PITTAS, BROADBILLS AND ASITIES

PITTAS, BROADBILLS AND ASITIES

Frank Lambert
and
Martin Woodcock

PICA PRESS
SUSSEX

RUSSEL FRIEDMAN BOOKS
SOUTH AFRICA

THE NETHERLANDS
AND BELGIUM

© 1996 Frank Lambert and Martin Woodcock
Pica Press (an imprint of Helm Information Ltd),
The Banks, Mountfield,
Nr. Robertsbridge,
East Sussex TN32 5JY

ISBN 1-873403-24-0

A CIP catalogue record for this book is available from the British Library.

All rights reserved. No reproduction, copy or transmission of this publication may be made without written permission.

No paragraph of this publication may be reproduced, copied or transmitted save with written permission in accordance with the provisions of the Copyright Act 1956 (as amended), or under the terms of any licence permitting limited copying issued by the Copyright Licensing Agency, 7 Ridgmount Street, London WC1 7AE.

Any person who does any unauthorised act in relation to this publication may be liable to criminal prosecution and civil claims for damages.

Published in Southern Africa by
  Russel Friedman Books CC
  PO Box 73,
  Halfway House
  1685,
  South Africa.

ISBN 1-875091-05-X
(Southern Africa only)

Published in the Netherlands and Belgium as *Dutch Birding Vogelgids 6* by
  Ger Meesters Boekprodukties
  Vrijheidsweg 86,
  2033 CE Haarlem,
  The Netherlands.

ISBN 90-74345-11-5
(Netherlands and Belgium only)

Editor: Nigel Redman
Copy Editing: Nigel Collar
Production and Design: Julie Reynolds

Computer graphics and typesetting by Fluke Art, Bexhill-on-Sea, E Sussex.
    Colour separation by Staples Printers, Kettering, Northants.
      Printed and bound by Hartnolls Limited, Bodmin, Cornwall.

Contents

		Page	Plate
Preface		1	
Acknowledgements		3	
Glossary		5	
Style and Layout of the Book		7	
Pittas, Broadbills and Asities: An Overview		13	
Phylogenetic relationships		13	
Classification		14	
Biogeographic history and distribution		17	
Evolutionary ecology of broadbills and asities		19	
Distinguishing characteristics of pittas and broadbills		19	
Food and foraging		19	
Social and breeding behaviour		21	
Nesting and care of young		25	
Migration and other movements		27	
Threats and conservation		28	
Colour Section		31	
Systematic Section		81	
Pitta			
Eared Pitta	*Pitta phayrei*	83	**1**
Blue-naped Pitta	*Pitta nipalensis*	85	**2/16**
Blue-rumped Pitta	*Pitta soror*	88	**2/16**
Rusty-naped Pitta	*Pitta oatesi*	92	**1/16**
Schneider's Pitta	*Pitta schneideri*	95	**3/16**
Giant Pitta	*Pitta caerulea*	97	**3/16**
Blue Pitta	*Pitta cyanea*	101	**4**
Banded Pitta	*Pitta guajana*	104	**5/16**
Bar-bellied Pitta	*Pitta elliotii*	108	**6**
Gurney's Pitta	*Pitta gurneyi*	110	**6/16**
Blue-headed Pitta	*Pitta baudii*	115	**6**
Hooded Pitta	*Pitta sordida*	117	**7/16**
Ivory-breasted Pitta	*Pitta maxima*	126	**8**
Superb Pitta	*Pitta superba*	128	**8**
Azure-breasted Pitta	*Pitta steerii*	129	**8**
Whiskered Pitta	*Pitta kochi*	131	**9/16**
Red-bellied Pitta	*Pitta erythrogaster*	133	**9/10**
Sula Pitta	*Pitta dohertyi*	140	**10**
Blue-banded Pitta	*Pitta arquata*	141	**11**
Garnet Pitta	*Pitta granatina*	143	**11**
Black-headed Pitta	*Pitta (granatina) ussheri*	146	**11**
Graceful Pitta	*Pitta venusta*	148	**11**
African Pitta	*Pitta angolensis*	150	**12**
Green-breasted Pitta	*Pitta reichenowi*	156	**12**
Indian Pitta	*Pitta brachyura*	159	**12**

		Page	Plate
Fairy Pitta	*Pitta nympha*	162	**13**
Blue-winged Pitta	*Pitta moluccensis*	166	**13/16**
Mangrove Pitta	*Pitta megarhyncha*	170	**13**
Elegant Pitta	*Pitta elegans*	174	**14**
Noisy Pitta	*Pitta versicolor*	177	**15**
Black-faced Pitta	*Pitta anerythra*	181	**14**
Rainbow Pitta	*Pitta iris*	183	**15**

SMITHORNIS

African Broadbill	*Smithornis capensis*	186	**17**
Grey-headed Broadbill	*Smithornis sharpei*	190	**18**
Rufous-sided Broadbill	*Smithornis rufolateralis*	193	**18**

CALYPTOMENA

Green Broadbill	*Calyptomena viridis*	196	**19**
Hose's Broadbill	*Calyptomena hosii*	201	**19**
Whitehead's Broadbill	*Calyptomena whiteheadi*	203	**19**

CYMBIRHYNCHUS

Black-and-Red Broadbill	*Cymbirhynchus macrorhynchos*	205	**21**

PSARISOMUS

Long-tailed Broadbill	*Psarisomus dalhousiae*	210	**20**

SERILOPHUS

Silver-breasted Broadbill	*Serilophus lunatus*	215	**23**

EURYLAIMUS

Banded Broadbill	*Eurylaimus javanicus*	220	**21**
Black-and-Yellow Broadbill	*Eurylaimus ochromalus*	224	**22**
Mindanao Wattled Broadbill	*Eurylaimus steerii*	227	**22**
Visayan Wattled Broadbill	*Eurylaimus samarensis*	229	**22**

CORYDON

Dusky Broadbill	*Corydon sumatranus*	231	**20**

PSEUDOCALYPTOMENA

African Green Broadbill	*Pseudocalyptomena graueri*	235	**18**

PHILEPITTA

Velvet Asity	*Philepitta castanea*	238	**24**
Schlegel's Asity	*Philepitta schlegeli*	242	**24**

NEODREPANIS

Common Sunbird Asity	*Neodrepanis coruscans*	245	**24**
Yellow-bellied Sunbird Asity	*Neodrepanis hypoxantha*	248	**24**

BIBLIOGRAPHY 255

INDEX OF SCIENTIFIC AND ENGLISH NAMES 269

List of Figures

	Page
Pitta head showing method of bill measurement	10
Pitta Topography	11
Nest of Giant Pitta	99
Roosting Blue-headed Pitta	116
Wing patterns of different individuals of Hooded Pitta	117
Raised crown and extended blue gorget of Azure-breasted Pitta	130
Raised posture of African Pitta	155
Heads and bills of Mangrove, Blue-winged, Fairy and Indian Pittas	163
Nest of Fairy Pitta	165
Fledgling Blue-winged Pitta	169
Nest of Mangrove Pitta	172
Rainbow Pitta upright posture, wing-spreading and bowing	184
Nest of African Broadbill	189
Nest of Green Broadbill	200
Nest of Long-tailed Broadbill	214
Nest of Silver-breasted Broadbill	219
Nuptial behaviour of Black-and-Yellow Broadbill	226
Nest of Dusky Broadbill	234
Schlegel's Asity at nest	244
Head of male Yellow-bellied Sunbird Asity	248
Bill and tip of first primary of Common Sunbird Asity and Yellow-bellied Sunbird Asity	252

*Dedicated to my parents, Francis and Joyce,
and to Anni, for encouragement, tolerance and support
during the long hours spent preparing this book.*

Frank Lambert

PREFACE

On my first visit to Asia, birding in the Himalayan region in 1978-9, seeing a pitta became something of an obsession, after one of my companions described having watched a Blue-naped Pitta at close quarters in Nepal. It was not until 2-3 months later that I finally saw this elusive species, in the hills of Assam. Here we also saw our first Long-tailed and Silver-breasted Broadbills and, when we returned home, my mind was set to see more of these beautiful birds. On a subsequent trip to Thailand I was somewhat disappointed with efforts to see more pittas and broadbills, with only brief views of a few species being obtained. Later, however, during four years in Peninsular Malaysia and another year in Borneo, my skills at finding these and other often secretive forest creatures were greatly enhanced by birding with Dennis Yong, to whom I am greatly indebted. Over that period and on many subsequent trips to Asia and Wallacea I have managed to track down and observe more than my fair share of these exquisite birds.

The greatest diversity of pittas and broadbills is to be found in the tall tropical rainforests of South-East Asia. Venturing into such habitat is, for many people, perhaps daunting, but an experience that should not be missed for anyone with a keen interest in nature. Observing pittas and broadbills in these forests can be very challenging, so when one is eventually seen well it is usually something to be remembered and certainly one of the highlights of any birding trip. During my years of research on tropical forest birds in Asia, I have had a few brief opportunities to dabble in the study of pittas and broadbills, by radiotagging a few individuals for short periods, and in doing so was very surprised to find that virtually no-one else had invested time in working on these marvellous birds.

Indeed, there have been no serious attempt by field biologists, until very recently, to investigate any species of pitta or broadbill in depth. Even now the only really detailed studies have been of Rainbow Pitta in Australia, Gurney's Pitta in Thailand and of asities in Madagascar. It is difficult to understand why such a diverse and alluring group of birds should have been so ignored, but perhaps there is an assumption that working with such elusive birds would be too difficult. However, I do not believe this to be the case, and sincerely hope that the publication of this book will stimulate a greater interest in studying this extraordinarily beautiful and fascinating group of birds.

ACKNOWLEDGEMENTS

Although the pittas and broadbills are a relatively small group of species, the research for this book has taken more than three years. This has entailed not only a wide search of published and grey literature, and an examination of countless museum specimens, but also consultation with many of the active field ornithologists and birders who have first-hand experience of these fascinating species in the wild. In particular, I am very grateful for the many contributions, both large and small, that derive from the field notes of such people. These contributors are identified in the text against the information that they so helpfully supplied (as *in litt.*, or verbally), but they all are also listed below.

The person deserving my greatest thanks, however, is undoubtedly Tim Inskipp, who, acting as both an ornithological encyclopedia and library, was able to provide me with numerous obscure references and papers that might easily have otherwise been overlooked or difficult to trace. At this juncture, I should also express my sincere thanks to Nigel Collar, who provided me with much encouragement in pursuing my earlier interests in tropical forest birds.

As mentioned above, museum specimens and libraries have been consulted to the extent that it was feasible, and many staff and associates at these important institutions deserve a mention for their assistance: Michael Walters, Peter Colston, Mark Adams, Effie Warr and Robert Prys-Jones (Natural History Museum, Tring); J. Phillip Angle, Louise H. Emmons and Mort Isler (Smithsonian Institution, Washington), René Dekker (National Museum of Natural History, Leiden), Ray Symonds (University Museum of Zoology, Cambridge), as well as the staff at the Sarawak Museum, Kuching; the Museum of Zoology, Bogor; Nguyen Cu at the museum of the Institute of Ecology and Biological Resources, Hanoi; and David Wells at the University of Malaya, Kuala Lumpur, who allowed me access to the nest record cards held in the Department of Zoology. It was not possible to visit all the museums that hold important specimens of pittas, but very valuable data held in some of these institutions were supplied by the following individuals, to whom I extend my thanks: Lloyd Kiff (Western Foundation of Vertebrate Zoology), Leon Bennun and Cecilia Gichuki (National Museums of Kenya), Michel Louette (Musée Royal de l'Afrique Centrale) and Mary LeCroy (American Museum of Natural History, New York). Specimens were lent to me for study at the British Museum by a number of institutions, and I must thank Jon Fjeldså (Zoological Museum, University of Copenhagen), David Willard (Field Museum of Natural History, Chicago), and staff of the Australian National Wildlife Collection (CSIRO) in Canberra for their assistance in this regard. I am also greatly indebted to Graeme and Carol Green for their kind hospitality during my many visits to the Natural History Museum at Tring.

A number of staff at BirdLife International in Cambridge, in particular Nigel Collar, Alison Stattersfield, Mike Crosby and Adrian Long, were also very helpful in supplying me with various pieces of useful information.

Text on the asities has been greatly enhanced by recent research by a number of field ornithologists, all of whom have generously provided me with unpublished manuscripts. In particular, I must express my gratitude to Richard O. Prum (Museum of Natural History, University of Kansas), Tiana Razafindratsita, Steve Goodman (Field Museum of Natural History, Chicago), Frank Hawkins, Roger Safford and Pete Morris for providing such information. Frank Hawkins must also be thanked for his useful advice concerning finding asities in Madagascar during my visit there in 1993. I am also very grateful to Udo Zimmermann (Northern Territory University) for providing me with the results of his recent doctoral studies on Rainbow Pittas, and to Frederick H. Sheldon for providing me with a copy of his unpublished annotated checklist of the birds of Sabah.

In addition to observations of birds in the wild, data on captive birds were sought from a number of sources. In particular, I would like to thank Robert S. Webster (San Antonio Zoo) for providing me with his enormously interesting observations of the displays and nesting habits of captive Green Broadbills.

Tape recordings of pittas and broadbills were very kindly supplied by a number of active sound recordists, including Bas van Balen, David Gibbs, Ole Jakobsen and Jelle Scharringa, whilst Richard Ranft at the British Library of Wildlife Sounds (National Sound Archive), in

London, also supplied me with recordings deposited there and assisted in making a number of sonograms.

In painting the plates and describing plumages and nests, the work was greatly enhanced by the provision of many photographs. To these photographers Martin Woodcock and I extend our gratitude: Louise H. Emmons, Simon Harrap, Simon Cook, Jelle Scharringa, Nigel Bean, John MacKinnon, John Howes, Kimura Tsutoma, Frank Hawkins, Pete Morris, Stig Jensen, Michael Poulsen, K. David Bishop, Adrian Long, Jonathan Eames, Dave Showler, Crawford Prentice, Mark Andrews, Simon Stuart, Uthai Treesucon, Phil Round, Charles Francis, Morten Strange and the BirdLife International Photo Library.

Assistance in identifying a number of food plants that were used by asities in Madagascar was kindly provided by Beverley Lewis (Missouri Botanical Garden), Don Reid and Frank Hawkins. Palms used by nesting pittas were kindly identified by John Dransfield (Herbarium, Royal Botanic Gardens, Kew).

Roland Wirth, Tamsin Humphries, Sarah Thomas and Radoslaw Ratajsczak assisted in the translation of a number of papers from German, Dutch and French.

Other individuals who assisted in the provision of their unpublished notes, or other relevant information, and to whom I extend my thanks are: Des Allen, Gary Allport, Paul Andrew, Mark Andrews, Dylan Aspinwall, Bas van Balen, Nigel Bean, Arnoud van den Berg, André Brosset, Stuart Butchart, K. David Bishop, Christopher Bowden, Hugh Buck, Tom Butynski, Clide Carter, Mike Chong, Anwaruddin Choudhury, Jan Christiansen, Major J.F.R. Colbrook-Robjent, Nigel Collar, Finn Danielsen, Pete Davidson, Geoffrey Davison, Will Duckworth, Steve Duffield, Guy Dutson, Jonathan Eames, Louise H. Emmons, Mike Evans, Lincoln Fishpool, Charles Francis, Brian Gee, David Gibbs, Alan Greensmith, Richard Grimmett, Simon Harrap, Frank Hawkins, Pete Hayman, Phil Heath, Derek Holmes, Jesper Hornskov, J. Houwing, John Howes, Ole Frode Jakobsen, Stig Jensen, Paul Jepson, Andrew D. Johns, Ben King, Olivier Langrand, Annabel Lee, Alan Lewis, Adrian Long, Joe Marshall, Ken Mitchell, Pete Morris, Yus Rusila Noor (Asian Wetland Bureau), Glenda Noramly, Richard Noske, Andrew Owen, Colin Poole, Michael Poulsen, Crawford Prentice, J. Rafidison, Rajang Rajanthan, Nigel Redman, Jon Riley, Craig Robson, Philip Round, Frank Rozendaal, Roger Safford, Ravi Sankaran, Jelle Scharringa, Thomas S. Schulenberg, Derek Scott, Dave Sargeant, Lucia Liu Severinghaus, Dave Showler, Robert Stjernstedt, Simon Stuart, Tony Stones, Morten Strange, Richard Thewlis, Rob Timmins, Uthai Treesucon, Don Turner, Shunji Usui, Jaime Pérez del Val, Filip Verbelen, David Wells, James Wolstencroft, Dennis Yong and Z. Dahaban.

Comments on the introductory section were kindly provided by Richard Prum, Nigel Collar and Nigel Redman. Various species accounts were reviewed by Phil Round, Richard Prum, Adam Gretton, Christopher Bowden, Craig Robson, Tim Inskipp, Carol Inskipp, Frank Hawkins, Udo Zimmermann, Lincoln Fishpool, Bas van Balen, Will Duckworth and David Wells.

Finally, I would like to take this opportunity to mention the many birders with whom I have shared, on many occasions, the experience of tracking down and observing pittas, broadbills and other shy forest birds during my frequent trips to Asia. In particular I extend my gratitude to Dennis Yong, John Howes, Marcel Silvius, Crawford Prentice, Reine-Marie Ramassamay, Agustinus Taufik, and members of the Malaysian Nature Society, for their companionship in the field and friendship during my years in Malaysia and Indonesia, as well as Richard Grimmett, Colin Winyard, Dick Filby, Nigel Redman, Richard Fairbank, Paul Andrew, Adrian Long, Ben King, Tony Greer, Annabel Lee, John Bowler, Jonathan Eames, Hugh Buck, Tim Fisher, Phil Round, Uthai Treesucon, Lindai Lee, Louise Emmons, David Bishop, Michael Lambarth and Sandra Fisher, whose companionship made my many visits to the forest that much more enjoyable.

GLOSSARY

Allopatric — Used of populations, species or taxa that occupy different and disjunct geographic areas.
Australasia — Used in this volume in its restricted sense, in that it includes Australia, New Guinea and other islands on the Sahul Shelf, but not the islands of eastern Wallacea.
CITES — The Convention on International Trade in Endangered Species of Wild Flora and Fauna. The Convention prohibits international trade in specimens of species included in any of the Appendices without the prior grant of a CITES permit. International trade in species in Appendix I, in particular, is subject to very strict and stringent rules.
Congeneric — Belonging to the same genus.
Dahomey Gap — The gap in distribution of lowland tropical moist forest vegetation in West Africa, in present-day Benin (see Moreau 1966).
Frugivore — A fruit-eater.
Gape — The mouth opening, formed in the angle between the two mandibles when the lower jaw is dropped.
Immature — In this book, immature is used to describe birds that approximate the adult in size and have clearly started to acquire adult plumage.
Indochina — Vietnam, Laos and Cambodia.
Juvenile — Recently fledged young which is not fully grown and, in many species, lacks most plumage colours of the adult.
Nectarivore — A nectar- and pollen-eater.
Parapatric — Used of populations whose geographical ranges are contiguous but not overlapping (so that gene-flow between them is possible).
Polygyny — Mating system in which a male mates with two or more females during the course of a breeding season.
Sunda Region — Peninsular Malaysia and the Greater Sunda Islands (Borneo, Sumatra and Java).
Sympatric — Used of populations, species or taxa that occupy the same geographic areas (though not necessarily the same habitat).
Taxon/Taxa — An expression to denote a taxonomic unit (/units) that is sufficiently different from other such units to be treated as a distinct, separate unit. A taxon includes all taxonomic units at subordinate rank (for example, *Pitta granatina* may be referred to as a taxon, in which case its subspecies are also included, but *Pitta granatina granatina* can also be called a taxon, but in this instance there are no taxonomic units of lower rank).
Wallacea — Oceanic islands lying between the Sunda Shelf of continental South-East Asia and the Sahul Shelf of New Guinea/Australia. This region includes Sulawesi, the Lesser Sundas, the Moluccas and intervening islands.
Wallace's Line — An imaginary line separating the Oriental and Australasian zoogeographical regions. The line passes between Bali and Lombok, Sulawesi and Borneo, and the Philippines and the Moluccas.

ABBREVIATIONS: Museum collections referred to in the text

Reference is frequently made to data collected from various museum collections. These museums are abbreviated in the text as follows.

BMNH — Natural History Museum, Tring
FMNH — Field Museum of Natural History, Chicago
IEBR — Institute of Ecology and Biological Resources, Hanoi, Vietnam
MZB — Museum of Zoology, Bogor, Indonesia
RMNH — Rijksmuseum van Natuurlijke Historie, Leiden
SarM — Sarawak Museum, Kuching
TIST — Thai Institute for Science and Technology, Bangkok
UM — University of Malaya (Department of Zoology), Kuala Lumpur (includes data from Nest Record Cards)
UMZC — University Museum of Zoology, Cambridge
USNM — United States National Museum, Smithsonian Institution, Washington DC

STYLE AND LAYOUT OF THE BOOK

This book follows the general style of other books in this series that deal with specific taxonomic groupings of birds. Although, like other volumes, there is strong focus on identification, this particular volume perhaps places a greater emphasis on general biology and ecology. There is also greater than usual attention to subspecies where appropriate, since, not only is the taxonomy of these species in relative infancy, but also the geographical variation shown by a number of presently recognised species is quite remarkable. It is hoped that by focusing on this variation, interest will be fostered in resolving some of the taxonomic questions that are raised in this volume.

Since there has been no previous detailed treatise on the pittas, broadbills and asities, an attempt has been made to draw together all the ecological data available from a large number of disparate sources. In researching the book, as well as consulting published literature, many ornithologists and active birders were consulted for their unpublished notes and information. None of the birds covered by this book could be described as well known, and many are so poorly documented that very little indeed is known about them. Hence the individual species accounts vary greatly in detail and length.

The first part of this volume attempts an overview of the taxonomy and phylogeny, distribution and general biology of the pittas and broadbills, and also provides some brief discussion of the conservation concerns associated with this group of birds.

The main part of the book is devoted to the species accounts. The taxonomy followed, discussed in more detail in the introductory section, is partly based on Sibley and Monroe (1990), but also on the work of Prum (1994), which has significantly advanced our knowledge of the broadbills and asities since the work of Sibley and Monroe was completed.

SPECIES NUMBERS

Each species in this volume has been given a number. These numbers have no taxonomic significance, but are used simply to assist the reader in linking species accounts and plate text with plate labels.

PLATES

Martin Woodcock's excellent plates depict all the known species of pitta, broadbill and asity. Where subspecies differ significantly, an attempt was made to illustrate these differences wherever possible. Similarly, juvenile and immature birds are shown alongside adults where this was possible, although, as with subspecies, this is entirely dependent on the availability of museum specimens or good photographs. For some species, such as Whiskered Pitta, no suitable juvenile material was available until a late stage in preparing the book, when a photograph became available, and it was not therefore possible to paint this individual next to the adult. Instead, it is depicted on the plate of immature pittas that was designed primarily as an aid to seeing more clearly the differences between juveniles which might be encountered in the same geographical area.

Plate captions provide a short synopsis of the range of each taxon depicted, and, for each individual illustrated, summarise diagnostic features.

SPECIES ACCOUNTS

For each species, an individual account is furnished. These accounts are subdivided into the sections described below. Within the species accounts, numerous references are provided to identify the corresponding source of information. Where this information is published or in press or in preparation, the full reference is provided in the References at the back of this book. Unpublished information is credited to the person who provided the information, as *in litt.* if they communicated in writing, or as 'verbally' if the information derived from conversation.

Metric measurements are used throughout this book. A number of abbreviations are used, as follows: mm millimetres, m metres, km kilometres, ha hectare (100ha = 1km^2), g grams, hr hour, s second; c. in front of a number is used to denote *circa* (i.e. approximately).

Alternative names that have been used for the species treated in the book are also provided at the beginning of the species accounts.

Taxonomy

A brief section discussing taxonomy is provided for a number of species, in particular those that are not formally described as species elsewhere. In the case of Black-headed Pitta of north Borneo, for example, a separate account is given in recognition that this taxon should probably be considered as a separate species from Garnet Pitta. In this particular instance, and in recognition that my view that Black-headed Pitta represents a valid species may not be accepted by all, and that other taxonomic views are held by others, a trinomial is used with parentheses to denote the species to which other taxonomists prefer to treat the taxon. The reasons why I believe such taxa to be good species are presented in this section.

Identification

In this section, a general description of the bird, including the most important identification features, are provided. Approximate size is also indicated by the provision of body lengths, from outstretched specimens (skins), measured from the bill tip to tail tip. Other biometrics, such as wing length, tarsus length and weight, are provided at the end of the Description.

The identification of pittas and broadbills in the field is usually relatively straightforward, but finding and observing them is a different matter. Views of pittas, in particular, are often rather fleeting, so it is useful to know which species to expect and what to look for in advance. Many pittas and broadbills are often very vocal, and knowing their calls will also aid greatly in finding these species.

For the majority of species, adults can usually be easily identified, but younger birds may pose more difficult identification problems, and so their diagnostic characters are also mentioned here. Unfortunately, young pittas are too poorly known to age with any precision, but they clearly fall into two rough categories. In this book, a distinction has been made between these categories by using the terms **juvenile** and **immature.** Both terms are rather vague in definition, but represent an attempt to distinguish between young birds that have recently left the nest and those that have left the nest several months previously but have not acquired full adult plumage. In general, birds defined as juveniles are not fully grown and, in many species, lack most plumage colours of adults. In contrast, immatures, as defined here, are approximately adult-sized and have often acquired plumage features that resemble those of an adult, usually making them at once distinguishable.

Within the identification section, and under the sub-heading **Similar species**, the diagnostic features that assist in distinguishing the species from other, often sympatric species, are also detailed. Similarities between species in vocalisations may be mentioned in this section, but these are mostly to be found in the section on Voice.

Voice

Most pittas and many broadbills have vocalisations that often provide the best or only evidence of their presence at a site. And, very often, the calls of pittas and broadbills provide the initial clue that enables a patient observer to find them. Although most species have distinctive calls, distinguishing between the superficially similar calls of some species may sometimes be difficult, and where relevant these differences are mentioned in the text. It should be noted that, at a distance, some of the disyllabic or trisyllabic calls typically given by a number of species of pitta may sound monosyllabic, particularly when one of the notes is much quieter than the others.

The descriptions are, of course, a poor substitute to listening to sound recordings, since they are merely the author's or other observers' interpretation, in words, of the sound that they hear. If possible, therefore, it is advisable to listen to appropriate recordings before venturing into the field. Many species of pitta, though not usually broadbills, readily respond to tapes of their calls, and this behavioural characteristic sometimes offers the easiest opportunity of observing them. Some species also respond to human imitations of their call (e.g. the Garnet Pitta, Blue-banded Pitta, Rusty-naped Pitta, Bar-bellied Pitta), so it is always worth bearing this in mind if difficulties arise in seeing a calling pitta.

Sonograms have also been provided for a number of species whose calls sound superficially similar. Sonograms were prepared by the British Library National Sound Archive using Avisoft-Sonograph Pro for Windows. The frequency and time axes are shown on each graph. The linear frequency scale on the vertical axis is calibrated in kiloHertz (kHz), and the linear timescale on the horizontal axis in seconds (s). The relative amplitude in decibels (dB) is roughly indicated by shades of grey, from white (low amplitude) to black (high amplitude).

Distribution

This section furnishes a detailed description of the global distribution of the species. For some species, however, it was more pragmatic to provide finer details of distribution under the relevant subspecies descriptions, within the section on Geographical Variation. **Maps** illustrate the distribution of all species, with divisions to show the distribution of individual subspecies, and, for migratory taxa, the known breeding and non-breeding ranges.

Unlike most other books that deal with taxonomic groupings of birds, an attempt has been made, where possible, to define the approximate present known range of a species, rather than simply blocking in a whole region, country or island, etc. Constructing maps in this way has involved assessing the extent of suitable habitat believed to remain within the known altitudinal range of the species. Clearly, such an approach has many difficulties, and there will undoubtedly be errors, but it is felt that by illustrating present range in this way, the reader will obtain a more realistic impression of how a particular taxon is distributed. For many species, the result of this exercise serves to emphasise the fragmented nature of present distribution, and hence the potential threats faced by many forest bird species. This approach also means that some parts of the historical range of a species may not be mapped, if the species or its habitat is no longer present.

Obviously, even with this method of mapping there is no guarantee that any particular species will occur throughout the range shown, since habitat and topography are rarely uniform over large areas, even though they may be mapped as such. Birds undoubtedly recognise many more habitats than cartographers or ornithologists. It should be also borne in mind that the altitudinal range of many species apparently varies within their range, often with topography, and this will also lead to inaccuracies in representing the real range of any taxon. Finally, it should be mentioned that for most species the exact limits of their range are unknown, and in such instances assumptions about the extent of a taxon's range were made using the information available.

The extent of remaining forest cover was ascertained from maps compiled by the World Conservation Monitoring Centre (Collins *et al.* 1991, Sayer *et al.* 1992), many of which are based on relatively old data (when considering the rate of deforestation in some areas). This must also, therefore, be borne in mind when interpreting the maps.

Geographical Variation

Many species of pitta and broadbill are polytypic, with more than one recognised subspecies. In this section, the distribution of these subspecies is described, along with a brief description of how the various subspecies differ from each other. In some instances, comparisons are made to the nominate subspecies which is described in full in the section headed Description. Certain measurements are also provided in this section if these are thought to be pertinent to the differentiation of particular subspecies.

A general preamble is often provided where it is thought necessary to justify the inclusion (i.e. recognition) or exclusion of particular subspecies, or to inform the reader of taxonomic issues that need resolving.

Habitat

Pittas and broadbills are all essentially forest birds. However, the types of forests that they occupy are less well documented, and in this section, details are given of all the habitat types that they are known to occupy. For some species, this varies, depending on which part of the species's range is considered. Similarly, the altitudinal range of many species is not uniform throughout their global range, and in this section details of known altitudinal distribution are also provided.

Status

The assessment of status is relative, and whilst one observer may feel a particular species is common, another may believe the opposite. For this reason, interpretation of this section requires considerable caution. Large forest birds such as Giant Pitta may occur naturally at low density, but since we know so little about it, it is very difficult to make judgements concerning its true status. Giant Pitta is also an example of a species that is very secretive and very quiet for most of the year, so that it may appear to be very rare until the breeding season, when it can be heard calling. At Danum Valley, Sabah, for example, observing this species is a relatively rare event, but at the right season it is possible to hear four or more birds calling simultaneously.

It should also be noted that, for a number of species, the only available comments on status were made a considerable time ago, and that various factors may have caused changes in their status since that time.

For species of conservation concern, the threats that they face, and the degree of threat, are documented in this section. Details of ranging behaviour, such as home-range size, of pittas and broadbills are virtually unstudied. However, where information of this nature is available, it is included in this section.

Food

In general this is a brief section that lists all the items of diet that are known to be used by any particular species. In the case of insectivores, it is often possible to say something about the kinds of prey items that are consumed. The range of prey items in the diets of insectivorous species is, however, likely to be far greater than indicated here. A simple reason for this is that it is usually impossible to identify prey items, except for large, obvious ones, in the field. Furthermore, and rather unfortunately, very few collectors ever examine stomach contents, and even when they do, the remains of soft-bodied invertebrates are usually all but impossible to identify.

Perhaps even less is known about the full range of fruits or flowers that are visited by frugivorous or nectarivorous species. Observers are usually less competent at identifying different species of plant than they are at distinguishing, say, a spider and a grasshopper, making the identification of plant species that provide fruit or nectar for various species of broadbill impossible, so the lists of plants that are important are generally far from complete.

Habits

The habits of pittas are relatively uniform, and most species conform to the general habits outlined in the introductory section of the book. For this reason, the section on habits is rather brief for many species. Social behaviour is poorly documented for all but a few species of pitta and broadbill, but for those where information is available, this is included in this section. However, habits related to breeding, such as information on nest construction and incubation, are generally addressed in the following section, on Breeding.

Breeding

The degree of knowledge of the breeding biology of species of pitta and broadbill varies considerably. At one end of the spectrum are species such as Schneider's Pitta, the nest of which has never been found, and Hose's Broadbill, for which no details of its nest have ever been recorded, whilst at the other are Black-and-Red Broadbill, Gurney's, Giant and Rainbow Pitta, the nests of which are well known. Hence the length of this section, in particular, varies significantly. Where there is a large amount of information on breeding, this section has sometimes been subdivided into three, to cover seasonality, nests and eggs.

Within this section details of breeding seasonality are given. For less well-known species, details of the dates and places where individual nests have been found, or fledged young observed or collected, are provided, since this information would otherwise be very difficult to access. For better-known species this level of detail is usually omitted since there is not enough space to document every nest, etc. that has been documented. Where information is available, this description of breeding seasonality is followed by an account of the sort of sites where the species makes its nest, a description of the nest and the materials used, and a description of the eggs. If known, information regarding parental involvement in nest construction, nest attendance and subsequent care of the young is also provided in this section.

Description

This section offers a detailed treatment of the plumage and bare-part coloration of the species. For polymorphic species, the description refers to one subspecies (usually the nominate), as indicated at the beginning of the description.

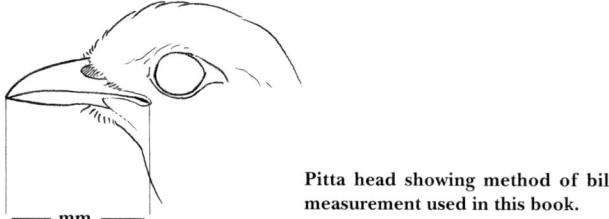

Pitta head showing method of bill measurement used in this book.

Measurements

In general, minimum and maximum measurements are given in metric units. Except for a few instances where specimens of particular subspecies were unavailable (in which case a reference to source is provided), all measurements were made by the author. With the exception of a few species for which there were insufficient specimens, measurements were taken from at least 15 skins of adult birds. Birds growing primaries, or in obvious moult, were not used to obtain measurements of wing or tail length.

The wing length is the flattened wing chord, and is an easy measurement to compare. Tail length is a more difficult measurement to take, and two people may frequently arrive at different lengths for the same specimens. It should also be mentioned that it is not always easy (on skins) to ascertain whether the tail is growing. In this volume, tail length refers to the distance from the base of the central feathers, where these emerge from the skin, to their tip. The bill length (mm) is the distance from the corner of gape to the tip (see figure above). For some species, in particular broadbills, the bill width (mm), which is the width of the bill at the gape, is also given. To compare bill sizes of Indian and Fairy Pitta, and subspecies of Red-bellied Pitta, measurements of bill depths (mm) are also provided, these being the maximum depth, which is usually at the culmen base, where the feathers of the forehead meet the upper mandible. Tarsus (mm) is a measurement of the distance from the joint at the proximal end of the tarsus (fused metatarsals) to its base, where it joins the middle toe (phalanges). Weight (grams) of the live bird derive from a variety of sources, including specimen labels, published values, and unpublished ringing data.

REFERENCES

This section of the book provides the details of all literature cited in the book.

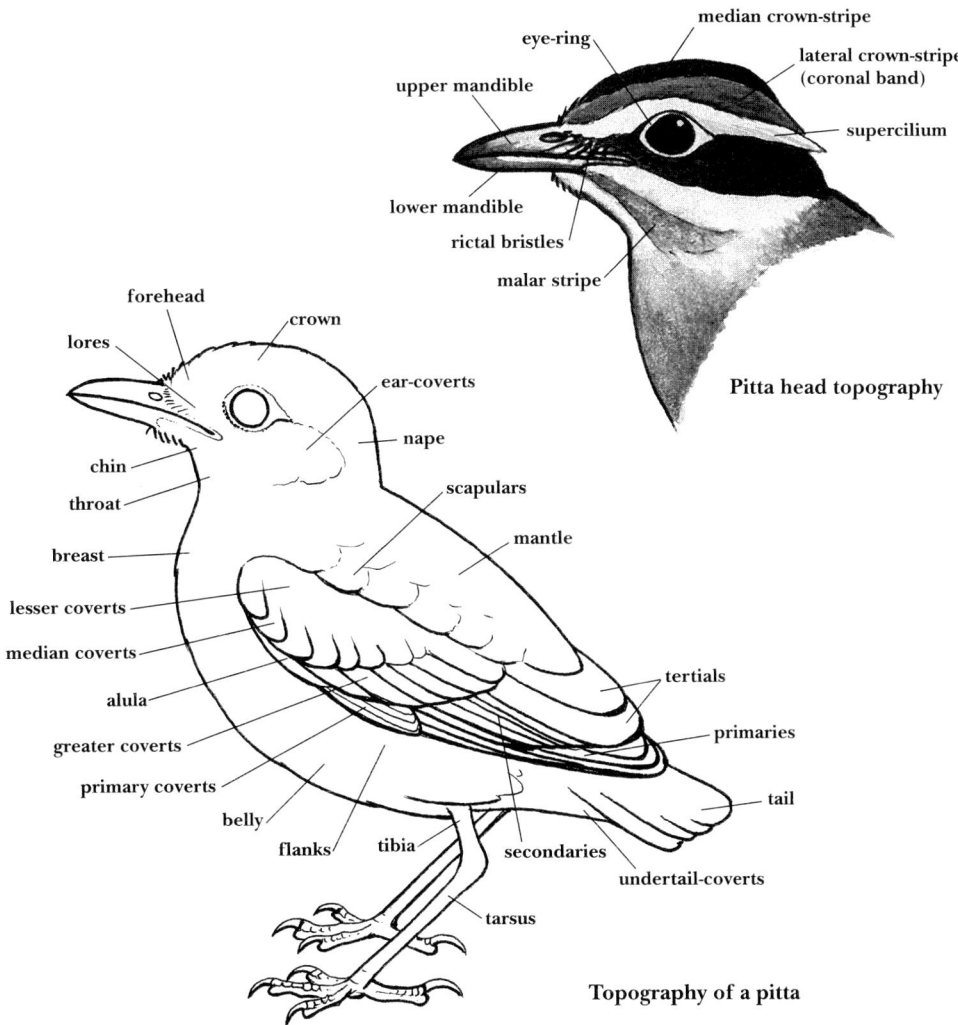

Pitta head topography

Topography of a pitta

PITTAS, BROADBILLS AND ASITIES: AN OVERVIEW

INTRODUCTION

The Old World suboscines comprise two relict families, the pittas, Pittidae, and the broadbills and asities, Eurylaimidae. This rather diverse but small assemblage of 51 species contains some of the most beautiful, shy and fascinating forest birds in the Old World.

The Pittidae is a relatively uniform family of medium-sized insectivorous terrestrial birds that is largely confined to the tropical and subtropical forests of Asia, Australia and intervening islands, with a few outlying species occurring on islands to the east of New Guinea and on mainland Africa. At the present time, all 32 species are in the genus *Pitta*. They are mostly brilliantly coloured and rather thrush-like in their behaviour, and as a consequence were originally called 'jewel-thrushes'. They are usually solitary or found in dispersed pairs, and do not flock with other species.

Although some pittas occur in more open habitats, most are shy inhabitants of the floor of forest, and observing them in this habitat provides a major challenge despite their colourful appearance. They have a well-deserved reputation for secretive, skulking habits, as noted in Ludlow and Kinnear (1937), who wrote of collecting a Blue-naped Pitta in Sikkim: '...it was scuttling about on the ground in the dense undergrowth. As I pressed the trigger I said to myself a rat'.

Broadbills and asities are, with few exceptions, also brightly coloured, but unlike pittas they are mostly sociable and none is terrestrial. They are a diverse group, with 19 species in 10 genera and five sub-families. Some species may join mixed bird flocks. They are much more diverse in appearance and ecology than pittas, with six genera in Asia and two in Africa. With the exception of sunbird asities, they are chunky, docile birds with large heads, broad flattened beaks and short legs. Broadbills are largely confined to the wetter parts of Africa and to South-East Asia, with a few species with ranges that extend through Burma and the foothills of the Himalayas in the Indian Subcontinent.

The four asities, all endemic to Madagascar, were previously included in their own family, but are now considered to be broadbills (Prum 1993). They are perhaps the most fascinating species within the Eurylaimidae. Although the two asities in the genus *Philepitta* have a somewhat similar appearance to other broadbills, the two sunbird asities in the genus *Neodrepanis* look very much like sunbirds, being small-bodied and agile, and having long, narrow, downcurved bills. All breeding male asities also have facial wattles, or caruncles, which have a structure that is apparently unique (Prum *et al.* 1994, R. Prum *in litt.*). The only other broadbills with eye wattles are those found in the Philippine archipelago, at the extreme limit of the range of broadbills in South-East Asia. Although the majority of broadbills and asities inhabit tall tropical and subtropical forests, some African species, such as African Broadbill and Schlegel's Asity, may inhabit regions where seasonally dry forests and savanna-like habitats predominate.

The pittas and broadbills are, taken together, an extremely interesting group of birds, with a diverse array of feeding and breeding habits, and a complex phylogeny that should be of great interest to evolutionary and biogeographic theory. Nevertheless, despite their obvious appeal, remarkably few detailed studies have focused on pittas or broadbills, and with the exception of Elliot's (1870) *Monograph of the Pittidae*, there has been no major treatise that deals exclusively with either family. This book seeks to fill that gap and to stimulate an interest in this fascinating and alluring group of birds. Although the present book is designed to be a comprehensive guide and handbook to the pittas and broadbills, our knowledge of many of these birds, in particular relating to taxonomy and ecology, is relatively poor. Even the nests and eggs of some species, such as Schneider's Pitta, remain unknown. This volume will hopefully assist in identifying such gaps in our knowledge, as well as providing an important source of information on many aspects of the biology and ecology of pittas and broadbills for years to come.

PHYLOGENETIC RELATIONSHIPS

There are two diverse lineages of passerine or perching birds. These are the oscines, with a complex vocal apparatus (syrinx), and the suboscines, with a simple, though variable, vocal apparatus. Whilst there are a number of families of suboscine passerine in the New World (the Americas), there are only two recognised families in the Old World (Africa and Asia). These are the Pittidae and Eurylaimidae of the infraorder Eurylaimides.

Historically, there have been various proposals about the relationship of pittas, broadbills and asities (e.g. Sclater 1872, Garrod 1876, 1877, 1878, Milne-Edwards and Grandidier 1879, Forbes 1880a, 1880b, Pycroft 1905, Amadon 1951, Sibley 1970, Olson 1971, Raikow 1987), and a comprehensive historical review of the taxonomy of the pittas and broadbills is provided by Sibley and Ahlquist (1990).

Recent evidence from DNA hybridisation has shown that the pittas are a monophyletic group and that they are most closely related to the broadbills (Sibley and Ahlquist 1990). However, whilst pittas and broadbills have long been classified as suboscine (Müller 1847), and the two asities in the genus *Philepitta* were placed in the suboscines in 1880 (Forbes 1880b), the taxonomic affinities of asities in the genus *Neodrepanis* were poorly understood until very recently. Indeed, the Common Sunbird Asity *N. coruscans* was placed with the sunbirds (Nectariniidae) for more than 75 years after its discovery (at that time, the existence of a second sunbird asity was unknown).

It was not until 1951 that examination of the syrinx of *Neodrepanis* showed that the sunbird asities were in fact suboscines, and that their remarkable morphological similarities with sunbirds was an extreme example of convergent evolution (Amadon 1951). They were subsequently placed in the then endemic Madagascan family, Philepittidae. Detailed work by R. Prum *et al.* (1994, and *in litt.*) on the structural colour production in the caruncle of asities has provided further compelling evidence that the two asity genera are related, whilst his recent research on syringeal and skeletal morphology has demonstrated that the asities represent a Madagascan lineage of the broadbills, and are therefore members of the Eurylaimidae (Prum 1993).

Thus, the pittas, broadbills and asities, despite their varied external appearance and considerable ecological differences, are phylogenetically closely related. In contrast, the New Zealand wrens (Acanthisittidae), once grouped with the Old World suboscines, are now considered to form a sister group to the oscine passerines (Raikow 1987) or to the entire passerines (Sibley *et al.* 1982, 1988, Sibley and Ahlquist 1990) and hence are not included in this book.

The phylogeny (evolutionary history) within the Old World suboscines has recently been the subject of several studies. These have relied on traditional forms of taxonomic comparison, such as an examination of anatomy, as well as more modern biochemical studies, based on recent advances in biotechnology. Both lines of research, most notably into hindlimb musculature (Raikow 1987), skeletal and syringeal morphology (Prum 1993), and DNA-DNA hybridisation (Sibley and Ahlquist 1990), indicate that the pittas are the most recent historical relatives (sister group) of a lineage that includes both the broadbills and asities. Based on the evidence from these studies, Prum (1993) was able to demonstrate not only that asities were members of the Eurylaimidae, but that the African Green Broadbill *Pseudocalyptomena graueri* of Zaïre and Uganda was more closely related to Madagascan *Philepitta* and *Neodrepanis* than to the other broadbills. Interestingly, the close morphological similarities between *Pseudocalyptomena* and *Philepitta* were recognised by Lowe (1931) when he realised that the former was a broadbill rather than an oscine flycatcher, as originally classified (the African Broadbill having originally been placed in the Muscicapidae).

Prum's (1993) study, and subsequent research on the sequence of molecular DNA in the asities and broadbills (Prum 1994, R. Prum *in litt.*), indicates that *Pseudocalyptomena* and the asities either evolved from a single lineage that diversified after Madagascar broke away from Africa during the Middle Jurassic (between 165 and 130 million years ago: Coffin and Rabinowitz 1987), or that the Asian broadbills dispersed to both Africa and Madagascar after they had separated. Whichever scenario is correct, Prum's (1993) hypothesis on phylogeny was that the ancestors of the asities were frugivorous, and that the nectarivorous habit of the asities, with consequent adaptations in bill morphology, evolved in response to resource availability on Madagascar. However, the recent realisation that the tongues of both *Philepitta* species show some adaptations to nectarivory means that nectar foraging is primitive to asities (*contra* Prum 1993), and that *Neodrepanis* species have evolved nectar specialisation from a more general (frugivore-nectarivore) diet (R. Prum *in litt.*). Whilst it is now clear that *Pseudocalyptomena* is a close relative of the asities, the other broadbills in Africa, of the genus *Smithornis*, have a very primitive syrinx which is most like that of pittas (Prum 1993).

CLASSIFICATION

DNA hybridisation evidence indicates that there are large genealogical distances between the species of pitta (Sibley and Ahlquist 1990), and that they should, like broadbills, probably be separable into several genera: indeed, as recently as 1982, Wolters (1982) reinstated six genera. Nevertheless, in view of the limited scope of the DNA studies that have been conducted to date, Sibley and Ahlquist (1990) proposed that all pitta species should be retained in a single genus until further studies are performed. It seems inevitable, however, that further elucidation of the genetic diversity within this superficially uniform group will lead to a taxonomic revision in the future.

Although it may be rash to predict some of the likely changes in classification that might occur as a result of further research, there are some clear candidates for taxonomic revision. For example, the very distinctive Eared Pitta *Pitta phayrei* may best be restored to its original monotypic genus, *Anthrocincla* (Blyth 1863), whilst other genera recognised by Whitehead (1893), *Hydrornis* (which included Blue-naped Pitta, Rusty-naped Pitta and Blue-rumped Pitta, and would perhaps also include Schneider's Pitta and Giant Pitta, which Whitehead put in the genus *Gigantipitta*) and *Iridipitta* and/or *Eucichla* (in which Whitehead included Blue Pitta, Bar-bellied Pitta, Gurney's Pitta, and Banded Pitta), could well be reinstated in future taxonomic revisions. In addition, the group of rather small, graceful pittas with scarlet or red underparts that inhabit the rainforests of the Sunda region, namely Garnet Pitta, Black-headed Pitta, Graceful Pitta and Blue-banded Pitta, and perhaps the Red-bellied, Whiskered and Sula Pittas of Philippines, Wallacea and Australasia, might reasonably also be placed in their own genus; Wolters (1982) kept this group in its own group (subgenus *Erythropitta*) within the genus *Pitta*.

In the following classification for pittas, Sibley and Monroe (1990) is followed (with the addition of one taxon, Black-headed Pitta, which has been elevated to species level in this volume), including their designation of superspecies groups, indicated in square brackets. However, alternative, provisional groupings, based on my examination of plumage, are also indicated by spaces in the sequence. The sequence followed by Sibley and Monroe (1990) differs significantly from that presented by Mayr (1979; Peters's sequence), but should not be considered definitive.

In contrast to the pittas, the taxonomy of the broadbills has recently been revised. From his detailed studies, Prum (1993) proposed that the broadbills and asities should be placed in five subfamilies of the Eurylaimidae. The classification and sequence presented below follows this proposal except that four species of *Eurylaimus* are recognised in this volume, rather than three.

Within the Eurylaimidae, the five subfamilies are arranged so that each is the sister group of the remaining subfamilies in the sequence. The relationships of the genera within the subfamily Eurylaiminae have not yet been resolved, whilst the genus *Corydon* was placed in this subfamily provisionally, requiring further investigation (Prum 1993). The Philippine wattled broadbills are unique amongst the broadbills in possessing a ring of fleshy wattles around the eyes, although this is shared with the Madagascan asities. For this reason they were originally placed in their own genus *Sarcophanops*, and although *Eurylaimus* is used here, following Prum (1993), it may be more appropriate to revert to this genus in the case of these very distinct species.

It seems inevitable that new DNA hybridisation studies or the use of new technologies designed to elucidate taxonomic relationships will lead to revisions of the sequence and status of taxa of pittas and broadbills. In particular, it should be emphasised that taxonomists have hardly explored the possibility that some of the many distinctive subspecies of pitta inhabiting the various islands between the Asian mainland and the New Guinea region may represent good species. The species of main concern in this respect are Banded Pitta *P. guajana*, Red-bellied Pitta *P. erythrogaster* and Hooded Pitta *P. sordida*. Hence, with many taxonomic questions still unanswered, the following sequence should be treated as provisional.

Order Passeriformes
Suborder Tyranni
Infraorder EURYLAIMIDES
Superfamily PITTOIDEA

Family **Pittidae**

> Genus *Pitta*
>
> | | |
> |---|---|
> | *Pitta phayrei* | Eared Pitta |
> | *Pitta nipalensis* | Blue-naped Pitta |
> | *Pitta soror* | Blue-rumped Pitta |
> | *Pitta oatesi* | Rusty-naped Pitta |
> | *Pitta schneideri* | Schneider's Pitta |
> | *Pitta caerulea* | Giant Pitta |
> | *Pitta cyanea* | Blue Pitta |
> | *Pitta guajana* | Banded Pitta |
> | *Pitta elliotii* | Bar-bellied Pitta |
> | *Pitta gurneyi* | Gurney's Pitta |
> | *Pitta baudii* | Blue-headed Pitta |

Pitta [*sordida*] *sordida*	Hooded Pitta
Pitta [*sordida*] *maxima*	Ivory-breasted Pitta
Pitta [*sordida*] *superba*	Superb Pitta
Pitta [*sordida*] *steerii*	Azure-breasted Pitta
Pitta kochi	Whiskered Pitta
Pitta [*erythrogaster*] *erythrogaster*	Red-bellied Pitta
Pitta [*erythrogaster*] *dohertyi*	Sula Pitta
Pitta arquata	Blue-banded Pitta
Pitta [*granatina*] *granatina*	Garnet Pitta ⎫
Pitta [*granatina*] *ussheri*	Black-headed Pitta ⎬ garnet pitta group
Pitta [*granatina*] *venusta*	Graceful Pitta ⎭
Pitta [*angolensis*] *angolensis*	African Pitta
Pitta [*angolensis*] *reichenowi*	Green-breasted Pitta
Pitta [*brachyura*] *brachyura*	Indian Pitta
Pitta [*brachyura*] *nympha*	Fairy Pitta
Pitta [*brachyura*] *moluccensis*	Blue-winged Pitta
Pitta [*brachyura*] *megarhyncha*	Mangrove Pitta
Pitta [*versicolor*] *elegans*	Elegant Pitta
Pitta [*versicolor*] *versicolor*	Noisy Pitta
Pitta [*versicolor*] *anerythra*	Black-faced Pitta
Pitta [*versicolor*] *iris*	Rainbow Pitta

Family **Euryaimidae**

    Subfamily Smithornithinae

        Genus *Smithornis*

Smithornis capensis	African Broadbill
Smithornis sharpei	Grey-headed Broadbill
Smithornis rufolateralis	Rufous-sided Broadbill

    Subfamily Calyptomeninae

        Genus *Calyptomena*

Calyptomena viridis	Green Broadbill
Calyptomena hosii	Hose's Broadbill
Calyptomena whiteheadi	Whitehead's Broadbill

    Subfamily Eurylaiminae

        Genus *Cymbirhynchus*

Cymbirhynchus macrorhynchos	Black-and-Red Broadbill

        Genus *Psarisomus*

Psarisomus dalhousiae	Long-tailed Broadbill

        Genus *Serilophus*

Serilophus lunatus	Silver-breasted Broadbill

        Genus *Eurylaimus*

Eurylaimus javanicus	Banded Broadbill
Eurylaimus ochromalus	Black-and-Yellow Broadbill
Eurylaimus steerii	Mindanao Wattled Broadbill
Eurylaimus samarensis	Visayan Wattled Broadbill

        Genus *Corydon*

Corydon sumatranus	Dusky Broadbill

Subfamily Pseudocalyptomeninae
    Genus *Pseudocalyptomena*
        Pseudocalyptomena graueri   African Green Broadbill

Subfamily Philepittinae
    Genus *Philepitta*
        Philepitta castanea   Velvet Asity
        Philepitta schlegeli   Schlegel's Asity

    Genus *Neodrepanis*
        Neodrepanis coruscans   Common Sunbird Asity
        Neodrepanis hypoxantha   Yellow-bellied Sunbird Asity

BIOGEOGRAPHIC HISTORY AND DISTRIBUTION

Little has been published concerning the biogeographic histories of pittas or broadbills. Olsen (1971) presented one of the first theories relating to the broadbills, based on an assessment of various morphological characters: however, his systematic work has since largely been shown to be incorrect.

Olson (1971) noted that none of the broadbill genera appears to be particularly closely related to any other, and that the rather scattered distribution of few, specialised species and the existence of supposed fossil eurylaimids in Bavaria (Ballman 1969) indicates a retreating family, so that modern broadbills perhaps represent the oldest remaining passerine stock. Olson (1971) therefore speculated that their suboscine ancestors arose in the Old World tropics early in the Tertiary and distributed themselves widely throughout the world in suitable habitats. When, sometime later, more advanced oscines evolved, the ancestral suboscine stock would have been largely replaced everywhere except in South America, which had been isolated during the Tertiary. This isolation permitted the persistence and radiation of suboscines in South America at the same time as suboscines in the Old World were being replaced by more advanced oscine species. As a consequence, relatively few, rather specialised suboscines survived in the Old World, whereas more than a thousand species remain in South America today.

Although Olson's (1971) scenario for the biogeographic history of broadbills may in part be correct, in particular the assertion that broadbills represent a retreating family, some of his other, more specific hypotheses have recently been largely disproved by new evidence. Indeed, the Bavarian broadbills upon which Olson based some of his ideas are probably just barely identifiable as passerines, and placed in broadbills on the basis of size (R. Prum *in litt.*). Hence, whilst Olson suggested that *Smithornis* was the most advanced broadbill, a study of the phylogenetic relationships of Old World suboscines has led to a very different notion, that *Smithornis* is the first distinct lineage within the broadbills (Prum 1993).

Prum's studies indicate a complex biogeographic history and evolutionary radiation of broadbills and asities. He proposes the following scenario leading to the modern-day distribution of the five subfamilies of the Eurylaimidae. Originally, a geographic division of an undifferentiated eurylaimid ancestor gave rise to the primitive genus *Smithornis* in Africa and to the common ancestor to all other genera in Asia. Following this event, diversification in Asia resulted in the evolution of *Calyptomena*, whilst the common ancestor dispersed or expanded into Africa/Madagascar. Isolation between Africa/Madagascar and Asian lineages then occurred. Subsequently, the ancestral African Green Broadbill (proto-*Pseudocalyptomena*) was isolated from Madagascan asities, leading to the evolution of modern-day asities on Madagascar and *Pseudocalyptomena* in Africa, whilst the Asian lineage underwent complex diversification to produce the five insectivorous broadbill genera that we find in Asia today.

Today, the centre of diversity of both pittas and broadbills is in the Sunda Region. In north Borneo, for example, 5-6 species of pitta and five species of broadbill coexist in some areas of hill forest. If contiguous areas of forest that extend from the lowlands to the mountains are considered together, a remarkable nine species of broadbill (of which two are endemic) and six resident (two endemic) and two migratory species of pitta can be found in relatively small areas. The number of pittas and broadbills occurring in Sumatra is similar, with seven or eight breeding species of pitta (two endemic) and seven broadbills. In contrast, there are only two species of pitta and four species of broadbill in the whole of Africa, and four species of asity on Madagascar. In the Afrotropics, no more than four suboscines are likely to be found at any site. In Madagascar, sites exist where three species of asity coexist, but in general no more than two are present.

There are no broadbills in Wallacea or Australasia, and no more than two species of pitta occur sympatrically anywhere to the east of Wallace's Line.

The biogeographic history of pittas has never been investigated in detail, even though John Whitehead made a provisional attempt to investigate this as early as last century, when he compared the adults and juveniles of various taxa to form hypotheses regarding the relationships between different species. For example, he noted that immature pittas of buff-breasted pittas (such as Blue-winged Pitta *P. moluccensis*) are similar to the adults, whilst those of other species such as Hooded Pitta, without buff breasts, tend to be more similar to the young of buff-breasted species than they are to their parents, suggesting that the buff-breasted species are more primitive (Whitehead 1893). Even the juveniles of well-differentiated species such as Red-bellied Pitta *P. erythrogaster* and Ivory-breasted Pitta *P. maxima* have buff on their breasts. The fact that African Pitta *P. angolensis* is also buff below further supports the idea that ancestral pitta stock was buff-breasted. Auber (1964), who studied the production of colours in pittas, provided evidence that the green upperparts of African pittas and many Asian and Australasian forms was also derived from an ancestor with green upperparts, although it should be noted that the African pittas differ from Asian species in the structure and pigmentation of their green feathers, suggesting early divergence in histological details despite retention of similar coloration. That the immatures of both African species very closely resemble that of Indian Pitta *P. brachyura* is also of interest. This also suggests close ancestral links, and leads to the suspicion that, like Indian and African Pitta, ancestral pittas may have been migratory: indeed, populations of six of the nine extant buff-breasted species are migrants.

The migratory habits of an ancestral pitta, or of several species of extant pitta breeding in northerly latitudes, may have resulted in the colonisation of new territory to the south and the subsequent evolution of new species. This seems one explanation for the occurrence of the buff-breasted Elegant Pitta *P. elegans* in Wallacea, Noisy Pitta *P. versicolor* in Australia, and Black-faced Pitta *P. anerythra* in the Solomon Islands, all of which may have derived from a migratory northern stock. Vagrant Blue-winged Pittas from continental Asia, for example, are known to have reached Sulawesi, Christmas Island and Australia. Two other northern, buff-breasted species, Indian Pitta *P. brachyura* and Fairy Pitta *P. nympha*, are also long-distance migrants.

Wallacea and Australia have, after all, been separated from what now constitutes the African continental land mass for at least 140 million years, and there has never been a land bridge in recent geological time that forest bird species from mainland Asia would have been able to cross to Australasia, although there is good evidence that Sulawesi was once joined to continental South-East Asia (Audley-Charles *et al.* 1981, Cranbrook 1981). Although the process of island-hopping could have been involved in the spread of pittas to Australasia, the idea that migrants were the main source of original colonisation seems just as plausible.

The other obvious candidate for colonisation of the Australasian and Wallacean region through migration is Hooded Pitta *P. sordida*. Its most northerly population, *P. s. cucullata*, is a relatively long-distance migrant, whilst populations of the nominate subspecies, in the Philippines, are known to have strong dispersive abilities. Although 'island-hopping' may be an explanation for the emergence of many distinct subspecies of Hooded Pitta on New Guinea and its satellites, it becomes difficult to explain colonisation in this manner when it is considered that only Sulawesi, in the whole of Wallacea, has a representative of this species. This absence of Hooded Pittas from most of the Wallacean islands strongly suggests that Hooded Pittas in New Guinea and its satellites are just as likely to have arisen through colonisation by a migratory form.

The origins of the small, red-bellied species of pitta remains unclear. Whitehead (1893) pointed out that evidence suggests that Blue-banded Pitta *P. arquata* (endemic to Borneo) is a link between the garnet pittas of the Sunda Region and the more easterly distributed Red-bellied Pitta *P. erythrogaster*. If we assume that similarity between juveniles is indicative of a closer relationship, this link seems highly probable, Whitehead's evidence being that whilst juvenile Blue-banded Pitta resembles that of sympatric Black-headed Pitta *P. (granatina) ussheri* and Garnet Pitta *Pitta (g.) granatina*, the adult more closely resembles adult Red-bellied Pitta.

Red-bellied Pitta is also clearly related to Whiskered Pitta *P. kochi*, and the existence of two clearly related sympatric species that differ considerably in size and altitudinal preference perhaps suggests that there were two waves of invasion into Luzon by the ancestral stock of these species. Red-bellied Pitta, unlike the garnet pittas or Blue-banded Pitta, has managed to disperse to many of the islands of Wallacea and Australasia, and very distinctive taxa have evolved on many of these islands. Whether these taxa are really representative of a single species remains unclear, and a detailed investigation into the *P. erythrogaster* group would be of great interest to both modern taxonomy and phylogeny.

EVOLUTIONARY ECOLOGY OF BROADBILLS AND ASITIES

Whilst it is not possible to reproduce Prum's (1993) detailed arguments here, available data suggest that the radiation in bill morphologies now apparent in modern-day broadbills and asities derived from a common ancestor with a wide bill and wide gape. The evolutionary transition in bill morphology, from the ancestral form, which was primarily adapted to insectivory, to one suited to frugivory (with the loss of a wide bill but retention of a wide gape), probably occurred twice. This evolutionary change apparently occurred independently, first in the genus *Calyptomena*, and second in the *Pseudocalyptomena* asity clade. Subsequent to this, the sunbird asities (*Neodrepanis*) lost the wide gape and the bill elongated as an adaptation to nectarivory and floral insectivory.

Within Madagascar, the evolution of the specialised nectarivorous feeding ecology from a frugivore or more general frugivore-nectarivore ancestor is maybe a consequence of a depauperate flora, or perhaps of strong seasonality, with periods of fruit scarcity. Fruit-eating is rare amongst birds in Madagascar, with only a few frugivorous genera and species present in the avifauna (e.g. *Treron*, *Alectroenas*, *Coracopsis*, *Philepitta* and *Hartlaubius*).

DISTINGUISHING CHARACTERISTICS OF PITTAS AND BROADBILLS

Within the passerines, pittas are defined by the following characters: short wings and tail; stout bill, long tarsus with the podotheca entire in front and smooth (bilaminate planta tarsi); temporal fossae extending across the occipital region of the skull, nearly meeting in the midline; syrinx bronchial (haploophone), lacking intrinsic muscles (Sibley and Ahlquist 1990); syrinx lacking pessulus (Prum 1993). In addition, both sexes of most species have colourful plumage.

Typical broadbills have a broad head with large eyes and broad gapes, and 11 primaries. Broadbills are separated from other passeriformes on the basis of the following (Sibley and Ahlquist 1990): the tendon of the flexor hallucis is connected by a vinculum to the tendon of the flexor profundus; toes three and four are joined at the base; podotheca with scutes anteriorly and small, six-sided scales posteriorly; 15 cervical vertebrae; 12 tail feathers; main leg artery the ischiatic. *Pseudocalyptomena*, and perhaps other broadbills, has paired carotids, in contrast to other passeriformes. *Calyptomena* broadbills are distinctive, not only in their specialised frugivorous diet (shared only with *Pseudocalyptomena* and *Philepitta*), but also in having a relatively short bill that is partially hidden by a tuft of antrorse loral plumes, and in possessing only ten primaries (Sibley and Ahlquist 1990).

Asities differ from other broadbills in having 12 tail feathers and in having the tarsus covered in scutes arranged in regular series and the syrinx encircled by a heavy uppermost bronchial ring (Sibley and Ahlquist 1990). At least one asity (*Philepitta castanea*) has a plantar vinculum as other broadbills, although *Neodrepanis coruscans* does not show this feature (Raikow 1987). *Philepitta* and all broadbills apart from *Smithornis* have pointed, bladelike spina externa of the sternum. In *Neodrepanis* this is weakly formed, but in nearly all passerines, and in *Smithornis*, it is strongly forked, with extensive lateral arms (Olson 1971, Prum 1993). Male asities possess a characteristic eye-wattle or caruncle that changes in size, becoming prominent and brightly coloured when breeding, but vestigial at other times. The vivid colours of the caruncles of Velvet Asity *Philepitta castanea* and of both sunbird asities *Neodrepanis* are produced by constructive reflection from arrays of collagen fibres, a mechanism that, as far as is known, is unique amongst animals (Prum *et al.* 1994, R. Prum *in litt.*).

Finally, it should be mentioned that all broadbills and asities, regardless of behavioural and ecological differences, have pendant nests with a side opening.

FOOD AND FORAGING

The pittas, broadbills and asities, between them, have developed a diverse array of feeding techniques, and depend on a rather wide range of food resources. Whilst pittas are relatively uniform in their feeding preferences and habits, the broadbills and asities are far more diverse.

Pittas

Pittas are terrestrial insectivores, their diet consisting almost entirely of invertebrates that are caught on the ground or on structures such as logs or boulders. Pittas most frequently forage by turning over leaf-litter and digging in damp soil, where they catch a wide variety of prey items that vary in size from small ants to large crickets and earthworms. Prey items are often battered on a nearby log or branch, and when young are being fed, such items are frequently tossed to one side whilst further prey are caught before four to five prey items are gathered up to deliver to the nest simultaneously.

Most observations of pittas suggest that they are quite active when searching for prey, and Gurney's Pitta has been observed jumping some 10cm into the air to take a spider from a web. Like many other rainforest birds, pittas may also take advantage of the opportunity to feed on emerging alate termites, which are a rich source of food. Baker (1926) observed Hooded Pittas feeding on flying termites by leaping into the air after them or pursuing them with remorseless energy: those that crept under leaves or moss were promptly exposed by kicking away the vegetation and being seized in the bill. During more leisurely feeding, pittas occasionally pause to cock their heads to one side, in a thrush-like manner, presumably whilst listening for prey. African Pittas have been observed to feed by standing motionless in deep leaflitter for 3-5 minutes whilst watching for prey and then hopping to a new spot 1-2m away (Rathbun 1978). If a potential prey item is spotted, the bird lunges at it and captures it in its bill.

Whether or not pittas are opportunists, eating any edible items that they come across, or more specialised, is unclear, since the diets of pittas have never been studied in the field. However, evidence suggests that there may be some degree of specialisation, at least when breeding. Studies of nesting Gurney's Pitta suggest that adults primarily catch annelid worms (earthworms) to bring to the nest. During two years of study at three nests of Gurney's Pitta, Round and Treesucon (1986) and Gretton (1988) found that 73-78.8% of identified food items were annelid worms. However, since annelids are easier to identify than other prey items the actual proportion may have been lower, and prey items less than 1cm long were certainly under-recorded. Nevertheless, it would appear that the great majority of items fed to fledgling Gurney's are annelid worms, indicating either that there is some degree of specialisation or that these prey were the easiest to procure at the time of nesting. In a similar study of Giant Pittas *P. caerulea*, at least a third of prey items fed to young were earthworms, and another third were snails (Round *et al.* 1989), perhaps indicating a preference to feed young on soft foods.

Earthworms (annelids) feature in the diets of many pittas, and are often highly sought-after. A Hooded Pitta kept by Harrison (1964) and fed liberally with various natural foods showed a strong preference for earthworms, consuming 986 of them, as well as 574 crickets, 28 caterpillars, 32 dragonflies and butterflies, five spiders and two wasps over a 15-day period. On one day, the food fed to this captive bird (which itself weighed 69.5g) was carefully weighed, and the bird consumed 83.5g. Harrison (1964) estimated that, when soft items were readily available, the bird's daily intake never dropped below 70g, but when soft items were limited and harder food was offered, much of the latter was left and daily intake dropped. Although the bird ate diverse prey items that were offered, including large, hard shiny beetles (one of 34g), and a large, fiercesome centipede (15cm long and weighing 99g: even the poison claws were consumed), it completely rejected hairy caterpillars (dead or alive) and lizards, including tiny geckos. It also showed little interest in spiders, wasps more than 2.5cm long, snails more than c.7mm across and small fish.

Apart from annelids, items fed to young in the nests of Gurney's Pitta that were studied by Round and Treesucon (1986) and Gretton (1988) included large winged insects, insect larvae, de-shelled snails, slugs, a locust, cicada, beetle, cockroach, large white butterfly and, on four occasions, a small frog. Although de-shelled snails were included in the nestling diet, no anvils have been found, and no shell-removing activity was noted during these studies. Other species of pitta that regularly eat snails, such as Giant Pitta and Noisy Pitta *P. versicolor*, have been shown to have 'anvils', these being suitable rocks where snail shells are regularly broken.

Seeds have been found in the stomachs of a number of pittas including Graceful, Garnet, Noisy and Rainbow Pitta. Whether fruits are eaten regularly is unclear, although this seems rather unlikely. It seems more plausible that pittas eat fruits off the ground that are infested with various insects or insect larvae. The most unexpected feeding record for any species of pitta was perhaps that reported by Becking (1989). Stomachs he examined of a pair of Banded Pittas that were found dead in east Java contained feathers of a nearby dead White-rumped Shama *Copsychus malabaricus* that had been trapped, indicating that the birds had been pecking at the corpse of this bird before becoming trapped themselves and dying. In this instance too, it is perhaps more likely that the pittas were pecking at maggots in the corpse rather than eating the dead shama's flesh. It should be noted, however, that 'field mice' have been recorded in the diet of Blue-naped Pitta and small snakes in the diet of Giant Pitta, indicating that larger pittas have the ability to feed on smaller vertebrates when opportunities arise.

Broadbills and Asities

In comparison to pittas, members of the Eurylaimidae exhibit a wide spectrum of feeding habits and diets, and display a variety of adaptations to specialised diets. A minority of species are specialised frugivores (*Calyptomena* and probably *Pseudocalyptomena*, as well as the *Philepitta* asities), with the remainder being insectivorous. However, as noted by Fogden (1970), the lack of a sharp cutting edge to the bill, such as those possessed by sympatric frugivorous barbets *Megalaima* sp., means that *Calyptomena* broadbills are unable to manipulate large fruits easily and therefore tend mostly to eat soft fruits, such as figs. The importance of figs in the diet of Green Broadbill is clearly demonstrated by the fact that they were recorded eating the figs of 21 species of *Ficus* in a 2.1km² study site in Peninsular Malaysia (see Table 2 in species account) (Lambert 1987, 1989b).

The insectivorous broadbills have bills that are long as well as broad, and often very strongly hooked at the tip, making them formidable predators of large insects such as orthopterans (crickets, grasshoppers, katydids) as well as enabling some species to tackle small vertebrates. The Black-and-Red Broadbill, for example, has been recorded eating freshwater crabs and on one occasion a small fish, whilst a Banded Broadbill collected by Davison contained a 10cm-long lizard (Baker 1926). The prominently hooked bill of the singularly dull Dusky Broadbill is disproportionately large and is reported to be the widest of any of the Passeriformes, having almost the appearance of the bill of a frogmouth (Podargidae). Although the diet of this largely canopy species is poorly known, the size of its bill suggests that it may specialise on very large prey items. Most species of insectivorous broadbills also very occasionally eat fruit, as indicated by the presence of seeds in specimens of a number of species.

The Madagascan asities are perhaps the most interesting of the Eurylaimidae in terms of dietary specialisation. Velvet Asity is apparently a specialised frugivore-nectarivore (with fruit being more important); Schlegel's Asity has a similar diet, but nectar and perhaps pollen may be relatively more important; whilst the two sunbird asities (*Neodrepanis*) have evolved in parallel to the true sunbirds and developed the very specialised adaptations associated with a nectarivorous or pollen-feeding diet, including the long, fine, curved bill and long, tubular tongue. Salomonsen (1965) reported that, in contrast, the tongue of Velvet Asity is unspecialised, but a recent examination of this organ by R. Prum (*in litt.*) shows otherwise: it is bifid near the tip, and the tip is brush-like. This, and an observation of Velvet Asities feeding at the flowers of mistletoes, strongly suggests that the species is partially nectarivorous. Nevertheless, the great majority of feeding observations to date are apparently of fruit-eating. Examination of the tongues of two specimens of Schlegel's Asity at BMNH confirm that this species also has a specialised tongue. Whilst the tongues of the specimens appear to be damaged, the tip is clearly brush-like, although the fine brushes are perhaps shorter than those of Velvet Asity. Furthermore, the tongue is attenuated, with a narrow section at the tip that appears to be formed by rolling of the tongue, to form a short tube. Whether this rolling is natural, or resulted during preparation of the specimens, is unclear. However, rolling of the tongue at the tip to form a fine tube might be envisaged to have arisen as an adaptation to nectarivory.

SOCIAL AND BREEDING BEHAVIOUR

Pittas

The social and behavioural biology of pittas is very poorly studied, and, with few exceptions, descriptions of behaviour derive largely from anecdotal observations.

Most pittas appear to be territorial, and Blue-winged Pittas have been netted on the same territory for at least five successive years (McClure 1974). Baker (1934) documented the nesting of a pair of Blue-naped Pittas, presumed to be the same birds, that returned for ten successive years to build a nest in almost exactly the same site.

Even migrant pittas on their wintering grounds appear to be territorial: ringing studies in Peninsular Malaysia have shown that wintering Hooded Pittas may return to the same territory in different years (Medway and Wells 1976), whilst marked African Pittas occupied the same non-breeding locality in Kenya during two successive years (Rathbun 1978). Indian Pittas are also known to be territorial in both breeding and wintering grounds (Henry 1971). Resident pittas may also aggressively defend their territories against conspecifics: in Sabah, two male Giant Pittas were observed in a long chase down a slope after a period of calling, and were seen briefly scuffling whilst locked together on the ground, before continuing the chase, in low flight, further down the slope (pers. obs.).

Although pittas are territorial, anecdotal observations of Gurney's Pitta suggest that unpaired males may rove widely and that pairs may move territory during the breeding season following an unsuccessful breeding attempt (Gretton 1988). In an intensively studied site in Thailand, the density of this endangered species was very low, being no more than 3.6-6.0 breeding pairs per km². Whilst Gurney's Pitta has been studied in detail (Collar *et al.* 1986, Round and Treesucon 1986, Gretton 1988, Round 1992, Gretton *et al.* 1993, in press) details of home range size of other species are scant. However, Keith *et al.* (1992) reported that 28ha of dense deciduous thicket in the Zambezi Valley contained six pairs of African Pittas, suggesting a population density in this habitat in the order of 43 birds per km². Although the exact size of territories occupied by African Pittas during breeding is unknown, during the non-breeding season marked African Pittas occupied extremely small home ranges of some 0.3-0.36ha at Gede Ruins, Kenya (Rathbun

1978). In contrast, two Black-headed Pittas that were radiotracked in Sabah, one for 1.5 months and the other for 34 days, ranged over areas of c.4.5-4.7ha (Lambert and Howes in prep.), suggesting a maximum population density in the region of 40-44 birds per km^2 in uniform habitat. Both birds favoured specific ravines, spending considerable periods of time in these areas. The two individuals tracked appeared to be in adjacent territories; neither was ever detected within the area used by the other.

Udo Zimmermann (in press) documents several social signals made by Rainbow Pittas in Australia, which he studied intensively during 1993-94. These included a 'bowing display', interpreted as being a social signal used to maintain territorial boundaries, and usually involving two birds. During the display, the birds alternate between an upright posture and bowing movement, made in slow motion. Growl-like calls accompanying bowing movements were not heard at other times. A second behaviour documented by Zimmermann was 'wing flicking', which is given as an alarm or to deter or distract the approach of potential nest predators. During wing flicking, the adult flicks its wings at short intervals, whilst simultaneously giving an alarm call. Occasionally, when an observer approaches a nest with young, nervous or highly alarmed adults perform a 'wing spreading' display, in which the wings are partly opened and held in a vertical position for c.3s. A final behaviour recorded by Zimmermann was a 'ducking posture' that was seen when several drongos dived to capture prey flushed by a foraging Rainbow Pitta: the pitta reacted by ducking, with its tail in the air, breast near the ground and wings slightly open, but the head stretched backwards with the bill pointing upwards. Another Australian species, Noisy Pitta, regularly performed a very similar ducking display in captivity, apparently as a threat (Norris 1964) and Blue-winged Pittas adopt a similar posture when threatened (pers. obs.). Captive African Pittas also adopt a low crouching posture with wings extended sideways and the bill pointed upwards, this being associated with aggressive behaviour (Sclater and Moreau 1933).

When approached by an observer, African Pittas may also crouch low with the breast feathers puffed out momentarily so that the feet are hidden, before stretching up to full height with the breast puffed out and the buff supercilia erect at the rear, to form 'horns' (Keith *et al.* 1992). Closely related Green-breasted Pittas have also been observed to raise the buff supercilia to form erect tufts if alarmed. Long pointed feathers at the tip of the supercilia are present in other species of pitta, although these may normally be flattened against the head, as in Elegant Pitta, but are sometimes permanently exposed as horns, as in Eared Pitta and Black-headed Pitta. Whether these horns have any social or behavioural function in these species is unknown. Certainly, the feathers of the head of a number of species seem to be important in providing social signals. For example, Blue-winged Pittas may raise the crown feathers into a high crest when handled, presumably as a response to danger (pers. obs.), whilst Azure-breasted Pittas have been observed to raise the crown feathers to form an impressive helmet-shaped crest whilst calling (pers. obs.). During handling, one Azure-breasted Pitta raised its crown whilst simultaneously erecting its inner wing-coverts to form a gorget of azure feathers that faced forward (J. Hornbuckle *in litt.*). Giant Pittas have also been observed to ruffle out their head feathers in the hand (M. Woodcock verbally).

The bowing behaviour noted above is not restricted to Rainbow Pitta. It has also been noted in Red-bellied Pitta (Beruldsen and Uhlenhut 1995, Zimmermann in press) and in Hooded Pitta (Lansdown *et al.* 1991). Details of this behaviour by these species are limited, but in Hooded Pitta bowing is accompanied by a typical alarm call, rather than a vocalisation that is apparently associated only with bowing. Garnet Pitta has also been seen to bow whilst uttering an unusual call (M. Chong *in litt.*), but in this instance the behaviour was believed to relate to a form of greeting associated with pair-bonding: whether this interpretation is correct or not is, however, unresolved. 'Head bobbing' by Hooded Pitta was also observed as a response to disturbance or as a reaction to alarm calling. Like Rainbow Pitta, Hooded Pitta flicks its wings in response to danger, or as a distraction display when it has young close by (Lansdown *et al.* 1991). Occasionally, Hooded Pittas combine wing flicking and bowing. In extreme alarm, they may perform a wing-spreading behaviour, similar to that observed by Zimmermann for Rainbow Pitta, but the wings are held out for some 30-45s. Wing flicking has also been noted in Blue-winged Pitta, Mangrove Pitta (Lansdown *et al.* 1991) and Indian Pitta (noted when a rival Indian Pitta enters the territory of another: Henry 1971).

Captive Hooded Pittas have also been observed performing various displays related to courtship and breeding. Birds in captivity are usually very aggressive towards each other and frequently fight, so that it is difficult to keep two birds, even a pair, together during the non-breeding season. However, with the onset of breeding, adults (of opposite sex) have been observed to chase each other around and intermittently perform a display in which both sexes stretch up to their extreme height and then suddenly lapse back to their normal posture whilst rapidly bobbing the tail up and down, and making a growling noise. During this period both adults were usually together for most of the day. Occasionally, one bird would also stand in front of the other and flick its wings, and the male was observed crouching whilst making the growling noise. The male made such crouching movements at one place for several consecutive days, and apparently collected nesting material, which littered the floor in the area where he had performed the display. Finally, after about one week of such behaviour, nest-building began (Delacour 1934, Lee *et al.* 1989).

In Africa, pittas have developed a set of displays related to breeding that have not been observed amongst pittas elsewhere, although Gurney's may perform a brief wing-flapping display when calling (Hume and Davison 1878, Gretton 1988). At the beginning of the breeding season, African Pittas perform a display in which they stand on a (near) horizontal branch and jump 25-45cm above the perch before parachuting back to the branch with a few rapid shallow wing-beats (Reichenow 1903, Chapin 1953). A distinctive call accompanies this display. As it jumps, the crimson belly is fluffed out (Ginn *et al.* 1989), whilst between jumps and bouts of calling the bird may sway on its hips from a horizontal to a vertical posture whilst puffing out the breast feathers to expose the red belly (Chapin 1953, Keith *et al.* 1992).

Recently hatched nestling Hooded Pittas have been seen to react to intruders in a way that is reminiscent of the ducking behaviour of adult Rainbow Pittas. When disturbed, the nestlings assumed a posture in which their rumps were exposed and head tucked down under the shoulders whilst their hedgehog-like quills bristled out towards the nest entrance (Lee *et al.* 1989). In contrast, nestling African Pittas are reported to flatten themselves on the bottom of the nest if alarmed, so that they are hard to see.

Broadbills

Unlike pittas, broadbills tend to be sociable birds, and many species occur in small flocks or join the mixed flocks of insectivorous birds that are so characteristic of tropical and subtropical forests. Little is known about the territoriality of broadbills. The impression gained when working at various study sites in South-East Asia is, however, that groups of broadbills may range widely and that different groups may overlap in range. Whether individual pairs leave flocks and defend territories whilst breeding is unknown, although there is no evidence to suggest that they do.

Whilst there have been no detailed ecological studies of any of the broadbills, a brief radiotelemetry study of Green Broadbills showed that this species roamed rather widely in the forest, and that different individuals overlap considerably in their ranging. Green Broadbill is, however, a specialised fruit-eater, and as such its ranging behaviour is likely to differ from that of its insectivorous relatives, since it is very much dependent on tracking fruit resources in the forest, which are rather patchily and often sparsely distributed. During the radiotelemetry study one bird ranged over an estimated 13ha in seven days, whilst another ranged within a 24ha area in only six days (Lambert 1989b). It is not unusual, during an early morning watch, to see a number of individuals of both sexes of this species, and of Whitehead's Broadbills, visiting a single fruiting tree.

More is known about the repertoire of displays exhibited by Green Broadbill than any other species of broadbill, although it should be noted that these displays have only been observed when pairs of birds have been kept together in captivity. In the wild, it should not be ruled out that male displays are performed in the context of some form of social gathering, such as a male lek.

Following nest construction by a female at the San Antonio Zoo, a male Green Broadbill was observed to display vigorously by ruffling head feathers and performing a deep head-nodding display in which the head moved below the level of the perch (Webster 1991). The perches appeared to have been specially chosen, and the males would occasionally approach them in an unusual, slow, butterfly-like flight. After several days of this behaviour, the male regurgitated food for the female and then copulated. Another display performed by male Green Broadbill in close proximity to females was observed after it had ricocheted around the zoo display in a very regular pattern. In this display, the male would sit in a hunched, neckless posture, exposing the tiny patches of bright yellow and black in front of the eye, and simultaneously repeatedly flash its wings whilst opening its mouth wide in a gaping display. Mating has been observed in one collection after such a gaping display. Ritualised courtship feeding, in which berries are regurgitated for the female, sometimes apparently in response to female solicitation, has been observed in collections on a number of occasions.

The most dramatic and astonishing male behaviour observed to date, observed by R. Webster at the San Antonio Zoo, is the beak pirouette display. In this display, the male Green Broadbill launches himself vertically to a tree limb and spins below it, hovering with the bill agape for some 5-10s whilst the female looks on (see species account for full details). To date, however, the female has never been observed to take any obvious interest in these male acrobatics (Webster 1991).

Although little is known about the African Green Broadbill, many authors have commented on the similarity between this presumed frugivore and the Asian green broadbills of the genus *Calyptomena*. Whether this species has any interesting displays comparable to those of the Asian green broadbills remains to be discovered. However, anatomical evidence suggests that this is a possibility, since Lowe (1931) remarked that the pectoralis major muscle of African Green Broadbill is remarkable for its size and thickness. These muscles are responsible for the wing's downstroke, and as noted by Webster (1991) these are the muscles that the Green Broadbill uses to remain in the spinning beak pirouetting display that has been observed in captivity.

Although behaviours associated with breeding are poorly known in other species of Asian broadbill, those of the three African *Smithornis* species are relatively well documented. The most notable behaviour, exhibited by all three species, is an elliptical display-flight in which the bird puffs out and exposes its white back feathers and emits a loud mechanical trill (caused by vibration of stiff, twisted, outer primaries). Rather intriguingly, they are reported to be performed by both sexes (Chapin 1953, Brosset and Erard 1986), although this perhaps needs verification in view of recent observations of what may be leks.

Although the elliptical flights of *Smithornis* broadbills have been interpreted as primarily having a territorial function, at other times they are clearly related to pair formation and as a prelude to copulation. Recently, some very interesting observations related to this display were made by M. Andrews (*in litt.*) on Mt Kupe, Cameroon. Here, Andrews observed four Grey-headed Broadbills in close association, two males, a female and an unsexed individual. The two males were observed making the characteristic elliptical flights from a low horizontal branch whilst the female looked on from a slightly higher perch. The males alternated in flying out in a tight circle, returning to the branch in quick succession and showing the white lower back continuously. Even whilst perched the males swivelled with puffed-out throats with their wings held low to accentuate the slightly raised white feathers of the lower back. The two males were also observed jumping up and down, occasionally doing this together in an alternate fashion, in a manner somewhat reminiscent of the displays of lekking Neotropical manakins (Piprinae). Rufous-sided Broadbills evidently show similar behaviour: Brosset and Erard (1986) observed three displaying in close proximity to each other without any suggestion of aggression, on several occasions. Similarly, Chapin (1953) observed two males that were sitting a few metres apart uttering short whistled sounds whilst a single female sat nearby: the males seemed to be excited since they were wagging their tails up and down and spreading the white patches of the back.

Whether such behaviour represents some form of male lek is presently unknown. In tropical forest bird species that have true leks there is usually strong sexual dimorphism, whilst all known lekking species with altricial young (i.e. dependent on their parents) have specialised diets of fruit, nectar or seeds. Furthermore, an important predisposing condition for lekking species is that males are emancipated from nest attendance and parental care. *Smithornis* broadbills are not known to exhibit any of these attributes, but in view of the fascinating observations of M. Andrews (*in litt.*), Brosset and Erard (1986) and Chapin (1953), and no evidence that males participate in incubation of eggs (see species accounts), further study of the breeding behaviour of these birds would be of great interest. It is conceivable that such research may show that the *Smithornis* broadbills are, like Neotropical *Mionectes* flycatchers and pihas *Lipaugus* spp., sexually monomorphic but polygynous lekking species.

Amongst the broadbills, but excluding asities (see below), the best candidates likely to have leks are the *Calyptomena* broadbills of Borneo and the Sunda region. The breeding systems of *Calyptomena* species are still poorly documented, but given that they are specialised fruit-eaters that show striking sexual dimorphism, and only females apparently build the nest and incubate eggs, it would not be surprising if they are found to be polygynous and to have male display arenas, or leks. Although no clear evidence of lekking has been demonstrated, the ritualised behaviours of Green Broadbills observed in captivity, and the documentation by A. Greensmith (*in litt.*) of two males head-bobbing together in the wild, suggest that further investigation of their breeding system might prove fascinating. Both of the *Calyptomena* species endemic to Borneo are little known, although the observation of at least three male Whitehead's Broadbills perched close to one another and calling in what appeared to be a competitive manner (pers. obs.) is certainly suggestive of lekking behaviour.

Asities

Asities differ significantly in appearance to other members of the Eurylaimidae, and recent research has shown that they also exhibit behaviours that have not been documented in any of their broadbill relatives. To date, only one species has been studied in any detail, but it seems likely that all four asities may have unusual social systems and breeding behaviour.

Recent detailed studies of Velvet Asity at Ranomafana National Park have led to exciting discoveries regarding the breeding behaviour of this species. During January and November 1994, Richard Prum and Tiana Razafindratsita (in prep.) made observations of territorial males, using videotape and sound recording equipment to document their research. Six individuals were colour-banded, three of which were subsequently observed as territorial males. These males were found to set up non-resource-based territories which were defended from other males, and in which they performed elaborate, secondary sexual displays. The display repertoire (see species account for details) included both intersexual courtship elements, such as the erect wing-flap display, which was performed during interactions between resident males and visiting females, as well as ritualised intrasexual competitive elements (gape and hanging gape displays) that were observed when males interacted. Male territories lacked the necessary resources for reproduction since (1) they were too small to include significant food resources; (2) they apparently lacked any individuals of the specific species of tree in which nests are built; and (3) several nests observed were not placed in or even near to any male territory.

One male was occasionally observed to associate with a pair of female-plumaged Velvet Asities that constructed a nest, indicating that some males may establish some pair-bond or social association with females (indeed, it should be noted that the observations of Prum and Razafindratsita may not provide a complete picture, since there is a male specimen in the University Museum of Zoology, Cambridge, that was perhaps shot on a nest last century, although the original label may have actually stated 'west' rather than nest). Despite this apparent behavioural plasticity, the vast majority of male Velvet Asities occupy territories that are not associated with nests and all nests observed were not associated with any male territories. The extreme sexual dimorphism exhibited by Velvet Asity is apparently the result of intersexual selection that is a consequence of the breeding system, with its male displays and polygyny. Other supporting evidence for this conclusion comes from the observation that only females apparently build the nest and care for fledglings.

As well as polygyny, aspects of the behaviour observed by Prum and Razafindratsita (in prep.) strongly suggest that there is cooperation in nest-building (as in some of the Asian broadbills), since two nests that they observed were attended by two female-plumaged birds.

The study by Prum and Razafindratsita clearly demonstrates that the numerous striking similarities between Velvet Asity and the lekking Neotropical manakins (Pipridae) are not coincidental, but relate to similarities in their breeding systems. Both Velvet Asity and the manakins are understorey frugivores that inhabit tropical rainforest. In addition, both exhibit strong sexual dimorphism, with ornamentation and delayed male plumage maturation. Such characteristics are well known in a number of bird taxa, including Neotropical cotingas, as well as the manakins and the birds-of-paradise and bowerbirds of the New Guinea region. As demonstrated by David Snow (1976), specialised frugivory by species in these groups has led to the release of males from parental care and the opportunity for female choice. These traits have then led to the evolution of lekking, intersexual display behaviour and elaborate sexual dimorphism. Although more research is required on all asities, it seems clear that Velvet Asity, at least, falls into this specialised category of frugivorous bird. The displays of Yellow-bellied Sunbird Asities recently observed by F. Hawkins (*in litt.* and in prep.: see species account) also suggest some form of unusual social behaviour. The observations of P. Morris and B. Wright of six male Schlegel's Asities singing along 150m section of trail (P. Morris verbally) is suggestive of dispersed lekking (Prum and Razafindratsita in prep.): the fact that males participate in nest construction (Hawkins 1994), however, makes this unlikely unless there is considerable behavioural plasticity associated with its breeding system.

NESTING AND CARE OF YOUNG

All species of pitta make domed nests that are placed on the ground or in vegetation within a few metres of the ground. In contrast, broadbills and asities construct woven globular hanging nests with side entrances that are suspended at various heights from the ground, or over water.

The construction of domed nests may be important in predator avoidance. Evidence for this is scanty, but Bell (1982b) showed that rainforest birds that built domed nests in southern Papua New Guinea, such as Hooded Pitta, were more successful than those of species that had open nests. Although nests of broadbills and asities are very conspicuous, their pendant nature and careful positioning over open spaces often provides some degree of protection from would-be nestling or egg predators. This is particularly true for those nests that have very long pendants or which are hung from the fine outer branches of a tree or palm. Long-tailed and Dusky Broadbill, for example, frequently hang their nests from the extreme tips of rattan spikes, making them more or less inaccessible to most predators. Baker (1934) once observed a Rhesus Macaque *Macaca mulatta* stretching out to the nest of a Long-tailed Broadbill, which the macaque could almost reach, but in its efforts to grasp the nest it tumbled into the stream below. Not all broadbills, however, suspend their nests from the tips of branches: the *Smithornis* broadbills of Africa and *Serilophus* and *Calyptomena* in Asia hang their nests by weaving the materials across a relatively broad section of a branch.

The habit of hanging nests above water may also contribute to protection from predators and, in its extreme form, Black-and-Red Broadbills may suspend their nests from sticks or other objects that protrude from the water itself. Such strategies, however, may also have a downside, since such nests are susceptible to destruction by flooding. In India, Long-tailed Broadbills have been observed to suspend their nests from telegraph wires stretched across forest, which presumably confers a very high degree of protection from predators. On the other hand, two nests of Black-and-Red Broadbill that were suspended from telegraph wires in Malaysia were both destroyed by strong winds. Nests of this species are, however, frequently destroyed: at least six of 17 nests (35%) that are documented on nest record cards (UM) were destroyed before breeding had been completed. Even for nests that survive destruction there are undoubtedly some that will fail to produce young because of predation or accidents which lead to the death of the nestlings or eggs.

Typically, a very high proportion of all nests in tropical forests fail, and this is largely due to predation (Snow 1976). Snakes, large lizards, birds, rats and other arboreal mammals all take their toll, although in the case of pittas and broadbills little is known about the species that regularly prey on their eggs and young, or the typical rate of survivorship of young to adulthood. The only species for which some data of this sort exist is Gurney's Pitta. Gretton (1988) followed the progress of three nests of this species in some detail, and found that on average only one young fledged per nest and the mortality rate was at least 72.7% when all known eggs and chicks were considered. Chicks were known to be lost through predation by a snake and, probably, through simply falling out of the nest. Data on Gurney's Pitta compiled by P. Round (*in litt.*) show that young definitely fledged from only four nests out of 13 that contained eggs or nestlings when discovered. In total, eight young fledged from these four nests, from a combined total of 36 eggs and young found in the 13 nests.

Nest materials used by pittas, broadbills and asities are diverse. Many pittas build their nest on a platform of sticks, and most include sticks in the outer covering, where usually mixed with dead leaves and other materials. Although many pitta nests are rather simple, in that they use dead leaves, leaf skeletons and twigs, they are often well concealed by the addition of moss, large leaves and other materials to the roof, this sometimes partially hiding the nest entrance.

Materials used by broadbills in Asia, with the exception of Green Broadbill, are generally more diverse than those used by pittas, particularly since many species construct a trailing 'tail' that dangles below the nest and presumably provides additional camouflage. Pieces of moss, bark, spiders' web and cocoons, among other miscellaneous materials, may be used in camouflaging broadbill nests. Green Broadbills are rather unusual in that their nest is usually made almost entirely of compactly woven coarse fibres, giving the nest a rather uniform appearance. Some nests of Silver-breasted Broadbill may also be made primarily from one material, although the long tail at the bottom is invariably adorned with miscellaneous material. The female Green Broadbill, unlike incubating birds on other broadbill nests on record, sits with her head protruding out of the nest hole.

Nests of African broadbills in the genus *Smithornis* are made from various plant fibres, dead leaves, green moss and twigs, and these are often woven together with the long black hair-like strands of fungi (*Marasmius* spp.). Asities, like other broadbills, construct hanging nests that are usually long and ragged, as well as being camouflaged. The materials used by asities are apparently diverse, and often include moss in the outer part of the nest structure. Schlegel's Asity may collect spiders' webs to hold the nest structure together.

In many accounts grass is mentioned as a nest material, and photographs often suggest that grass is indeed used. However, careful examination of museum collections suggests that the fine grass-like fibres that are often present, and indeed may form the primary component of the nests of some species (e.g. Green Broadbill and African Broadbill), is in fact derived from stripping the dried and rotting (skeletonised) leaves of monocotyledonous plants such as those of various palms or bamboo (which is actually a grass). In the case of nests of African Broadbill, these fibres may be extracted from rotting pieces of wood, the bark of which is still attached in one nest example in BMNH. In appearance, these fibres resemble grass, and reference to grass as a material used in nest construction may sometimes therefore be erroneous.

Although Skutch (1987) makes no mention of pittas or broadbills in his worldwide survey of cooperative breeding and related behaviour, it seems to be quite clear that certain Asian broadbills have helpers at their nests. A number of observers have documented more than an individual pair of broadbills participating in nest-building, and there is good evidence that both Dusky Broadbill and Long-tailed Broadbill have helpers, at least at the nest-building stage. Meyer de Schauensee (1928), for example, reported that ten Dusky Broadbills were building a nest that he collected in Thailand, whilst Phillipps (1970) observed five Long-tailed Broadbills attending a nest in Borneo. In Madagascar, Prum and Razafindratsita (in prep.) observed two female-plumaged Velvet Asities building nests, which also suggests that some form of cooperative breeding may occur.

As mentioned above, Velvet Asity is also an exception to the general rule within the Old World suboscines in that the male does not participate in nest construction or parental care. As far as is known, the only other exceptions are the two species of sunbird asity (but not Schlegel's Asity), and Green Broadbill. It also seems likely that the two other species of green broadbill in Borneo (*Calyptomena* sp.) and perhaps the frugivorous African Green Broadbill also fall into this category, although nest construction by these species, as well as the three species of African *Smithornis* broadbills, is poorly documented.

Both sexes of pittas, and both sexes of most broadbills, are known to incubate eggs. Following hatching, both sexes also feed the nestlings. There are few known exceptions to these rules, these being the Velvet Asity and probably the sunbird asities and *Calyptomena* broadbills, although data are scant.

The nestling period of pittas is generally rather short, with some species fledging after less than two weeks. Hence young pittas leave the nest when still relatively small. Skins of juvenile Blue-winged Pittas that had apparently fledged, for example, were only about two-thirds the overall head and body length of an adult (Round and Treesucon 1986).

Giant Pittas are known to brood their nestlings at night, even when they are 13 to 14 days old, and only two days from leaving the nest (Round *et al.* 1989). Gurney's Pitta young are brooded almost continuously by the female, and occasionally by the male, for the first three days after hatching, and then approximately half the daylight hours for the next one to three days. The female also broods the chicks at night for the first seven days after hatching (Gretton 1988).

The degree of dependency of fledged juveniles on adults is not well documented. Captive Giant Pittas, however, can feed themselves 24 days after hatching (McKelvey and Mille 1979), whilst Everett (1974) reported that captive juvenile Indian Pittas were fully independent within five days, suggesting that the ability to become independent of parents is rapidly acquired by pittas. Broadbills, in contrast, may be more dependent on their parents, with the juveniles of insectivorous Asian species usually staying with parties of adults that presumably include their parents. Observations of Banded Broadbills in Sarawak showed that whilst about 70-80% of food was provided to young birds by their parents 13 weeks after fledging, this had dropped to around 20-30% by the time they were 20 weeks old (Fogden 1972).

MIGRATION AND OTHER MOVEMENTS

Pittas

Despite their secretive habits, and preference for relatively dense vegetation, a number of pitta species are well-documented migrants or partial migrants (i.e. only part of the population is migratory). With the exception of one subspecies of Hooded Pitta, all of these pittas are 'buff-breasted' species, the best known of which are Indian, Blue-winged, African and Fairy Pitta. Some populations of Noisy Pitta are also migratory, and there is good evidence that two subspecies of Elegant Pitta migrate, although little is known about these movements. Details of the migratory patterns of these species are provided in the relevant species accounts.

Whilst these aforementioned species are true migrants, several species of pitta also apparently make regular but rather local movements. In the Philippines, for example, large numbers of Red-bellied Pittas have been trapped at Dalton Pass, Luzon, but next to nothing is known about the destination of such birds, or even whether such movements are regularly made by all individuals in the population. Evidence suggests, however, that these birds do not move very far, and the movements may therefore perhaps be best described as dispersive. This species may also be a partial migrant in its Australian range, moving north to New Guinea during the austral winter (Draffan *et al.* 1983, Coates 1990), although evidence for this is still inconclusive. The reported movements of Blue Pitta and Blue-naped Pitta in the north of their range are apparently largely altitudinal, in response to unfavourable weather conditions or food availability. Unfavourable conditions are probably also responsible for the apparent nomadic movements of Red-bellied Pittas in the highly seasonal forest of southern New Guinea (Coates 1990).

As far as is known, pittas are nocturnal migrants and, as a consequence, they are among the species that are frequently attracted to lights at particular sites along their migration routes. Although nocturnal migration may enable pittas to avoid predation by diurnal raptors, this attraction to lights has unfortunately enabled man to exploit relatively large numbers of at least one species, Indian Pitta, by using lights to attract the birds.

On the other hand, the attraction to lights has enabled some studies of migrating pittas, the best known of which are those that have been carried out at Fraser's Hill in Peninsular Malaysia (Medway and Wells 1976). These studies have shown, for instance, that both Hooded and Blue-winged Pittas move through Malaysia at a very precise time, within a relatively narrow range of variation from year to year, and that both species complete their wing moult on the breeding grounds, prior to migrating. The studies have, however, perhaps thrown up more questions than they have answered: why, for example, does the migration of Blue-winged Pitta start, peak and end earlier than that of Hooded Pitta, and why has the latter been consistently caught in approximately double the numbers of Blue-winged Pitta (Wells 1990a)?

There is some evidence to suggest that migrating Blue-winged and Hooded Pittas may form loose flocks. Batchelor (1959), for example, reported that a large flock of Blue-winged Pittas landed on his boat off the coast of Sarawak on 15 October. Evidence that Hooded Pittas migrate in flocks is more anecdotal, deriving from observations of waves of these birds flying past lights on migration, and the fact that, often, small numbers will be caught in appropriately set nets almost simultaneously (pers. obs.). During migration, falls of Hooded Pitta sometimes occur, perhaps supporting the idea that migrating pittas may flock. F. Verbelen (1993, *in litt.*) reports encountering such a fall in Kaeng Krachen National Park, Thailand, in mid-April 1992. During this period, at least 30-35 Hooded Pittas were observed on a 1.5-2km stretch of jeep track. The birds were mostly very tame and evidently exhausted, and, sadly, at least eight birds were killed by the few cars that used the track.

Broadbills and Asities

There is no evidence that any species of broadbill is migratory, although several may be rather nomadic or make altitudinal movements in response to food availability or prevailing weather conditions. In South-East Asia, although poorly documented, there is some evidence that Green Broadbills, like other specialised frugivores in the region, are nomadic. In December 1968, four Green Broadbills were netted shortly after daybreak or just after dark in montane forest at Fraser's Hill (Wells 1970). Since this species does not usually occur in montane forest, it seems likely that these birds were moving in response to food shortages at lower altitude, in their normal range. Further north, both Long-tailed and Silver-breasted Broadbills apparently move to lower altitudes in harsh winters, presumably when cold conditions make foraging on invertebrate prey very difficult. Movements by other Asian species have not been documented, but in Madagascar the Yellow-bellied Sunbird Asity may be an altitudinal migrant, although in this instance it is not known whether this is in response to the availability of suitable flowers to feed at, or to seasonality in temperature and rainfall.

THREATS AND CONSERVATION

Upon the discovery of Gurney's Pitta, Hume (1875b) wrote that 'no more beautiful or interesting addition to our Indian Avifauna has been made for many a long day...'. Today, this once relatively common creature is probably one of the rarest species of bird in Asia, and the only known population is seriously threatened by habitat loss. The plight of Gurney's Pitta has received much attention and publicity, and as a consequence pittas are relatively well known to conservationists as well as to the general public. Gurney's Pitta, however, is not alone in being under threat. A number of other species of pitta and broadbill can certainly be considered as threatened, and at the lower taxonomic level of subspecies some taxa may well be in more serious trouble than Gurney's Pitta.

In a recent review of the world's threatened bird species (Collar *et al.* 1994), BirdLife International listed seven species of pitta, two broadbills and one asity that they considered to be globally threatened, based on draft IUCN criteria (Mace and Stuart 1994) and the species recognised by Sibley and Monroe (1990, 1993). These were Gurney's Pitta, considered to be Critically Endangered; Yellow-bellied Sunbird Asity, considered to be Endangered; Schneider's Pitta, Superb Pitta, Azure-breasted Pitta, Whiskered Pitta, Fairy Pitta, Black-faced Pitta, African Green Broadbill and Wattled Broadbill, all of which were considered to be Vulnerable. If BirdLife accept the arguments, presented in the species accounts, that the wattled broadbills of the Philippines represent two rather than one species, then the threat categories for these will need reevaluation.

Habitat Loss

Pittas and broadbills with small global ranges (e.g. Gurney's Pitta, Superb Pitta, African Green Broadbill, Wattled Broadbill) or those that breed in areas where suitable habitat has disappeared over much of their former range (e.g. Azure-breasted Pitta, Fairy Pitta, Yellow-bellied Sunbird Asity) are increasingly threatened by loss of habitat. Most species presently have large enough ranges that habitat loss is not of immediate concern, although in the longer term significant changes in habitat availability are likely to affect these and other species of pitta and broadbill.

Fortunately, the majority of species of pitta and broadbill are apparently very adaptable, and are often able to survive in habitats that have been altered by man. Indeed, a number of pittas, including Gurney's, may even prefer secondary growth and regenerating forests to primary forest (Round 1992). Nevertheless, the exact habitat requirements of the majority of species, and their degree of tolerance to habitat disturbance, remain unknown. Although, for example, studies of the effects of logging on birds in Sabah have shown that relatively good numbers of pittas and broadbills can survive in regenerating forests (Lambert 1992), these individuals may be highly dependent on the smaller fragments of original habitat that are invariably found in logging concessions. If such fragments were also removed, it is conceivable that some less tolerant species might disappear. Clearly, until the habitat requirements of threatened pittas and broadbills are better known, it would be a pragmatic conservation strategy to assume that they need relatively intact original habitats. To this it should be added that, as a rule, once forest areas have been disturbed by man, the level of secondary threats, such as fire or hunting, very often increases.

The Bird Trade and Hunting

Although some species of pitta are trapped for the trade in live birds, this cannot be considered a significant threat when compared to habitat loss. However, in instances where little habitat now remains, such as the case with Gurney's Pitta in peninsular Thailand, any trade may be of significance. In peninsular Thailand, trade in pittas and other birds is common, and imitation of calls are used to trap Banded Pittas in mist-nets, whilst fledglings of this species and of Blue-winged and Hooded Pitta are taken from nests (pers. obs.). In the past these methods were also used to trap Gurney's Pitta, but the species is now relatively well protected, and capture for trade is probably no longer a threat, although there are occasional instances where snares set for other species could inadvertently trap and kill a non-target species such as Gurney's Pitta (Round 1992).

Very little information exists on the number of pittas in international trade. Many that are taken from nests probably die before they have the chance to enter any international trade. In 1992, the Malaysian government proposed the inclusion of the Pittidae in Appendix II of CITES. If accepted, this would have led to the monitoring of pittas in international trade, and hence enhance our knowledge of the numbers and species involved in trade, but in the event the Conference of the Parties rejected the proposal. According to the document prepared for the conference (IUCN 1992), very few pittas are recorded in international trade, and of these, none is considered to be threatened. However, since most pittas are not currently listed in the CITES Appendices (the exceptions being Gurney's and Whiskered in Appendix I; Banded and Fairy in Appendix II: WCMC 1995), the real level of trade is undocumented. Inclusion of pittas on Appendix I effectively bans international trade between Parties to CITES, except in unusual circumstances, whilst inclusion in Appendix II requires export permits and documentation of trade. One important stipulation of Appendix II is that export permits are granted only in instances where the Scientific Authority of the member state has advised that such an export will not be detrimental to the survival of that species. This is designed to ensure that the harvest level (i.e. trapping) is sustainable. Due to lack of data on the effects of harvesting, however, it is very unusual that such an assessment can be made, and exports are, strictly speaking, often made in contravention of the convention.

Hunting of pittas for food may be of more significance than capture for trade. Their relatively large size makes them an obvious target for local hunters, and in some parts of their global range special traps are used to capture pittas and other terrestrial birds. For example, in Liberia, local people trap African Pitta in snares set in openings at intervals along barricade fences erected in the forest that may run for many tens of metres (Allen 1930). Van Tyne (1933) reported that hunters in Tonkin, Vietnam, were very fond of the meat of Blue-rumped and Blue-naped Pittas, which they trapped using simple box traps baited with live crickets and frogs. In Vietnam, trapping of terrestrial birds, including pittas, for food is still a widespread practice, although other, more lethal trapping methods are often employed (Lambert *et al.* 1994, J. Eames verbally). In the highlands of Luzon, Whiskered Pittas are caught by local people (Jakobsen and Andersen in prep.), also for food. In some cases, pittas may also be caught for other purposes, the most unlikely of which is perhaps the case of Blue-banded Pittas, which were snared and skinned in Sarawak to make toys for small boys (Smythies 1957). It is not known, however, whether this practice continues today.

Some pittas are also subject to trapping for food on migration. Fairy Pittas are regularly trapped in Taiwan (Severinghaus *et al.* 1991); in the Barail Range of Assam, Hooded Pittas attracted to lights are knocked out of the sky with poles (Choudhury 1986); and in Peninsular Thailand, migrant Blue-winged Pittas are caught in coastal mangroves with snares made from fishing line (J. Howes *in litt.*). Migrant species, fortunately, tend to have large clutch-sizes, and, in view of the relatively long days and long breeding seasons in the higher latitudes where they occur, may even have more than one brood each breeding season. However, the relative rarity of Fairy Pitta on its known breeding grounds and the fact that many areas in its range have been seriously affected by habitat modification and habitat loss is cause for concern, and it is possible that Fairy Pitta is seriously threatened, despite its relatively large range. The Vulnerable threat category allocated to this species by Collar *et al.* (1994) perhaps understates the threats faced by this species, and needs careful reconsideration.

Inadequate Taxonomic Knowledge

Whilst taxonomy *per se* may not immediately spring to mind as a conservation problem, it may be of significance in the context of how conservation priorities are set by leading conservation organisations, such as BirdLife International (formerly ICBP). Firstly, threat assessments for birds are invariably carried out at the species, rather than subspecies, level. Secondly, regional priorities in South-East Asia and Wallacea have, in the last few years, relied to a great extent on the identification of Endemic Bird Areas, based on the geographic range of the world's restricted-range species (ICBP 1992). Rather few endangered species, and usually only those that are considered to be critically threatened, such as Gurney's Pitta, have been the focus for conservation projects. Considering the very limited financial and human resources available to deal with the conservation problems faced by birds, this is clearly an efficient approach, and it would be difficult to justify priorities based on concerns for individual subspecies. However, the whole process of priority setting relies on the correct allocation of taxa to species level.

In South-East Asia and Wallacea, where the majority of pittas and broadbills reside, taxonomy is still inadequately known, and there are many subspecies, including those of pittas and broadbills, that might well be recognised as good species if sufficient relevant data were available. Indeed, it should be noted that if the phylogenetic species concept (Cracraft 1983, 1987) was used instead of the biological species concept usually applied by ornithologists, many subspecies of pitta dealt with in this volume would undoubtedly be recognised as species. The case for adopting the phylogenetic species concept is provided by Hazevoet (1995), and the case against doing so will shortly be provided by D. Snow and (separately) by N. Collar (N. Collar *in litt.*).

The problem of taxonomy is of particular concern with respect to the subspecies of Red-bellied Pitta *Pitta erythrogaster* and Hooded Pitta *P. sordida*. Some of the taxa presently included in these widespread species are very distinctive, and several also have very restricted ranges (e.g. *P. e. cyanonota*, which is endemic to Ternate, *P. s. sanghirana* of Sangihe Island and *P. s. rosenbergii* of Biak). However, because they are not recognised by traditional taxonomists as separate species, but instead belong to well-known wide-ranging species, they are not usually considered in conservation planning or priority setting. Hence, a detailed review of the taxonomic status of these species, even using traditional taxonomic criteria, might make a valuable contribution to conservation, if it means that the conservation needs of presently unrecognised species are investigated as a consequence.

PLATES
1-24

PLATE 1: EARED AND RUSTY-NAPED PITTAS

1 **Eared Pitta** *Pitta phayrei* **Text and map page 83**

South-East Asia from Burma and Thailand to South Yunnan, China and Indochina.

Both sexes have broad whitish-buff lateral head-stripes that can protrude as ears at the back of the head, chestnut-rufous upperparts, and bold black and rufous-buff speckling on the wing-coverts.

- **1a** **Adult male** Black central crown to nape, broad black face-patch and malar stripe. Underparts rich fulvous-orange with black speckling on flanks.
- **1b** **Adult female** Differs from male in having black parts of the head replaced with dark brown, more heavily marked underparts and paler, buff rather than pink undertail-coverts.
- **1c** **Immature** Dull rufescent-brown above with a distinct buffy-white supercilium and ear-coverts from an early age and broken buffy wing-bar. Bill tip dries pale, but colour uncertain.

4 **Rusty-naped Pitta** *Pitta oatesi* **Text and map page 92**

See also Plate 16.

Mountains from Burma to Indochina and Peninsular Malaysia.

Crown and nape rusty, with black line behind the eye. Upperparts green, tinged rusty in the female, underparts fulvous. Often has pink suffusion to throat and blue-tinged rump. Immatures are dark brown with white spotting on the wing-coverts and whitish underparts spotted and streaked darker.

- **4a** **Adult male** (*oatesi*; Burma to Thailand, north-east Laos and western Yunnan) Rump blue or bluish-green (green in female). Amount of pink suffusion on throat varies considerably. Head more uniform fulvous than other subspecies, with fulvous merging into mantle.
- **4b** **Adult male** (*deborah*; Peninsular Malaysia) Darker than other subspecies, with darker green mantle and upperside of tail. Male had very bright pink wash to entire underparts, very pink throat and blue rump. Slightly smaller than other subspecies.
- **4c** **Adult male** (*castaneiceps*; south-east China, central Laos and north-west Vietnam) Green rump. Chestnut-rufous crown clearly demarcated from green mantle. Deeper buff below than *oatesi*, and usually more prominent pink suffusion to throat and ear-coverts and broader black streak behind ear-coverts.
- **4d** **Adult male** (*bolovenensis*; southern Laos on Bolovens Plateau; possibly also in southern Vietnam) Similar to *castaneiceps* but has bright blue rump to lower mantle and variable blue wash to upper mantle, back and inner upperwing-coverts.

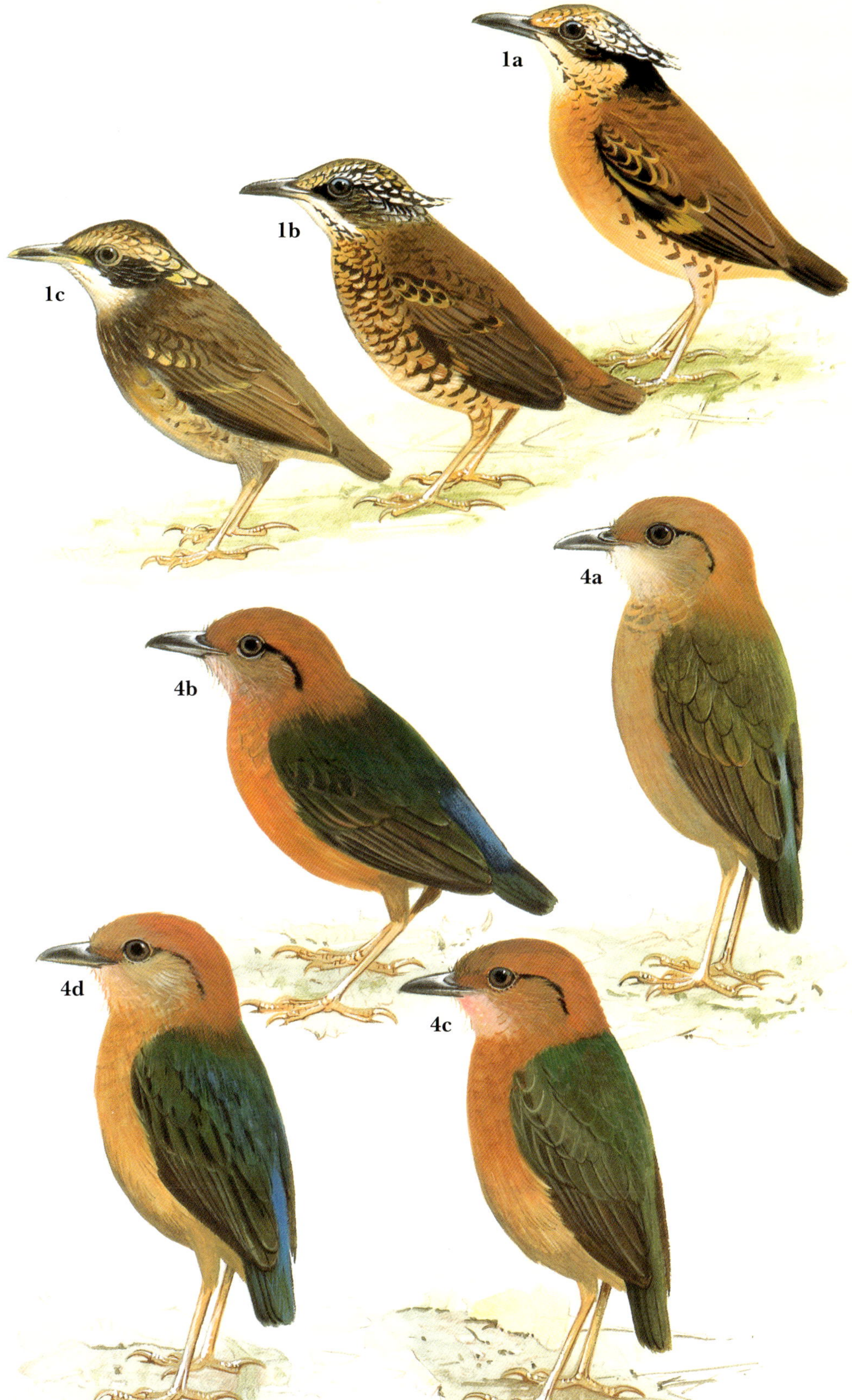
1a
1b
1c
4a
4b
4d
4c

PLATE 2: BLUE-RUMPED AND BLUE-NAPED PITTAS

3   **Blue-rumped Pitta** *Pitta soror*   **Text and map page 88**

See also Plate 16.

Range disjunct, from the Yao Shan range in south-eastern China, throughout Indochina to the Khao Soi Dao range of south-east Thailand, and on Hainan.

Upperparts green, with a blue rump. Generally dull purple or lilac on the forehead and ear-coverts, with a bright orange-rufous supercilium. Colour of nape depends on subspecies. The underparts vary in colour from dull fulvous to a deep orange-buff, often with a pinkish tinge on the breast. Bill pale.

- **3a**   **Adult male** (*tonkinensis*) Green crown and nape, faintly tinged blue. Upperparts tinged blue. Strong pink suffusion below. Underparts paler than those of male *soror*, and more strongly suffused with pink.
- **3b**   **Adult female** (*tonkinensis*; Guangxi [Yao Shan] in China, south to central Tonkin in north-west Vietnam) Dull green crown and nape concolorous with mantle, with no blue tinge.
- **3c**   **Adult female** (*douglasi*; Hainan) Similar to female *tonkinensis*, but lacks any black scaling on breast, is more purple on sides of forehead and more strongly suffused with pink below. Male is similar in appearance to *soror*, but with blue of head and nape replaced with green, and less purple on forehead.
- **3d**   **Adult male** (*soror*; Vietnam in southern Annam and Cochinchina) Crown and nape blue, forehead and forecrown lilac to dull purple.
- **3e**   **Adult female** (*soror*) Duller than male, with green crown.

2   **Blue-naped Pitta** *Pitta nipalensis*   **Text and map page 85**

See also Plate 16.

Southern Himalayas, from central Nepal east to Bhutan and north-east India, south to Bangladesh and northern Burma (*nipalensis*), and in Indochina (*hendeei*).

Head fulvous with bright blue hindcrown and nape in the male, and green nape in the female. The underparts are fulvous and upperparts green, often faintly tinged with blue on the lower mantle and rump.

- **2a**   **Adult male** (*nipalensis*; southern Himalayas) Nape blue, often extending onto the upper mantle. Rump green, sometimes mixed with blue.
- **2b**   **Adult female** (*nipalensis*) Nape green. Usually duller pink hue on underparts than in male.
- **2c**   **Immature** (*nipalensis*) Dark brown with whitish throat and speckling on the sides of head, buffy belly; marked with large, well-defined buff streaks on the crown, buff spots on the back and spotted with white on the breast. Similar to juvenile Rusty-naped Pitta (see also Plate 16).

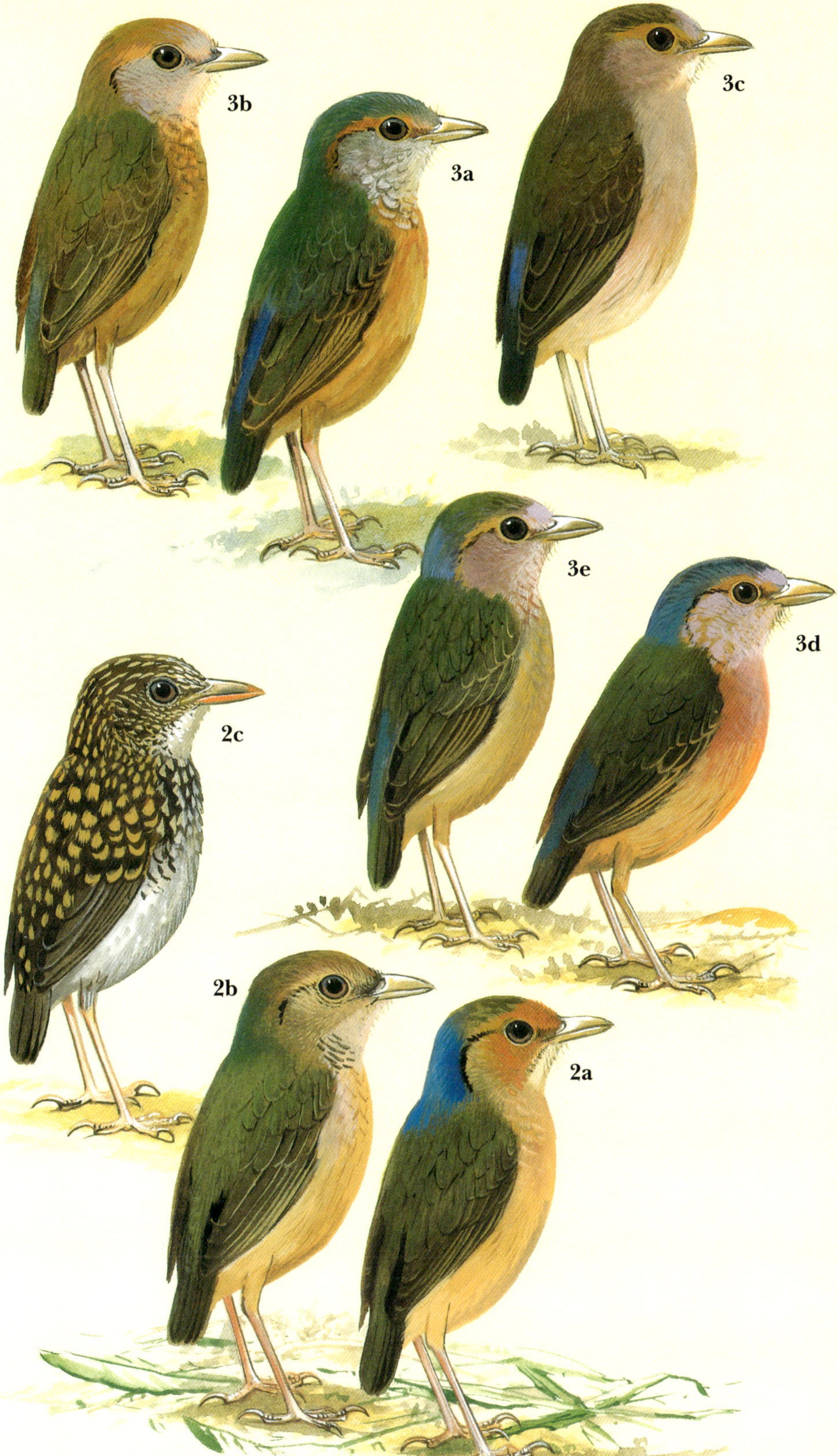
3b
3a
3c
3e
3d
2c
2b
2a

PLATE 3: GIANT AND SCHNEIDER'S PITTAS

6 Giant Pitta *Pitta caerulea* **Text and map page 97**

See also Plate 16.

Peninsular Thailand to Malaysia, Sumatra and Borneo.

A large, robust species, with a diagnostic black necklace.

- **6a** **Adult male** (*caerulea*; Tenasserim and Peninsular Thailand to Peninsular Malaysia) Upperparts entirely blue with a black hindcrown and nape.
- **6b** **Adult female** (*caerulea*) Upperparts rufous brown with black-scaled greyish crown and pale blue rump and tail.
- **6c** **Juvenile** (*caerulea*) Dark brown with creamy-buff to fulvous-buff ear-coverts and crown. Coarse mottled brown and creamy-buff pattern on the nape and hindcrown. Upperparts unmarked dark brown with blue tail. Underparts dark brown, marked with creamy-buff or fulvous-buff that becomes more prominent with age. See also Plate 16.
- **6d** **Adult female** (*hosei*; Borneo) Differs from nominate subspecies in the richer, chestnut crown to nape, which is scaled, rather than barred. Males are similar to male *caerulea*.

5 Schneider's Pitta *Pitta schneideri* **Text and map page 95**

See also Plate 16.

Mountains of Sumatra.

A medium-large species with blue upperparts and chestnut-rufous forehead and crown to nape. The breast and belly are fulvous, with narrow black edges to the feathers of the upper breast, sometimes forming a broken line.

- **5a** **Adult male** Bright dark blue back, rump and tail. Forehead and crown to nape bright chestnut-rufous, separated from the blue mantle by a narrow black line.
- **5b** **Adult female** Differs from the male in less bright coloration and in lacking the black collar on the hind-neck.
- **5c** **Immature** Rufous-brown plumage with white throat, buff scaling on crown and rufous on ear-coverts. Wing-coverts marked with buffy or white spots. See also plate 16.

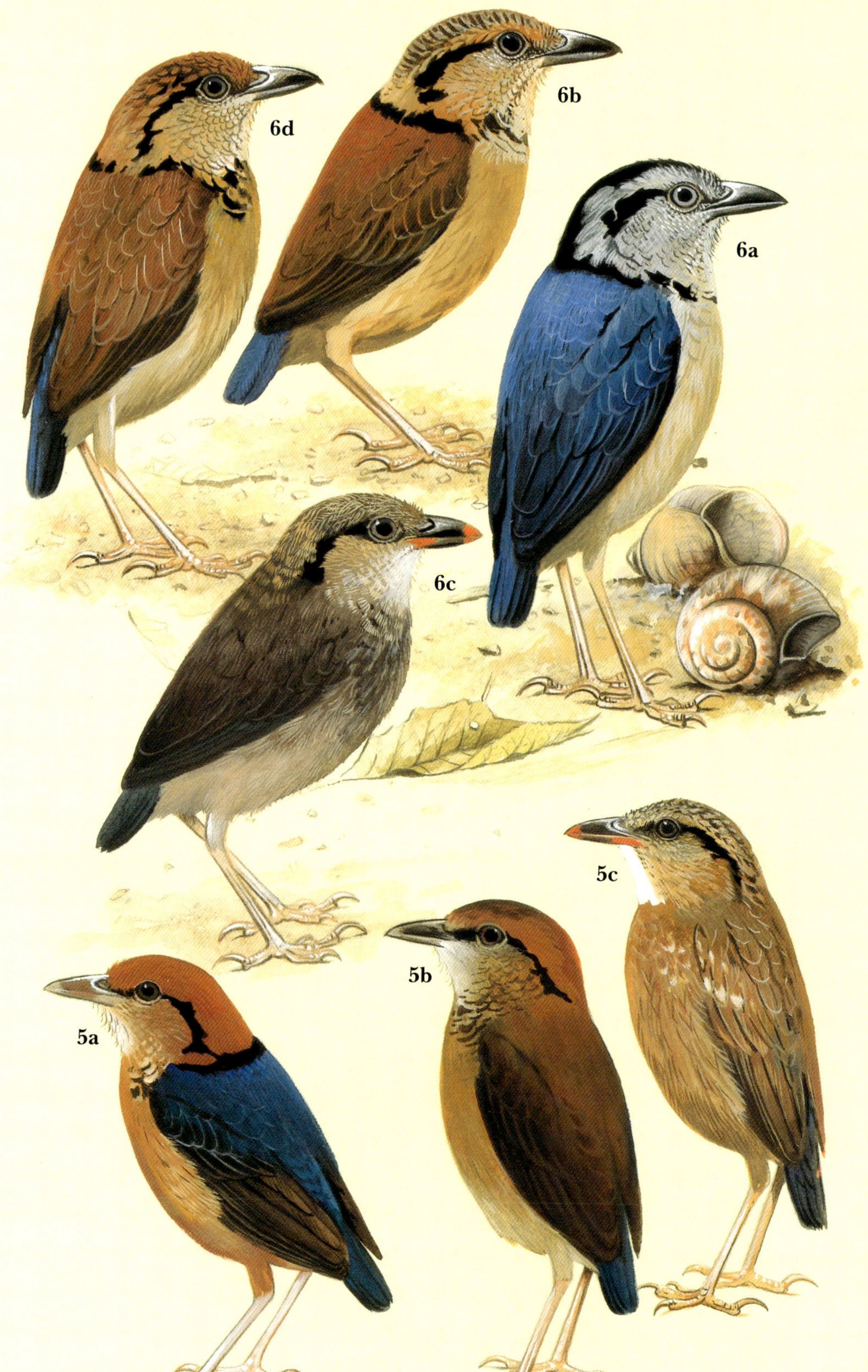
6d
6b
6a
6c
5c
5b
5a

PLATE 4: BLUE PITTA

7   Blue Pitta *Pitta cyanea*   **Text and map page 101**

A medium-sized pitta that ranges from north-east India to Burma, Thailand and Indochina.
Upperparts blue, or olive-brown in females of most subspecies, with scarlet or orange hindcrown. Pattern of underparts very variable, being more or less barred and spotted with black.

- **7a** **Adult male** (*willoughbyi*; mountains in central Laos and southern Annam, Vietnam) Dark individual. Generally brighter than *cyanea*, particularly on the breast, which is usually well marked with red tones.
- **7b** **Adult male** (*willoughbyi*) Pale individual.
- **7c** **Adult female** (*willoughbyi*) Upperparts olive-brown with variable amount of blue on upper mantle, with some individuals also having scattered blue feathers on back. Underparts more or less tinged with red tones.
- **7d** **Adult male** (*cyanea*; north-east India, south-east Bangladesh, Burma, Thailand, Yunnan and north-east Laos; ? north-west Vietnam) Usually lacks red tones to breast.
- **7e** **Adult female** (*cyanea*) Upperparts olive-brown, head dull.
- **7f** **Immature** (*cyanea*) Dark brown with bold rufescent-chestnut streaks on the crown and supercilium, and broader, duller rufous-buff streaks on the mantle. Breast to belly dark brown prominently marked with large rufous-buff spots, but these are rapidly replaced by black-barred or spotted whitish-blue feathers. Tail blue. Bill has bright red base.
- **7g** **Adult male** (*aurantiaca*; mountains of Cambodia and south-east Thailand) Nape and sides of head yellowish-orange.

7a

7c

7b

7d

7g

7e

7f

PLATE 5: BANDED PITTA

8   **Banded Pitta** *Pitta guajana*   **Text and map page 104**

Peninsular Thailand to Malaysia and the Greater Sunda Islands.

The broad bright yellow to orange supercilium, black crown and mask, rich chestnut-brown upperparts with a broad white stripe across the wing-coverts and blue tail make this a most distinctive species. Subspecies vary considerably. Taxonomy needs investigation: may comprise more than one species.

- **8a**   **Adult male** (*guajana*; east Java and Bali) Bright yellow supercilium, a conspicuous violet-blue breast-band and fine violet and yellow or yellowish-buff bars from breast to undertail-coverts. See also *P. g. affinis*.
- **8b**   **Adult female** (*guajana*) Golden-brown crown and golden-buff supercilium, a narrow black band bordering white throat and dull yellowish and brown bars on underparts.
- **8c**   **Adult male** (*irena*; peninsular Thailand and Tenasserim to Peninsular Malaysia and Sumatra) Supercilium yellow in front of the eye, becoming orange, usually behind the eye, and joining to form bright fiery orange on the nape. Underparts purple with a blackish border between the purple and the white throat and prominent orange bars on the sides of the breast.
- **8d**   **Adult male** (*irena*) Individual with restricted orange in supercilium.
- **8e**   **Adult female** (*irena*) Head and upperparts duller than male. Narrowly barred below with dark brown and pale orangy-pink to yellowish-buff. Crown colour variable, usually black but often mixed with chestnut feathers.
- **8f**   **Adult male** (*schwaneri*; Borneo) Underparts and supercilium of the male are yellow, with narrow purple bars from the breast to undertail-coverts and purple centre of the belly.
- **8g**   **Adult female** (*schwaneri*) Supercilium golden-buff becoming yellow at rear. Crown varies from entirely chestnut to entirely black or may be chestnut at the forehead and becoming black at the rear.
- **8h**   **Juvenile** (*schwaneri*) Head pattern reminiscent of adult, with buffy supercilium and dark mask; crown dark, variably mixed with chestnut-buff speckles. Upperwing-coverts tipped with buff (white in other subspecies). The breast is streaked at first but becomes more barred with age.

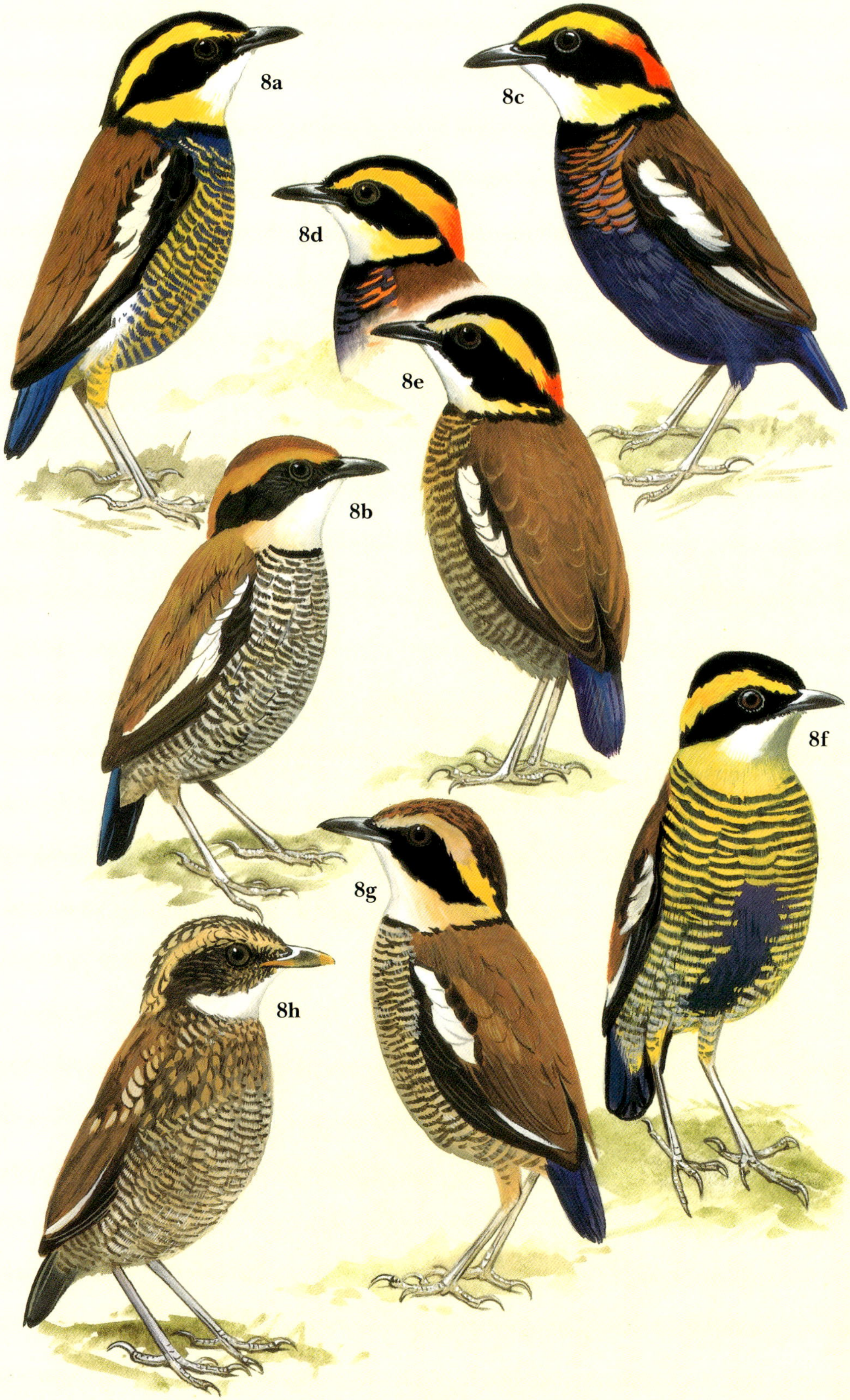
8a
8c
8d
8e
8b
8f
8g
8h

PLATE 6: BAR-BELLIED, GURNEY'S AND BLUE-HEADED PITTAS

9   **Bar-bellied Pitta** *Pitta elliotii*   **Text and map page 108**

Vietnam, Cambodia, Laos, east and possibly south-east Thailand.

The black mask, green upperparts, blue tail and fine yellow-and-black bars of the underparts are shared by both sexes.

- **9a**   **Adult male** Crown blue-green, throat pale blue. Blue belly-patch.
- **9b**   **Adult female** Dull ochraceous-buff head, throat and upper breast, and yellow centre to the belly.
- **9c**   **Immature** Crown streaked with black. Sides of forehead and supercilium unmarked buff. Throat white; breast mixed green and pale brown.

10   **Gurney's Pitta** *Pitta gurneyi*   **Text and map page 110**

Tenasserim and peninsular Thailand.

Adults have rufous-brown wings and mantle and blue tail.

- **10a**   **Adult male** Vivid dark ultramarine-blue hindcrown and nape, yellow underparts with black belly and black bars on the flanks.
- **10b**   **Adult female** Ochraceous-yellow crown, blackish-brown eye-patch and finely barred, black-and-yellow-buff underparts.
- **10c**   **Juvenile** Buffy speckling on the crown and underparts, but largely unmarked dark brown mantle and back, and turquoise-blue tail.

11   **Blue-headed Pitta** *Pitta baudii*   **Text and map page 115**

Endemic to Borneo.

Both sexes have rich reddish-brown mantle and wing-coverts, a bold white bar in the wing and a blue tail.

- **11a**   **Adult male** Bright blue crown to nape. The black mask and white throat contrast with the dark purple-blue underparts.
- **11b**   **Adult female** Head and underparts rich fulvous-buff.

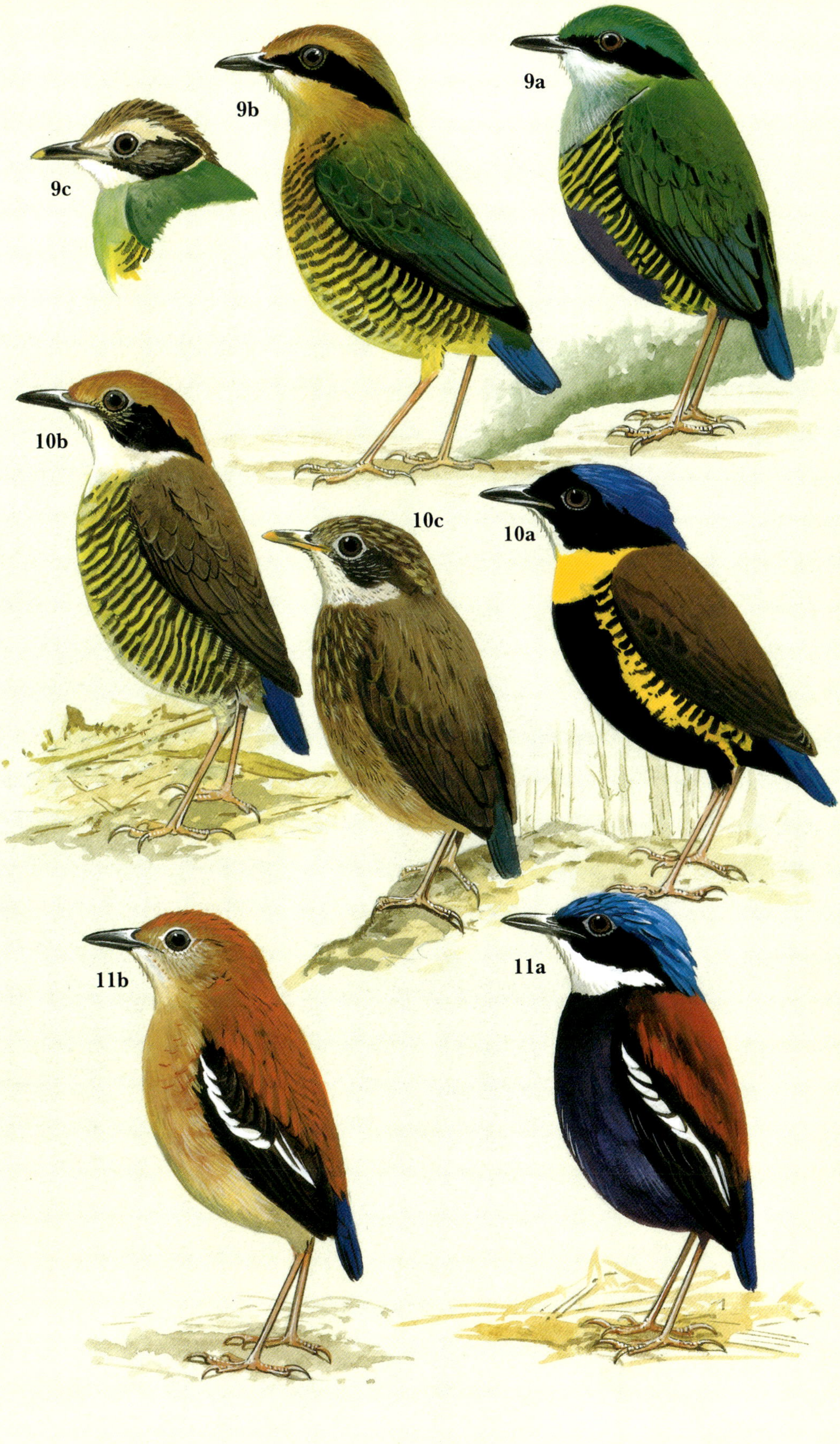
9c
9b
9a
10b
10c
10a
11b
11a

PLATE 7: HOODED PITTA

12 Hooded Pitta *Pitta sordida*   **Text and map page 117**

See also Plates 13 and 16.

One of the most widely distributed pittas, with twelve subspecies ranging from the Himalayas through South-East Asia, the Philippines and Greater Sunda Islands to Sulawesi and the New Guinea region.

Most subspecies have a black head, and all have a green body with a variable amount of azure-blue on the upperwing-coverts, an azure-blue rump and red to scarlet undertail-coverts. Subspecies differ primarily in the colour of the crown and tail, extent of white in the primaries, the extent of black on the upper belly and the amount and shade of the blue or violet on the underparts.

- **12a** **Adult** (*cucullata*; the Himalayas, north-east India, Burma, Yunnan, Thailand and Indochina. Northern populations move to the Sunda Region in the winter) Crown to nape rich dark chestnut-brown. Body paler green than other subspecies. Large white wing-patch. See also *P. s. abbotti* (Nicobar Islands) and *P. s. bangkana* (Bangka and Belitung Islands). Flying bird illustrated on Plate 13.
- **12b** **Immature** (*cucullata*) Plumage dark, lacking streaking or spotting. Head reminiscent of adult. Upperparts unmarked dark brown; underparts unmarked dark buffy-brown. Bold whitish wing-bar. Juvenile illustrated on Plate 16.
- **12c** **Adult** (*forsteni*; Sulawesi) Large, lacks white in the primaries. Tail green.
- **12d** **Adult** (*sordida*; Philippines except Palawan) Underparts pale greenish with blue suffusion; upper belly black, lower belly and undertail-coverts bright red. Tail black, narrowly tipped with blue. See also *P. s. palawanensis*.
- **12e** **Adult** (*novaeguineae*; New Guinea, including the West Papuan Islands) Flanks bluish-green; glittering pale green or greenish-yellow patch across the upper breast. White in wing is confined to a few small spots in the central primaries. Usually has a large black patch on lower breast, surrounded by indigo-blue. See also *P. s. hebetior* (Karkar Island) and *P. s. goodfellowi* (Aru Islands).
- **12f** **Juvenile** (*novaeguineae*) Head black with whitish throat. Body plumage dusky. Lacks any sign of pale wing-flash.
- **12g** **Adult** (*muelleri*; Greater Sunda Islands) Similar to *sordida*, but less extensive or no black on belly.
- **12h** **Adult** (*rosenbergii*; Biak Island) Extensive, deep violet-blue flanks; lacks black on the lower breast. Centre of belly and undertail-coverts scarlet.

12a

12b

12c

12d

12f

12e

12g

12h

PLATE 8: IVORY-BREASTED, SUPERB AND AZURE-BREASTED PITTAS

13 Ivory-breasted Pitta *Pitta maxima* **Text and map page 126**

Endemic to the north Moluccan Islands, Indonesia.

- **13a** **Adult** (*maxima*; Halmahera, Bacan, Kasiruta, ? Mandioli, ? Obi) The head, throat and upperparts are black with a glittering azure-blue wing-covert patch. Underparts silvery-white with crimson belly. In flight, the broad white band across the primaries is distinctive.
- **13b** **Adult** (*maxima*) In flight.
- **13c** **Adult** (*morotaiensis*; Morotai) The blue in the wings is less extensive and of a darker colour, with dull blue replacing the green in the leading edge of secondaries. The white wing-patch of *morotaiensis* is larger and more prominent. Larger than the nominate subspecies.
- **13d** **Adult** (*morotaiensis*) In flight.

14 Superb Pitta *Pitta superba* **Text and map page 128**

Confined to Manus Island, Admiralty Islands.

- **14** **Adult** Glossy jet-black with upperwing-coverts largely iridescent pale turquoise-green and a red patch on the belly. The only pitta on Manus.

15 Azure-breasted Pitta *Pitta steerii* **Text and map page 129**

The Philippines, on the islands of Mindanao, Samar, Leyte and Bohol.

- **15** **Adult** Unmistakable, with black head, green upperparts with prominent azure-blue wing-patch. Pale blue breast and flanks unique.

13a

13b

13c

13d

14

15

PLATE 9: WHISKERED AND RED-BELLIED PITTAS

16 Whiskered Pitta *Pitta kochi* **Text and map page 131**

See also Plate 16.

Endemic to Luzon, where largely montane.

- **16a** **Adult** Significantly larger than Red-bellied Pitta. Dark olive-green above with orange-rufous rear of crown and nape. Bold pale malar stripe. Breast, edges to greater coverts, tail and uppertail-coverts are slaty-blue. Sides of upper breast bright olive; belly to undertail-coverts scarlet.
- **16b** **Juvenile** Illustrated and described on Plate 16.

17 Red-bellied Pitta *Pitta erythrogaster* **Text and map page 133**

See also Plate 10.

Widely distributed on islands from the Philippines to Australasia, with 24 recognised subspecies (but taxonomy requires investigation).

The blue breast and red belly are common to all subspecies.

- **17a** **Adult** (*erythrogaster*; Philippine archipelago, except for Palawan Province) Crown to nape and sides of head bright pale rufous, throat darker brown becoming black on upper breast. Centre of breast and collar blue; rest of mantle and sides of the breast green.
- **17b** **Juvenile** (*erythrogaster*) Dull brown above with blue on rump and tail. Whitish throat, and buffish-white underparts with dark brown or black scaling on breast.
- **17c** **Adult** (*inspeculata*; Talaud Islands, north-east of Sulawesi, Indonesia) Dark chestnut head; entire upperparts blue. Narrow dark blue breast-band. Red of underparts extends onto lower breast.
- **17d** **Adult** (*cyanonota*; Ternate, Indonesia) Lacks green in plumage, like *inspeculata*, but rear crown much brighter. The throat is pale buffy-brown and the whole breast is pale blue. Some birds have a narrow blackish band separating the red and blue.
- **17e** **Adult** (*rubrinucha*; Buru, Indonesia) Similar to *erythrogaster*, but lacks the blue collar and has a duller brown head marked with pale blue on the centre of the rear crown and the rear of the ear-coverts, and a distinctive reddish-orange nape-patch.

17a

17b

16a

17c

17d

17e

PLATE 10: RED-BELLIED AND SULA PITTAS

18 Sula Pitta *Pitta dohertyi*   Text and map page 140

Sula Islands of Indonesia.

    18   **Adult** The sides of the face and entire throat are black, extending as a broad nuchal collar. The nuchal collar is bordered below by a pale blue band, which contrasts with the green back. The blue breast-band is separated from the scarlet-orange belly by a distinct blackish-blue band. The entire upperwing-coverts and outer edges and tips to the secondaries are pale blue with some green tinge in the tertials, and, unlike those of subspecies of Red-bellied Pitta, the entire upperparts shine.

17 Red-bellied Pitta *Pitta erythrogaster*   Text and map page 133

See also Plate 9.

    17f   **Adult** (*celebensis*; Sulawesi, Manterawu Island and the Togian Islands, Indonesia) Differs from *erythrogaster* (Plate 9) in having darker, redder crown and nape, with a variable amount of pale blue on the central crown; narrower blue nape-band; more extensive blue on the breast, separated from the red belly by a broad black breast-band.

    17g   **Adult** (*rufiventris*; North Moluccan islands, Indonesia) Similar to *macklotii*, but paler, duller rufous crown and nape and greyish throat.

    17h   **Adult** (*macklotii*; Western and southern New Guinea; Islands of Misool, Salawati, Batanta, Waigeo and Yapen; and Cape York Peninsula, Australia) Most similar to *celebensis*, but throat darker, crown to nape bright orange-rufous and lacks pale blue on head and upper mantle. The width of the black breast-band varies considerably.

    17i   **Adult** (*gazellae*; New Britain and Umboi Island, Bismarck Archipelago) Differs from *macklotii* in having brighter orange-red sides of the rear head and nape, and in the possession of a few blue feathers on the crown. Dark breast-band rather indistinct.

    17j   **Adult** (*extima*; New Hanover Island, Bismarck Archipelago) Similar to *macklotii*, but nape brighter orange-red and lacks black breast-band.

    17k   **Adult** (*meeki*; Rossel Island, Louisiade Archipelago) Similar to *macklotii*, but paler, duller rufous crown and nape and greyish throat.

    17l   **Adult** (*finschii*; Fergusson and Goodenough Islands, D'Entrecasteaux Archipelago) Upperparts and broad breast-band entirely blue. Nape dull chestnut, concolorous with the crown. Differs from *cyanonota* (Plate 9) in larger size and head colour.

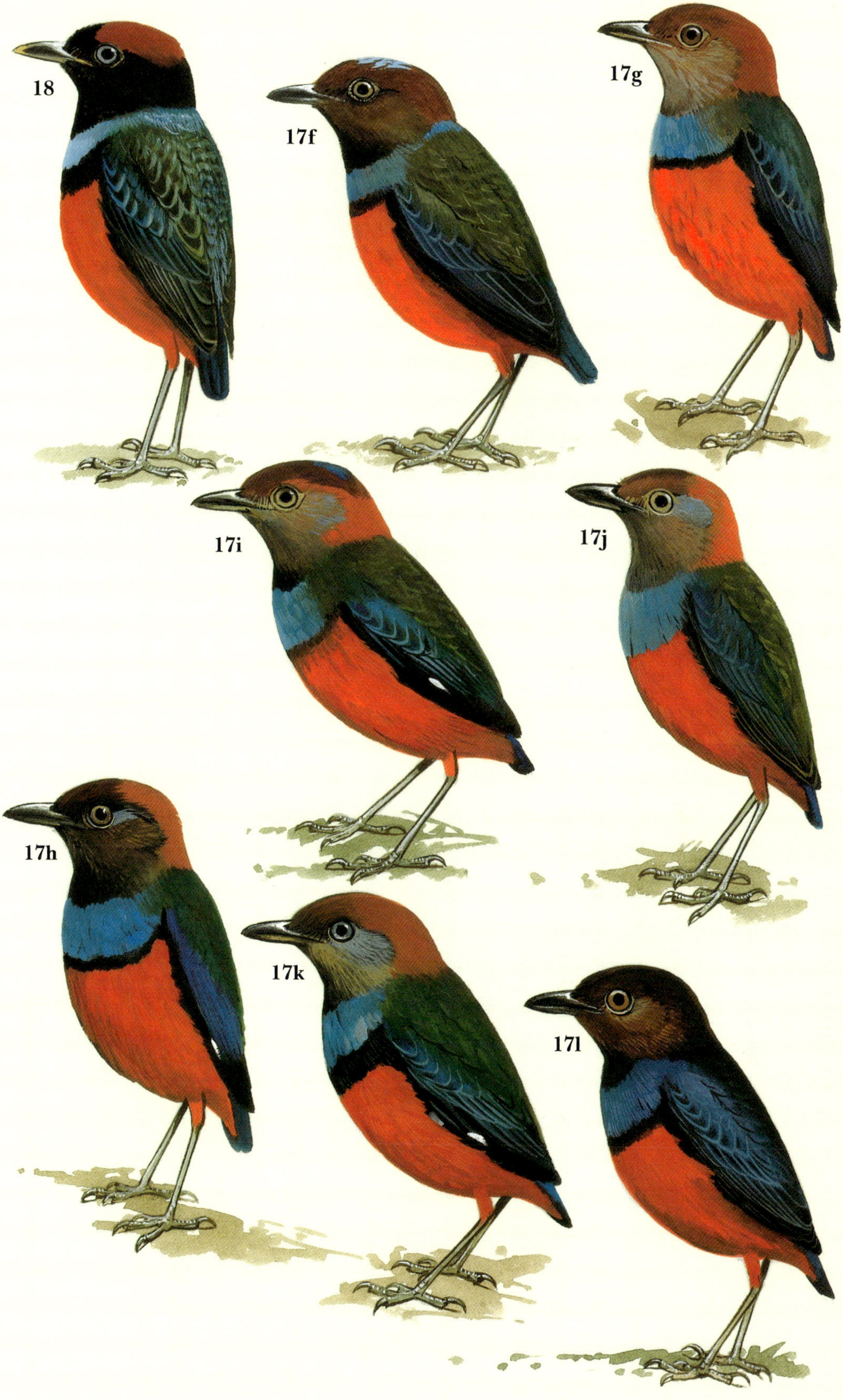
18
17f
17g
17i
17j
17h
17k
17l

PLATE 11: BLUE-BANDED PITTA AND THE GARNET PITTA GROUP

19 Blue-banded Pitta *Pitta arquata* — Text and map page 141

Endemic to Borneo.

- **19a** **Adult** Unmistakable, with largely orange-red head and scarlet underparts, prominent narrow azure stripe behind the eye and narrow band across the breast, and olive-green upperparts.
- **19b** **Juvenile** Brown with paler belly and red on flanks.

20 Garnet Pitta *Pitta granatina* — Text and map page 143

Peninsular Thailand through Malaysia to Borneo and northern Sumatra.

Dark purple-blue upperparts, bright scarlet crown to nape and belly to undertail-coverts, iridescent pale blue bend of wing and stripe behind the eye.

- **20a** **Adult** (*coccinea*; Sumatra, Peninsular Malaysia and Peninsular Thailand) Greater amount of red scaling on the breast and more red on the crown than *granatina*. Most individuals have only the forehead black.
- **20b** **Adult** (*granatina*; Borneo, except the north) Most individuals have the entire forehead and forecrown black.
- **20c** **Juvenile** (*granatina*) Young birds are uniform dark chocolate-brown, gradually acquiring red on the nape and belly.

21 Black-headed Pitta *Pitta (granatina) ussheri* — Text and map page 146

Endemic to north Borneo.

- **21a** **Adult** Unmistakable, with black head and breast, crimson belly, and prominent narrow pale blue post-ocular stripe. The upperparts are purple-blue with an iridescent azure-blue patch at the bend of the wing.
- **21b** **Juvenile** Uniform dark brown, obtaining red on the belly with age.

22 Graceful Pitta *Pitta venusta* — Text and map page 148

Hills and mountains of Sumatra.

- **22** **Adult** Head, breast and upperparts maroon (though this may appear dark brown in the field), with a long, fine azure-blue post-ocular stripe that extends to the nape. The breast is maroon, tinged with red and merging into the bright scarlet of the belly to undertail-coverts. The wings are dull purple with an indistinct dark azure-blue line in the outer edge of the closed wing.

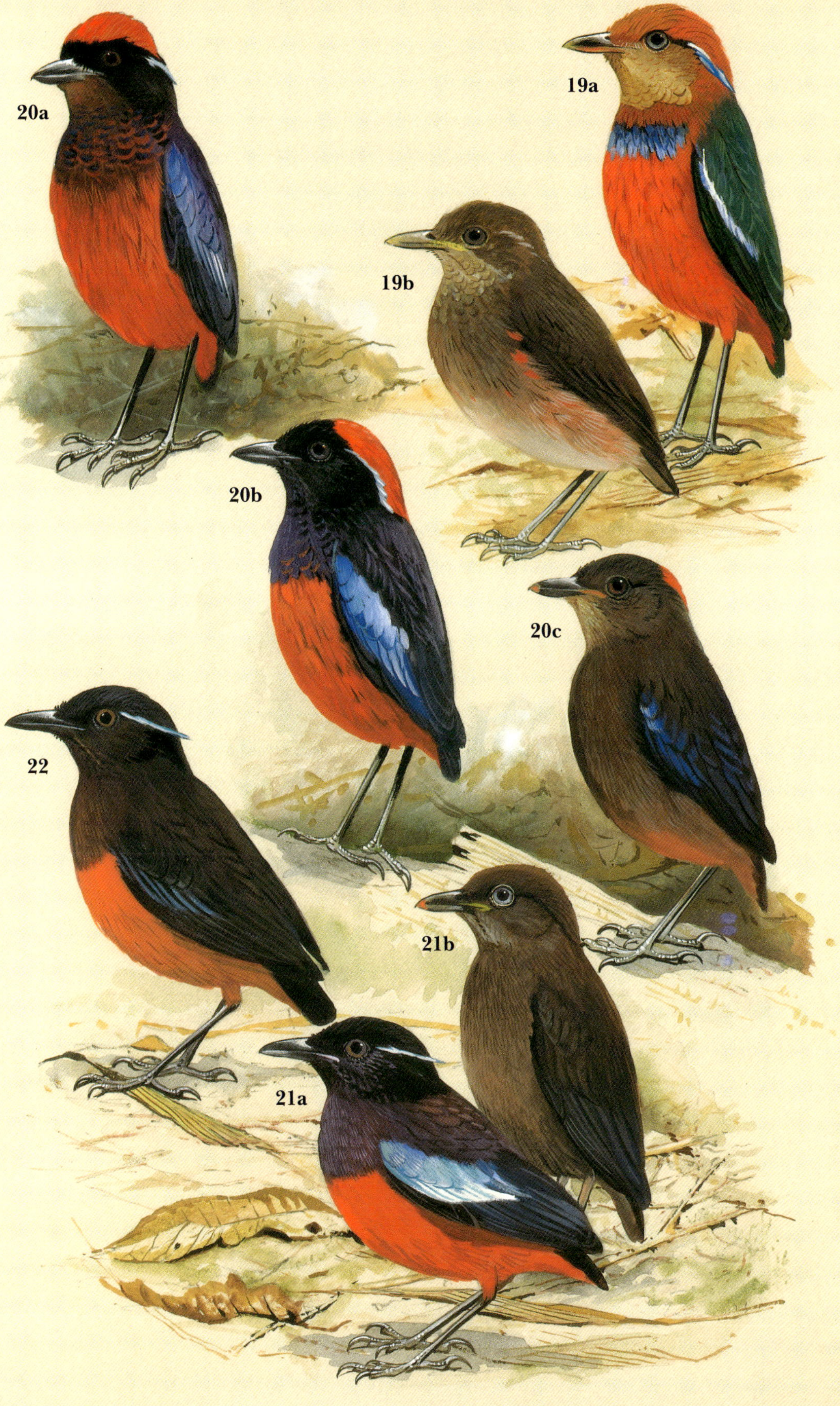
20a
19a
19b
20b
20c
22
21b
21a

PLATE 12: AFRICAN, GREEN-BREASTED AND INDIAN PITTAS

23   African Pitta *Pitta angolensis*   **Text and map page 150**

A medium-sized pitta that is widely distributed in moist woodlands of eastern Africa and in the lowland coastal forests of West Africa.

Subspecies differ markedly in size. Prominent broad golden-buff supercilium. The upperparts are green with an azure-blue rump. The wings are darker with a line of azure-blue feather-tips across the innermost coverts, bordered below by feathers that may be marked with azure-blue or deep violet, depending on subspecies. Below, the throat and breast are white, with a distinctive pink wash, the breast and flanks are a cinnamon-buff, whilst the belly to undertail-coverts is deep red. A white wing-patch is conspicuous in flight.

- **23a**   **Adult** (*pulih*; Sierra Leone and Guinea to northern and western parts of Cameroon) From Mt Nimba (see species account). Small with extensive, neatly defined red on under parts and usually no or few blue tips in outer greater coverts. Supercilium bicoloured. Underparts have greenish sheen.
- **23b**   **Juvenile** (*pulih*) Reminiscent of adult but much darker, especially below. Usually lacks azure-blue wing-spots and has pink undertail-coverts and belly. Bill orange-red to red with black band near tip.
- **23c**   **Adult** (*longipennis*; East Africa, where migratory) Violaceous tinge to the azure blue of rump and prominent violet tips to the outer medium and greater wing-coverts. Supercilium usually rather uniform. Red restricted to centre of belly and undertail-coverts. Larger than West African subspecies.
- **23d**   **Adult** (*longipennis*) In flight.

24   Green-breasted Pitta *Pitta reichenowi*   **Text and map page 156**

The Congo basin and parts of southern and western Uganda.

- **24a**   **Adult** Breast and upperparts green, throat white, belly to undertail-coverts red, rump azure-blue. Breast often glossed with golden-yellowish. Upperwing marked with 2-3 lines of conspicuous azure-blue spots; lower spots may be deeper blue.
- **24b**   **Juvenile** Dark brown above with dark ochre supercilium and pinky-white throat. Wings marked with one to two lines of azure-blue spots. Breast dark ochre-brown, washed with olive, belly and undertail-coverts pink. Bill distinctly bicoloured.

25   Indian Pitta *Pitta brachyura*   **Text and map page 159**

Foothills of the Himalayas to southern India and Sri Lanka. Migratory.

- **25a**   **Adult** Broad buffy supercilium, white and pointed at nape. Upperparts green, under parts pale cinnamon-buff with scarlet undertail-coverts. Rump and wing-patch iridescent azure-blue.
- **25b**   **Juvenile** Bold supercilium but generally subdued plumage; underparts with the lower parts smoky-black and the scarlet on the belly faint or hardly visible. Bill orange-red to red with black band near tip.
- **25c**   **Adult** In flight. Bright azure-blue patch in wing and rump; white patch in primaries.

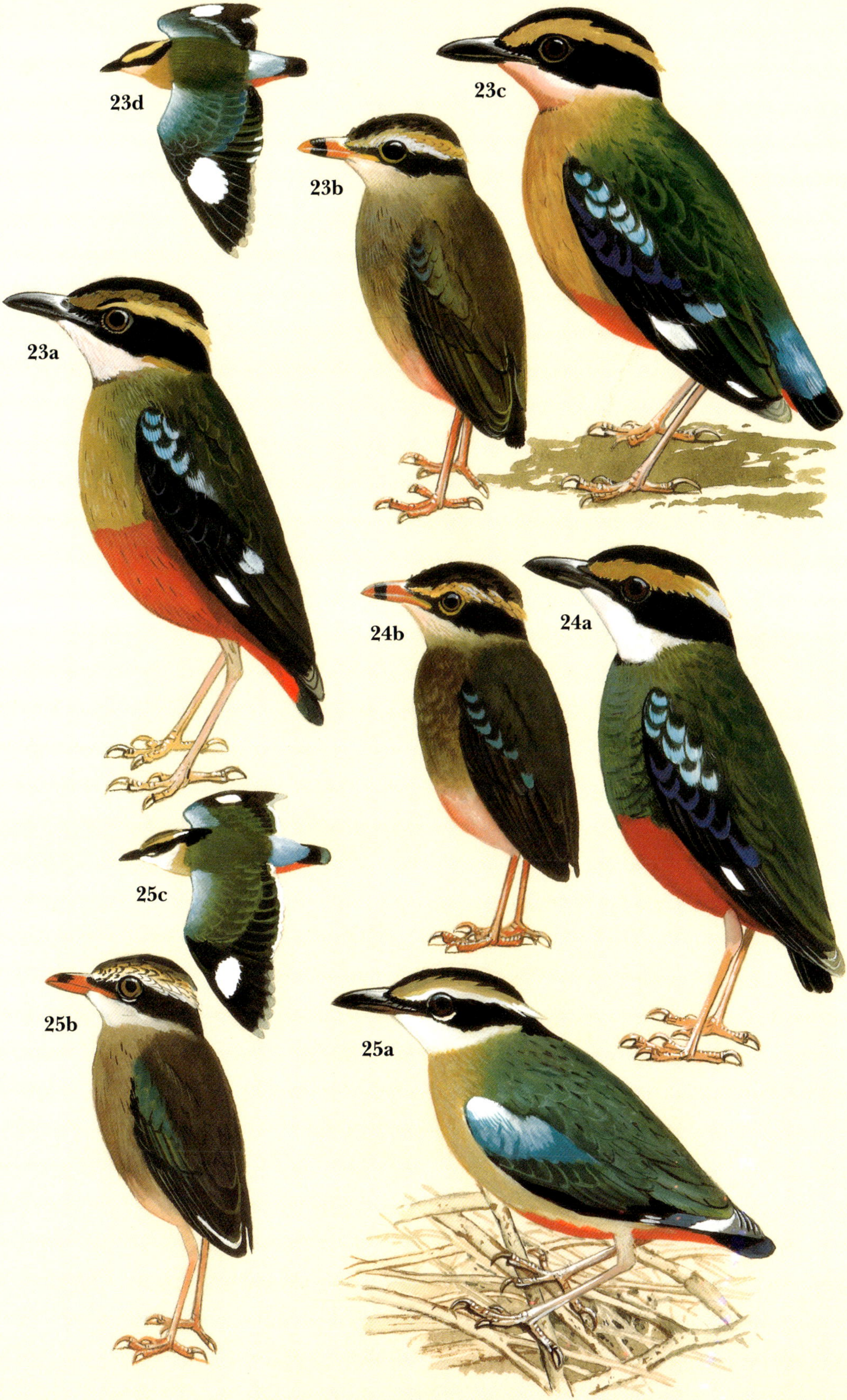
23d
23c
23b
23a
24b
24a
25c
25b
25a

PLATE 13: MANGROVE, BLUE-WINGED AND FAIRY PITTAS

28   **Mangrove Pitta** *Pitta megarhyncha*   Text and map page 170

       A mangrove specialist that occurs along coasts, deltas and along tidal parts of rivers from the Bay of Bengal to Peninsular Malaysia, and Sumatra.

          28a   **Adult** Massive bill, white chin and brown coronal bands. Upperparts dull greenish with bright glossy violet-blue wing-coverts. Underparts buff with crimson vent.

          28b   **Immature** Massive bill evident from young age. Much duller above than adults, usually with only a suffusion of red on vent.

27   **Blue-winged Pitta** *Pitta moluccensis*   Text and map page 166

       Northern parts of South-East Asia, migrating to the Sunda Region.

          27a   **Adult** Broad buffy-brown lateral coronal bands that contrast strongly with black centre of crown. Upperparts dull greenish with bright glossy violet-blue wing-coverts. Underparts buff with crimson vent.

          27b   **Immature** Duller than adult, with salmon-pink vent. Bill tip and base reddish.

          27c   **Adult** In flight. Wing largely blue with conspicuous large round white wing-patch.

26   **Fairy Pitta** *Pitta nympha*   Text and map page 162

       Breeds in eastern China, Japan, Korea and Taiwan, and migrates to Borneo during the winter. Some birds probably winter in southern China. Migrants reach Vietnam.

          26a   **Adult** Broad bright chestnut to chestnut-orange coronal bands contrast strongly with creamy-buff supercilium. Upperparts green with azure-blue wing-patch. Underparts creamy-buff with dark red undertail-coverts.

          26b   **Adult** In flight. Small white wing-spot.

12   **Hooded Pitta** *Pitta sordida*   Text and map page 117

       See also Plate 7 and 16.

       The northern, migratory subspecies is shown because of potential confusion (in flight) with Blue-winged or Fairy Pitta during migration.

          12   **Adult** (*cucullata*) In flight. Rich chestnut-brown crown, green body. Wings with relatively small white spot and azure-blue patch at bend.

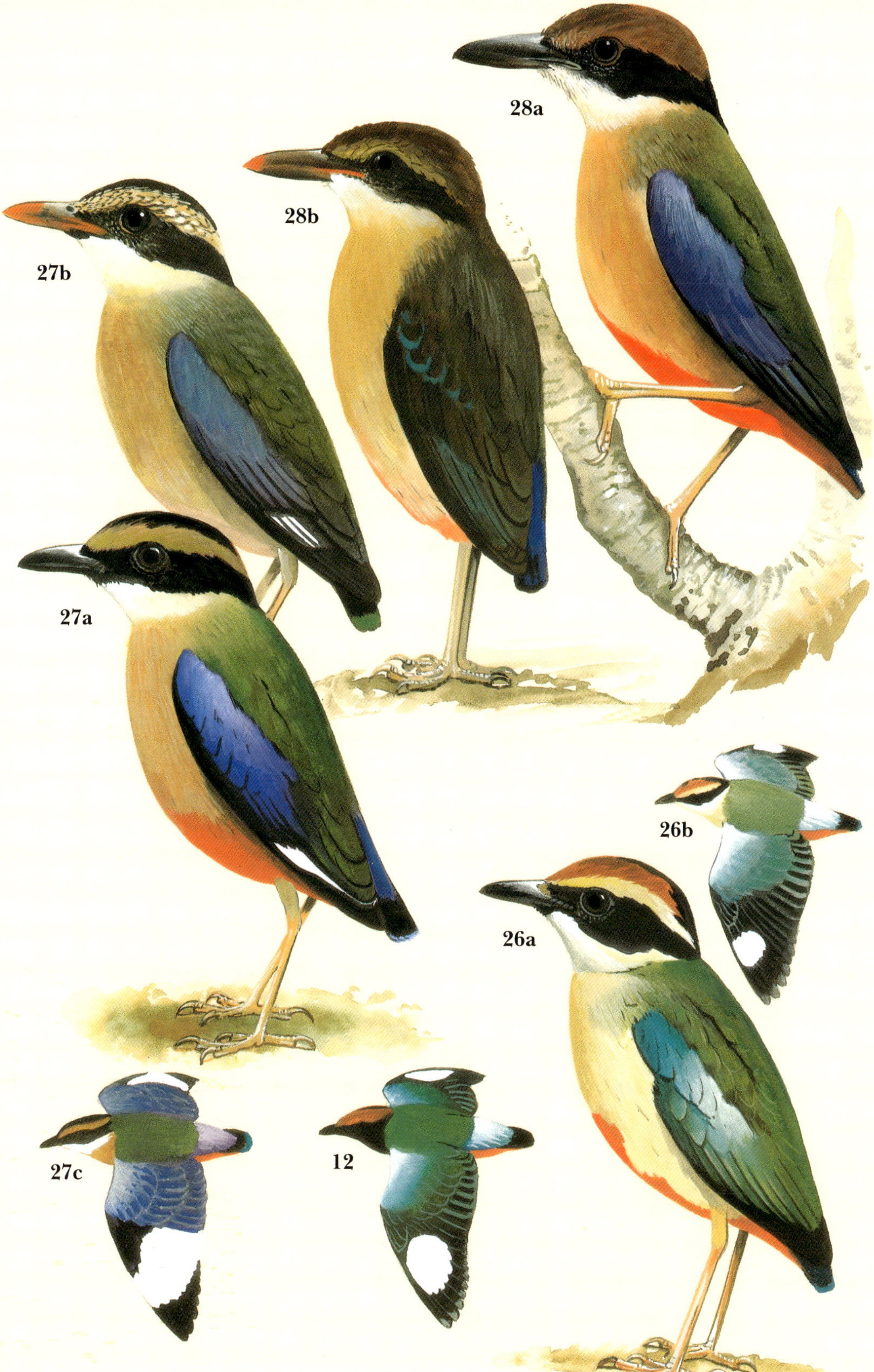

28a
28b
27b
27a
26b
26a
27c
12

PLATE 14: ELEGANT AND BLACK-FACED PITTAS

29   **Elegant Pitta** *Pitta elegans*   **Text and map page 174**

Widely distributed on islands of Wallacea, Indonesia. Two subspecies (nominate and perhaps *vigorsii*) may be migratory, but migratory patterns poorly known.

All subspecies have green upperparts with azure-blue wing-patch and rump, black crown bordered by contrasting supercilium, and cinnamon-buff underparts with a variable amount of red or scarlet on the belly and undertail-coverts.

- **29a**   **Adult** (*concinna*; Lombok, Sumbawa, Flores, Adonara, Lomblen and Alor as well as smaller offshore islands) Supercilium broad, golden-buff in front of the eye, but narrowing and becoming white behind the eye and strongly tinged with blue near tip. Throat black, extending onto the upper breast as a point. Large black patch on belly centre.
- **29b**   **Juvenile** (*concinna*) Chestnut-buff forehead extending as a broad coronal band, becoming bluish-white behind the eye. Plumage darker and duller than that of adult.
- **29c**   **Adult** (*virginalis*; Tanahjampea, Kalaotoa, Kalao in Flores Sea) Pale cinnamon-buff supercilium. Some black on centre of belly.
- **29d**   **Adult** (*elegans*; Breeds on islands of central Lesser Sunda arc, apparently migrating northwards to various islands in the Moluccas, Sula Islands, and small islands north of Sulawesi) Similar to *virginalis*, but supercilium narrower and buff. Tail narrowly tipped green.
- **29e**   **Adult** (*vigorsii*; Primarily on small islands in eastern Lesser Sunda arc, including Tanimbar, and between Seram and the Kai Islands, Banda, Kaledupa [south-east Sulawesi] and probably Seram) Supercilium becomes pale blue behind the eye. Throat white, centrally buff. Tail tip bright green.
- **29f**   **Adult** (*maria*; Sumba) Differs from nominate in having more extensive, paler red on the belly and a blue tinge to the posterior part of the supercilium. White in wing confined to a small spot.

31   **Black-faced Pitta** *Pitta anerythra*   **Text and map page 181**

Northern Solomon Islands of Bougainville, Choiseul and Ysabel.

Head black with a chestnut crown or a chestnut band across the rear of the crown. The upperparts are a dark deep green, marked with a conspicuous shining azure-blue patch on the wing-coverts.

- **31a**   **Adult** (*pallida*; Bougainville) Largest and palest subspecies, with flanks and belly to undertail-coverts almost white. Most birds have a band of chestnut across the nape, but sometimes lacking.
- **31b**   **Adult** (*nigrifrons*; Choiseul) Forehead and forecrown black, crown rich chestnut-brown becoming buffier on nape. Underparts deep ochre.

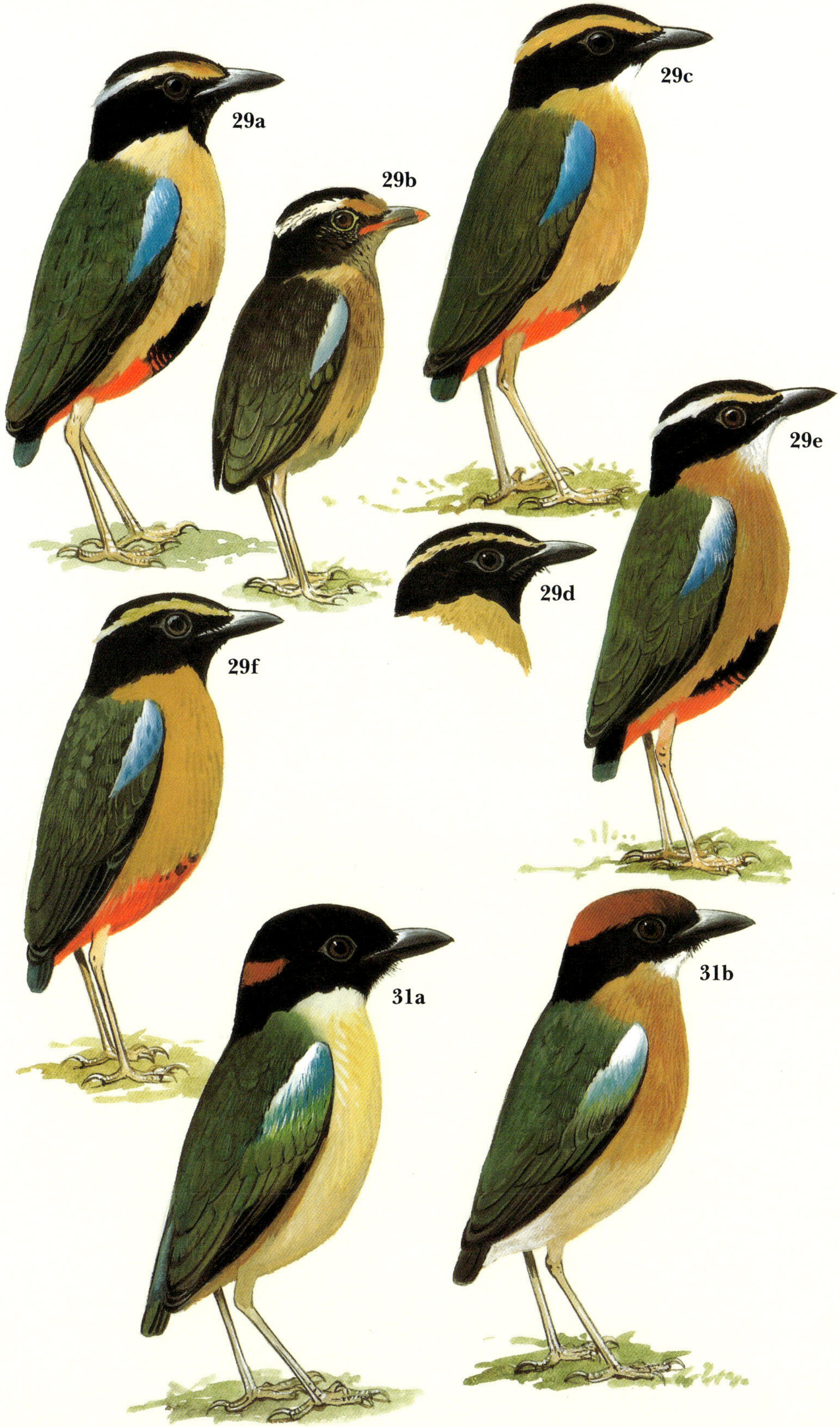
29a
29b
29c
29d
29e
29f
31a
31b

PLATE 15: AUSTRALIAN PITTAS

30 Noisy Pitta *Pitta versicolor*   **Text and map page 177**

Coastal New South Wales to Cape York and islands in the Torres Strait. Migrants visit southern New Guinea.

Black head with a chestnut crown that is streaked black in the centre. Upperparts green with bold shining azure-blue wing-covert patch and uppertail-coverts. Underparts buff with black throat and belly patch, and red vent.

- **30a**   **Adult** (*versicolor*; southern Queensland and New South Wales, and islands off the east coast. During the austral winter, some birds may visit Cape York Peninsula) Large. Underparts yellowish-buff. Large white spot in wing.
- **30b**   **Adult** (*simillima*; eastern Cape York Peninsula south to northern Queensland and on larger islands in the Torres Strait. Migrants visit southern New Guinea) Small, sometimes only half the size of nominate. Underparts rich cinnamon-buff.
- **30c**   **Immature** (*simillima*) Duller than adults. Lack black on throat and chin and azure-blue on the wings. Extent of black on abdomen is reduced and undertail-coverts are pinky-red. Bill black with an orange tip.
- **30d**   **Adult** (*versicolor*) In flight.

32 Rainbow Pitta *Pitta iris*   **Text and map page 183**

Confined to Top End region of the Northern Territory, north-west Australia, and perhaps in the coastal region of the Kimberley Region of Western Australia.

- **32a**   **Adult** Head and underparts black except for chestnut-brown post-ocular coronal band, often meeting on nape, and pinky-red undertail-coverts. Upperparts bright green, marked with a large azure and violet-blue wing-covert patch.
- **32b**   **Adult** Showing variation in head pattern.
- **32c**   **Immature** Distinguished from adults by duller coloration and white edges to the feathers of chin. Orange gape and reddish-brown legs.

30a

30b

30c

30d

32b

32a

32c

PLATE 16: JUVENILE AND IMMATURE PITTAS

This plate depicts four sets of juvenile and immature pittas that may be confused in the field, as well as juvenile Whiskered Pitta (painted from a photograph obtained after completion of Plate 9).

GROUP 1 - Blue-naped, Rusty-naped and Blue-rumped Pittas

See also Plates 1 and 2.

2   **Blue-naped Pitta** *Pitta nipalensis*   Text and map page 85
    Juvenile (*nipalensis*; specimen from upper Burma) Buffy-white streaks on crown. Buffy or occasionally creamy-white spots on the upperwing-coverts. Breast spotted with buff to buffy-white.

4   **Rusty-naped Pitta** *Pitta oatesi*   Text and map page 92
    Immature (*oatesi*; specimen from northern Thailand) Whiter on ear-coverts than in immature Blue-naped Pitta. Streaks on crown and spots on upperwing-coverts whitish. Breast spotted with white.

3   **Blue-rumped Pitta** *Pitta soror*   Text and map page 88
    Juvenile (*soror*; central Annam) Large buffy-orange spots on upperwing-coverts. Crown marked with bold, broad buff streaks.

GROUP 2 - Schneider's and Giant Pittas

See also Plate 3.

5   **Schneider's Pitta** *Pitta schneideri*   Text and map page 95
    Juvenile (Sumatra) Whitish spotting on crown contrasting with orange spotting on nape, and sometimes on upper mantle. Bolder white-and-cream scaling on breast and flanks than in Giant Pitta. Spotting on upperwing-coverts.

6   **Giant Pitta** *Pitta caerulea*   Text and map page 97
    Juvenile (*caerulea*; specimen from Malacca) Uniform creamy to buffy spotting on crown to nape. Lacks spotting on upperwing-coverts.

GROUP 3 - Banded and Gurney's Pittas

See also Plates 5 and 6.

8   **Banded Pitta** *Pitta guajana*   Text and map page 104
    Juvenile (*irena*; specimen from Peninsular Thailand) Prominent white bars across the wing.

10 **Gurney's Pitta** *Pitta gurneyi*   Text and map page 110
    Juvenile (Peninsular Thailand) Lacks white wing-bars.

GROUP 4 - Hooded and Blue-winged Pittas

See also Plates 7 and 13.

12 **Hooded Pitta** *Pitta sordida*   Text and map page 117
    Juvenile (*cucullata*; South-East Asia) White band across the upperwing-coverts. Lacks any trace of blue in wing.

27 **Blue-winged Pitta** *Pitta moluccensis*   Text and map page 166
    Juvenile (South-East Asia) Dull blue in wings but no white in upperwing-coverts.

Whiskered Pitta

See also Plate 9.

16 **Whiskered Pitta** *Pitta kochi*   Text and map page 131
    Juvenile (Luzon) Blackish-brown with white spots on the breast and flanks, and distinctive pale moustachial streak.

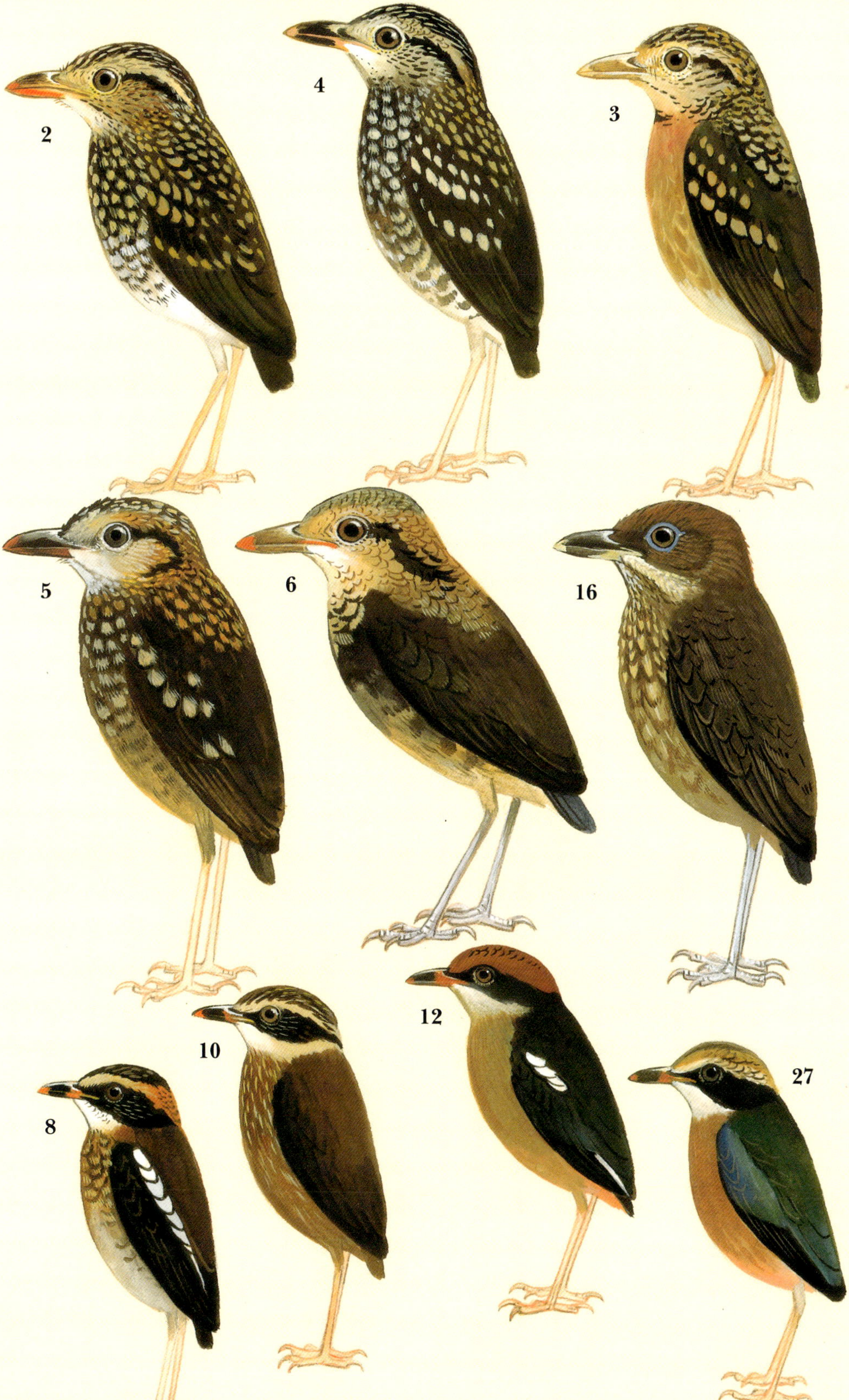
2
4
3
5
6
16
10
12
8
27

PLATE 17: AFRICAN BROADBILL

33 African Broadbill *Smithornis capensis*   Text and map page 186

Widely distributed in the tropics of southern, central and eastern Africa with isolated populations from central Ghana to Sierra Leone, in Cameroon, Gabon, Central African Republic and Angola.

- **33a** **Adult male** (*capensis*; South Africa in coastal Natal and southern Zululand) Forehead more or less buff, ear-coverts greyish, narrow grey nape-band and grey tinge to rufous-brown mantle. Underparts silky-white boldly streaked with blackish-brown on sides of throat, breast and flanks.
- **33b** **Adult female** (*capensis*) Crown grey with black streaking.
- **33c** **Adult** (*capensis*) In flight. Conspicuous white back-patch.
- **33d** **Adult male** (*albigularis*; north Malawi, north Zambia, Zaïre and in west [?central] Tanzania) Upperparts greyish. Breast washed yellowish-buff. Ear-coverts grey, streaked white.
- **33e** **Adult female** (*medianus*; highlands of central Kenya and north-east Tanzania) More uniform, less streaked mantle and rump than other subspecies. Grey wash across upper mantle. Crown blackish-grey, usually with indistinct paler grey scaling.
- **33f** **Juvenile** (*medianus*) Closely resembles an adult female, but very short-winged and short-tailed.
- **33g** **Adult male** (*camarunensis*; Cameroon, Gabon, Central African Republic). Mantle to rump and wings rich rufescent. Underparts heavily streaked and strongly washed with buff, especially on breast.
- **33h** **Adult female** (*camarunensis*) Crown varies from black, almost lacking streaks, to heavily streaked black and rufous.
- **33i** **Adult female** (*camarunensis*) Showing variation in crown.

33b

33a

33d

33f

33e

33c

33i

33h

33g

PLATE 18: AFRICAN BROADBILLS II

47 **African Green Broadbill** *Pseudocalyptomena graueri*   Text and map page 235

A small broadbill endemic to montane forests in eastern Zaïre and western Uganda.

- **47** **Adult** Sexes are similar. Green upperparts and belly, pale blue throat and undertail-coverts. The head is marked by a buffish crown, finely streaked with black, a black eye-stripe and a black streaked moustachial area. Immatures are duller, with green undertail-coverts.

35 **Rufous-sided Broadbill** *Smithornis rufolateralis*   Text and map page 193

Central and West Africa.

A very small broadbill. Both sexes have silky-white underparts with black streaking and a conspicuous bright orange patch on the sides of the breast. White spots in the wing form a broken bar and there is a hidden white mantle-patch that shows in flight.

- **35a** **Adult male** (*budongoensis*; central and north-east Zaïre and western Uganda) Head black, upperparts rufescent-brown with extensive black on mantle.
- **35b** **Adult female** (*budongoensis*) Head dark greyish-brown, tinged rufous and indistinctly streaked darker; upperparts rufous-brown usually lacking black; orange sides of breast duller than in male.
- **35c** **Adult female** (*rufolateralis*; West Africa from Liberia and Cameroon to eastern Zaïre) Head dark brown.
- **35d** **Adult male** (*rufolateralis*) In flight, showing exposed white back-patch.

34 **Grey-headed Broadbill** *Smithornis sharpei*   Text and map page 190

The Congo basin, where present in extreme east and extreme west, and Bioko Island.

Crown and ear-coverts blue-grey, breast marked by a prominent broad orange band, divided by white in the centre. Rest of underparts are more or less white with bold black streaks; the upperparts are rufescent-brown with a few black patches on the back and mantle and hidden silky-white feather bases. Wings are browner than the mantle and back, and lack any pale bars or spots.

- **34a** **Adult male** (*sharpei*; Bioko) Crown blue-grey with diffuse streaks; underparts marked with broad orange breast-band and blackish streaking.
- **34b** **Adult female** (*sharpei*) Less extensive and duller orange on breast than male. Crown colour varies from dark grey to blue-grey.
- **34c** **Adult** (*sharpei*) In flight, showing exposed white bases to back feathers.
- **34d** **Adult male** (*zenkeri*; Cameroon and northern Gabon) Crown and ear-coverts sootier than nominate; underparts typically with denser, bolder black streaking.

35b

35a

47

35c

35d

34c

34b

34a

34d

PLATE 19: *CALYPTOMENA* BROADBILLS

36   Green Broadbill *Calyptomena viridis*   Text and map page 196

South-East Asia from Tenasserim to Peninsular Malaysia, Borneo, Sumatra and various offshore islands.

This is the smallest species of the genus. In flight the plump body, short tail and broad-based but pointed wings are distinctive.

- **36a** **Adult male** (*viridis*; Borneo, Sumatra, Nias, Batu Islands, Lingga Archipelago and North Natuna Islands) Iridescent dark green with prominent black ear-patch and broad black bars and patches on the wings.
- **36b** **Adult female** (*viridis*) Uniform green, paler than the male, with a lime-green eye-ring.
- **36c** **Adult male** (*siberu*; Mentawai Islands) Largest and darkest subspecies. Males have a distinctive bright blue wash to plumage, particularly noticeable on throat and belly to undertail-coverts.

37   Hose's Broadbill *Calyptomena hosii*   Text and map page 201

Hills and mountains of Borneo.

Both sexes have well-defined round black spots on the wing-coverts.

- **37a** **Adult male** Bright iridescent green above with black spots on nape and behind ear-coverts. Underparts distinctive, with extensive indigo-blue on the breast and deep blue on the belly to undertail-coverts.
- **37b** **Adult female** Paler and more olive-green above than male, with black spots confined to the wing-coverts. Prominent lime-green eye-ring. Underparts lime-green, notably paler and yellower than the upperparts, with sky-blue on vent.
- **37c** **Immature male** Similar to female, but has dark feathers on the lower nape. Immature males have less extensive blue on the underparts and lack most of the black head markings.

38   Whitehead's Broadbill *Calyptomena whiteheadi*   Text and map page 203

Montane forests of Borneo.

Both sexes have a prominent black throat-patch, and black markings on the upperparts and wing-coverts. Paler green than other *Calyptomena* species.

- **38a** **Adult male** Very bright. Prominent black spot behind eye and on nape. Bold black streaks on underparts.
- **38b** **Adult female** Slightly smaller and duller than males, lacking the black head-spots, and are a uniform, duller green below with no black markings.

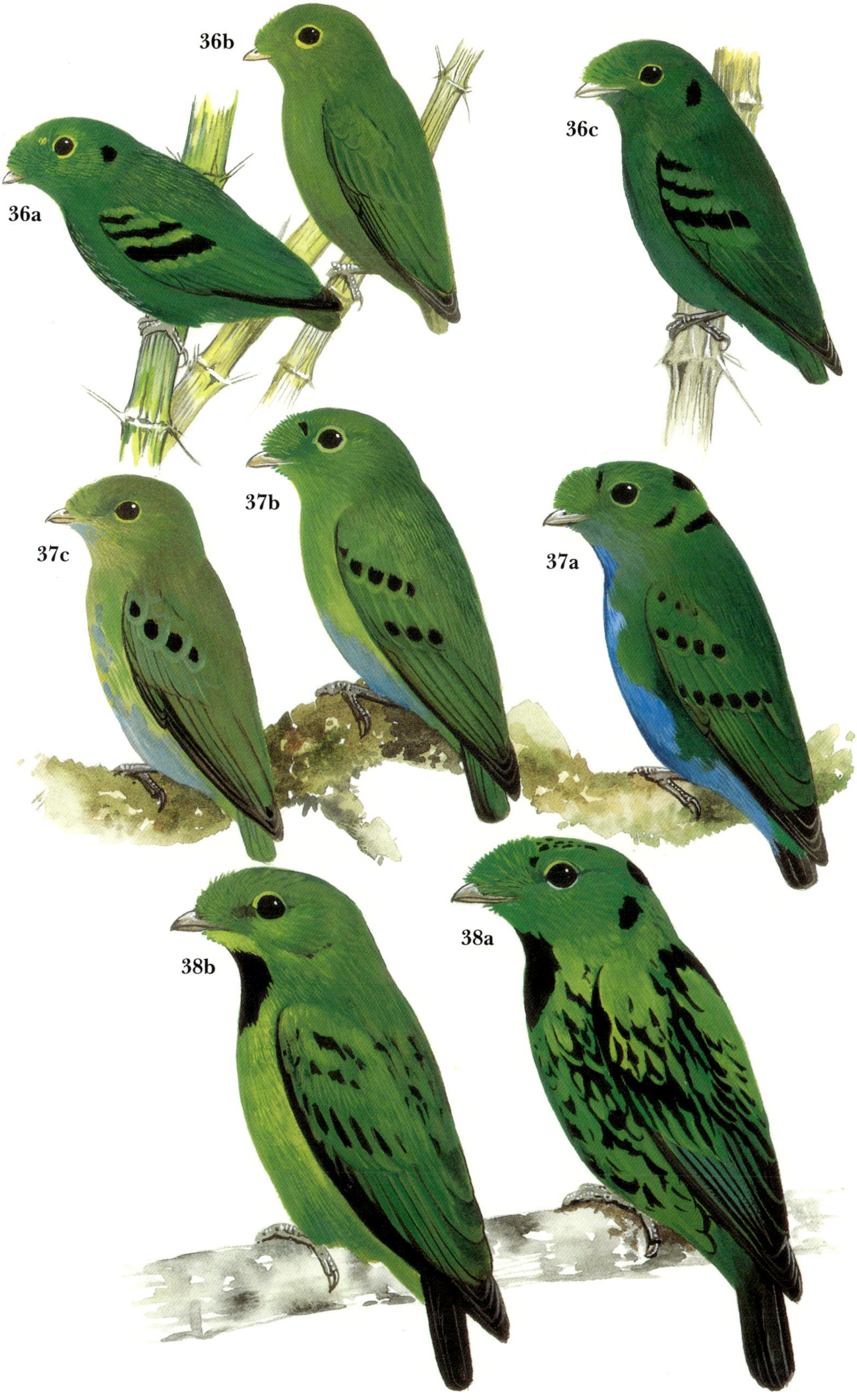
36b
36c
36a
37b
37c
37a
38b
38a

PLATE 20: DUSKY AND LONG-TAILED BROADBILLS

46 Dusky Broadbill *Corydon sumatranus* Text and map page 231

Forests from Indochina and Tenasserim to Malaysia, Sumatra and Borneo.

Sexes are similar. This is a large, blackish, thickset, gregarious species, with an obvious massive pinkish to purplish bill and orbital skin. Throat pale, usually contrasting with rest of plumage. The small white patch in the primaries is usually visible in the field, but the orange or flame-coloured streaks on the mantle are not.

- **46a** **Adult** (*brunnescens*; Borneo and the North Natuna Islands) Sooty black upperparts that are distinctly tinged with olive-green, particularly on the rump and uppertail-coverts. Throat-patch usually deep rufous-brown.
- **46b** **Adult** (*laoensis*; Indochina, Burma and Thailand south to Tenasserim and Trang) Dark blackish-brown with more or less contrasting whitish throat.
- **46c** **Juvenile** (*laoensis*) More uniform than adult, with no clearly defined pale throat.

40 Long-tailed Broadbill *Psarisomus dalhousiae* Text and map page 210

A long-tailed, unmistakable species that is widely distributed from the eastern Himalayas throughout South-East Asia to Sumatra and Borneo.

Sexes are very similar. Plumage green with striking black cap, bright yellow throat, face and collar, and long, graduated tail that is blue above. Top of crown blue. In flight, the wings appear blue, and the white patch on the underside of the blue flight feathers is prominent. Head patterns between individuals vary greatly: the degree of this variation is shown in the illustrations.

- **40a** **Adult male** (*psittacinus*; Peninsular Malaysia and Sumatra) Shorter-winged but longer-tailed than nominate. Collar is usually a mixture of yellow and white.
- **40b** **Adult female** (*psittacinus*) Usually differs from male in having a concealed or partially concealed yellow band across the nape.
- **40c** **Adult male** (*dalhousiae*; Himalayas to north-east India and south-east Bangladesh, Burma, northern Thailand, northern Indochina and southern China) Tends to have little white in collar.
- **40d** **Adult male** (*dalhousiae*) Variant with blue underparts (such birds are usually from the Himalayas to northern Burma).
- **40e** **Immature** (*dalhousiae*) Head differs from adult in having green crown but pale yellow on the lores and rear of ear-coverts; chin and throat greenish-yellow.

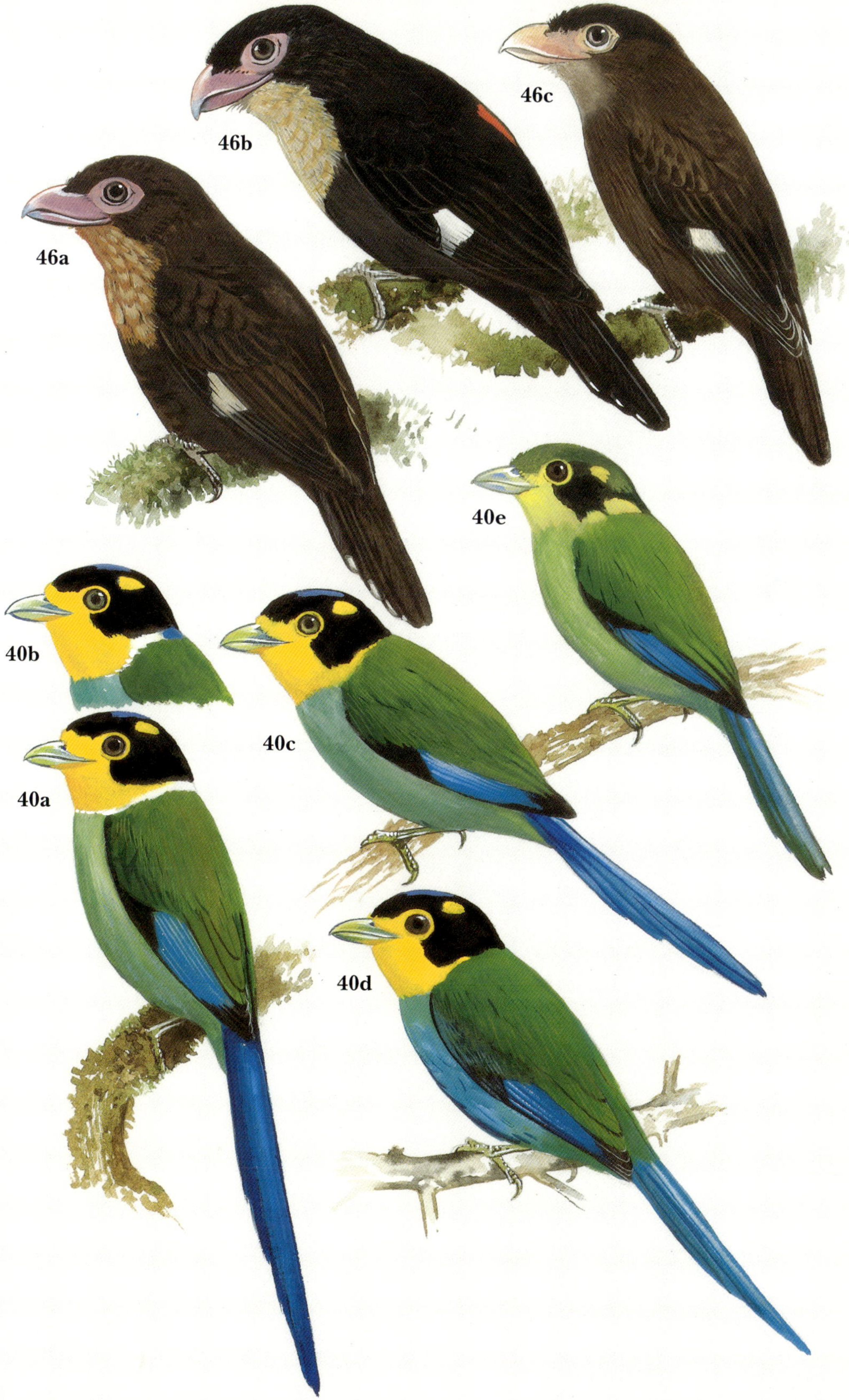
46b
46c
46a
40e
40b
40c
40a
40d

PLATE 21: BLACK-AND-RED AND BANDED BROADBILLS

39 **Black-and-Red Broadbill** *Cymbirhynchus macrorhynchos*   Text and map page 205

Indochina and Burma to Malaysia, Sumatra and Borneo.

Sexes are similar. Black above with a bold, long white wing-flash and maroon rump; deep maroon below with black breast-band. Blue-and-yellow bill conspicuous. Usually near water.

- **39a** **Adult** (*macrorhynchos*; Borneo) White on the tail almost obsolete. Lacks crimson spots in secondaries.
- **39b** **Juvenile** (*macrorhynchos*) Browner than adults, but has a white wing-patch, black breast-band and patches of maroon on the rump and, usually, on the central belly and undertail-coverts.
- **39c** **Adult** (*affinis*; western Burma) Crimson spots on the innermost secondaries. Considerably smaller than nominate, with more conspicuous white wing-spot, broader white tail-bars and narrow black edges to the crimson feathers of the rump.

42 **Banded Broadbill** *Eurylaimus javanicus*   Text and map page 220

Indochina to the Sunda Region.

Relatively large, with unique combination of purple-maroon head and underparts, and yellow streaking and spotting on the dark upperparts.

- **42a** **Adult male** (*harterti*; Indochina, Burma, Thailand, Peninsular Malaysia, Sumatra and the Riau Archipelago) Large, with maroon upper mantle and more or less purple undertail-coverts. Distinct black breast-band.
- **42b** **Adult female** (*harterti*) Lacks the breast-band of male.
- **42c** **Juvenile** (*harterti*) Pale brown head and upperparts marked with yellow. Large yellow spots on wings. Recently fledged birds have sooty band across upper breast.
- **42d** **Adult male** (*javanicus*; Java) Small with brown upper mantle and pure yellow undertail-coverts. Lacks black breast-band typical of males of other subspecies and is paler below.
- **42e** **Adult female** (*javanicus*) Undertail-coverts pure yellow. Head and throat glossed with grey.

39a
39b
39c

42b
42a
42d
42c
42e

PLATE 22: BLACK-AND-YELLOW AND WATTLED BROADBILLS

43 **Black-and-Yellow Broadbill** *Eurylaimus ochromalus*   **Text and map page 224**

Sunda Region from peninsular Thailand and Tenasserim to Malaysia, Borneo and Sumatra.

A small species with black head and upperparts, separated by a white collar, and conspicuous bold yellow markings on the back and wings; the breast is vinaceous pink, fading into pale yellow on the belly and undertail-coverts. Bill brightly coloured.

- **43a** **Adult male** (*ochromalus*; most of range) Complete black breast-band.
- **43b** **Adult female** (*ochromalus*) Broken breast-band.
- **43c** **Juvenile** (*ochromalus*) Underparts greyish-white with no breast-band. Pale yellow supercilium.

45 **Visayan Wattled Broadbill** *Eurylaimus samarensis*   **Text and map page 229**

Philippines; islands of Leyte, Samar and Bohol.

Mottled grey nuchal collar, purple mantle and prominent white and lilac bar across the tertials and secondaries. Adults have a conspicuous pale blue wattle around eye.

- **45a** **Adult male** Underparts pink.
- **45b** **Adult female** Underparts pure white.
- **45c** **Juvenile** Head dark brown, washed with grey on crown and with a whitish collar extending across sides of neck. Upperparts mixed olive and grey-brown with rufescent tinge in patches. Breast and flanks greyish-brown; belly and undertail-coverts white.

44 **Mindanao Wattled Broadbill** *Eurylaimus steerii*   **Text and map page 227**

Philippines; islands of Mindanao, Basilan, Malamaui, Dinagat and Siargao.

Both sexes have dark grey mantle separated from purple crown by a white collar, and a prominent narrow white and yellow bar across the secondaries and tertials. Adults have a broad pale blue wattle.

- **44a** **Adult male** (*steerii*; Zamboanga peninsula, Mindanao) Underparts pink.
- **44b** **Adult female** (*steerii*) Underparts pure white.
- **44c** **Immature** (*steerii*) Differ from adults in having white throat (this gradually becoming black), olive-green crown and wash to upperparts and ill-defined wing-bar with buffish-pink rather than white. Wattle reported to be yellow.

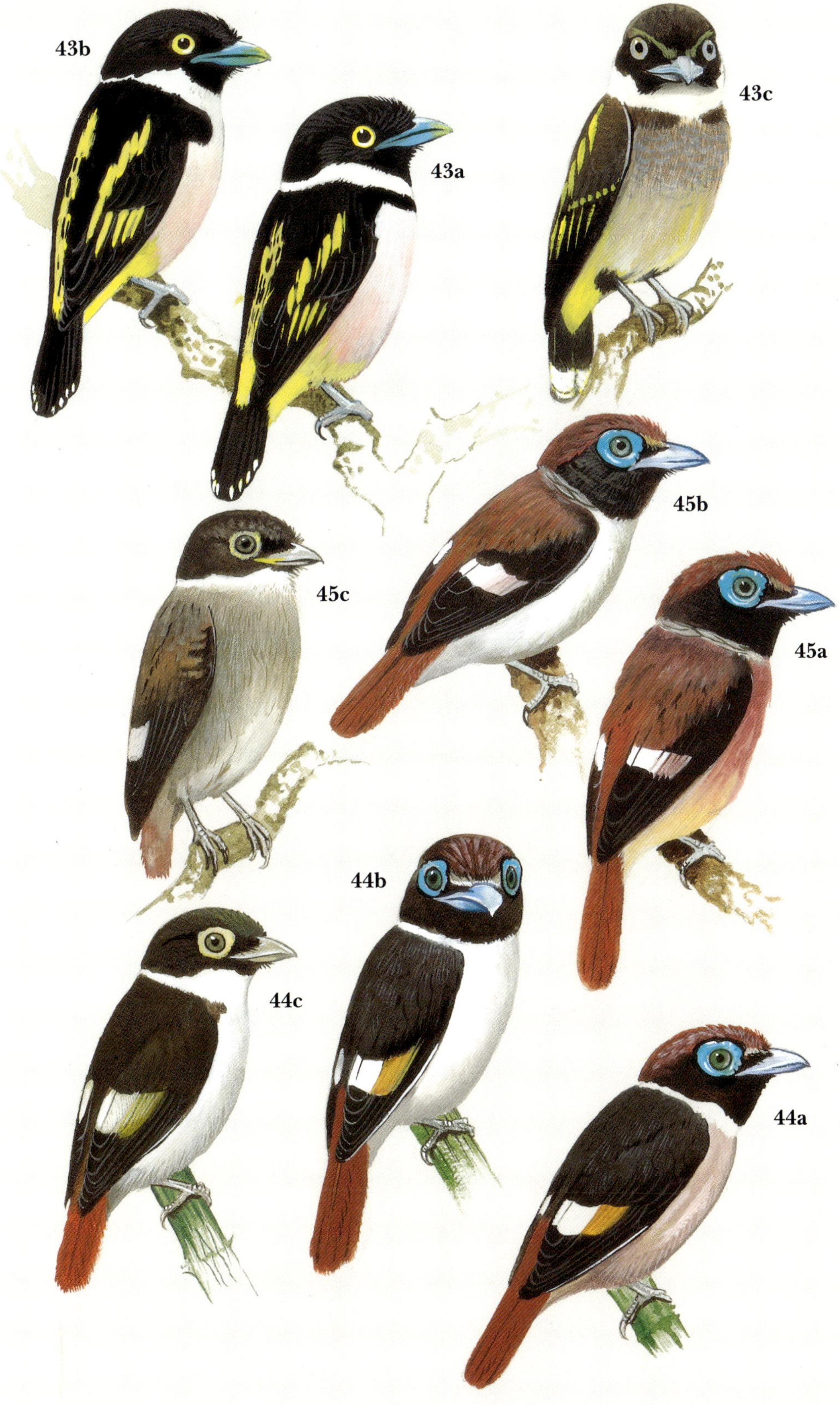
43b
43c
43a
45b
45c
45a
44b
44c
44a

PLATE 23: SILVER-BREASTED BROADBILL

41   Silver-breasted Broadbill *Serilophus lunatus*   Text and map page 215

A small, highly gregarious species that is widely distributed from the Himalayas to Indochina and the Sunda Region.

The plumage is rather subdued, with greyish underparts and grey to grey-brown upperparts. The distinctive features are the broad black supercilium and striking blue, black and rufous markings on the wing, and a more or less rufous rump. The tail is black, with white-tipped outer feathers. Females differ from males in having a fine, often broken, silver band across the upper breast.

- **41a** **Adult male** (*lunatus*; southern Burma and north-west Thailand to northern Tenasserim) Lores usually rusty; ear-coverts, crown and mantle ferruginous, mixed with some grey on mantle. Small area of forehead ashy in most birds. Uppertail-coverts and rump bright pale rufous.
- **41b** **Adult male** (*rubropygia*; Nepal to north-east India, hill tracts in Bangladesh, and the upper Chindwin valley, Burma) Narrower, darker blue secondary bar than other subspecies, and broad dark blue tips to outer webs of flight feathers. Crown, nape and upper mantle pale grey; rest of upperparts darker grey, tinged brownish. Lores dark grey. Supercilium less strongly marked than in other subspecies, and underparts generally greyer.
- **41c** **Adult female** (*rubropygia*) Silver breast-band broken in centre.
- **41d** **Adult male** (*rothschildi*; Peninsular Malaysia) Lores pale ashy, sides of head and ear-coverts greyish. Extensive rufous on upperparts. Females tend to have broader white breast-bands than nominate.
- **41e** **Juvenile** (*rothschildi*) Closely resembles adult, but has very short wings and tail.
- **41f** **Adult female** (*intensus*; Sumatra) Silver breast-band complete.
- **41g** **Adult male** (*polionotus*; mountains of Hainan) Lores black; sides of head, ear-coverts, crown and nape pale ashy, tinged with darker olive-brown; scapulars and upper back ashy-grey. Breast dark.

41a

41c

41b

41e

41d

41f

41g

PLATE 24: ASITIES AND SUNBIRD ASITIES

48 **Velvet Asity** *Philepitta castanea* Text and map page 238

Wet forests of eastern and northern Madagascar.

A short-tailed, plump species with a medium length, slightly curved bill.

- **48a** **Adult male breeding** Velvety black with a prominent deep lime-green wattle over and in front of the eye. Yellow shoulder-patch rarely visible in the field.
- **48b** **Adult male non-breeding** Black with prominent narrow yellow scales on body and spots on forehead to nape. Wattle absent or much reduced in size.
- **48c** **Adult female** Olive-green with prominent yellow eye-ring and moustachial streak and indistinct yellow supercilium in front of the eye. The upperparts are unmarked, the underparts scaled with yellow and green, except for the undertail-coverts.

49 **Schlegel's Asity** *Philepitta schlegeli* Text and map page 242

Seasonally dry forests of western Madagascar.

- **49a** **Adult male breeding** Black head with conspicuous large bright blue and pale green wattle. Underparts bright yellow, upperparts bright olive-green with yellow and black markings on the mantle and back.
- **49b** **Adult male non-breeding** Similar to female, but with vestigial wattle.
- **49c** **Adult female** Bold yellow eye-ring and olive-green upperparts, usually with narrow, short, yellowish streaks on the crown to nape. Virtually unmarked yellow belly and undertail-coverts. Immatures are indistinguishable from adult females.

50 **Common Sunbird Asity** *Neodrepanis coruscans* Text and map page 245

Wet forests of eastern Madagascar.

This is a tiny species with a distinctive, long, strongly downcurved bill, reminiscent of a sunbird. The tail is very short, hardly extending past the wing tips. The underparts are yellow-olive with dark greyish to blackish markings on the breast.

- **50a** **Adult male breeding** Head and upperparts bright dark metallic blue. Long rectangular pale azure-blue wattle extends over and behind the eyes. Greater coverts and secondaries yellowish. Primaries have diagnostic narrow yellow outer margins.
- **50b** **Adult male non-breeding** Metallic blue confined to tail, rump, uppertail-coverts and innerwing-coverts. Wattle much reduced in size.
- **50c** **Adult female** Olive-green above, paler, more mottled olive-grey below with bright yellow flanks and undertail-coverts.

51 **Yellow-bellied Sunbird Asity** *Neodrepanis hypoxantha* Text and map page 248

Montane forests in eastern Madagascar.

Underparts are entirely bright yellow in all adult plumages. Compared to Common Sunbird Asity, the bill is shorter and less strongly curved.

- **51a** **Adult male breeding** Metallic dark blue head, upperparts and tail. Large, rather rectangular, green-and-blue wattle over and behind the eyes. Fringing on the primaries, secondaries and greater and medium upperwing-coverts is iridescent blue and concolorous with rest of the upperparts.
- **51b** **Adult male non-breeding** Has an undeveloped wattle that may not be visible; slightly duller yellow below than breeding male with no metallic blue on head and mantle.
- **51c** **Adult female** Greenish crown and upperparts, yellow cheeks and underparts.

48c
48b
48a
49c
49b
49a
50c
50b
51c
50a
51b
51a

SYSTEMATIC SECTION

PITTA

The 31 or 32 species of pitta are currently all included in a single genus. However, morphological variation within this genus suggests a basis for subdivision, and future taxonomic work may well result in recognition of other genera. Whilst some species are monotypic, those with large ranges or an island distribution are frequently represented by a number of subspecies, some of which may be good species; further taxonomic study is required. Pittas are shy, terrestrial inhabitants of forests in Africa, Asia, Wallacea and Australasia. They are short-tailed, strong-billed, rather plump and generally brightly coloured, with simple, but often far reaching, calls. Twelve species are sexually dichromatic and males and females can be separated in the field. Most species are resident, but some populations of six species are migrants and evidence suggests that two other species make regular seasonal movements.

1 EARED PITTA
Pitta phayrei   Plate 1

Anthocincla Phayrei, J. Asiat. Soc. Bengal, 31 (1862), p. 343. Tounghoo, Burma.

Alternative name: Phayre's Pitta

FIELD IDENTIFICATION Length 20-24. This is a rather elusive, well-camouflaged species that occurs from Burma to Indochina. Both sexes have broad whitish-buff to ochraceous lateral crown-stripes that become white above the ear-coverts and can protrude as white ears at the back of the head, chestnut-rufous upperparts and bold black and rufous-buff markings on the wing-coverts. Males have black central crown to nape, broad black face-patch and diffuse black malar stripe. The underparts are a rich fulvous-orange with black speckling on flanks and pink or fulvous-pink undertail-coverts. Females differ from males in having the black parts of the head dark brown and supercilium duller. The underparts are a paler buff and are more heavily marked with black spots and scales. Immatures are rufescent-brown above and have a distinct buffy-white supercilium from an early age. The throat is white, whilst the ear-coverts, breast and flanks are dark grey-brown. The belly and undertail-coverts are usually a mixture of grey-brown and buff or rufous-buff.
Similar species The Eared Pitta is unlikely to be confused with any other pitta in its range, except perhaps with immature birds. Immature Blue-naped, Rusty-naped and Blue-rumped Pittas have buff or rusty spotting above, whilst both immature Rusty-naped and Blue-naped Pitta are distinctly marked with white below. In contrast, immature Eared Pitta has a distinct buffy-white supercilium from an early age but lacks distinctive spotting on the upperparts and is not spotted or scaled with white below. Immature Blue Pitta has a blue tail.

VOICE The call of Eared Pitta is a drawn-out whistle, *whee-ow-whit*, with emphasis on the first part of the first note and the middle note, which may sometimes sound as one. This call is most frequently heard at dusk. The distinctive alarm of this species is a brief whine, recalling a dog (Lekagul and Round 1991). This call may also be used as a contact note (P. Round *in litt.*).

DISTRIBUTION Eared Pitta is distributed from southern Yunnan in China, northern Burma, south through Tenasserim and east through Thailand to Indochina. It has also recently been reported from Hainan.

The most northerly record from Burma is reported to be from the Upper Madaya reserved forest in the Mandalaya district (Smith 1943, Smythies 1986). It also occurs in the Pegu Hills, Shan Hills, Karen Hills and Karenni. In adjacent Thailand, Eared Pitta has been found in the hills and mountains of the north, west and southeast, as well as in the central plains, but it has not been found in northeast Thailand, and its range does not extend to the peninsula (Round 1988, Lekagul and Round 1991).

In Indochina, it is found in northern Vietnam (Tonkin), Laos and Cambodia. The distribution of Eared Pitta in Cambodia is poorly documented. However, Kloss collected it in south-east Thailand on the coastal border with Cambodia (Robinson 1915a), whilst Engelbach (1936, 1938) found it at low altitudes on the slopes of Bokor and B. King found it nearby at Kep, Kampot (Thomas 1964); Eared Pitta is therefore probably well distributed in the hills of the Elephant and Cardamom Mountains of southern Cambodia.

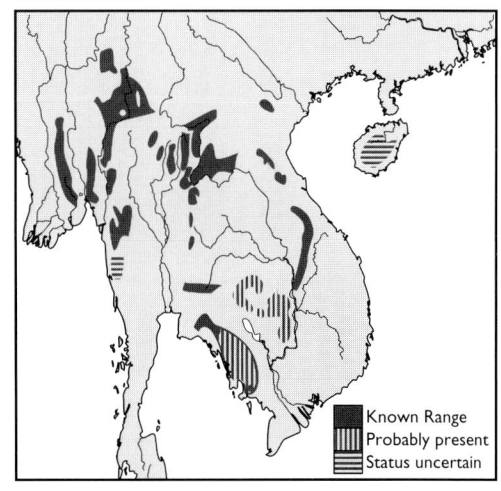

Distribution based on approximate extent of forest remaining within the altitudinal range of the species. Former range more widespread. Presence in Cambodia and southern Laos requires confirmation.

Delacour and Jabouille (1927, 1931) found Eared Pitta at Saravane, on the northern edge of the Bolovens Plateau, Napé, in central Laos near the Vietnamese border, and Xiengkhouang in central northern Laos, whilst Engelbach (1932) collected it at Attopeu to the east, and at Thateau to the north of the Bolovens Plateau. J.W. Duckworth (*in litt.*) found the species at Phou Khao Khouay near Vientiane, Laos. In Vietnam, Delacour and Jabouille (1931) found Eared Pitta at Muongmoun near the border with China in north-west Tonkin, and around the Babé lakes and Backan in north-east Tonkin. It also occurs in the central part of Tonkin, where present in Cuc Phuong (J. Eames *in litt.*) and has been found in central Annam at Kon Tum (Sa Thay, close to the border with north-east Cambodia: Bô Khoa Hoc *et al.* 1992), suggesting that it has been overlooked in other parts of the Annamitic chain in northern and central Vietnam.

Cheng (1987) reported that, in China, Eared Pitta is known only from Xiao Mengyang in the southern part of Xishuangbanna (= Hsi Shuang Pan Na), Yunnan, on the border with Laos to the east of the Mekong River. The only record from Hainan is of a single bird that was apparently observed in Jiangfengling Nature Reserve on 10 July 1990 (J. Christensen *in litt.*); the presence of this species on Hainan needs confirmation.

GEOGRAPHICAL VARIATION No subspecies are recognised, although northern birds apparently become slightly deeper in colour from west to east.

HABITAT Eared Pitta is a bird of dense evergreen forest, mixed deciduous forest and bamboo, also occurring in areas of secondary forests and, in some localities, in scrub (Round 1988) and in rather dry forest on limestone (P. Round *in litt.*). In southern Laos, R. Thewlis (*in litt.*) found it in areas which were a mosaic of deciduous trees, bamboo, rattan palms and lianes. This species is frequently found on steep slopes and, in comparison to Blue Pitta, Eared Pittas appear to prefer drier microhabitats in areas of sympatry.

Baker (1926) recorded the altitudinal range of this species in Burma and Thailand as around 460-1,830m. However, it has been collected at c.425m in the Mandalaya district (Smythies 1986), whilst Bingham (1903) found it at 450m in the southern Shan States of Burma. Although Eared Pitta occurs from the plains to over 1,800m in Thailand, it is most frequently found below 900m, and in Tenasserim the species is reported to be confined to the plains and low hills (Hume and Davison 1878, Round 1988). The possible record from Hainan (see Distribution) was in dense undergrowth on a forested slope at 915m in the mountains of Jiangfengling Nature Reserve (J. Christensen *in litt.*).

STATUS Hume (1875a) and Hume and Davison (1878) reported that, even last century, Eared Pitta was rare in the Pegu Hills of Burma and very rare in Tenasserim, whilst Oates (1883) believed it to be excessively rare throughout Burma. Smythies (1986) concurred with this view, stating that it was rare everywhere in Burma. Delacour and Jabouille (1929) described the status as rare in Indochina. Beaulieu (1944) was unable to find this species in northern Laos, and believed it to be very rare. However, it may be locally common in suitable habitat in parts of Indochina, since it was found to be common at Nam Khueng, north-central Laos (Delacour and Greenway 1940), was the commonest pitta at sea level near the coast in south-west Cambodia (Engelbach 1938) and has regularly been heard and observed at Cuc Phuong, Tonkin (J. Eames verbally). Cheng (1987) describes the status in Yunnan, China, as very rare. The status on Hainan is unclear.

FOOD Nothing is recorded of the diet of Eared Pitta. However, a bird watched foraging at close range grabbed food items rapidly as it uncovered them in the leaf-litter, suggesting that it was eating relatively mobile invertebrates (B. Gee *in litt.*).

HABITS Eared Pitta is a very poorly known species. However, the relatively short tarsi and slightly decurved bill suggest significant differences in niche to other pittas (P. Round *in litt.*). It is often tame but unobtrusive, and usually encountered in pairs. R. Thewlis (*in litt.*) reports that the members of a pair, followed for nearly six hours over a two-day period during January, stayed within 5m of each other during the whole period of observation. Sometimes both birds would stand motionless for periods of up to ten minutes. Under such circumstances their camouflage makes them almost invisible on the forest floor. P. Round (*in litt.*) notes that Eared Pittas generally cover less ground when feeding than any other species of pitta in Thailand, and will stay under the same bush, feeding, for prolonged periods.

In Thailand, an Eared Pitta has been observed digging vigorously into a large log on the forest floor for about twenty minutes, whilst two Orange-headed Ground Thrushes *Zoothera citrina* fed close by. The thrushes apparently benefited from the pitta's activities, feeding in the debris dug out by the pitta (D. Scott *in litt.*). A bird watched by B. Gee (*in litt.*) turned over leaf-litter as if foraged, rapidly grabbing prey items that were disturbed. Smith (1943) observed this species foraging amongst dead leaves on the bank of a small rocky stream in thick forest.

Eared Pittas will perch up in low trees or on vines in response to playback of the territorial call (P. Round *in litt.*).

BREEDING Breeding seasonality and habits are very poorly known, with few published details. Bingham (1903) found a nest in the headwaters of the Meple River, north Tenasserim, on 21 April 1881 (BMNH). Baker (1926, 1934) reported that the nest was found at 1,525m, although he provided the wrong locality and there may therefore also have been an error in his report of altitude. Apart from this, the only other record relating to breeding is of an immature that was collected in northern Thailand on 28 October (BMNH). The nest found by Bingham was a small, compact, oven-shaped structure, made on the ground at the foot of a tree. The entrance was on one side and situated so that it faced down the slope on which the nest was built, and had a firm little platform of sticks leading up to it. The nest was constructed of leaves, roots and grass whilst the nest cavity was lined with fine black roots. The nest contained four eggs which were glossy white, spotted with purplish-black, primarily at the broader end. The eggs, reported to be broad, rather pointed ovals and of the usual hard glossy white of most pittas' eggs, measured 27.4-27.9 x 21.6-23.3mm.

DESCRIPTION

Adult male Head patterns of individuals vary considerably, with no two birds being alike. Centre of crown black, broadening and extending backwards to nape; sides of forehead and sides of crown, including superciliary area, rufous or rufous mixed with white, the feathers being tipped with black, giving a bold scaled appearance: this feather tract extends to the sides of nape, the feathers from above and behind the eye usually being white, or pale rufous mixed with white and occasionally cream; the rearmost feathers at the sides of the crown are elongated and pointed, usually white with some subterminal black markings, and protrude beyond the head as 'ears'. Lores usually blackish, sometimes rufescent-brown; cheeks and ear-coverts black with variable amount of horizontal rufous streaking, these streaks usually absent behind the ear-coverts, where the black is unmarked and extends across nape. Mantle, back, scapulars, rump and uppertail-coverts chestnut-rufous, often with a few feathers at the top of mantle with black feather edges; uppertail rufescent, often with faint, dull, olive tinge. Undertail brownish-grey. Upperwing-coverts black with bold, broad rufous-buff tips to lesser and median coverts and terminal or subterminal patch in outer web of greater coverts; greater coverts unusually long and entirely black with exception of the innermost feather (i.e. the most visible on closed wing), which has inner web rufous-buff along entire length except for tip. Primaries dark brown, becoming paler at tip, and basally rufous-buff on outer web of outermost primary, both webs of primaries 2-5 and outermost webs of innermost primaries (where usually duller), these rufous-buff areas forming a bold patch in the closed wing; the rufous-buff patch is shortest on outermost primary and longest on primary 7, where extending about one quarter of length up feather. Secondaries dark brown, paler brown at tip, with broad rufescent tinge to outer web, brightest at edge; innermost secondaries and scapulars rufescent on both webs. Secondaries basally greyish on underwing (inner webs). Underwing-coverts pale buff mixed with dark brown. Centre of throat white, bordered by orangy-buff sides of throat, this area marked with variable amount of black scaling formed by narrow black tips to feathers, and appearing like an ill-defined moustachial streak; breast deep orange-fulvous to orange-buff, this colour extending along flanks and onto belly, where usually slightly paler. Underparts marked with variable amount of sparse black spotting, formed by subterminal spots to feathers, sometimes almost absent with only a few feathers of flanks being spotted, but usually spotted on flanks and also breast. Undertail-coverts unmarked pinky-buff to pink. Iris dark brown, bill black, legs pale brown or fleshy. As noted by (Hume 1875a), the bill is longer and more slender (compressed) than in most pitta species.

Adult female Centre of crown to nape dark brown, the feathers often darker terminally; sides of forehead and sides of crown dull buffy-white, the feathers scaled with narrow black edges and often dull chestnut nearer tip; feather tract from above and behind eye (extending to nape), white, mixed with black formed by black subterminal bands and edges; lores rufescent, the colour extending across cheeks and ear-coverts, where usually mixed with black. Rest of upperparts similar to male, though never with black on upper mantle; wings as male except that the buffy-rufous bases of primary feathers are much reduced and sometimes not visible in closed wing. Centre of throat white, sides creamy-white to buffy-white and marked with blackish scaling and spotting, these black marks being most intense in moustachial region, and often forming an ill-defined moustachial streak. Rest of underparts dull fulvous-rufous, often with more orange tones on breast, but never as orange as in male, and more intensely marked than male, with blackish subterminal spots that extend across entire breast and along flanks, although usually absent from belly centre. Undertail-coverts dull buffy-pink. The black markings of the lower flanks are often crescent-shaped, giving the appearance of broad scales. Bare parts as male.

Immature Forehead sides pale buff; centre of forehead to crown and nape dark brown, bordered by narrow band of rufous feathers from above eye to sides of nape, where broadening and becoming paler and buffier, with a few elongated but rounded, mainly buff feathers, forming 'ears' as in adult. Mantle to rump and tail, and scapulars, dull rufescent-brown. Innerwing-coverts dark brown with bold buffy notches at tips forming bar; greater coverts dark rufescent with indistinct small buffy notch at tips of innermost feathers; primary coverts blackish, innermost with paler rufous-buff outer web to basal two-thirds; secondaries dark brown with dull rufous edges to outer web except on innermost 1-2 secondaries, which are dull rufescent. Primaries dark brown with hidden buffy-cream bases. Throat white; ear-coverts, cheeks and lores blackish-brown, usually with indistinct rufous feathers on ear-coverts. Breast dark brown, usually with a few sparse narrow rufous streaks; flanks dark brown to greyish-brown, with rufous-orange to rufous-buff feathers of adult developing on flanks (and often with a few black subterminal spots); centre of lower breast becoming buffy-cream; belly unmarked buffy-rufous, extending to lower flanks and undertail-coverts. Iris brown. Bill tip, cutting edge and basal two-thirds of lower mandible dries yellow, but colour in life unrecorded; rest of bill dark. Legs paler than those of adult.

MEASUREMENTS Wing of male 103-107, of female 98-103; tail 47-53; bill length 34.5-39; tarsus 28-33; weight ?100 (male in TIST).

2 BLUE-NAPED PITTA
Pitta nipalensis   Plates 2 & 16

Paludicola Nipalensis Hodgson, 1837, J. Asiat. Soc. Bengal, 6, p.103: Nepal.

FIELD IDENTIFICATION Length 22-26. Blue-naped Pitta is a medium-large species with a range that extends from the Himalayas to Indochina. The head is largely fulvous to rufous-buff, including the forehead, and usually with a deeper rufous-orange colour in the post-ocular region. A black line starting behind the eye that bends sharply down behind the ear-coverts is not always visible. Males have a bright blue hindcrown and nape, whilst the female has a green nape. The upperparts are green, often faintly tinged with blue on the lower mantle and rump, whilst the underparts are fulvous. Males generally have a brighter pink hue on the underparts than females. Subadults are similar to adults but have a row of large creamy spots across the upperwing-coverts and a scaly, black gorget across the upper breast. Immatures are dark brown with whitish throat and buffy belly, and are marked with large, well-defined buff streaks on the crown, buff spots on the back and white spots on the breast.

Similar species In Tonkin and China (and perhaps Laos), where Blue-rumped Pitta is sympatric with Blue-naped Pitta, males of the former are easily distinguished since the subspecies there lacks blue on the nape but has a blue rump, in contrast to male Blue-naped Pitta. Females are much more similar, but can best be distinguished by the coloration of the throat, ear-coverts and forecrown, which is dull lilac in Blue-rumped Pitta and fulvous to rufous-buff in Blue-naped Pitta, and by the colour of the crown, which is green in female Blue-rumped Pitta but not in Blue-naped Pitta (which has the green confined to a neat patch on the nape in Indochina). Rusty-naped Pitta, which also occurs in Indochina, and is sympatric with Blue-naped Pitta in Arakan, and perhaps in the lower Chindwin River valley of Burma (Baker 1934), lacks the blue or green nape of Blue-naped Pitta.

Immatures are similar to immature Blue-rumped and Rusty-naped Pittas. However, whilst immature Blue-naped has buffy or occasionally creamy-white spots on the upperwing-coverts, those of Blue-rumped are buffy-orange, and the crown of Blue-rumped is marked with bolder, broader streaks. Immature Rusty-naped Pitta differs from immature Blue-naped Pitta in being much whiter on the ear-coverts (buff in immature Blue-naped), in having whiter streaks on the crown and whiter spots on the upperwing-coverts, and in having the breast spotted with white rather than buffy-white.

VOICE The call of this species has been described as a magnificent double whistle (Ali and Ripley 1983), most often made in the early mornings and late evenings. It is reported to be similar to that of Rusty-naped Pitta.

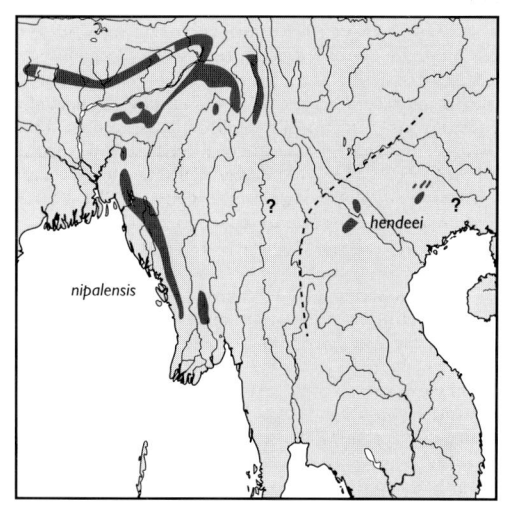

Distribution based on approximate extent of forest remaining within the altitudinal range of the species. Former range more widespread. Range in Burma may be more widespread than presently known.

DISTRIBUTION Blue-naped Pitta ranges from the foothills of the Himalayas in central Nepal east through Darjeeling, northern West Bengal, Sikkim, southern and western Bhutan and north-east India (including Meghalaya, Assam, Mizoram and Arunachal Pradesh), and southwards through the Chin Hills to the Arakan Hills of western Burma and Chittagong Hill Tracts of south-east Bangladesh.

It is widespread in north-east India, and Hume (1888), for example, obtained specimens from Manipur, northeast Cachar (southern Assam), Sadiya (north-east Assam/Arunachal Pradesh), Dibrugarh district (north-east Assam), the Khasi hills (Meghalaya), and the Dafla hills (eastern Arunachal Pradesh). Baker (1934) reported its occurrence at Hill Tippera, and it therefore probably also occurs in the Indian state of Tripura, the hills of which continue as the Chittagong Hill Tracts of Bangladesh. Inglis *et al.* (1920) found it in the plains of West Bengal, to the south of Darjeeling and western Bhutan, during winter.

A separate population occurs to the east, in Indochina, where it has been found in Yunnan (China) and in northern Laos and northern Vietnam. In Vietnam, Blue-naped Pitta occurs throughout Tonkin and in extreme northern Annam (Delacour 1930). In Laos, it has only been recorded from the extreme north, where known from Bountai (Tranninh Province) (Jabouille 1927, Bangs and van Tyne 1931, Delacour and Jabouille 1931), and presumably near Napé in central Laos, where R. Timmins and J.W. Duckworth (*in litt.*) found one for sale in a market.

Cheng (1987) suggests that this is an accidental species in China, where he reported its occurrence only in Hekou (Hokow) in south-eastern Yunnan Province in April. Yang Lan (1983), however, reported that there is a population in southern Yunnan and south-west Guangxi. Its occurrence in Guangxi seems probable in view of the distribution in adjacent north Vietnam (Tonkin).

Although Meyer de Schauensee (1946) reports collecting a specimen of Blue-naped Pitta on Khao Soi Dao, south-east Thailand, the locality and stated wing length (122mm) suggests the true identity of this bird to be Blue-rumped Pitta *P. soror flynnstonei.*

GEOGRAPHICAL VARIATION Two subspecies are recognised.

P. n. nipalensis (S Himalayas, from C Nepal east through W Bengal and Darjeeling to Sikkim and Bhutan; NE India; SE Bangladesh, and W and NW Burma. Birds in the Shan States of Burma are probably also this subspecies) Blue of nape frequently extends onto back and mantle. Wing 116-129.

P. n. hendeei (Indochina in Tonkin, Vietnam, N Laos, and S Yunnan and SW Guangxi, China) Similar to *nipalensis*, but much smaller, and blue is confined to a neat patch on the nape. Wing 107-114.

HABITAT AND RANGE This species inhabits a wide variety of habitats. It occurs in primary and secondary tropical and subtropical forests, including overgrown clearings, and is also to be found in various types of degraded habitats. Indeed, Baker (1926, 1934) asserted that Blue-naped Pittas prefer dense secondary growth and, next to that, open bamboo jungle without much undergrowth. He also noted that this species was frequently found in the bamboo-dominated secondary growth that springs up in hill cultivation when deserted. Baker (1934) found nests in dense evergreen undergrowth, in deciduous forests, and in bush and scrub growth on rocky hillsides.

In Nepal, Blue-naped Pittas frequent damp gullies in subtropical or tropical forests, usually with dense undergrowth (Inskipp and Inskipp 1991), but in north-east India they can be encountered on fairly dry ridge-tops as well as on steep slopes above rivers and streams, provided there is dense cover (pers. obs.). In Burma, Blue-naped Pitta has been found in teak forest where there is a dense undergrowth of dock leaves mixed with *Eupatorium odoratum*, an introduced weed (Smythies 1986).

Blue-naped Pitta has been found at low altitudes in the plains of north-east India, but is usually a bird of higher elevations, occurring up to 2,150m (Ripley 1982, Ali and Ripley 1983). Seasonal altitudinal movements may account for some records at lower elevations (see Movements), although Baker (1934) noted that it occasionally bred in the foothills in Dibrugarh, Assam. Ali and Ripley (1948) found it in the Mishmi Hills of eastern Arunachal Pradesh at 200-700m, whilst in the central region of Arunachal Pradesh it has been found at 425-1,830m (Baker 1913, Betts 1956). In Bhutan, it has been found at 640-c.1,000m

(Ludlow and Kinnear 1937, Ali *et al.* in press). In Sikkim, Blue-naped Pitta has been found up to 1,525m around Gopaldhara and near Mangpu (Stevens 1925). In northern Burma the species was found in rainforest in the vicinity of the Chindwin River, in rolling country below 200m (Mayr 1938), and at 1,400m on Mt Victoria in the Chin Hills (Stresemann and Heinrich 1939). Smythies (1986) noted that it occurs to around 1,830m in Burma.

In Indochina, the subspecies *hendeei* has been collected at altitudes of 150-365m in Tonkin and at 580m in Laos (Bangs and van Tyne 1931), but also occurs at higher altitudes: at Tam Dao in Tonkin, it occurs at c.900m (pers. obs.). It is well distributed in the mountains and hills of Tonkin (Delacour and Jabouille 1931) and Delacour (1930) also found it in limestone hills in extreme North Annam. In Tonkin, Vietnam, Delacour (1929a) found this species at the same localities as Blue-rumped Pitta.

STATUS Throughout much of its range in the Indian subcontinent, Blue-naped Pitta is rather a local bird, and may be scarce, as in Nepal (Inskipp and Inskipp 1991). During the nineteenth century, however, J. Cripps found it to be fairly common in dense forests on the borders of Assam and Arunachal Pradesh (Hume 1888). Baker (1934) reported that it was common in Assam, Manipur, the Lushai Hills (Mizoram), Hill Tippera (i.e. Tripura State, India) and in the Chittagong Hills of Bangladesh, although noting that it was commonest between 460 and 915m. Harvey (1990) reported that it was a rare breeding resident in north-east Bangladesh.

Hopwood (1912) found it to be uncommon in Arakan, Burma, but Baker (1934) reported it to be common in the northern Arakan Hills. However, its skulking habits and camouflaged plumage may in part account for the paucity of records: in other parts of its range it has been found to be common. Delacour and Jabouille (1927) found *hendeei* to be a common species in Tranninh Province, northern Laos, whilst Delacour and Greenway (1940) reported that it was very abundant in the vicinity of Louang Phrabang, north-central Laos. Delacour and Jabouille (1929) found it to be fairly common in Tonkin, Vietnam, although rarer than sympatric Blue-rumped Pitta. In recent times, however, forest in parts of its range, particularly in Tonkin, has been extensively cleared, so that in such areas it must now be confined to scattered localities.

MOVEMENTS Blue-naped Pittas apparently make altitudinal movements. Stevens (1915) noted that in Assam the species was distributed throughout the plains of the upper Brahmaputra during cold weather, whilst Inglis *et al.* (1920) found it commonly on the plains of Bengal Jalpaiguri in winter, but it has not been recorded here as a breeding species.

FOOD According to Baker (1926) and Ali and Ripley (1983), Blue-naped Pittas eat a variety of insects, such as ants and small beetles, as well as worms, grubs, lizards and even field mice. Delacour and Greenway (1940) found small snails and ants in stomachs of birds from Laos. In Tonkin, local people catch Blue-naped Pittas for food using live crickets or frogs as bait (Bangs and van Tyne 1931), suggesting that these are naturally part of their diet.

HABITS This pitta is very skulking, usually hopping away rapidly if disturbed, sometimes taking refuge by flying up to the low branches of small trees. A bird flushed into a tree in Assam proceeded to walk up and down the horizontal branch for a few minutes before flying to a more distant perch some 10m from the observer (pers. obs.). Blue-naped Pittas probably roost in small trees, and may also call from them in the early morning and evening.

This species is very well camouflaged as it feeds in leaf-litter under damp, shady cover. Its feeding actions are very thrush-like, moving a short distance, stopping to listen and probe the leaf-litter and then moving again. Ali and Ripley (1948) observed this species feeding on insects in cow manure, by picking it apart. Baker (1926) noted that the species often fed in pairs.

BREEDING In the Indian subcontinent, breeding is seasonal, occurring between April and August, with the majority of eggs being laid in May and June (Baker 1926, 1934, Harvey 1990, Inskipp and Inskipp 1991). Eggs in BMNH were collected in Darjeeling, north Bengal and Sikkim between 16 May and 16 June (from five clutches); from the Khasi and North Cachar Hills between 17 April and 2 August (from 29 clutches) and near Dibrugarh and Sadiya in the upper Brahmaputra valley on 14 May and 4 June. Baker (1926) noted that eggs found at the end of August may be from second broods. Cripps collected a juvenile in Arunachal Pradesh on 1 September 1879 (Hume 1888). A recently fledged bird was collected at Cherrapunji, Assam, on 28 May (USNM) and an immature was collected in Assam on 8 October (FMNH). In Sikkim, breeding birds have been found in June, and immature birds collected in June and August (Stevens 1925, BMNH).

The nest is a large, loosely constructed, oblong-domed structure, with the entrance at one end. It is usually situated on the ground or up to 2m above, in a tangle of bushes or brambles, or a clump of bamboo, the fork of a small sapling or on a platform of branches and debris on a bush (Hume 1880, Baker 1926, 1934). Rarely, nests are situated at higher levels: Baker (1934) found them up to 6m above the ground. Baker (1926) noted that most frequently the nest is built on the ground in bamboo, scrub or thin secondary growth, and, more rarely, in forest.

Nests in trees are usually built in stout forks of two or more upright boughs or in among a tangle of branches, and often such nests are built on platforms of decaying vegetation and other material placed in a convenient position on which a nest can rest. In the case of two nests on such platforms that were found in Sikkim, the entrance faced the only part of the scrub by which the nest could be approached, affording the sitting adults a good view and easy access of escape at the first sign of danger. In Sikkim, a nest built on a square platform that was more than 50cm wide was about 19.7cm high and 22.9cm wide within which the cavity was some 14cm in diameter and about 2.5cm deep from the lower edge of the entrance. The entrance was about 9.5cm in diameter (Baker 1934).

Nests of *nipalensis* that have been found on the ground were made almost entirely of bamboo leaves, with a few roots and 'grass', and lined with roots. Nests that are built above the ground have been noted to contain a greater variety of materials, including twigs, roots, fern-fronds and weed stems that are used to bind together the bamboo leaves, and may be lined with bamboo or coarse roots and 'grass'. The nest material may be so rotten and loosely held together that the nest falls apart when handled (Hume 1880, Baker 1934).

Three to seven eggs, but more usually 3-5, are laid by Blue-naped Pittas in the Indian subcontinent (Baker 1926,

BMNH). The eggs are broad ovals, with the average size of a sample of 100 eggs of the nominate subspecies being 29.5 x 23.4mm and the range of sizes being 26.1-32.6 x 21.8-25.6mm (Baker 1926). These eggs are glossy white, variably, but usually sparingly, marked with inky-purple to reddish-brown, blackish-brown, lilac or lavender-grey (Hume 1880, Oates 1883, Baker 1895). Eggs are sometimes tinged with pink. In the majority of eggs, markings consist of spots and small irregular blotches, whilst in others the blotches are larger and more or less intermingled with short broad streaks and straggling lines and other kinds of marks. Some eggs may be almost unmarked, whilst others have numerous markings. Typically, markings are more numerous towards the broader end (Baker 1895).

Both sexes share the tasks of nest-building, incubation and feeding the young. Baker (1934) reported that incubation takes about 17 days. Baker also reported that one pair of this species, presumed to be the same, returned year after year to build a nest in almost exactly the same site. Over a period of ten years, this pair had two broods every year, the first eggs being laid on 4-6 May, with brood size always four.

There are no published details on breeding biology or seasonality in Indochina.

DESCRIPTION *P. n. nipalensis*
Adult male Forehead and crown rufous-buff, with variable amount of green on crown; supercilium orange-fulvous, lores black mixed with fulvous; black post-ocular stripe beginning well behind eye and bending sharply downwards behind the ear-coverts; nape-patch deep blue, variable in extent and in most birds extending onto the mantle, and sometimes onto the rear crown. Rest of upperparts dull green, often with a golden sheen formed by mixing of orange. Some birds have blue feathers on the rump, giving a blue tinge. Primaries brown with narrow rufous leading edge to all but outermost two feathers, the rufous becoming more rufous-buff on innermost feathers; secondaries brown with outer margins of feathers largely dull green except for the extreme edge, which is narrowly rufous-buff. Innermost two tertials dull green. Flight feathers distinctly greyish from below. The outermost five primaries are marked with a hidden whitish to buffy-white bar across the inner margin near the feather base, this being narrowest on the outermost primary. Underwing-coverts dark greyish-brown with buffish to rufous-buff markings, densest on inner part. Bend of wing rufescent with narrow dark brown feather-tips. Tail feathers green above, greyish-brown below. Chin and throat pale buff; rest of underparts fulvous to orange-buff. The depth of colour of the underparts varies considerably, with many individuals having a fulvous-pink wash on the throat and foreneck. Many birds have blackish scaling on the upper breast formed by narrow black tips and broad black bases to feathers. Individuals from Assam, Manipur and Meghalaya frequently have black centres to the scapulars and feathers of the mantle, rump and uppertail-coverts. These black feather centres are rarely present in birds from Nepal and Sikkim. Eye light to dark brown; bill brown, paler at tip; legs reddish-slate to fleshy-pink.
Adult female Differs from male in having the blue nape replaced by an area of bright green, though this is often suffused with a blue wash. The underparts of females are less pink below than males, and, in series, females tend to have slightly paler underparts, though there is individual variation. Iris reddish-brown.

Nestling Baker (1926) describes the nestling as being dark brown above and pale fulvous-pink below, with fine black scaling, most noticeable on the breast and flanks, formed by narrow black bases to the feathers. Feathers of the upperparts have pale fulvous centres, but blackish edges.
Immature Head boldly marked: crown to nape dark brown with striking buffy-white or creamy-white streaks, tinged brown near the edges, most dense on the forehead; feathers of superciliary tract whitish with narrow brown edges; lores buffy-white; ear-coverts buffy-rufous with white shaft-streaks to some feathers. Throat white, becoming buffier with age; upper breast brown with buffy-white tips to feathers forming distinct large spots, these usually also extending along flanks. The rufous feathers of the breast and belly gradually replace the spotted areas, with the upper breast feathers the last to be replaced. Undertail-coverts mixed white and fulvous. Feathers or mantle to rump dark brown with bold, rich buffy subterminal spots, mixed with some greenish-bronze feathers in older birds. Older immatures have lower mantle and back green, with bronzy tinge to feathers of mantle, and may have creamy rather than buffy spotting on mantle. Upperwing-coverts dark brown, with dark buffy to buffy-orange subterminal spots and dark brown terminal edges, and innermost feathers tinged with bronze. Tertials bronzy-green. Secondaries pale brown on inner web, pale bronzy-green tinged on outer web. Primaries brown with broad rufous-bronze edge to outer web. Underside of primaries paler, with short white band across inner web at base of outermost four feathers. Uppertail-coverts brown with bold buffy to creamy spots at tip. Tail green with bronzy sheen and small, narrow, pointed rufescent patch at centre of tip. Iris hazel-brown; legs and feet yellowish-pink to dull reddish. Lower mandible and tip orange-red; rest of bill dusky brown.
Subadults are similar to the adult but have a row of four large cream spots across the edge of the greater coverts, a gorget across the upper breast formed by black scales, and more prominent scaling on the sides of the neck than an adult. Bill yellowish-pink turning blacker with age.

MEASUREMENTS Wing 116-129; tail 56-68; tarsus 46-53; bill 27-30; weight 110-132 (BMNH).

3 BLUE-RUMPED PITTA
Pitta soror Plates 2 & 16

Pitta (*Hydrornis*) *soror* Wardlaw-Ramsay, 1881, Ibis, p.496: Saigon, Cochinchina.

Alternative name: Blue-backed Pitta, Blue-headed Pitta

FIELD IDENTIFICATION Length 20-22. This is a medium-large species of pitta with a restricted range, occurring only in Indochina and south-east Thailand and on Hainan. The upperparts are green to bright green (depending on subspecies), with a distinctive blue lower back, and rump. Head colour varies according to subspecies, but is dull purple or lilac on the forehead and ear-coverts (although a subdued pink in some parts of the range, such as Tonkin), with a bright orange-rufous supercilium or post-ocular stripe which is usually bordered by a black line that extends around the ear-coverts. Males of most subspecies have blue napes (green with blue tinge in Tonkin and Hainan), this patch sometimes extending onto the crown and often the upper mantle; the nape of females is dull blue to green, depending on subspecies. The underparts vary in colour from dull fulvous to a deep orange-buff, often with a pinkish tinge on the breast. The bill is pale. In flight, the blue rump and uppertail-coverts are quite conspicuous, contrasting sharply with the green mantle. Immatures have coarse streaking or spotting from the forehead to crown, and rufous-buff droplet-like patches on the back and rump. Upperwing-coverts of immatures are marked with large, buffy-orange subterminal spots, the breast is pinky-buff to whitish with a few darker feathers mixed in (sometimes forming a gorget) whilst the flanks tend to be more rufous. The undertail-coverts of immatures are usually pale buffy-white, even in birds with breast colour similar to adults.

Similar species Most similar to Rusty-naped Pitta and Blue-naped Pitta. Where sympatric (China and Tonkin, and possibly adjacent Laos), Blue-rumped Pitta is relatively easy to distinguish from Blue-naped Pitta, being a much brighter bird with distinctive plumage differences. Blue-naped Pittas in Indochina usually have no blue on the rump. Males have a neat blue patch on the nape, in contrast to male Blue-rumped Pittas in this part of their range, which have blue rumps but no bold blue patch on the nape, making differentiation of males easy. Both sexes of Blue-rumped Pitta have a distinctive lilac or purple forecrown (fulvous to rufous-buff in Blue-naped) and bold orange lores and post-ocular stripe (often also the supercilium) that contrasts with the rest of the head coloration. Females can best be distinguished by these differences in head coloration, as well as by the colour of the crown, which is green in female Blue-rumped Pitta. In Indochina, the green is confined to a neat patch on the nape of female Blue-naped Pitta. Blue-naped Pitta is not known to occur in parts of Indochina where different subspecies of Blue-rumped Pitta (with blue napes) occur. Blue or green on the crown and nape are the best features for distinguishing adult Blue-rumped Pitta from adult Rusty-naped Pitta: both sexes of the latter have chestnut crown to nape.

Immature Blue-rumped Pittas are very similar to immature Blue-naped and Rusty-naped Pittas; for differences see under respective species accounts.

VOICE Several calls are described, but all are given infrequently. In Tonkin, this species gives a single full note *waeoe* or *weeya*, with only a slight inflection (Robson *et al.* 1989). This call, which is generally repeated at intervals of at least 7s, is probably a territorial call, although it seems to be rarely made. Other calls described are a a sharp, breathless *tew*, similar to the first syllable of the call of Rusty-naped Pitta, and a longer, falling-tone, mellow *tiu* (Legakul and Round 1991). This call, or something similar, is the most frequently heard in the Annamese lowlands of Vietnam: it is a quiet, rather frog-like note (*ppew* or *eau*), repeated at short intervals (pers. obs.). This call may function as an alarm note (C. Robson *in litt.*). When approached by an observer, an adult male with dependent spotted young repeatedly gave a single, sharp, explosive note, *hwip* or *hwit*, from a perch on a horizontal vine some 30cm off the ground (pers. obs.).

DISTRIBUTION The range of Blue-rumped Pitta is centred on Indochina. It is known from the Yao Shan range in south-eastern China (Mt Yaoshan, Guangxi Zhuang Autonomous Region [Kwangsi]: Cheng 1987), and patchily through Indochina to the Khao Soi Dao range of southeast Thailand, and on the island of Hainan. In Vietnam, Blue-rumped Pitta has been recorded from northernmost Tonkin to northern Cochinchina, with no confirmed records from south of the Mekong delta. Records suggest that the range is disjunct, with no records from most of Dac Lac and Gia Lai-Kon Tum Provinces of Vietnam or from the Mekong delta and abutting Mekong-Bassac river system (Delacour 1929, Rozendaal 1988, 1993). Robinson and Kloss (1919) reported that they had found Blue-rumped Pitta at 915m and at 2,290m in the Langbian Mountains (Dalat) of South Annam. However, the birds that they identified as this species at 2,290m were undoubtedly Rusty-naped Pittas and the record from 915m involved a young bird that was collected (specimen untraced): confirmation of its identity and therefore of the presence of Blue-rumped Pitta in the Langbian Mountains is therefore needed.

In Cambodia, Blue-rumped Pitta is known only from the Elephant Mountains near the south coast (Thomas 1964), whilst in Laos it has only been recorded in the extreme south (Salter 1993, Thewlis 1993), in the hills abutting and to the north-east of the Bolovens Plateau (Engelbach 1932), and in central Laos at Nam Kading (R. Timmins and J.W. Duckworth *in litt.*). Elsewhere in Laos, Timmins *et al.* (1993) did not find it in the forests of Phou Xang He, Delacour and Greenway (1940) failed to find it in the vicinity of Louang Phrabang, and Beaulieu (1944) did not record it anywhere in the northern province of Tranninh. However, its presence in north-west Tonkin (Bangs and van Tyne 1931) suggests that it is likely to occur in parts of northern Laos and perhaps also in adjacent China, although there is now little forest left in this region. On the island of Hainan, Blue-rumped Pitta is found in the Seven Finger Mountains.

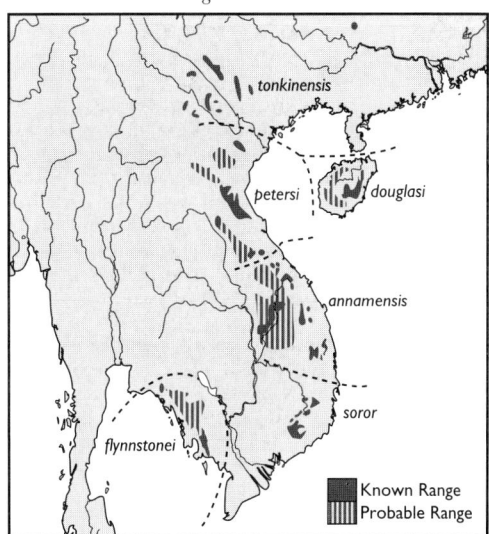

Map shows known distribution based on approximate extent of forest remaining. Other areas where suitable habitat remains at appropriate altitudes are also shown but confirmation of occurrence in these areas is required.

GEOGRAPHICAL VARIATION The apparent disjunct range of Blue-rumped Pitta has led to the description of six subspecies, although, as pointed out by Rozendaal (1993), the large distributional gaps in Vietnam are likely to be attributable to limited collecting activity. The species shows strong clinal variation in the colour of the nape and upper mantle. Hence, whilst six subspecies are recognised here, some of these may be lumped in the future, in particular those in Vietnam. Size and the colour of the crown and nape are the main features used to separate presently recognised subspecies.

P. s. soror (Vietnam in S Annam and Cochinchina between the Mekong River and the Tay Nguyen Plateau) Crown and nape blue, forehead and forecrown lilac to dull purple. Female duller than male, with green crown (wing 110-113).

P. s. annamensis (Vietnam in C Annam and S Laos) Lilac crown and face, rather pale blue rump, and usually larger and bolder black patch at rear of ear-coverts than *soror*. Size similar to *petersi* (wing 108-118).

P. s. petersi (formerly *P. s. intermedia*) (Vietnam in N Annam and SE Tonkin, and C Laos) Intermediate between *soror* and *tonkinensis* in size, with crown and nape pale blue, slightly tinged green (wing 115-122).

P. s. tonkinensis (S China, south to C Tonkin, Vietnam) Green crown and nape, faintly tinged blue. Upperparts tinged blue, underparts tend to be paler than those of *soror*, and more strongly suffused with pink. Large size (wing 117-126).

P. s. flynnstonei (SW Cambodia and SE Thailand) Differs from *soror* in larger size, being almost as large as *tonkinensis* (wing 120-125).

P. s. douglasi (Hainan, in the Seven Finger Mountains) Small (wing 112-114). Male has purple forecrown, green crown and upperparts with blue restricted to rump and occasionally some feathers of the lower mantle. Blue tinge to nape and upper mantle. Purplish ear-coverts and throat. Breast strongly washed with pink. Female similar to *tonkinensis*, but lacks any black scaling on breast, is more purple on sides of forehead and more strongly suffused with pink on throat and breast.

HABITAT Blue-rumped Pittas are tolerant of a wide range of habitat types. In Vietnam, they inhabit evergreen forest and areas of bamboo, although they are not found in pure bamboo stands (Robson 1991). In Tonkin and North and Central Annam, they survive well in secondary and logged evergreen forests (Robson *et al.* 1989; pers. obs.). Delacour (1930) found them in limestone hills in extreme North Annam. Thewlis (1993) found this species in primary and degraded semi-evergreen forests along river courses in southern Laos, but not in deciduous areas.

In the Annamese lowlands, two nests were found in dark, damp, flat areas of primary riverine forest with a dense understorey dominated by saplings of broad-leaved trees (Lambert *et al.* in press). Adults with dependent young were also observed on steep dry slopes dominated by fan palms *Licuala* sp. and amongst tangles of vines in dark areas of forest on limestone cliffs and outcrops.

Blue-rumped Pitta occupies a very wide altitudinal range. Delacour (1929) found it common to at least 915m in Indochina. Nguyen Cu (1990) reported its occurrence at 1500-1,900m in north Vietnam, although its presence at 1,900m has been questioned (C. Robson *in litt.*). In Tonkin, *tonkinensis* has been found at 300-1,700m (Rozendaal 1990). In North Annam, *petersi* occurs at altitudes as low as 50m, and in Central Annam it occurs at 30-700m (Robson *et al.* 1989, 1991; pers. obs.). In Cochinchina, *soror* has been found below 170m (Beaulieu 1932, Rozendaal 1990). In Laos a specimen was collected at 90m (Delacour 1929a), and it has been observed below 100m on the Xe Kong Plains and elsewhere at altitudes up to 850m (Engelbach 1932, R. Timmins and J.W. Duckworth *in litt.*). In south-east Thailand, however, *flynnstonei* is known only from evergreen forests at 900-1,670m (Rozendaal 1993). In adjacent Cambodia, it has been found at 915m near Bokor in the Elephant Mountains (Delacour 1929), and presumably also occurs throughout the Cardamom Mountains.

In Tonkin, Vietnam, Delacour (1929a) found this species in the same localities as Blue-naped Pitta. Rozendaal (1990) found Blue-rumped Pitta and Rusty-naped Pitta sympatrically in the Hoang Lien Son range of Vietnam.

STATUS Delacour (1929) described the status of *soror* as common, and of *petersi* as numerous in parts of North Annam, whilst Kinnear (1929) found that *tonkinensis* was fairly common in north Tonkin. In south-east Thailand, this is now an uncommon or rare species (Round 1988), but in parts of Vietnam, such as in Cuc Phuong National Park, Tonkin, it is still common (Robson *et al.* 1989), and *annamensis* is reported to be common in southern Laos (Thewlis 1993). Blue-rumped Pitta is also relatively common throughout the forests of the Annamese lowlands, although generally less abundant than sympatric Bar-bellied Pitta (pers. obs.). Engelbach (1932) reported that *soror* was not rare on the slopes north-east of Bolovens Plateau although not found there above 800m. On Hainan, *douglasi* is said to be very rare (Cheng 1987).

MOVEMENTS None recorded.

FOOD In Tonkin, local people were reported to catch this species using traps baited with live crickets and small frogs (van Tyne 1933), suggesting that these items usually form part of their diet. C. Robson (*in litt.*) has observed Blue-rumped Pittas eating snails, which are smashed on rocks. The sound of snail shells being broken on rocks is not uncommon in areas where this species occurs, suggesting that they may form an important component of its diet. No other information pertaining to diet has been reported.

HABITS This is a shy and very secretive species, hopping off rapidly or flying when approached, although adults with young are often very approachable. Mud stuck to the bills of birds observed in Annam suggest that they sometimes probe the soil whilst feeding (pers. obs.). Birds flying off from an observer invariably land on the ground. Adults appear to be solitary, although occasionally pairs may be encountered in good foraging sites. Blue-rumped Pittas call rather infrequently, from the ground, even during the breeding season (pers. obs.). R. Timmins (*in litt.*) observed this species calling from a 20cm high termite mound in Laos.

BREEDING The only nests of this species that have been found were in primary forest in the lowlands of Vu Quang Nature Reserve, central Vietnam, on 4 and 6 June 1994 and in logged forest in Ha Tinh Province on 24 June 1995 (Lambert *et al.* in press, J. Eames *in litt.*). A recently fledged juvenile was observed in central Annam on 6 July 1994, whilst older fledglings were observed at various sites and on various dates from 13 June to mid-July. Immature *annamensis* with spotted head, nape, upper mantle and wing-coverts, have been collected in Central Annam on 21 July; another collected on 24 December was more advanced with the plumage of its head like that of an adult. Björkegren collected five young birds between 25 July and 13 August (Eames and Ericson in prep.). Taken together, these observations therefore suggest a rather short breeding season in Vietnam, with nesting apparently occurring in May-June or July. In Hainan, a fully fledged immature, with buff spotting at the tips of the greater and median upperwing-coverts, was observed at Jiangfengling on 6 July (J. Christensen *in litt.*).

The nest found at Vu Quang on 4 June contained three eggs, but was predated on 8 June, when the nest was found to be partly destroyed and two eggs had gone (the remaining egg was cold). The nest discovered on 6 June contained three recently hatched young. Only females were seen incubating the eggs and young of these nests (Lambert *et al.* in press).

The two nests found in 1994 were situated within 100m of each other, in primary forest, one on the flat floodplain of a large river at 30m altitude, the other in a flat area of forest adjacent to a stream at c.50m altitude (the two areas separated by a cliff). The latter nest was situated about 2.4m off the ground in an unusual narrow fold in the trunk of a large tree, formed at the point where two buttresses intersected. The base of the nest was built primarily of twigs, many of which were branched, measuring up to 15-20cm long and mostly 4-6mm wide, though one twig was 8mm wide. Mixed in with the twigs were a few rootlets and dead leaves. The roof and sides of the nest were of dead leaves, the majority being from broad-leaved trees, but mixed in were 1-2 long strips of palm leaf. The rear and sides of the nest cavity were lined with leaves, whilst the nest cup was made with black rootlets of c.1-2cm diameter. The total height of the nest was c.22cm, with an entrance hole 8cm high and 6cm wide. The second nest, containing three pulli, was situated in a 4.5m tall understorey tree with spiny, holly-like leaves and trunk diameter of c.8cm at breast height. The nest was at the base of the canopy of the tree, where placed on a fork at c.2.3m off the ground and within c.30cm of an intact but disused wasps' nest that hung from the same tree. The nest base was made of large twigs, the biggest measuring 74cm long and 1cm wide. These long twigs extended forward from the base to form a platform in front of the nest hole, on which the adults could land. The nest was 22cm high and 20cm wide externally, with the platform extending some 15cm out from the entrance. The entrance was 8.5cm wide and 9.5cm high, and the distance from the entrance hole to the rear of the nest cavity was 15cm. The nest sides and roof were constructed of the dry leaves of broad-leaved trees, with only one being skeletonised. The nest cavity contained a nest cup made of fine flexible twigs and rootlets up to 16cm in length (Lambert *et al.* in press).

The nest found by J. Eames (*in litt.*) on 23 June was in a 4.5m-tall rattan palm, and contained three chicks. It was situated amongst the branching fronds of the rattan, and was an almost spherical dome made with dry dicotyledonous and rattan (probably *Calamus* sp.) leaves and twigs. The nest was some 24cm wide, 25cm high and 30cm deep with a nest entrance 1.3m above the ground. The entrance was 9cm wide and 8cm high, whilst the chamber was 17cm deep and 13cm wide. A 7cm-wide platform jutted out c.6cm in front of the nest entrance. The nest was under observation for c.4.5 hours over one morning, during which time an adult was observed to sit motionless on the nest platform for more than three hours but no feeding visits were observed. The pulli were estimated to be 2-3 days old and weighed 11, 17 and 17g, suggesting asynchronous hatching.

The eggs found in 1994 were pinky-white with large chocolate-brown speckles and blotches that were concentrated at the broad end and smaller spots near the narrow end. One egg measured 29 x 22mm.

A recently fledged juvenile that was observed with one parent at the base of a jagged limestone outcrop on 6 July raises the possibility that this species could sometimes build its nest in crevices in limestone (pers. obs.).

DESCRIPTION *P. s. soror*

Adult male Forehead lilac-purple, becoming progressively more blue towards rear of crown; rear of crown and nape bright blue, occasionally purple-blue; supercilium dark orange, narrowest in front and over eye, broadening behind and in many birds extending as a narrow orange border to the rear of the ear-coverts. There is a variable amount of black behind the eye, most prominent at the rear end of the supercilium. Eye large, with narrow but distinct white eye-ring. Mantle and back bright green, or a mixture of green and blue; rump blue; uppertail-coverts and tail green; upperwing-coverts green, concolorous with mantle, sometimes with rufous feather edgings. Some individuals have black centres to the green feathers of mantle, back and uppertail-coverts. Primaries brown with paler outer edge; outer secondaries brown with green outer half of outer web and rufous outer edge, inner secondaries similar but with green tips and variable amount of green on the inner web. Hidden white bar across the base of the primaries formed by a white patch in the inner webs. Throat and ear-coverts purplish-pink, paler than forecrown; rest of underparts deep orange-fulvous, with strong pink wash to feathers of breast; undertail-coverts slightly paler, and some individuals may have whitish belly to undertail-coverts. Some males have narrow black bands across the feathers of the upper breast, giving a scaly appearance. Iris dark brown; bill pinkish-white to pinkish-horn with the distal third whitish-horn; legs pinky-horn to fleshy brown.

Adult female Forehead purple becoming green on the crown and blue on nape; nape colour variable with some individuals having green feathers mixed in with the blue; orange supercilium usually narrower, less prominent than that of male. Back, mantle and uppertail-coverts green, slightly duller in colour than those of male; variable amount of blue on rump. Underparts as male but paler, with duller pink tinge to throat. Most females show some scaling on the upper breast. Bill yellowish-white with basal two-thirds of upper mandible greyish-horn.

Subadult (*soror*) Distinguished from adult by the possession of one or more conspicuous buff spots in the greater coverts, and usually has belly and undertail-coverts distinctly paler than the breast. Slightly younger birds (*douglasi*) are reported to have buff spots on the greater and median upperwing-coverts, greyish throats with pinkish tinge and fine black spotting, brown forecrown and a very short eye-stripe that is hardly more than a black spot behind the eye (J. Christensen *in litt.*).

Immature Feathers of forehead to nape buffy-white with dark brown edges and tips, giving coarse streaked or spotted appearance; this pattern extending onto the mantle, back, rump and uppertail-coverts. Feather tract from upper lores to above ear-coverts, passing narrowly above eye, lack the dark edges; lores brownish; ear-coverts buffy with pinkish sheen, this colour extending onto the throat, where buffy-pink feathers mix with a few darker brown ones. Centres of feathers of mantle and back richer, more rufous-buff than those of crown. Some green feathers usually present on mantle and blue feathers on lower back or rump. Tail dull green with blue tinge. Breast pinky-buff with a few darker feathers mixed in, the dark feather-bases at top of breast sometimes being exposed; sides of breast, flanks and sides of belly shows less pink and are more rufous; undertail-coverts pale buffy-white. Upperwing-coverts dark brown with large, buffy-orange subterminal spots, though some inner greater coverts may lack spots; primary coverts brown with buff on outer edge; primaries pale brown with buff to olive-buff tinge to outer webs; secondaries brown with olive to green tinge to outer web; four outermost primaries with narrow whitish bar across inner web near base; underwing-coverts buffy-orange, mixed with some darker brown feathers. Bill pale, but lower mandible darker than upper.

MEASUREMENTS Wing 110-113; tail 50-56; bill length 31-33, bill width 13.5-16; weight (one female *annamensis*) 102.

4 RUSTY-NAPED PITTA
Pitta oatesi Plates 1 & 16

Hydrornis oatesi Hume, 1873, Stray Feathers, 1, p.477: Toungoo district, Upper Pegu [Burma].

Alternative name: Fulvous Pitta

FIELD IDENTIFICATION Length 21-25. This is a distinctive medium-large species of pitta that inhabits mountains from Burma to Indochina and Peninsular Malaysia. The head is mostly chestnut-rufous, brightest on crown to nape, and with paler, pink-tinged throat and ear-coverts, and black post-ocular stripe that extends around the ear-coverts. The upperparts are green, tinged rusty in the female and, in Indochina, with a variable amount of blue on the rump and back. The underparts are deep fulvous, often with a pink suffusion to the throat. Rump colour varies with subspecies and sex from blue to green: in most parts of its range, blue on the rump is indicative of a male, although in the Bolovens Plateau of Laos and perhaps adjacent Vietnam females may also have blue on the rump. Immatures are blackish-brown with bold white streaks on the crown, bold buff droplets and spots on the upperparts and wing-coverts and a dark brown breast spotted with white.
Similar species Several sympatric pittas of mainland Asia are similar to Rusty-naped Pitta. Blue-naped Pitta and Blue-rumped Pitta are superficially very similar, but these differ in having blue or green hindcrown and nape, whilst Blue-rumped Pitta also differs in having a distinctive blue lower back and rump. Juvenile and immature Rusty-naped Pittas are confusingly similar to those of several other species. Immature and juvenile Blue-naped Pittas are almost identical in pattern of markings, but similar-aged Rusty-naped Pitta has white rather than buffy-white streaks on crown, white rather than buff to buffy-white spotting on breast, and whiter spots on the upperwing-coverts. Juvenile Blue-rumped Pitta differs in having broader, bolder buffy streakings on the crown and large, bold buffy-orange spots on the wing-coverts. Young Rusty-naped Pitta can be distinguished from young Blue Pitta by its large bill with extensive orange-flesh or pink coloration, and by the brownish, rather than blue, tail. Note that some calls of Blue Pitta are also similar to those of Rusty-naped Pitta. Juvenile Eared Pitta is safely distinguished by the virtual lack of markings on its upperparts, with the exception of a long buffy supercilium.

VOICE The main call is a loud, breathless double note, *chium whit* or *chow-whit*, the first note rather truncated, the second rather explosive and inflected. In Malaysia, Rusty-naped Pitta calls seasonally, with peak calling in the period March-May. At other times of the year, calls are rarely heard, although the species may respond readily to playback or imitation of its call. During the peak breeding season, the *chium-whit* call may be uttered every 4-5s with up to 25 calls being given in a row. In Vietnam, a quiet rather frog-like call, very similar to the *ppew* call of Blue-rumped Pitta, has also been heard (J. Eames verbally).

An alarm note recorded in the vicinity of a nest in Malaysia was described as a loud, emphatic, woodpecker-like, metallic *check* or *weck* that is sometimes run into a rattle when highly agitated. M. Chong (*in litt.*) described the agitated call of a female in Malaysia, apparently trying to lure a juvenile away from him, as a single loud *chek*, repeated regularly. On the ground near the nest a soft musical churr, *chur-r-r-r-t*, *wer-r-r-r-r-rt* or *tur-r-r-r-rt*, may also be made (King 1978). In northern Thailand, another call, probably given by a female in a breeding territory, has been described as an explosive, liquid, falling-tone *poouw* (Round 1983). Immatures have been heard to utter an explosive, rather woodpecker-like alarm call, *tchick*.

The voice of the this species was erroneously described as *bong-bong* by Deignan (1945), based on an identification made by local people.

DISTRIBUTION The Rusty-naped Pitta is a submontane to montane species that is found throughout much of Burma, and east through Thailand to Indochina. Another population occurs in the Central Range of Peninsular Malaysia, where it is known in the Main Range from Mount Telapa Burok (Negri Sembilan State, 2°50'N, 102°03'E), north to the Genting Highlands, Fraser's Hill, and the Larut Hills (Maxwell's Hill) (D. Wells *in litt.*, C. Prentice *in litt.*). A nest of Rusty-naped Pitta obtained in Perak, in the extreme north of Malaysia (Baker 1926), was presumably that of *deborah*.

In Burma it occurs from the Myitkyina District in the north-east and the southern Shan States through the eastern Pegu Hills, Karenni, the Karen Hills and northern and central Tenasserim to south-west Thailand. Its range extends eastwards through northern Thailand and north-east Laos and western Yunnan.

Cheng (1987) documents the existence of two subspecies in Yunnan, China, although *oatesi* is only known from one female specimen from Yingjiang in the west; *castaneiceps* has been found in southern Xishuangbanna and south-east Pingbian. In Vietnam, its distribution is apparently disjunct, with a population in north-west Tonkin and another on the Langbian Peaks (Dalat) and southern Annam in southern Vietnam (Delacour and Greenway 1940b), but apparently none in the intervening mountains of Central and North Annam. In adjacent Laos, Rusty-naped Pittas occur on the Bolovens Plateau (including the Dong Hua Sao National Biodiversity Conservation Area) and northwards through the central and northern highlands (R. Timmins and J.W. Duckworth *in litt.*).

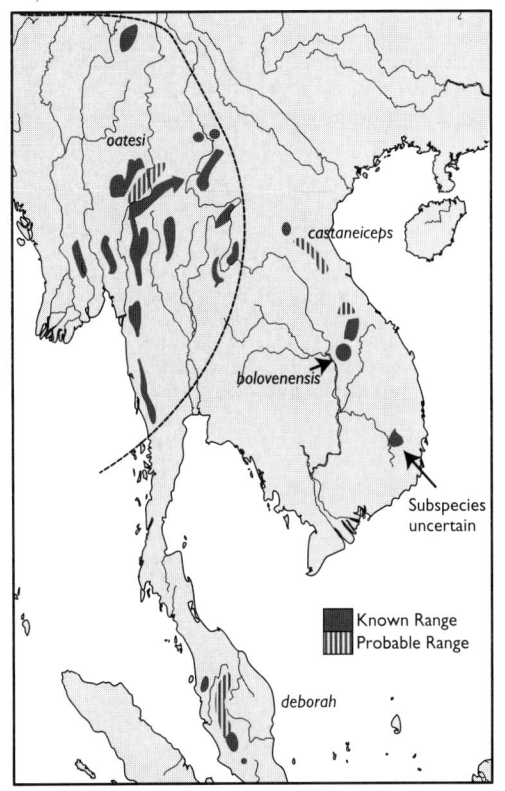

Map shows known distribution based on approximate extent of forest remaining within the altitudinal range of the species. Other areas where suitable habitat remains are also shown but confirmation of occurrence in these areas is required.

GEOGRAPHICAL VARIATION Four subspecies are described.

P. o. oatesi (Burma to SW Thailand; and eastwards through N Thailand and NE Laos and W Yunnan) Male rump with little blue or bluish-green, sometimes lacking blue entirely; female rump green. Head more uniform fulvous than other subspecies, and with the colour of crown fading onto mantle. Upperparts slightly more olive than other subspecies. Amount of pink suffusion on throat varies considerably.

P. o. castaneiceps (SE China to C Laos and Vietnam in NW Tonkin. Birds in southern Annam and on the Langbian Plateau, Vietnam, are also believed to belong to this subspecies) Both sexes have green rump. Border of the chestnut-rufous of lower nape and the green of mantle well defined. Usually very narrow orange edges to feathers of upperparts. Slightly smaller than nominate (wing 112-118), with darker, more chestnut tones on crown and nape, deeper buff underparts, and, usually, more prominent pink suffusion to throat and ear-coverts and broader black streak behind eye.

P. o. bolovenensis (SC Laos, where known only from the Bolovens Plateau. Birds from the Langbian Peaks, Dalat, in southern Vietnam may also be referable to this subspecies) Similar to *castaneiceps*, but has bright blue rump to lower mantle and variable blue wash to upper mantle. Males also have blue wash on back and inner upperwing-coverts. Female differs from male in having blue usually confined to the upper and middle rump.

P. o. deborah (Peninsular Malaysia) Darker than other subspecies, with darker green mantle and upperside of tail. Male had very bright pink wash to entire underparts, very pink throat and blue rump. Slightly smaller than other subspecies (wing 107.5).

HABITAT The Rusty-naped Pitta inhabits the understorey of forests in the submontane to montane zones throughout its range. Compared to the similar and partially sympatric Blue-naped Pitta, it tends to occupy denser parts of the forest interior, where it often frequents ravines (Baker 1926). In Peninsular Malaysia it is most frequently encountered on slopes and in steep ravines with a dense understorey of saplings, rattans and other palms, although it also occurs in more open areas, such as paths and along rocky streambeds, at dawn and dusk (pers. obs.).

In Thailand, nominate *oatesi* has been recorded at c.500-2,590m in evergreen forests, wooded second growth, deciduous bamboo brake (one record) and in an overgrown fruit orchid on Doi Pui, at 1,400m (Round 1988, Lekagul and Round 1991, P. Round *in litt.*). Hume and Davison (1878) found the species at altitudes of 760-1,525m in Tenasserim, whilst further north in Burma this species has been collected at 1,370-1,615m in the southern Shan States (Rippon 1901, Bingham 1903, Meyer de Schauensee 1934) and Smith *et al.* (1943) reported that it occurs at 760-1,220m in Karenni. Oates (1882) found this species to be common along certain streams in the evergreen forests of the Pegu Hills, Burma. In Peninsular Malaysia, *deborah* has been recorded in primary montane rainforest above 1,100m (Wells 1990b).

All records of *castaneiceps* are from above 1,000m, up to 1,600-1,700m (Delacour 1930, Bangs and van Tyne 1931, Delacour and Jabouille 1931, Rozendaal 1990). Most observations have been made on steep slopes, and this subspecies evidently favours such terrain: Delacour (1930) found it above 1,525m in a mountainous limestone area in Tonkin.

Although Rozendaal (1988) reported that Rusty-naped Pitta (race *bolovenensis*) and Blue-rumped Pitta *P. s. soror* seem to replace each other altitudinally on the Bolovens Plateau in Laos, Engelbach (1932) found both species at similar altitudes there, whilst Rozendaal (1990) found both species (*P. o. castaneiceps* and *P. s. tonkinensis*) occurring sympatrically at c.1,600-1,700m in the Hoang Lien Son Range of Vietnam. On the Langbian Plateau, Vietnam, Robinson and Kloss (1919) collected pittas at c.2,290m which were undoubtedly this species, although they were identified at the time as Blue-rumped Pitta.

STATUS Round (1984) described its status as rare in suitable habitat on Doi Pui, Thailand. Oates (1882) found Rusty-naped Pitta to be a locally common species in the Pegu Hills of Burma, and Wardlaw-Ramsay (1877) stated that it was a common species in Burma. Baker (1926), however, noted that, whilst common in the Pegu Hills, it was comparatively rare in the Arakan Hills. Calling at Fraser's Hill, Peninsular Malaysia, suggests that in suitable habitat *deborah* is relatively common with densities probably in the order of at least five pairs per km^2 (pers. obs.). In Tonkin, Delacour (1930) found this species to be abundant in the vicinity of Chapa, above Lookay on the Red River. Engelbach (1932) found *bolovenensis* common on the Bolovens Plateau above 500m. In China, where two subspecies have been recorded from Yunnan Province, the status is described as very rare (Cheng 1987).

FOOD Hume and Davison (1878) found insects, ants, grubs, slugs and small shells of snails in specimens from Tenasserim that they examined, whilst Oates (1883) noted that it eats worms and large insects.

HABITS Rusty-naped Pitta is exceptionally skulking, even for a pitta. If flushed this species does not usually alight in trees, but lands on the ground before rapidly hopping away. Observations suggest that it feeds primarily by turning over leaves in the upper part of the leaf-litter: at Fraser's Hill, Peninsular Malaysia, it has been observed feeding amongst very wet, rotting vegetation and household debris thrown down from a house in a deep dark ravine as well as on drier parts of steep slopes (pers. obs.).

In Peninsular Malaysia, a female Rusty-naped Pitta was observed apparently defending its eggs from a Green Magpie *Cissa chinensis* that was about to inspect the inside of the nest. The magpie promptly left when the female pitta alighted adjacent to the nest and adopted an unusual, presumably threatening, posture in which the neck was stretched out (S. Duffield *in litt.*).

BREEDING Specimens of the nominate subspecies with enlarged gonads have been collected in northern Thailand in early April and early September, whilst very young birds have been collected in September (Williamson 1918, BMNH) and observed in mid-August (P. Round *in litt.*), and a well-grown immature on 4 December (USNM). In Burma, a nest with four eggs was discovered on 7 April 1913 and, in the southern Shan States, a spotted immature was collected on 3 February (Meyer de Schauensee 1934). Baker (1926) found nests in Burma in March, April and May whilst eggs from nine clutches collected in Tenasserim and Arakan, Burma (BMNH), were collected between 11 February and 22 May. Björkgren collected an immature *castaneiceps* in Tonkin on 28 January (Eames and Ericson in prep.). Engelbach (1932) collected an immature *bolovenensis* at 600m on the Bolovens Plateau, Cambodia, on 5 July, whilst another, well-grown immature was collected in December in the Langbian Peaks (USNM). These records suggest the Rusty-naped Pitta is primarily a wet season breeder, at least in this part of its range. Further south, a nest was obtained in Perak, Malaysia in January (Baker 1926). Peak calling appears to occur, in Malaysia, between March and May, suggesting that this is the main breeding season, which is after the main period of heavy rains.

The race *deborah* was first described by King (1978) who, on 8 April 1977, flushed an adult male off a nest at Fraser's Hill, Peninsula Malaysia. However, Baker (1926) reported receiving a nest with adults from Perak, Malaysia, the clutch of which is in BMNH, having been collected on 2 January 1906. The nest discovered by King (1978) was situated in the leaf axil of a *Caryota* palm 3m off the ground, was globular and contained two eggs (one infertile). Another nest with eggs was found at Fraser's Hill on 3 March 1992 in primary forest. The same nest contained two young on 8 May and these fledged on 27 or 28 May, indicating a minimum fledgling period of 19-20 days (D. Showler *in litt.*, B. Gee *in litt.*). The nest was in a 7m tall rattan palm (J. Dransfield *in litt.*), probably of the genus *Calamus* or *Daemonorops*, situated where the leaf axils converged at about 1.7m off the ground, and was supported by the long thin spines of the palm. At Fraser's Hill, juvenile Rusty-naped Pittas have been observed with an adult, in different years, on 29 April, 9 June and 17 June. Another juvenile was observed there on 30 July (H. Buck *in litt.*, M. Chong *in litt.*, J. Howes *in litt.*).

The nest of *deborah* is a rough dome shape, measuring some 18-28cm in length, and slightly less in width, with an entrance hole of c.10cm diameter, partially concealed by broad, large dead leaves which form the roof. Usually, the nest is largely constructed of layers of dead leaves and leaf skeletons of various broad-leaved trees plus some palm leaflets, dead fern stems and a few wood rootlets. The base is of large twigs, mostly 20-30cm long and 5-8mm wide. The nest chamber, measuring up to 13cm from the entrance to the back, contains a neat, 3cm-deep cup of c.10.5cm diameter. The cup of the nest discovered by King (1978) was constructed of black plant fibre, thought to be from palms and fungal hyphae, whilst that described by D. Showler (*in litt.*) was constructed from fine grass-like material.

Although confirmed nests of *deborah* have been found only in palms, M. Chong (*in litt.*) discovered nests thought to belong to Rusty-naped Pitta at Fraser's Hill that were situated at the base of 1-2m-tall, overhanging rocks on steep slopes. These nests, discovered on 29 April and 10 May, were close to calling Rusty-naped Pittas, and similar in construction to the nest of Mangrove Pitta. Both were large domes, some 35-39cm long, 19-20cm wide with a 15cm-long platform in front of the entrance. The entrance to one nest was 8cm wide by 10.5cm high, whilst the chambers were 21-22cm deep, c.5.5-6.5cm high and c.11cm wide. Assuming that these nests were indeed built by pittas, the dimensions, in particular the size of the entrance hole, suggests that they belonged to Rusty-naped rather than Giant, the only other potential species to breed at this site. It is interesting to note that these two nests, situated on the ground, were clearly longer, with a deeper nest cavity, than nests found in palms, and both possessed platforms, a feature not recorded in front of nests in palms. Nest form may therefore vary considerably according to where the nest is built. Both nests were reported to be primarily built with dead leaves, though the wall at the entrance to one was propped up with short twigs, whilst the platform in front of the entrance hole was made of small dead twigs and branchlets (M. Chong *in litt.*).

Clutch-size in Burma varies from 3-6, although most frequently 4-5, whilst nests in Malaysia have contained 2-4 eggs (BMNH). Two eggs of *deborah* described by King (1978) measured 29.3 x 25.0mm and 30.5 x 25.3mm. They were slightly glossed white with purple-brown speckles and dense irregular-shaped chestnut-red speckles over the broad end, and the latter sparsely distributed elsewhere. Baker (1926) gave average measurements for a sample of 44 eggs of the nominate subspecies as 28.5 x 24.3mm with the range in size being 25.9-31.3 x 23.0-25.2mm.

DESCRIPTION *P. o. oatesi*
Adult male Crown to nape rufous, sometimes with sides of forehead and forehead paler buff; distinct long black post-ocular stripe bends down behind ear-coverts; lore colour variable, darker than forehead, basally black, tipped with dark orange, and can appear all rufous or all black. The upperparts and tail are dull, dark green, although the rump is usually suffused with blue. Some birds have a

streak of black in central tips of feathers of mantle, rump and inner greater coverts. Upperwing-coverts green, mixed with golden-rufous; primaries and outer secondaries brown, strongly tinged green on outer web; inner secondaries tinged green on both webs. Underwing-coverts white with buff tinge. Hidden creamy-white to white bar across base of inner webs of primaries 1-4 or 1-5. The chin and throat are whitish, usually suffused with pink; rest of underparts rufous, becoming deeper buff on the belly and undertail-coverts. Iris olive-brown to dark brown; bill blackish-brown, paler at base; legs salmon-brown.

Adult female Differs from the male in having the green upperparts suffused with rufous, deeper rufous underparts and a narrow zone of indistinct blackish scaling on the lower throat and upper breast, formed by feathers with basal two-thirds black and the rest pale, and occasionally very narrow black tips.

Juvenile Forehead, crown, nape, upper mantle to rump and greater upperwing-coverts blackish-brown, and boldly marked with buff and white: bold, narrow white streaks (formed by white centre of feathers) from forehead to nape; bold buff droplet-shaped spots on centres of mantle feathers, these becoming rounder and whiter on the median and outer greater coverts, back, and rump. Flight feathers similar to those of adult, but with a duller green tinge. Throat white, with some dark basal feathers often showing through; ear-coverts whitish with some indistinct blackish-brown streaking; breast and flanks blackish-brown with bold white spots formed by white tips to feathers; centre of belly and undertail-coverts mostly white with dark bases of feathers sometimes showing through, particularly on flanks. The first sign of adult plumage is usually the appearance of green feathers on the mantle. Iris dark brown; legs pink to pinkish-brown; bill black with commissure, gape and tip pink or pinkish-horn.

MEASUREMENTS Wing of male 113-129; tail 59-72; bill 35.5-37; tarsus 50-56; weight c.99-135, (*deborah*) 116.

5 SCHNEIDER'S PITTA
Pitta schneideri Plates 3 & 16

Pitta schneideri, Hartert, 1909, Bull. Brit. Orn. Club 25, pp. 9-10.

FIELD IDENTIFICATION Length 20.7-23.0. Schneider's Pitta is a medium-large, sexually dimorphic species that is endemic to the mountains of Sumatra. Males have a bright dark blue back, rump and tail. The forehead and crown to nape are bright chestnut-rufous, and, in males, this is separated from the blue mantle by a narrow black line. The sides of the head are a duller fulvous colour, bordered by a bold black line that extends back from the eye and is broadest at the rear. The throat is white, and the breast and belly are fulvous, with narrow black edges to the feathers of the upper breast, sometimes forming a broken line. The back, mantle, scapulars and upperwing-coverts of the female are rufous-brown, less chestnut than the head, but the rump and tail are blue. The underparts of the female are similar to those of the male.

Juveniles are blackish-brown with variable amounts of buff and whitish edgings in the plumage, giving a scaly appearance: the crown is streaked with whitish, the nape prominently marked with rufous spots, and the dark brown underparts are scaled with white, although some markings on the breast may be orange-fulvous. The upperwing-coverts are marked with buffy or white spots, whilst the throat and sides of the forehead are whitish, and the ear-coverts densely spotted with rufous.

Similar species Giant Pitta is sympatric with Schneider's Pitta and, although Giant is larger, it may be confused. Adults can easily be distinguished by the colour of crown to nape, which is chestnut in Schneider's Pitta. Juvenile and immature birds are more difficult to distinguish, but Schneider's has whitish spotting on crown contrasting with orange spotting on nape, and sometimes on upper mantle (uniform creamy to buffy spotting in young Giant); spotting on upperwing-coverts (absent in Giant); and much bolder, white-and-cream scaling on breast and flanks. The spotting on the nape becomes buffier with age, whilst the underparts become fulvous, with juvenile feathers restricted to the breast.

VOICE Described as a loud whistle not unlike that of a whistling thrush *Myiophoneus* by Robinson and Kloss (1936), but more recent observations, also on Mt Kerinci, suggest that the call is very quiet. Hurrell (1989) described it as a low, rather soft, double whistle, quite drawn out, rising on the first note and falling on the second, with a slight pause between. At close range it is distinctly tremulous but deceptively soft, and individuals are often much closer than their calls might suggest. Schneider's Pitta may call all year, since they have been heard in late December (pers. obs.) and mid-August (Hurrell 1989). However, observers have noted that the call appears to be given only during a brief period immediately after dawn. Hurrell (1989) noted that it was given in sequences of up to eleven calls, and that these were repeated every 5-6s.

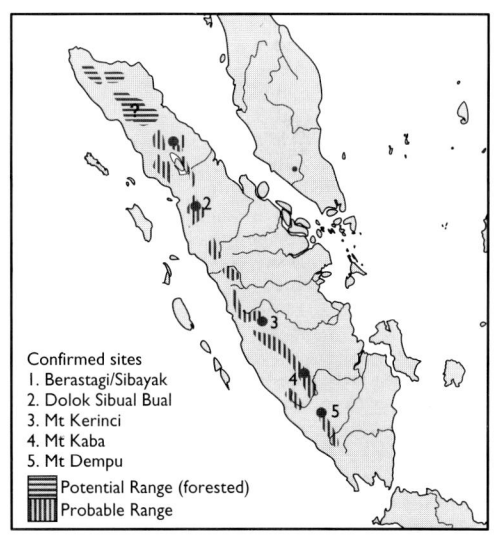

Confirmed sites
1. Berastagi/Sibayak
2. Dolok Sibual Bual
3. Mt Kerinci
4. Mt Kaba
5. Mt Dempu

Potential Range (forested)
Probable Range

Known sites are indicated together with areas of montane forest where the species may occur. Occurrence in northern part of island needs confirmation.

DISTRIBUTION Schneider's Pitta is endemic to the mountains of Sumatra, Indonesia. It is known from few localities, although it is perhaps more widespread than records suggest. It was first discovered during the last years of the nineteenth century, and most specimens were collected between 1914 and 1918. There were no more sightings of the species, which had become something of an enigma, until rediscovered in 1988 (Hurrell 1989). The distribution of Schneider's Pitta remains poorly known, but it may be largely confined to the Barisan Range and Batak Highlands. The type specimen was collected on Mt Sibayak (03°15'N), in the Batak Highlands, in the early part of this century, and Schneider's Pitta has recently been observed here (Holmes in prep.) and has been heard near this mountain, at Berastagi (Collar *et al.* 1994). In the Barisan Range, it is known from the vicinity of Mt Kerinci, Mt Kaba and Mt Dempu (4°02'S). There is also a recent unconfirmed record from Dolok Sibual Bual Nature Reserve, to the north of other known localities in the Barisan Range, and to the west of the Batak Highlands (Collar *et al.* 1994).

GEOGRAPHICAL VARIATION None recorded.

HABITAT Schneider's Pitta is a bird of submontane and lower montane forests, which is probably now confined to isolated mountain peaks. On Mt Sibayak, for example, it occurs in relict montane forest with dense undergrowth (Holmes in prep.). Most birds have been collected or observed between c.900 and 1,800m. Nevertheless, Schneider's Pittas occur at considerably higher elevations, since a specimen was collected on Mt Dempu at 2,300m, and another has been observed on Mt Kerinci at about 2,375m. Within its submontane and montane forest habitat, this species has been found in dense cover in dark, damp ravines within primary forest as well as in dense cover within tree-fall clearings (B. Gee *in litt.*).

STATUS In 1914, Schneider's Pitta was reportedly a common bird in the vicinity of Mt Kerinci, below 2,135m (Robinson and Kloss 1936). Nevertheless, even before 1917, Jacobson reported that the forest had been cleared in the vicinity of Mt Kerinci up to c.1,000m, and that virgin forest on Mt Dempu started only at c.1,200m (Robinson and Kloss 1924). Today, agriculture extends, in places, to 1,800m on Mt Kerinci in the Kerinci-Seblat National Park, and evidence suggests that encroachment is continuing (Hurrell 1989). Hence, although Schneider's Pitta lives at relatively high altitude, much of the forest in the altitudinal range from which it is known has been modified considerably, or cleared for agriculture. Furthermore, all records to date have been from primary forest, and there is as yet no evidence that this species can survive in degraded areas. Schneider's Pitta is therefore one of the few pittas which is considered to be threatened, and is classified as such by Collar *et al.* (1994), who consider it to be Vulnerable.

HABITS Schneider's Pittas are usually recorded in pairs, feeding by vigorously turning over fallen leaves. Although often shy, they may sometimes hop along paths in front of an observer.

FOOD Robinson and Kloss (1936) stated that, on Mount Kerinci, this species fed primarily on a species of large cockroach. Hurrell (1989) observed a male with what appeared to be a large grey millipede (*Myriapoda* sp.).

BREEDING Jacobson collected immatures on Mt Kaba at 1,200m in primary forest (Robinson and Kloss 1924). Although he does not state when he collected there, it was during a period when heavy deluges of rain were frequent. In 1914, Robinson and Kloss (1918a) collected immatures at 1,430m and 2,225m between 25 March and 14 May. In the Padang Highlands, young were collected at the base of Mt Singgalang in June 1834 (Müller and Schlegel 1840, Rozendaal 1990), and at Surian on the eastern slopes of the Bukit Barisan range (Büttikofer 1887, Rozendaal 1990). An immature was collected on Mt Dempu on 28 July (MZB). In 1991, two immatures were observed on Mt Kerinci in mid-June (Heath 1992), in the following year another immature was observed on 15 July (S. Duffield *in litt.*), and in 1994 a juvenile was seen in late July (Holmes in prep.). No nest or eggs of this species have ever been found.

DESCRIPTION
Adult male Forehead to nape dark chestnut-rufous, sometimes with very fine indistinct black edges to feathers. Side of head marked by a bold black eye-stripe, most prominent behind the eye, that broadens at the rear of the ear-coverts; ear-coverts fulvous, sometimes with very narrow blackish feather-tips, giving scaly or dirty appearance. A narrow black band across the top of the mantle separates the chestnut-rufous of the nape from the bright dark blue of mantle to rump and tail. Tertials brownish with olive tinge to outer web; inner lesser coverts blue, concolorous with mantle; median coverts and outer lesser coverts olive-brown sometimes tinged rufous; greater coverts brown, tinged olive; primary coverts blackish. Primaries brown; secondaries brown edged with dull greenish-buff. Tertials brown, usually with indistinct blue margins at tip of outermost two feathers. Outermost four primaries have hidden white band across the base of innermost webs. Underwing-coverts blackish-brown with extensive rufous-buff on feathers nearest body. Throat white, often tinged with pink, and becoming buff on the lower throat. A broken black band across the upper breast is formed by black basal two-thirds to feathers; rest of underparts dark fulvous. Iris hazel or sepia-brown; legs grey with purplish or purplish-flesh tones; bill brown with paler, pinkish base and paler tip.
Adult female Similar to male but lacks the black band across the upper back, and upperparts are dark rufous-brown, merging into richer, bright chestnut-rufous of nape; only the tail and uppertail-coverts are blue. Also lacks the dull blue tinge to tertials. Bill greyish-horn, paler at gape and tip.
Juvenile and immature Head well marked, but pattern differing with age and sex. Younger birds have fine central spots of buffy-white, most dense on the forehead, less so on the blackish-brown crown, with spots becoming more rufous towards the nape; ear-coverts and cheeks greyish, mottled in appearance and bordered above by contrasting dark streak; feathers of sides of neck white to rufous, with dark terminal bands giving scaly appearance. The spots on the nape and sides of the head become more rusty with age and form a rusty spotted band across the nape and sides of the head. Juvenile females may have forehead and forecrown, nape and sides of hindcrown dark chestnut. Upperparts dark brown, usually with some buffy spotting on the upper mantle in younger birds. Recently fledged birds have most feathers of the rump and uppertail-coverts brown marked with small whitish apical spots, but these are usually mixed with some blue-tipped feathers, and the rump becomes bluer with age. Jacobson noted that juveniles assume the chestnut and blue of the upper parts in patches (Robinson and Kloss 1924), the first blue feathers on the mantle usually being blue on the inner web and greenish on the outer web. Lesser and

median coverts brown, more or less marked with 2-3 rows of prominent whitish-buff spots in the apical parts of the feathers, these becoming more buff with age. Tail dull dark blue. Tertials, secondaries and primaries brown, remiges appearing greyish from below; underwing-coverts mostly brown with some white feathers and spots near the edges. Chin and throat white; rest of underparts dull blackish-brown with some buffy-rufous feathers on the upper breast and prominent broad white spots in apical parts of feathers of breast, flanks and belly that become denser towards the belly, giving a scaly appearance. In older birds, these feathers are gradually replaced with orangy feathers that grow in patches. Iris dark; legs and feet purplish-flesh; bill black with vermilion markings on tip and basal half of lower mandible (Robinson and Kloss 1918).

MEASUREMENTS Wing 117-125; tail 47-52; bill 37.4-41.2; tarsus 49.6-55.7; weight unrecorded.

6   GIANT PITTA
Pitta caerulea   Plates 3 & 16

Miothera caerulea Raffles, 1822, Trans. Linn. Soc. London, 13, p.301. Sumatra.

Alternative name: Great Blue Pitta

FIELD IDENTIFICATION Length 28-29. This is a large, robust, massive-billed pitta of the Sunda Region, ranging from Peninsular Thailand to Malaysia, Sumatra and Borneo. Males are unmistakable, with a black crown to nape, bright blue upperparts and a distinctive black necklace across the creamy-buff to pale fulvous-buff underparts. A black eye-stripe breaks up the greyish to grey-brown sides of the head. Females differ considerably from males, but are equally distinctive. Their upperparts are chestnut-rufous with a black band across the upper mantle and a black-scaled (Borneo only) or black-barred dull dark rufous crown, and blue rump and tail. The underparts are similar to those of the male. Juveniles are dark brown with distinctive head markings; the ear-coverts and crown are creamy-buff to fulvous-buff with coarse mottled brown and creamy-buff pattern on the nape and hindcrown. The upperparts, from the mantle to rump, are an unmarked dark brown, although the tail is blue. The underparts, except the white throat, are dark brown, marked with creamy-buff or fulvous-buff which becomes more prominent with age.

Similar species In Sumatra, Giant Pitta may be found in the same areas as the closely related, but smaller, Schneider's Pitta. Adult male Schneider's Pitta is easily distinguished by the combination of bright chestnut-rufous crown and nape and blue upperparts, separated by a black band. Females of the two species are more similar, but that of Schneider's Pitta has a much brighter, unmarked chestnut-rufous crown and a black band across the top of the mantle, lacking in female Giant Pitta. Juveniles are much more likely to be confused and, indeed, specimens have been misidentified in the past (Rozendaal 1990). However, whilst juveniles of these two species are superficially similar, Schneider's differs, in having whitish spotting on the crown that contrasts with dull orange spotting on the nape; spotting on the wing-coverts; and much bolder white-and-cream scaling on the breast and flanks.

In contrast, juvenile Giant Pitta has the head to nape marked with virtually uniform cream or buff, lacks extensive heavy spotting on the underparts, and lacks spotting on the wing-coverts. Juvenile Giant Pitta lacks any chestnut or rufous tinge to spots of nape and upper mantle.

In Peninsular Malaysia, Giant Pitta may occur sympatrically with Rusty-naped Pitta and has been confused. Adults are, however, easily distinguished by size and upperpart coloration; juveniles can also be distinguished easily, since juvenile Rusty-naped Pitta is well marked with buff and white spotting on both upperparts and underparts. These markings are absent in young Giant Pitta.

VOICE The commonest call of the Giant Pitta is a loud, slow, mournful, disyllabic whistle which is probably associated with territoriality, best transcribed as *hwoo-er* or *whee-er*, lacking the explosive quality of most other pitta calls. The two notes sound very similar, descending in pitch and with a downward inflection. In Thailand, only males have been observed making this whistle, usually from the ground or a fallen log. Both adults and juveniles have been heard to utter a short discordant whistle, *phreew*, of 'tuneless quality', and it is suspected that this call may act as an alarm or contact call. Early in the breeding season this call is made primarily by males.

Normally, calling is seasonal. At a site in Peninsular Thailand, the mournful double whistle of Giant Pitta was only heard in April and the latter half of June, even though observers were present from mid-March to early August: the onset of the rainy season appears to be the peak time for calling (McLoughlin 1988). In Sabah, this species was regularly heard calling between October and March, but in August none would even respond to playback of the call (pers. obs.). Although calling can occur at any time of the day, most calls are made during the first and last hour of daylight. Calling periods are usually short, lasting several minutes at most, but very occasionally males may call for periods of up to 50 minutes.

Although Giant Pittas are rarely heard, they regularly respond to tapes during the breeding season. Under these circumstances, calls may be made from low trees, rather than from the ground. In Thailand, Giant Pittas have been noted to make a short whistle, but including the occasional lower-pitched note, in response to imitations of their call (McLoughlin 1988).

DISTRIBUTION Confined to the Sunda Region, where present in Sumatra, Borneo, Tenasserim, Peninsular Thailand and the northern half of Peninsular Malaysia. On the mainland, Giant Pittas have been found northwards to Maprit at 10°50'N in Peninsular Thailand (Baker 1919), and in Tenasserim apparently as far north as Mt Nwalabo (14°2'N, 98°28'E: Baker 1926, Robinson and Chasen 1939). In Peninsular Malaysia, the most southerly records are from Selangor State in the west and Pahang in the east (Medway and Wells 1976). In Sumatra, Giant Pittas have been recorded only from Lampung Province in the southernmost part of the island and in the foothills of the Barisan Range, near Bengkulu, South Sumatra Province (Rozendaal 1990). A recent sight record from Mt Leuser, Aceh Province, almost certainly also relates to Giant Pitta (Holmes in prep.), and this species is probably more widespread in Sumatra than presently known. In Borneo, Giant Pitta may be confined to the northern half of the island, where it is known from Sabah and Sarawak, Malaysia, and the upper Mahakam River, East Kalimantan (Büttikofer 1900, Smythies 1957).

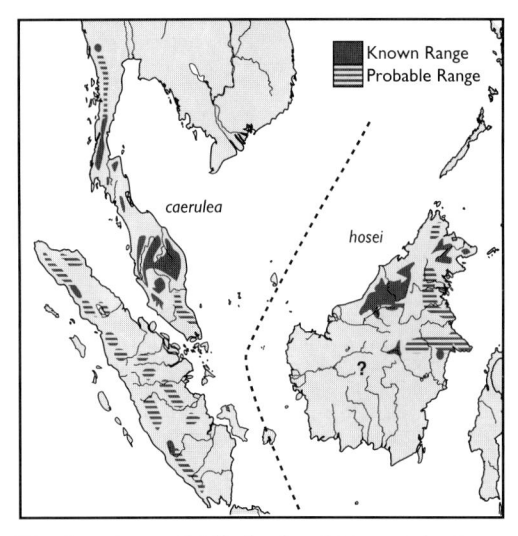

Map shows known distribution based on approximate extent of forest remaining within the altitudinal range of the species. Other areas where suitable habitat remains are also shown but confirmation of occurrence in these areas is required.

GEOGRAPHICAL VARIATION There are two subspecies of Giant Pitta, and it is the females which are more different.

P. c. caerulea (From Tenasserim, Burma, south through Peninsular Thailand to Peninsular Malaysia and Sumatra) Crown of female barred.

P. c. hosei (Borneo) Male similar to *caerulea*. Female differs from nominate female in having a richer chestnut crown to nape patterned with scales, and in having darker ear-coverts. Immatures have sparser, more buff spotting and scaling on crown to nape than immatures of nominate subspecies.

HABITAT Giant Pittas inhabit lowland, hill and submontane rainforests and disturbed forest habitats in the Sunda Region. Gibson-Hill (1949) noted that in Peninsular Thailand and Peninsular Malaysia, this species is most frequently associated with dense, usually swampy lowland forest, but more recent evidence suggests that Giant Pittas range widely in the lowlands and hill forests of mainland South-East Asia. Whilst most frequently recorded from relatively low altitudes, Giant Pitta has apparently been recorded as high as 885m on Khao Luang and 915m on Khao Phanom Bencha in Peninsular Thailand (Meyer de Schauensee 1946, Round *et al.* 1989), although there is some doubt about this record (P. Round *in litt.*). In Sumatra, a probable sight record was as high as 700-800m on Mt Leuser.

In Sabah, north Borneo, Giant Pitta appears to be locally common in both primary and selectively logged forests (Lambert 1990, 1992). Indeed, there is some indication that the species survives in many areas away from primary forest. One bird killed by a car in Sabah was in an area of scrub and rubber, whilst a number of sightings have been made in overgrown rubber estates and thickets (Sheldon, in prep.). As with the nominate subspecies, *hosei* has mostly been recorded in the lowlands and in hill forests at moderate altitudes. A.D. Johns (*in litt.*) found this species at the Maliau Basin at c.1,200m, but elsewhere in Sabah Giant Pitta has not been found above 200m (Sheldon in prep.). In Sarawak, most records are likewise from the lowlands, but there is one record from Mt Murud at c.900m (Smythies 1957). Although not recorded from Brunei (Mann 1987) it is likely that the species will be found there.

STATUS Gibson-Hill (1949) noted that Giant Pitta was moderately common in Perlis and northern Kedah, but further south, in Peninsular Malaysia, it is scarce, being known only from Larut in Perak, near the Krau River and in Malacca. Although Giant Pittas are rarely heard or seen, they may be more common than records suggest, at least in the parts of their range where good habitat remains. In the vicinity of Danum Valley, Sabah, for instance, at least four birds were occasionally heard calling simultaneously in an area of forest that had been logged nine years previously, whilst in nearby primary forest, there may have been as many as 2-3 males per km^2 judging by the numbers heard calling between mid-October and early March (Lambert 1990, 1992). Elsewhere in Sabah, A. Johns (*in litt.*) reported that the species was relatively common in the Maliau Basin. However, Fogden (1976) reported that Giant Pitta was a rare species throughout Sarawak.

In some parts of their range, Giant Pittas may now be rare as a consequence of forest clearance. In Thailand few areas of lowland forest persist, and the species may have disappeared from large areas that were formerly occupied. Nevertheless, observations from Thailand have shown that even relatively small patches of forest and scrub can support breeding pairs of Giant Pitta: one nest was located in a 2ha patch of secondary growth that was at least 2km from the nearest patch of primary forest. There is therefore reason to be optimistic about the survival of Giant Pittas in man-altered habitats despite the on-going clearance and degradation of lowland forests throughout its world range. Nevertheless, in the more seasonal parts of its range, such as Peninsular Thailand, the availability of water may limit occupancy of such habitats (P. Round *in litt.*). It may even be that advanced successional regrowth and logged-forest habitats are as favoured by this species as areas of pristine forest (Round *et al.* 1989). Last century, Giant Pitta was reported to have been rare in the evergreen forests of extreme southern Tenasserim (Hume and Davison 1878), but there is no recent information on its status in Burma.

In Sumatra, one of the three areas from which Giant Pittas have been recorded, Lampung Province, has been largely deforested. It seems likely, however, that Giant Pittas must have been overlooked at many localities in this large island: sites where it has so far been recorded are far apart.

MOVEMENTS Oates (1883) noted that Davison had found Giant Pitta in Tenasserim only from March to July and that it appeared to be partially migratory, but there is no good evidence to support this assertion.

FOOD Giant Pittas have been recorded feeding on a variety of invertebrates, as well as occasionally eating smaller vertebrates such as frogs, and even small snakes (Robinson and Chasen 1939). At a nest studied in Thailand, Round *et al.* (1989) were able to identify half of the 131 prey items that were brought to the nestlings during over 52 hours of observation. Earthworms and snails were the most common food items that were identified, each forming at least a third of the prey fed to the young. Indeed, many of the unidentified food items were also suspected of being snails and worms, and may have actually accounted for as much as 90% of prey fed to the young. Large insects or insect larvae (including an orthopteran and a probable mantid, and one 2cm long larva that was probably a beetle), and at least one frog, were also fed to the young, and together accounted for about a third of the identified items.

Snails are removed from their shells by smashing them on rocks. Suitable rocks may be used as 'anvils', where the shells of snails are regularly broken. Snail shells collected from such an anvil near the Giant Pitta nest studied in Thailand were identified as belonging to the genus *Cyclophorus*, and had been broken at the posterior part of the body whorl, although the flesh was probably extracted from its usual aperture. The snails that Giant Pittas eat are often large and robust; those of the *Cyclophorus* described above were 5-6cm in diameter. In Borneo, large spiders and trilobite beetles have been recorded in the diet of Giant Pitta. Davison reported that his specimens from Tenasserim had been feeding largely on large black ants (Oates 1883).

HABITS Despite its size, the Giant Pitta is one of the most elusive species of pitta, and is difficult to observe even when calling. Males are fiercely territorial, and react very aggressively to other males in their territories during the breeding season. In Sabah, on 3 March 1990, a male was observed chasing another male down a long slope in primary forest after a period of calling that suggested that the two individuals were initially at least 100m apart. The two males were seen briefly scuffling together on the ground, before continuing the chase, in low flight, further down the slope.

The female is very well camouflaged, so that only the bright shining blue tail is seen easily in dark understorey. Females have been observed to freeze and stretch their body up high to look at intruders rather than immediately taking flight when disturbed. In contrast, males usually hop away rapidly or fly off when disturbed: when flushed off the ground the wings sometimes clap, sounding rather like an Emerald Dove *Chalcophaps indica* (pers. obs.).

Giant Pittas often feed in a thrush-like manner, flicking over leaves and twigs in damp places with the bill and cocking the head sideways, presumably to listen for prey. Feeding is not restricted to such damp areas, however, since Giant Pittas have occasionally been observed feeding among the stones of dry, rocky streambeds. In Sabah, a male was found roosting about 3m above the ground on a horizontal liane loop, its head tucked inside the plumage and all visible feathers fluffed up so that the appearance was of a colourful round ball (L. Emmons *in litt.*). Medway (1972) caught a male roosting in undergrowth on Mt Benom, Peninsular Malaysia. In the hand this species may ruffle out its head feathers (M. Woodcock verbally).

BREEDING Although the nesting habits of this species is relatively well known from studies conducted in Thailand, there are few breeding records. Mainland breeding records consist of: the nest containing two eggs that was studied in detail, which was found in Peninsular Thailand on 22 July (Round *et al.* 1989); a nest with two eggs found in Krabi Province on 21 July (A. Pierce *in litt.* to P. Round); a half-grown nestling collected in Perlis, in the north of Peninsular Malaysia, in early November (Robinson and Kloss 1924); a recently fledged juvenile collected on 4 November in Burma (BMNH); a fledgling collected in Trang, Peninsular Thailand, on 20 August (Riley 1938, USNM); and two juveniles that were observed in Peninsular Thailand on 17 June (McLoughlin 1988, Round *et al.* 1989). The latter had probably fledged from one of four abandoned nests that are described below, this having been discovered on 24 June. These records, and those of nests documented below, suggest that the breeding season in mainland South-East Asia may be rather long, lasting from May to November, and hence coinciding with the south-west monsoon. In Borneo, Hose collected a fledgling on Mt Dulit in October (BMNH). There are no breeding records from Sumatra.

Four abandoned nests and one occupied nest of Giant Pitta that have been discovered in Peninsular Thailand were in secondary habitats. One of these nests was in an isolated 2ha patch of secondary scrub; the rest were in secondary forest fragments. The occupied nest was discovered on 22 July when it contained eggs, and subsequently monitored (Round *et al.* 1989), providing a wealth of information about the breeding biology of this species.

Nest of Giant Pitta

On 2 August, the eggs had already hatched.

All of the five Thai nests mentioned above were built in the forks of small palms, four in rattans of the genera *Calamus* or *Daemonorops*, and the other in a *Licuala* palm (Round *et al.* 1989). The four nests in secondary forests were placed close to the ground, three at c.1m or less above the ground, and one at c.2m off the ground. The nest discovered in secondary scrub was c.3m off the ground. A sixth nest from Thailand was found 1.2-1.5m off the ground in a *Daemonorops grandis* palm (A. Pierce *in litt.* to P. Round). Observations of captive birds show that both sexes participate in nest-building (McKelvey and Mille 1979). Two of the nests that were discovered in Thailand by Round *et al.* (1989), one the occupied nest, were in good condition and described. The nest is a dome, slightly higher than wide, with a lateral entrance and abutting flattened platform. The abandoned nest was c.30cm in diameter, whilst the occupied nest was 32cm high with a horizontal diameter of 21cm. The entrance holes of the two nests were c.10 x 15cm and 12 x 14cm respectively. Both nests were placed on a base of sticks and made of leaves, but the types of leaf differed. The abandoned nest was made principally from bamboo and palm leaves, whilst the occupied nest, including the platform, was constructed of dead leaves from a diversity of broadleaved trees, small twigs and a few green leaves; these components were rather loosely held together. The former nest was lined with rootlets.

Evidence suggests that the clutch-size of Giant Pitta in Peninsular Thailand is normally two. The eggs measure c.30 x 35mm, and are off-white, marked with fine brown speckles at the broad end, which may be denser, forming a diffuse band, away from the apex. In captivity, nesting Giant Pittas also have a clutch-size of two, but eggs are reported to be rounder, measuring c.26 x 32mm, and weighed c.10g fresh weight (McKelvey and Mille 1979).

Both adults incubate the eggs, feed the young and remove faecal sacs. At the nest studied by Round *et al.* (1989), the observers recorded 167 visits by adult Giant Pittas to the nest during an observation period of 52 hours 16 minutes, made over a 12-day period. The period between visits varied between about 30s and one hour, and the average time between visits was 18 minutes 47s, although the data suggest that visits were almost twice as frequent during the afternoon. Feeding continued even during periods of heavy rain. The average feeding visit lasted 36.7s, but varied considerably, ranging from a mere 7s to 157s. The male visited the nest slightly more frequently than the female, making 93 (55.7%) of the 167 visits. The adults were rarely observed visiting the nest simultaneously, with one notable exception, when the female spent c.1.5hrs brooding the chicks, then five or six days old, in the early morning. The female also brooded the young at night, even when they were 13 to 14 days old, and only two days from leaving the nest.

Both nestlings in a nest discovered by A. Pierce (*in litt.* to P. Round) in Peninsular Thailand on 21 July were predated: one was observed to be eaten by a snake on 24 July whilst the second disappeared later the same day.

Giant Pitta has been successfully bred in captivity, and some additional information on breeding biology derive from such breeding. At San Antonio Zoo, the first egg was laid some 10-18 days after starting to construct the nest, eggs hatched after 15-16 days in incubators, whilst hatching occurred c.12 hours after pipping. Artificially fed juveniles became active (hopping around like an adult) after some 20 days, and were self-feeding 24 days after hatching. When eggs were removed from a nest, a new nest was constructed and a second clutch laid (McKelvey and Mille 1979).

DESCRIPTION *P. c. caerulea*

Adult male Feathers of forehead grey, sometimes tinged with fulvous, with broad black tips and bands across the centre, forming narrow bars; crown to nape black; sides of forehead and long supercilium, extending onto rear of nape, pale grey to greyish-white, often faintly tinged with pale blue, and separated from ear-coverts by narrow black post-ocular stripe and very narrow black band across grey lores. Ear-coverts grey, sometimes with dull fulvous suffusion, and appearing narrowly scaled due to very narrow black edges to feathers. The black on the nape usually extends around from nape to form a narrow black breast-band; rest of upperparts and tail bright deep blue. Upperwing-coverts, scapulars and tertials blue with brown bases; primaries and primary coverts blackish-brown, secondaries blackish with blue tips and edges to outer webs, these broadening on innermost feathers where extending across entire outer web. Primaries 2-4 have small white to whitish-grey patches in the inner web at the base. Underwing-coverts blackish-brown marked with dull fulvous-buff bars at the bend of the wing and in the outermost section. Throat white to creamy, often tinged pinky-buff and normally separated from rest of underparts by an irregular-edged black band across the top of the breast (formed by black bases and basal sides to feathers). Rest of underparts creamy-buff to pale fulvous-buff. Some birds lack the black breast-band entirely. Iris grey to grey-brown, legs pink tinged with grey, bill black.

Adult female Feathers of forehead to nape very variable in colour, being creamy-brown to chestnut-brown and finely barred with black; lores usually similar in colour, but more or less mixed with black; feathers of superciliary tract usually unbarred. There is a black post-ocular stripe of variable width that bends down and broadens out behind the ear-coverts, sometimes joining black breast-band. The barred feathers of nape are separated from rich chestnut-rufous mantle, back, upperwing-coverts and upper rump by a black band of variable width (from virtually non-existent to 2cm broad); lower rump, uppertail-coverts and tail blue; rest of plumage as male, but marginally darker. Iris dark hazel; bill black, legs pink tinged purplish.

Juvenile Forehead and crown mostly creamy-buff, with dark brown feather-edges giving a more or less scaled appearance and with buff extending onto mantle, as patches, on many individuals; rest of upperparts unmarked dark brown, with blue tail. Throat whitish, ear-coverts creamy-buff separated from crown and nape by blackish-brown post-ocular stripe and narrow brown line across lores; breast and flanks brown with variable amount of creamy-buff edges to feathers, these forming narrow band of bold scaling on breast of some birds, but forming indistinct, narrow and sparse scaling on others; belly to undertail-coverts fulvous-buff mixed with brown, becoming buffier towards the undertail-coverts. Flight feathers brown, with dull bluish tinge to outer webs of secondaries developing in males at an early age. Feet purplish-grey; bill blackish with orange-red tip and orange to red gape. Bills of subadult males are reported to be entirely red (B. Gee *in litt.*). Hatchlings are purplish-black with yellowish-white rictal areas and soles of feet and bright crimson gape

(McKelvey and Mille 1979).

MEASUREMENTS Wing 142-155; tail 53-58; bill 39-40; tarsus 52-54; weight 198-202 (*hosei* 206-207 male).

7 BLUE PITTA
Pitta cyanea   Plate 4

Pitta cyanea Blyth, 1843, J. Asiat. Soc. Bengal, 12, p. 1008: Arakan, north-west Burma

FIELD IDENTIFICATION Length 19.5-23. Blue Pitta is a medium-sized pitta that is found from north-east India to Burma, Thailand and Indochina. The male is distinctive, with scarlet hindcrown and nape contrasting with blue upperparts, a broad black eye-stripe and moustache and fine black spotting and barring on the pale bluish-white underparts. Females are much duller, with olive upperparts, tinged with blue, and less obvious (occasionally lacking) orange on the nape, but they share the blue tail of the male. Markings on the underparts tend to be less dense than in the male, but these markings are very variable in both sexes. Immatures are dark brown above with bold rufescent-chestnut streaks on the crown and supercilium, and broader, duller rufous-buff streaks on the mantle that extend onto the upperwing-coverts. The throat is white and the breast to belly dark brown prominently marked with large rufous-buff spots, but these are rapidly replaced by black-barred or spotted whitish-blue feathers. The tail is blue. Bill of immature is black with a bright red base.
Similar species Adults are distinctive within their range and unlikely to be confused with other species of pitta. Juveniles, however, are more likely to be confused since they resemble the juveniles of other sympatric species. However, juvenile Blue Pitta obtains the distinctive black-and-white barred and spotted feathers on the underparts at an early age, and when these are present (usually on the breast and flanks) they are diagnostic. The blue tail also serves to separate Blue Pitta from juvenile Eared Pitta, Rusty-naped Pitta and Blue-naped Pitta. Dense streaking on the crown to mantle and breast of juvenile Blue Pitta also serves to distinguish it from juvenile Eared Pitta, whilst the lack of spotting on its upperwing-coverts distinguish it from juvenile Rusty-naped, Blue-rumped and Blue-naped Pitta, all of which have distinct spotting. Juvenile Rusty-naped Pitta also differs in having white, rather than buff, markings on the head and underparts. The cheeks of juvenile Blue Pitta are largely white, in contrast to those of juvenile Bar-bellied Pitta, which is otherwise quite similar. Juvenile Gurney's Pitta has indistinctly marked upperparts, narrow streaks rather than large spots on the breast and flanks, and solid black ear-coverts.

VOICE The call of Blue Pitta, probably given by both sexes (P. Round *in litt.*), is a clear drawn-out fluty double whistle *pleeow-whit*, the first note with a sliding quality, the second sharp. A shorter version of this call, *priaw-wit*, is also occasionally heard when excited, as is a rasping, squeaky, harsh alarm note, *skyeew*. In Thailand, calls are made throughout the year but the species is reported to be most vocal between April and October (Round and Treesucon 1983).

DISTRIBUTION The Blue Pitta is distributed from the eastern Himalayas to Burma, Thailand, Indochina and Yunnan, China.

In the Indian subcontinent, Blue Pitta is known from the Indian states of Arunachal Pradesh, Assam, Nagaland, Manipur, Mizoram and probably Tripura; north-east Bangladesh (Srimangal); the Chittagong Hill Tracts of south-eastern Bangladesh; the Chin, Pegu and Arakan Hills, Arakan, southern Shan States and Karen Hills of Burma, and south to Tavoy in Tenasserim. Although listed as a resident in Bhutan by Baker (1926), Ali and Ripley (1970) and Ripley (1982), there are no confirmed records from here (Ali *et al.* in press). Breeding has not been confirmed in Bangladesh, but is strongly suspected since the species is known to be present in the north-east in May and June (Harvey 1990).

Blue Pitta is distributed throughout most of Thailand except for the central plains – though perhaps formerly present: Baker (1934) reported it to be a plains breeder in Siam – and the southern peninsula. In the mountains of western Thailand it occurs southwards at least to Prachuap Khiri Khan province and Khao Luang (11°40'N) in the peninsula, and P. Round (*in litt.*) heard the *pleeow-whit* call of what must have been a pair of Blue Pittas in logged forest on the bank of the Khlong Sok River, Surat Thani Province (c.8°56'N). The nominate subspecies is replaced by *aurantiaca* in eastern Thailand in the San Kampheang Range, Nakhorn Ratchasima Province (Dickinson and Chaiyaphun 1968).

Blue Pitta is widely distributed in Indochina, although in China it is known only from the southern part of Xishuangbanna (Hsi-Shuang-Pan-Na), in southern Yunnan Province (Cheng 1987). In Vietnam it occurs from Tonkin south through Annam to the Langbian Plateau, and throughout much of Laos, from the Dong Hua Sao National Biodiversity Conservation Area (R. Timmins and J.W. Duckworth *in litt.*) and the Bolovens and Cammon Plateaus to northern Laos. The distribution in Cambodia is poorly known, and it has not been found in the Cardamom Mountains, despite its presence in south-east Thailand in the same range of highlands.

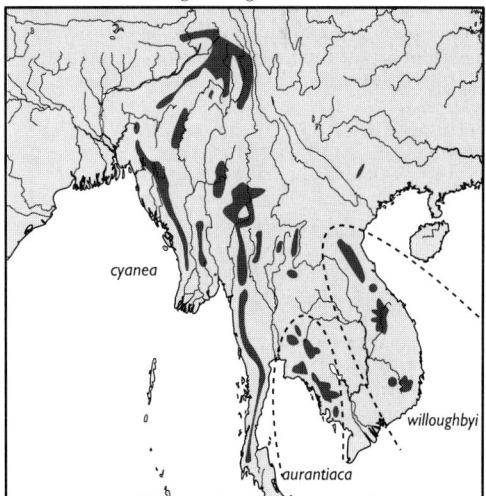

Distribution based on approximate extent of forest remaining within the altitudinal range of the species. Former range more widespread.

GEOGRAPHICAL VARIATION Three subspecies are recognised, and within these there may also be considerable variation.

P. c. cyanea (Known from the lower Himalayas in the north-eastern states of India, the Chittagong Hill Tracts of Bangladesh, and hills of Burma south to Tenasserim and to 11°40'N in Peninsular Thailand, and east to Yunnan, China, and NE Laos. Birds in west Tonkin, Vietnam, probably belong to this subspecies) Birds from Khao Yai, Thailand, are reported to be somewhat intermediate in the nape colour between *cyanea* and *aurantiaca* (P. Round *in litt.*).

P. c. aurantiaca (Mountains of Cambodia and SE Thailand in San Kampheang Range) Distinguished from *cyanea* and *willoughbyi* by the colour of the nape and sides of head which are yellowish-orange rather than scarlet (wing 114-115).

P. c. willoughbyi (Mountains of C Laos to S Annam, Vietnam) Brighter than *cyanea*, particularly on the breast, which is usually well marked with red tones (wing 110-116 in specimens from Laos, wing 120 in a bird from Dran, South Annam). Female has brownish upperparts with olive wash and variable amount of blue on mantle. Females with solid blue on upper mantle usually also have scattered blue feathers on back.

HABITAT In Burma, this species is reported to occur in damp, dark ravines in areas of mixed tree and bamboo in evergreen forests, although Oates (1883) noted that the habitat occupied by Blue Pitta is generally more open than that of many of its congeners. It has been recorded up to 2,000m in the Indian subcontinent (Ripley 1982), although more often encountered at lower altitudes. In parts of Burma, the species is found at altitudes as low as 60m, but it ascends to at least 1,830m (Smythies 1986).

In Thailand, the nominate subspecies inhabits dense vegetation of the forest floor in mixed deciduous, hill evergreen and montane evergreen forest, bamboo and sometimes secondary growth, from the plains to 1,070m in the peninsula. In northern Thailand it is mostly encountered above about 800m, and up to at least 1,680m (Deignan 1936, Meyer de Schauensee 1946, Round 1984), whilst at Khao Yai National Park Round and Treesucon (1983) found it commonly in dry evergreen forest at 700-800m. In eastern Thailand it has been collected at 610-1,070m (Meyer de Schauensee 1934, 1946). Deignan (1945) found it in swampy glades thickly covered with gingers and moist groves of thorny palm scrub.

In Indochina, where Blue Pitta is widely distributed, Delacour and Jabouille (1929) found this species to be primarily montane, with records of *willoughbyi* from localities at 800-1,200m in the Bolovens, Cammon (central Laos) and Langbian Plateau, although they failed to find *aurantiaca* at 1,000m in the Cardamom Mountains. In west Tonkin, Vietnam (probably the nominate subspecies), it has been observed on boulder-strewn slopes with a dense herbaceous understorey in primary evergreen forest at 940m (pers. obs.). However, Blue Pitta does occur at much lower altitudes in parts of its Vietnamese range, and J. Eames (*in litt.*) has found this species in the lowlands of Annam, in limestone hills within the Phong Nha Historical and Cultural Site. In Laos, Delacour and Jabouille (1931) reported that Blue Pitta had been collected from Napé and Xiengkhouang at c.1,000m.

STATUS In the Indian subcontinent and northern Thailand, Blue Pitta is a rather rare or uncommon and local species (Baker 1895, Deignan 1936, 1945, Ali and Ripley 1983, Round 1984), although it may formerly have been locally commoner in north-east India, since Baker (1934) stated that it was common in hills of the Brahmaputra River valley. Round and Treesucon (1983), however, report that Blue Pitta is a common but secretive species in Khao Yai National Park, Thailand. In view of the very shy nature of this species it may be commoner than is presently recognised in other parts of its range. This would explain why Wiles (1979) found it rare in lowland bamboo forest in south-west Thailand, since P. Round (*in litt.*) reports that it is still relatively common in suitable habitat in both west and south-east Thailand.

Thompson *et al.* (1993) consider the Blue Pitta to be a rare summer visitor to Bangladesh, where the only recent record was of birds suspected to be breeding during May and June 1988 in the West Bhanugach Reserve Forest, Srimangal, in the north-east (Harvey 1990). Hume and Davison (1878) reported that, in Tenasserim, this species was nowhere very common, whilst Delacour and Jabouille (1929) described the status of *willoughbyi* in Laos and Annam as rare. In Cambodia, Engelbach (1936, 1938) found *aurantiaca* to be common at c.400m near Bokor in the Elephant Mountains and frequently heard it, particularly above 700m. On the Bolovens Plateau, he reported that it was abundant above 800m (Engelbach 1932). The status in Yunnan, China, is described as very rare (Cheng 1987).

MOVEMENTS In the Indian subcontinent, the nominate subspecies is said to be a seasonal local migrant (Ripley 1982): Baker (1926), for example, states that Blue Pitta occurs to at least 1,525m in the breeding season, but that it is normally a bird of the plains. Thompson *et al.* (1993) considered that Blue Pitta is a rare breeding visitor to Bangladesh.

Delacour and Jabouille (1925) reported that a live Blue Pitta was caught at the lighthouse of Cap St Jaques, at the extreme southern tip of Cochinchina. However, these authors later failed to mention this record in *Les Oiseaux de l'Indochine Française* (Delacour and Jabouille 1931) although they mention that Blue-winged Pitta was caught at the lighthouse, suggesting that there they had confused the species in their 1925 publication. Clearly, reported movements need further confirmation. Differences in abundance or apparent seasonal absence in parts of the range of Blue Pitta may merely be a reflection of the birds' shy, quiet habits at certain times of the year.

FOOD Ali and Ripley (1983) list ants and other insects, grubs and land snails as food of the Blue Pitta in the Indian subcontinent, whilst Baker (1926) believed that they feed principally on ants and termites, although also eating other insects, insect larvae and worms.

HABITS Blue Pittas typically feed in leaf-litter, turning over leaves and occasionally digging their bills into damp soil. Usually this species is found in pairs, although the birds may not feed in close proximity. This can be a particularly shy species and difficult to observe, rapidly hopping away, rather than flying, at the slightest sign of danger.

J. Marshall (verbally) discovered a Blue Pitta in Burma that had been killed and was being eaten by a Collared Owlet *Glaucidium brodiei*, a predator that is much smaller than the pitta.

BREEDING In the Indian subcontinent and in Burma, breeding is seasonal, occurring principally in May and June, and occasionally in July (Baker 1926). Eggs in the possession of Baker (1934), all collected in the North Cachar and Khasi Hills of India, had been taken from 6 May to 26 July (BMNH). In Burma, a nest containing five eggs was discovered on 23 July 1903 and two nests with eggs were found at Kaukarit (Kawkareik) in Tenasserim in late May (Hume 1880, Oates 1883). Three clutches from Tenasserim in BMNH were collected between 4-26 May. Immatures have been collected in northern Thailand on 24 August (Baker 1919) and on 24 October (USNM), whilst one was observed with adults on 5 September, and an immature was collected in Tenasserim on 9 September (BMNH). Herbert (1924) collected a nest with five eggs in central Thailand on 14 June. At Khao Yai National Park, southern Thailand, three nests were discovered by Round and Treesucon (1983) in 1982: one with four eggs on 7 July, a male brooding or incubating on 21 August, and a nest with two nestlings on 12 September. In Thailand, therefore, Blue Pitta seems to breed mostly in the middle to late wet season, in contrast to most insectivorous birds, which have breeding seasons spanning the late dry season and wet season between February and June (Round and Treesucon 1983). This is also apparently somewhat later than breeding in the Indian subcontinent.

Most of the nests found in the Indian subcontinent by Baker (1934) were in very wet evergreen forest, and in most cases in forest in which the ground was broken up by steep precipitous ravines and outcrops of rocks and boulders, the latter all covered by the most luxuriant growth of moss, ferns and orchids. In contrast to the nests of other species of pitta (Rusty-naped and Blue-naped), Baker (1934) often found nests of Blue Pitta built on old stumps, the tops of rocks and on steeply sloping banks, and very occasionally on the ground almost, or entirely, in the open. However, other nests were found in scrub and bamboo in areas with ample undergrowth. Some nests found by Baker (1934) were not concealed in any way, perched conspicuously on the top of a stump or rock, but nevertheless so unlike a nest in appearance that they were hard to recognise as such.

In India, the nest is described as being similar to, but more compact than, nests of Rusty-naped and Blue-naped Pitta, being a rather loosely assembled oval ball of leaves, including those of bamboo, and roots, with an entrance hole at one end. Baker (1926, 1934) reported that, where bamboo was available, bamboo leaves formed the favourite material, birds often travelling far to obtain them, although bracken, grass, roots, moss, lichen and other leaves are used in considerable proportions, and the lining is usually of roots and leaves.

The nest collected for Herbert (1924) in Thailand was dome-shaped with a side entrance and said to have been situated on the ground at the base of a bank. It was primarily constructed of bamboo leaves and situated close to a clump of bamboo. In sharp contrast, the three nests found by Round and Treesucon (1983) in Khao Yai were situated 2.5-4.0m off the ground, resting on a clump of an epiphytic fern (*Asplenium nidus*) and hence abutting the side of a tree trunk which varied in diameter from 13-50 cm. Leaf-litter and other plant debris accumulating in these ferns formed the platform on which the Blue Pitta nests were found. These nests were similar in shape to that collected by Herbert (1924), but constructed of dry sticks, leaves or grass rather than bamboo. The base of Herbert's (1924) nest was made of wet leaves that had been matted together.

The Thai nests found by Round and Treesucon (1983) were about 25-30cm in diameter, whilst in India the nest is about 27.5 x 21cm in size. Herbert's (1924) nest was built on a platform that was c.30.5cm across and 7.6cm deep. Clutch-size in India, Burma and Thailand is probably usually 4-5 (BMNH), although Baker (1926) reported nests in India with as many as seven eggs, whilst one nest found at Khao Yai contained just two nestlings. These Thai nestlings were being brooded by the female, but the male probably also cares for the young since the two other nests discovered at Khao Yai, one of which contained eggs, were incubated/brooded by a male.

The eggs are glossy white, marked with spotting and blotching varying in colour from pale reddish to deep purple-black. Eggs may also be marked with streaks, particularly on the large end. These markings are generally more numerous than those on the eggs of sympatric Blue-naped and Rusty-naped Pittas and they are most similar in appearance to those of Hooded Pitta (Hume 1880, Baker 1926, 1934). The markings are usually distributed over the entire surface of the eggs, although in some eggs they may be restricted to the broader end. Fifty eggs of the nominate subspecies measured by Baker (1926, 1934) varied in size from 24.0-28.2 x 20.1-22.1mm, with an average size of 27.6 x 20.9mm. Eggs in the clutch of five found by Herbert (1924) were described as being broad ovals, and were clearly more round than those measured by Baker, since their average size was 24.8 x 20.8mm.

DESCRIPTION *P. c. cyanea*
Adult male Crown-stripe from centre of forehead black, widening on top of head and bordered by broad grey to pale grey-brown supercilium that mixes with reddish feathers behind eye and becomes more or less solid red on nape, where forming large patch. Eye-stripe, including lores, black, extending to sides of neck, where broadening, and separated from black moustachial area by greyish-white to buffy-white streak from sides of throat across ear-coverts; the moustachial area more or less scaly, with white basal feathers. Rest of upperparts uniform dark blue, including rump and tail. Some individuals have black triangular streaks in feathers of mantle, wing-coverts, uppertail-coverts and, occasionally, tail. Upperwing-coverts as mantle; tertials blue, secondaries dark brown with blue outer edges to distal half, narrowest on outer secondary and broadest on innermost; primaries dark brown. Throat white with variable amount of dark brown streaking, mostly in centre. Moustachial stripe broadens out at top of breast side. Breast, belly and flanks pale blue marked with numerous black spots and broad bars, and the blue often washed with buff or pinkish tones, particularly on breast. The pattern of the black markings on the underparts is highly variable, both in terms of size and of their intensity. Sides of breast, flanks, and sides of belly tend to be most prominently marked with bold black bars, but in some individuals these areas appear spotted rather than barred. Centre of breast and belly is usually spotted with black, though this may be lacking in some individuals. Underwing-coverts brown, with variable number of white axillaries. Underwing marked by white bar at base of primaries; outer six primaries have broad white bar across inner web, extending onto the outer web of primaries 2-7 and forms a narrow white bar that is visible in flight. Undertail-coverts white, sometimes with buffy-orange wash, occasionally spotted or barred with black. Iris dark brown or dark reddish-brown, eye-ring slaty, bill black, legs fleshy-plumbeous and feet soles yellowish-white.

Adult female Head pattern and coloration as male, except that the crown-stripe is often dark chestnut-brown rather than black and the throat whiter, less frequently tinged with buff. Upperparts dull olive-brown, usually with variable amount of blue on mantle and back, sometimes restricted to the top of the mantle, but otherwise extending to back and on rump. Tail and uppertail-coverts blue, often mixed with black streaks in centre of uppertail-covert feathers. Underpart pattern similar to that of male, but tends to be more barred and less spotted; the spots dark brown rather than black and the blue background coloration much paler and almost white overlaid with dull pinky-buff, especially on breast. Undertail-coverts white and usually tinged buff and marked with some dark brown bars. Wings as adult male, except that outer webs of secondaries have dull olive-green rather than bluish colour. Bill basally red.

Immature Forehead and central crown black, with long, broad rufescent-chestnut streaks that become denser on nape and are formed by chestnut feathers with black edges and bases; these feathers dominate the superciliary area, forming a more or less continuous chestnut band, marked with black scales, from side of forehead to nape; behind eye this band is bordered below by paler, buffier feathers. Lores and streak behind eye dark brown; ear-coverts dark brown mixed with white. Mantle dark brown with broad dark chestnut-rufous streaks extending onto upperwing-coverts. Flight feathers brown. Adult feathers replace juvenile with unmarked olive-green (female) or blue (male) feathers from lower back upwards. Tail and some uppertail-covert feathers blue (as adult). Throat white, boldly scaled with dark brown (tips to feathers) in malar region and on sides of neck. Breast to belly dark brown with bold buffy oval spots and streaks, these gradually replaced by white feathers boldly spotted and barred with black or dark brown. Undertail-coverts more or less unmarked white, occasionally tinged buff. Iris dark brown; bill light vermilion, turning to dark horn in patches; legs and feet light purplish-white.

MEASUREMENTS Wing 106-121; tail 57-62; bill 29.5-32.5; weight 99-120.

8 BANDED PITTA
Pitta guajana   Plates 5 & 16

Turdus Guajanus P. L. S. Müller, 1776, Natursyst., suppl., p. 146; based on "Merle de al Guiane" of Buffon (Daubenton), 1765-1780, Planches Enlum., 3, pl. 355: "Guajana"; corrected to Banjuwangi, eastern Java, by Kloss, 1926, Journ. Mal. Br. Roy. Asiat. Soc., 4, p. 161.
Alternative name: Blue-tailed Pitta

TAXONOMY The four taxa included here under *Pitta guajana* show considerable differences in plumage and were originally described as three separate species, with the Banded Pittas on Borneo and on Java both treated as endemic species. Whitehead (1893) and Riley (1938) were evidently convinced that these taxa represented 2-3 distinct species. Further investigations into the taxonomy of this group are required.

FIELD IDENTIFICATION Length 21-24. This is a medium-sized, sexually dimorphic species occurring in lowlands and hills from Peninsular Thailand to Malaysia and on the Greater Sunda Islands. Subspecies differ considerably. The broad bright yellow or yellow-and-orange supercilium, black crown and mask, white to yellow-and-white throat, and rich chestnut-brown upperparts with a broad white stripe across the wing-coverts and blue tail make this a most distinctive species. These common features can be used to identify both sexes of all subspecies apart from females on Java, which differ in having a golden-buff supercilium and rich golden-brown crown. Females of all subspecies, and males on Java, have finely barred underparts from breast to undertail-coverts, with blackish-brown and more or less yellow, yellowish-buff or orange-pink bars. Males have dark violet-blue on the underparts with fine orange or yellow bars on the breast and flanks, depending on subspecies. Juveniles have the distinctive white (buff in Java and Borneo) wing markings and blue tail of the adult, but are otherwise brown with buffy supercilia. The breast is streaked at first but becomes more barred with age.

Similar species Adults are unlikely to be confused with any other species in their range, the prominent white wing markings being diagnostic, except on Borneo, where Blue-headed Pitta also has such markings. In other respects, however, Blue-headed Pitta differs considerably and adults cannot be confused, the male of the latter with shining blue crown and reddish back, and the female with unmarked fulvous-buff head. Juvenile Banded Pittas also have prominent white bars across the wing (buff on Java and Borneo), enabling separation from juveniles of all other sympatric pittas, including juvenile Gurney's, which is otherwise very similar. Furthermore, immature female Banded Pittas acquire barred underparts from an early age, which enables separation from immatures of most sympatric species. Differences between recently fledged Banded Pittas and similar-aged Blue-headed Pittas on Borneo are not fully documented, but immature Blue-headed Pitta can apparently be distinguished by the almost uniform chestnut-brown crown: immature Banded Pitta has a blackish crown densely streaked or spotted with chestnut-rufous.

VOICE Two quite different calls are given throughout most of the range of this species. One is an abrupt, rather explosive, falling-tone *pouw* or *poww*, the other a short, low-pitched, slightly wavering *pprrr* or *kurrr*. The former call is relatively loud, but the latter tends to be quieter. Neither carries far in the forest understorey, and birds giving this call are often much closer than the calls might suggest. Tape-recordings suggest that the call made by birds in Sumatra is a softer *hwow* than the more explosive *pouw* made by birds in Peninsular Malaysia. Bornean birds also sound different, the *pouw* call being replaced by a less explosive *shewo*. In Peninsular Malaysia, calling is rather seasonal, with birds mostly silent around the turn of the year, starting to call in early March and continuing to at least June (pers. obs.). In Thailand, two adults gave very soft, hollow *whup* notes when approaching a nest with chicks, causing the chicks to raise their heads. On another occasion, a female gave a soft moaning *who-oo* when the male was on the nest, feeding chicks (P. Round *in litt.*).

DISTRIBUTION The Banded Pitta is a Sundaic species that is distributed from south of around 10°30'N in Peninsular Thailand (and probably the extreme south of Tenasserim, Burma) south to Johore in Peninsular Malaysia and the Greater Sunda Islands of Java, Sumatra and

Borneo.

The mainland population is widely distributed in suitable habitats in Peninsular Thailand and Peninsular Malaysia, but Banded Pitta has apparently never been recorded in Singapore (Medway and Wells 1976). In Sumatra it has been widely recorded, from Aceh Province in the north to Lampung in the south (van Marle and Voous 1988). The distribution in Borneo is poorly documented, and there is, for example, only one record from Brunei (Mann 1988). Banded Pittas are widely distributed on Java and the adjacent island of Bali, although their distribution on these islands is now limited by serious habitat fragmentation.

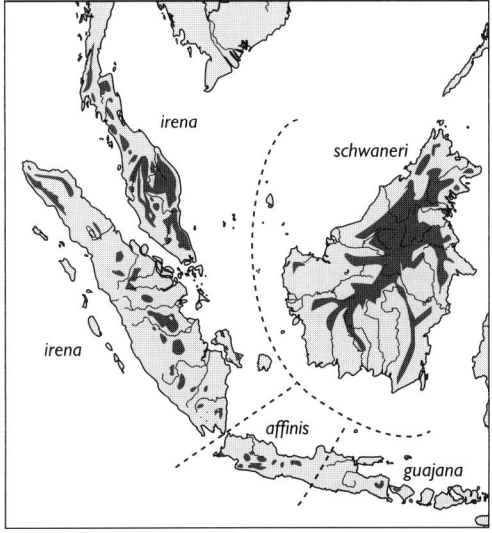

Distribution based on approximate extent of forest remaining, but the species is absent from the higher parts of mountains within the mapped range. Former range more widespread.

GEOGRAPHICAL VARIATION Six subspecies have been described, but only four are recognised here. The subspecies differ primarily in the coloration of the head and underparts.

There has been some confusion about the distribution of the taxon that was described as *Pitta irena ripleyi* by Deignan (1946a), based on an examination of only four birds. Deignan (1946a) gave the distribution of this subspecies as being from Peninsular Thailand south to Malacca and Pahang in Peninsular Malaysia, but later authors (Medway and Wells 1976, Mayr 1979) restricted the range of *ripleyi* to Peninsular Thailand, whilst birds in Peninsular Malaysia and Sumatra were treated as one subspecies, *irena*. Examination of series of specimens in BMNH suggests that variation among individuals is considerable, particularly in the features used by Deignan (1946a) in describing *ripleyi* (extent and distribution of scarlet in the supercilium, colour of the central abdomen in the male, and coloration of the mantle), and in this account *ripleyi* is not recognised, although there is a tendency for birds from Peninsular Thailand to have more orange-red in the supercilium than birds from Sumatra.

The other subspecies that was listed by Mayr (1979) but is not recognised here is *bangkae*, described as an endemic to the island of Bangka, south Sumatra. This subspecies is believed to have been described from a mislabelled specimen, probably originating from West Java (van Marle and Voous 1988).

As noted under Taxonomy, the subspecies *irena* and *schwaneri* differ considerably from each other and from the two subspecies on Java (*guajana* and *affinis*), and may represent distinct species rather than conspecific forms.

P. g. guajana (E Java and Bali) Blue gorget of male 10.5-17.5mm wide. Males have bright yellow supercilium, a conspicuous violet-blue breast-band and fine violet and yellow or yellowish-buff bars from breast to undertail-coverts. Females have golden-brown crowns and golden-buff supercilium, a narrow black band bordering white throat and dull yellowish and brown bars on underparts.

P. g. affinis (W Java) Very similar to *guajana*, differing in usually having a narrower blue band across the breast (6-10mm, but sometimes up to 16mm wide) and in slightly shorter bill, tail and tarsus (measurements of male: tail 62.8-69.8; bill 28.4-29.8mm; tarsus 41.1-42.9). Immatures may have the entire crown, including supercilium area, buff. Juveniles have buff spotting in upperwing-coverts.

P. g. schwaneri (Borneo) Underparts and supercilium of the male are yellow, with narrow purple bars from the breast to undertail-coverts and purple centre of the belly. Females have golden-buff supercilium that becomes yellow at rear, and chestnut, black or black mixed with chestnut forehead that becomes black on crown or nape. Upperwing-coverts of juveniles and immatures are tipped with buff, not white. Wing 96.5-106; tarsus 37-41; bill 29-30; tail 59-70.

P. g. irena (Sumatra, and from Peninsular Thailand and Tenasserim to Peninsular Malaysia) The supercilium of both sexes is yellow in front of the eye, becoming orange, usually behind the eye, and joining to form bright fiery orange on the nape. Males are purple below with a blackish border between the purple and the white throat and prominent orange bars on the sides of the breast. Females are narrowly barred below with dark brown and pale orangy-pink to yellowish-buff. The crown colour of females varies from all black to black mixed with chestnut. Juveniles and immatures have bold, pure white spotting on the upperwing-coverts. Typically smaller-billed than other subspecies. Wing 103-107; tail 59-71; tarsus 37-41; bill 26.5-30.

HABITAT Banded Pitta is principally a bird of primary forest, although also present in some selectively logged areas, particularly where taller stands of trees remain. Birds on Java have, however, been more often recorded in other habitats and may be more tolerant of habitat modification. It does not usually survive in areas of scrub, as do several other sympatric pitta species, and in parts of its range, such as Peninsular Thailand, is generally confined to the forest interior. Occasionally, it strays into more degraded habitats, and has been heard calling in a thick understorey of gingers (Zingiberaceae) within low secondary forest in Peninsular Malaysia (pers. obs.). In Peninsular Thailand, it also occurs in secondary forest, but access to permanent water seems to be a limiting factor (P. Round *in litt.*).

Although *irena* is known from sites throughout Sumatra, it appears to be absent from the extensive areas of swamp forest in the eastern lowlands. Major surveys in the swamp forests of Berbak Game Reserve (Silvius and

Verheugt 1986, Hornskov 1987) and in the Padang Sugihan Wildlife Reserve (Nash and Nash 1985) have failed to document the presence of this species, although it may have been overlooked. Robinson and Chasen (1939) also noted that this species avoids swamps in Peninsular Malaysia, even though it can regularly be found in riverine floodplains of lowland rainforest that are prone to deep flooding most years (pers. obs.).

In Sumatra, *irena* has been recorded to at least 750m (Kloss 1931a), and in Peninsular Malaysia Gibson-Hill (1949) reported its occurrence at 1,525m in the Cameron Highlands, although this record is highly questionable (D Wells *in litt.*). It is rarely encountered above the flat lowlands and low hills in the peninsula (pers. obs.). In Thailand, although not known from above about 610m (Robinson 1915) there is little area within its range that lies above this altitude.

In Borneo, Banded Pittas (*schwaneri*) occur mostly in the hills and mountains, including forests on calcareous hills (Pfeffer 1960), to at least 1,675m, although they are present locally in lowland areas down to 50m (Smythies 1957, SarM). In Sabah, recent records are all from below 800m (Sheldon in prep.) with the exception of records from 1,200m in the Maliau Basin (A. Johns *in litt.*). One of Whitehead's native collectors, however, apparently obtained two specimens at 1,525m on Mt Kinabalu last century (Sharpe 1889). Bornean birds are most frequently encountered along drier ridge-tops in tall Dipterocarp forest.

In west-central Java, Banded Pittas occur to at least 1,220m, whilst there is a record of Banded Pitta in mossy montane forest at 2,450m on Mt Gede, west Java (P. Heath *in litt.*). This is probably, however, exceptional; birds in west Java are most frequently encountered from sea level to 950m (Kuroda 1933; pers. obs.). Hoogerwerf (1948) found *affinis* at all sites that he visited on mainland west Java, but in east Java the nominate subspecies was found at only two of the sites he visited: above 1,300m in the Tengger Highlands and at 500m and above in the Idjen (=Ijen) Highlands. Kuroda (1933), who found *affinis* as low as 300m, states that it was fairly common in coffee plantations near Lawang 'where the absence of undergrowth allows free movement'. Bas van Balen (*in litt.*) found *affinis* in pure plantation forest, comprising primarily dipterocarps and pine trees, on Mt. Walat. Specimens from Java have also been collected in teak forests, and Vorderman (1882) reported that this was an abundant species in Salak *Salacca edulis* gardens in Jakarta. In Bali, Banded Pittas occur in riparian forest in Bali Barat National Park (R. Gregory-Smith unpublished), although B. van Balen (*in litt.*) notes that the habitat here is seriously degraded.

STATUS Banded Pittas are generally common in appropriate habitats throughout much of their range, although recent records suggest that they are rather local in distribution in Sumatra (D. Holmes *in litt.*). Nevertheless, some populations may become threatened in the near future. In Thailand, for example, the habitat of Banded Pitta is fast disappearing, and it is still occasionally caught for the illegal bird trade (Round 1992), a practice that was previously common. Trappers imitate the calls of the species to lure them into mist-nets, or take immature birds from the nest just prior to fledging. However, the inclusion of this species in Appendix II of CITES means that international trade can only occur with the issue of CITES export permits, and legal trade is monitored by countries belonging to the Convention, which include all its range states.

In Java, the extent of forest at appropriate altitudes has declined significantly, although Banded Pitta is able to survive in small forest patches (pers. obs.). It should be noted, however, that the species has disappeared from the Bogor botanical gardens (Diamond *et al.* 1987), which is now completely isolated from surrounding woodlands (although trapping may be partly responsible for its disappearance from this site, rather than the suitability of habitat). Whilst Sclater (1863) described Banded Pitta as being rare in south-east Borneo, it is a relatively common species in suitable forested habitats in Sabah (pers. obs.).

Round and Treesucon (1986) estimated that there were about 10 pairs of Banded Pittas in a lowland forest patch of 1.6 km² that was contiguous with forest on hill slopes (P. Round *in litt.*) in Peninsular Thailand; gives this a population density in the order of 12-13 birds per km².

FOOD In Borneo, specimens that have been examined have contained ants, a cockchafer, cockroaches, platyhelminths (flatworms) and various 'soft' insects (SarM, Smythies 1981, Sheldon in prep.). Specimens from central and eastern Borneo examined by Pfeffer (1960) contained terrestrial snails and insects and one bird had been eating small orange berries. Javan birds have been recorded as eating scarabid beetles, ants (Ponerinae), termites (workers of *Coptotermes gestroi*), caterpillars, earthworms and snails (Kuroda 1933, Sody 1989). In Peninsular Malaysia, Banded Pitta has been observed eating snails (J. Christensen *in litt.*), and Robinson and Chasen (1939) noted that Ridley found piles of snail shells in limestone areas that had been broken by birds that he attributed to this species.

Sody (1989) mentioned finding feathers in the stomach of Banded Pitta in Java, and, although this might seem strange, Becking (in Sody 1989) also found feathers in the stomachs of a pair of Banded Pittas in east Java. This pair of Pittas died after becoming stuck to bird lime surrounding a nearby dead White-rumped Shama *Copsychus malabaricus*, which was being used as a decoy by bird trappers and was tied to the ground. The Pittas had evidently been feeding on the dead shama, or perhaps on invertebrates in the corpse.

HABITS Calling suggests that the members of pairs of Banded Pittas feed apart but keep in close contact. Despite their bright colours, Banded Pittas can be very elusive at certain times of the year, hopping into dense vegetation at the first sign of intruders. At other times they may be more confiding, although not as confiding as many sympatric species. They feed on paths relatively rarely, usually keeping within the forest understorey. When startled they tend to fly some distance before alighting and hopping away rapidly. However, Banded Pittas have been observed in some exceptionally open situations, such as in a wide dry streambed and on a logging road in heavily logged forest.

BREEDING
Seasonality In Malaysia, recently fledged young have been found at Pasoh on 20 March (Wells 1975), and in the Krau Game Reserve between 3 November and 1 December (Wells 1990b, pers. obs., BMNH). In Peninsular Thailand, a nest with three eggs was found on 10 June (Robinson 1915), another with eggs on 11 July and a nest with two

young on 16 September (P. Round *in litt.*). Two recently fledged Banded Pittas were collected there on 12 September, one of which had some blue on the belly, and older juveniles have been collected on 19 September and 5 October (BMNH, Baker 1919). In Sumatra, one egg was collected in September 1915 (Beaufort and Bussy 1919 in van Marle and Voous 1988) and a recently fledged bird was collected on 2 May. These records suggest that Banded Pittas of subspecies *irena* breed all year round.

In Borneo, a nest was being built at the base of a rattan palm at 1,130m in the Hose Mountains on 2 April, while fledglings have been collected on Mt Ensuan at 305m on 11 May and at Tutoh on 1 January (Fogden 1965). Specimens with enlarged sexual organs have been collected in Sabah between 15 March and 4 August (Sheldon in prep.). Older immatures have been collected in Sarawak between 18 September and 1 November (SarM, BMNH), in south-east Borneo on 7 March, and at unspecified locality on 6 September (USNM). In Borneo, therefore, evidence suggests that breeding is more seasonal, occurring principally in the driest months and perhaps extending to the beginning of the wet season.

In Java, nests have been discovered from January to June, with peak egg laying in February-April (Kuroda 1933, Hoogerwerf 1949, Hellebrekers and Hoogerwerf 1967). Fledglings have been collected at the end of February in central Java and on 30 September at an unknown locality. Recently fledged juvenile *affinis* have been collected in west Java between 15 February and 20 May, whilst slightly older immature birds have been collected between mid-March and late April (MZB, RMNH).

Nests Nests are invariably placed above the ground. In Peninsular Thailand, P. Round (*in litt.*) found one nest against the bole of a large *Saraca indica* tree, on a boss of the trunk c.3m off the ground, and another nest in a small broadleaved but thorny understorey tree 2.5-3m off the ground. A nest described by Robinson (1915) as a large globular mass of dead leaves and fibres, was c.2m off the ground in a sapling. Two nests of *affinis* collected on Java (RMNH) are approximately 19 and 24cm in diameter. The largest of these was constructed with a diversity of dead broad leaves and grass stems of 1-2mm diameter and up to 40cm in length that were curved around the dome. Also within the structure were a few monocotyledonous leaves, including some from a rattan-like palm and some leaf skeletons. Some large pieces of split vine stem formed the base of the nest. The other nest, likewise, was primarily built of leaves (mostly of broad-leaved plants) but had no grasses in the structure. There were a few strips of larger leaves, small fern fronds and short pieces of rotted twig in the outer structure. The nest was lined with a fine mass of branching black roots, some of which were up to 1.5mm in diameter. B van Balen (*in litt.*) discovered a nest with three eggs in western Java in January. It was in a rattan c.80cm from the ground, and was c.30cm in diameter. The inner part of the nest was made of fine roots and twigs, whilst the outer parts were built entirely of coarse twigs and palm leaves. A nest from Thailand had the roof and sides made predominantly from pieces of broad leaves, including a few that were skeletonised, and small twigs (pers. obs.).

Eggs Clutches from Java usually comprise 3-4 eggs, though nests have been found with five and two eggs (Kuroda 1933, Hoogerwerf 1949). Eggs from Peninsular Thailand are broad blunt ovals, moderately glossy white and thickly spotted with dark purplish-brown spots and streaks, particularly at the broader end (Robinson 1915). Eggs from Java are white or creamy-white covered with scattered, well-defined irregular speckles and dots that are sepia to black, with black predominating. The underlying mottling, in some eggs forming the predominant markings, vary in colour from light lavender-grey to dark ashy. In most examples, markings are concentrated at the blunt end. In rare cases the markings are very faint or missing completely (Hellebrekers and Hoogerwerf 1967). Ninety-seven eggs from Java varied in size from 24-29.1 x 20-23.2mm (Hoogerwerf 1949, Hellebrekers and Hoogerwerf 1967). Three eggs from Peninsular Thailand measured 24.2-25.2 x 20.7-21.0mm (Robinson 1915). Hellebrekers and Hoogerwerf (1967) noted that eggs of Banded Pitta on Java differed from those of Hooded Pitta in that the markings are less numerous but better defined, and lack or have very little scribbling.

The subspecies *irena* has been bred in captivity. Clutch-sizes have varied from 3-5, perhaps occasionally two, with an interval of some 27hrs between the laying of two eggs. Incubation, shared by both sexes, took 13 days and the young fledged after a further 15 days. The young started to feed themselves after only four days, although both parents, in particular the female, still fed them on occasion (Vernon 1974).

DESCRIPTION *P. g. guajana*

Adult male Crown to nape black, usually joined at nape to broad black band across side of head from lores to ear-coverts and the sides of neck; broad bright yellow supercilium, broadening above or behind the eyes and extending to the nape, where pointed. Some birds have a few yellow feathers in the black of the nape but the yellow never joins across the nape. Mantle, back, upper rump, scapulars and tertials rich chestnut-brown; lower rump, uppertail-coverts and tail deep bright blue, often slightly brighter on the uppertail-coverts and with some chestnut-brown edges to feathers of lower rump; throat white, strongly washed with pale yellow which becomes deeper and is almost pure yellow at throat sides and often at edge of breast. Breast and throat are separated by a narrow, deep blue to violet band that has a very narrow, but variably wide, black upper border. The rest of the underparts are narrowly barred with violet-blue, or dark brown bars tinged with blue, and intervening yellow, with a variable amount of orange-yellow mixed with the latter: these bars formed by each feather having four narrow dark bands, the basal (hidden) bands generally being dark brown and those nearer the feather-tips with blue or violet tones, although some feathers of the belly and most of those of the undertail-coverts and feathers of the thighs may have all the bands dark brown. Some bars of the belly and flanks can be almost pinkish-yellow. Upperwing-coverts blackish with the median coverts marked with oblong-shaped white lower webs that broaden towards the tip and are often bordered by blue on the feather-edges or the inner part of the white. These white markings form a prominent band or double band across the closed wing. The outermost greater coverts may be marked with blue on leading edge, but lack white. Primaries dark brown, usually with a pale bluish tinge to the distal third of the outer edge of all but the outer two feathers, and with a narrow, hidden, white wing-bar on the underwing formed by white bases to the inner web of the outer five primaries. Secondaries dark brown, with prominent white leading edge formed by white distal third to half of a variable number (usually 3-

4) of outer feathers, and occasionally with blue edge to the innermost 1-2 feathers. Underwing-coverts mixed brown and white. Eye dark brown; bill black; legs purplish-pink.

Adult female Mantle to tail and wings as male, but mantle and back may be slightly duller and paler. Forehead to crown and nape golden chestnut-brown (with variable amount of dark mottling formed by dark bases of feathers showing through) merging to more or less pale golden-buff colour on sides of forehead and, usually, golden-buff superciliary area. Lores, feathering at base of bill, ear-coverts and sides of neck black, forming broad band across side of head (but not extending above eye). The nape and crown may be of similar, but slightly darker, colour to mantle, but rear of superciliary area usually contrasts with darker mantle. Ear-coverts often streaked with pale chestnut-brown. Throat white often tinged with creamy-buff or yellowy-buff, particularly at sides; narrow black band across upper breast; rest of underparts narrowly barred with brown and yellowish-white; occasionally some of the dark bars are tinged dull blue, especially on the lower breast, flanks and belly centre, whilst the pale bars are variously tinged yellowish-buff or a deeper yellow, and even mixed with pale orange-pink, so that the underpart colour can appear almost yellow in some individuals but yellowish-white in others.

Immature Crown dark, variably mixed with chestnut-buff speckles formed by pale centres to feathers that become darker and richer chestnut with age. Feathers of superciliary area streaked buffy-white, becoming more dense and richer, more golden-buff with age and, in immature male, yellow. Ear-coverts dark brown with fine paler streaks. Mantle and back brown, becoming more chestnut with age; tail dark blue, as adult; wings dark brown with prominent, large buffy-white spots formed by tip and inner web of greater coverts and becoming whiter with age. Throat whitish, mixed with brown that disappears with age; breast mottled buffy-brown, lower breast and belly to undertail-coverts more uniform dark buff, becoming barred with age, the bars being dark brown and buff in the female but showing progressively more blue in the male. Legs pale bluish. Base of bill and extreme tip usually orange.

Fledgling (*irena*) Culmen blackish, the rest of bill, gape and lining of the mouth flaming orange; legs and feet purplish-brown (Medway and Wells 1976).

MEASUREMENTS Wing 103-108; tail 69-71.5; bill 32.3-33.8; tarsus 40.2-44.0; weight 93-106 ('Java'), 60-79.6 *schwaneri*, 75-97 *irena*.

9 BAR-BELLIED PITTA
Pitta elliotii   Plate 6

Pitta elliotii Oustalet, 1874, Nouv. Arch. Mus. Hist. Nat. [Paris], 10, bull., p. 101, pl.2: Cochinchina.

Alternative name: Elliot's Pitta

FIELD IDENTIFICATION Length 19.5-21. Bar-bellied Pitta is endemic to lowland and hill forests in Indochina. The black mask, green upperparts, blue tail and fine yellow-and-black bars of the underparts are shared by both sexes. Males have blue-green crowns, a pale blue throat and dark blue belly-patch. Females have a dull ochraceous-buff head, throat and upper breast, and yellow centre to the belly. Immatures rapidly obtain adult body plumage, but the black-streaked crown renders easy distinction from adults.

Similar species Within its range, Bar-bellied Pitta is distinctive. Whilst males are clearly different from all other species, the female could be confused with Gurney's Pitta, although these species are allopatric in distribution. However, female Gurney's Pitta has brown rather than green upperparts. The lack of white in the wing serves to separate Bar-bellied Pitta from Banded Pitta at all ages.

VOICE The usual call of Bar-bellied Pitta is a trisyllabic, almost human-like whistle, variously described as *tu-wi-whil*, *per-ur-wu! hwt-whit-too* or *chawee-wu*, given at short intervals. The first two notes are ascending and the last descending, whilst the first note is shorter than the following. At a distance, only the last two syllables are generally heard, although birds with young occasionally give a two-note mellow whistle, *hhwee-hwha*, which is as long as the three-note call. In alarm, this species makes an explosive, rather shrill, descending note, *ppeu* or *ppew*, which is somewhat woodpecker-like, and reminiscent of the alarm of Gurney's Pitta. Females with young may utter a shorter version of this call.

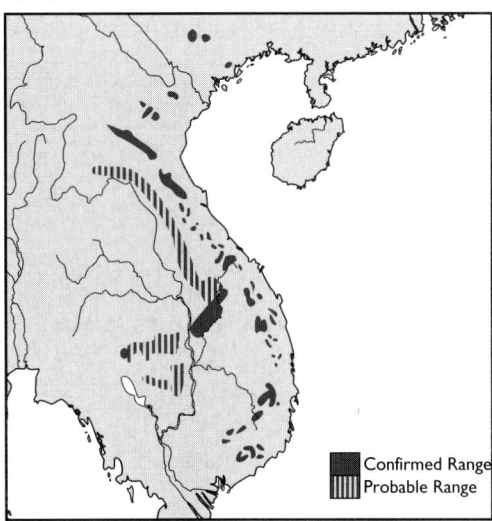

Distribution based on approximate extent of forest remaining within the altitudinal range of the species. Range probably more widespread: confirmation of presence in parts of Laos and Cambodia is required.

DISTRIBUTION Bar-bellied Pitta is endemic to Indochina, where present in Cambodia, Laos, Vietnam (Delacour and Jabouille 1940) and extreme east Thailand. The species may occur, or have occurred, in south-east Thailand (Lekagul and Round 1991), although this is questionable. The distribution of Bar-bellied Pitta in Cambodia is very poorly known, the only records coming from Angkor (Siem Reap), to the north of Tonle Sap (Thomas 1964). Despite fairly intensive collections in the forests of the Elephant Mountains, near Bokor, Kampot, no specimens or observations have been recorded for this area, and Bar-bellied Pitta may therefore be absent from the lowlands and hills of southern Cambodia: if this is the case then records from south-east Thailand are all the more questionable, and perhaps derive from market-purchased specimens that may be attributable to trade skins from elsewhere (Round 1988, Rozendaal 1989b). Elsewhere in Thailand, the only record is a bird that was said to have been 'collected by workers' in November 1983 in the Yot Dom Wildlife Sanctuary (c.14°20'N, 105°00'E) in the Dangrek Range along the northern border of Cambodia (P. Round *in litt.*). Forest at Yot Dom is more or less continuous with that in southern Laos.

In Laos, Bar-bellied Pittas occur from the extreme south to at least Phou Khao Khouay, near Vientiane, including the lower slopes of the Bolovens Plateau (Engelbach 1932). The distribution of Bar-bellied Pitta in northern Laos is unclear, although it seems to be absent from most areas. It was not found in the province of Tranninh by Beaulieu (1944), nor recorded in the forests bordering the Mekong River in the vicinity of Louang Phrabang (Delacour and Greenway 1940). In Vietnam, it is known from Backan in north-east Tonkin south to Anbinh, Cochinchina (Delacour and Jabouille 1931, Beaulieu 1932, Bô Khoa Hoc *et al.* 1992). The presence of Bar-bellied Pitta in western Tonkin is unconfirmed.

GEOGRAPHICAL VARIATION None described.

HABITAT The Bar-bellied Pitta is a bird of semi-evergreen and evergreen forests in the lowland and hills. Beaulieu (1932) found this species below 170m in Cochinchina. Elsewhere in Vietnam, Robson *et al.* (1989, 1992, 1993) found Bar-bellied Pittas in a variety of forested habitats, including logged forest and secondary growth, up to at least 800m. Delacour (1930) found it in limestone hills in extreme North Annam, and it is also found in forest on limestone hills in Central Annam, as well as in logged areas dominated by bamboo (pers. obs.). In the Annamese lowlands of Vietnam, this species is most frequently observed in areas of evergreen understorey where rocks, including protruding jagged limestone, are a prominent feature of the terrain (pers. obs.). In southern Laos, this species is found along river courses in primary semi-evergreen forests, including areas that have been logged, provided a good canopy remains, but it is apparently absent from deciduous and dry dipterocarp forests (Thewlis 1993, Duckworth *et al.* 1993). The species was often in areas with bamboo in the understorey, and appeared to favour land that was flat or gently sloping (R. Thewlis *in litt.*).

STATUS Delacour and Jabouille (1929) found Bar-bellied Pittas to be particularly abundant in Central and North Annam, whilst Wildash (1968) reported that it was common in parts of southern Vietnam. Though recently considered threatened by Collar and Andrew (1988), Bar-bellied Pitta has now been shown to be widespread and not uncommon in suitable habitat: it is therefore no longer considered as globally threatened although still listed as a near-threatened species by BirdLife International (Collar *et al.* 1994). Parts of its range have been largely deforested, such as most of Cochinchina, so its most southerly population in Vietnam is now probably that in Nam Bai Cat Tien National Park, southern Annam (Robson *et al.* 1992). Similarly, little forest at appropriate altitudes occurs in Tonkin, and the species is probably there confined to lowland areas along the bases of the mountains on the border with Laos and to the few small lowland protected areas such as Cuc Phuong National Park.

Rozendaal (1989a, 1989b) and Robson *et al.* (1992) found the bird to be common in both primary and logged forests and in other secondary forest formations in Vietnam, but in Thailand Bar-bellied Pitta must be a very rare and localised species that is endangered by forest clearance.

In 1994, sightings and calling suggested that the density in suitable forested parts of the Annamese lowlands must have been in the order of at least 20-30 birds per km^2 (pers. obs.). Different males were observed by R. Thewlis (*in litt.*) within 75m of each other in semi-evergreen forest, suggesting that the population density in suitable habitat in southern Laos is also high: Duckworth *et al.* (1993) reported that it was common and ubiquitous in primary semi-evergreen forests of southern Laos, but appeared to be rarer in degraded areas. In nearby riverine forest, the species was noted to be uncommon (Thewlis 1993).

MOVEMENTS None recorded.

FOOD Nothing is published on the diet of this species. An adult male feeding an immature was observed carrying a 2cm-long green caterpillar and earthworms (J. Eames *in litt.*). The stomach of a male contained the heads of termite soldiers and termite parts (pers. obs.).

HABITS This is typically a shy species. During and after the breeding season, Bar-bellied Pittas often forage as loosely associated pairs that keep in contact regularly by calling. Though very vocal in the breeding season, they are often only glimpsed briefly as they hop away from an observer. Foraging birds peck at the ground and turn over leaves whilst searching for food. A female was observed in central Vietnam repeatedly digging in the soil and leaf-litter whilst a Large Scimitar Babbler *Pomatorhinus hypoleucos* fed in a similar manner alongside (pers. obs.).

BREEDING Very little is recorded about the breeding biology of Bar-bellied Pitta. Both parents apparently build the nest (J. Wolstencroft pers. comm. to C. Robson *in litt.*), as a male was observed collecting twigs on 8 May whilst the female collected leaves. In Vietnam, recently fledged young were observed by C. Robson (*in litt.*) on 11 May in the Kon Cha Range (700-800m): this record indicates that nest-building may start as early as the first half of April. During a visit to central Vietnam from mid-May to early July, juvenile Bar-bellied Pittas, some of which were being fed by adults, were regularly observed (pers. obs.). Most juveniles observed had already obtained adult body plumage but retained immature feathers on the head and wing-coverts, and had probably fledged at least 3-6 weeks earlier. Nests have only been found in the second half of May (C. Robson *in litt.*, Nguyen Cu verbally). Hence, present evidence suggests that the breeding season is rather short, with birds nesting between early to mid-April and mid-June.

A nest containing three eggs was discovered in Vietnam on c.21 May (C. Robson *in litt.*, Nguyen Cu verbally). The nest was situated in the crown of a small tree about 5m off the ground, at the edge of a rather open area in degraded forest. The understorey in the adjacent area had been cleared by woodcutters (Nguyen Cu verbally). This nest was a slightly flattened dome with the front sloping backwards so that the entrance was at c.45° angle to the horizontal and hence half-facing the sky (see photograph in Rozendaal 1989a). It was built primarily from twigs of various sizes and dead leaves.

DESCRIPTION

Adult male Forehead and crown to nape bright golden-green, more or less tinged blue and merging into pale ultramarine-blue lower border that becomes more prominent at the rear, forming a narrow point of ultramarine-blue at sides of nape; broad black band across lores, sides of head and ear-coverts extending to sides of nape; mantle to rump and uppertail-coverts, and upperwing-coverts, green, more or less glossed with blue tones but darker than colour of head; tail dark bright ultramarine-blue. Some individuals have black wedges in tips of feathers of upperparts. Extent of blue gloss above varies considerably. Chin and centre of throat white, becoming pale greenish or pale blue-green on sides and on lower throat, and merging into pale yellowy-green of upper breast. Lower breast, flanks and sides of belly bright yellow, boldly but narrowly barred with black; patch on centre of belly, occasionally extending to lower breast, deep violet-blue; undertail-coverts deep dark blue with black bases or deep dark blue with green central parts to feathers and black base. Lower belly sometimes yellow with black bars. Bars on sides of belly may have purple sheen. Primaries blackish-brown with pale dark bluish leading edge to distal half of inner primaries. Distal half to three-quarters of secondaries broadly edged with blue: colour same as tertials and similar to rest of upperwing-coverts. Underwing-coverts brownish with some white feathers mixed in. Eyes dark brown; bill black; legs pink.

Adult female Forehead and sides of crown to nape rich ochraceous-buff, centre and rear of nape to nape green with yellowish sheen and mixed with variable amount of ochraceous-buff. Broad band from lores to back of head black with some ochraceous streaking on ear-coverts and lores. Rest of upperparts similar to those of male, but with blue sheen. Throat pale buffy, becoming deeper buff on breast; rest of underparts yellow, slightly duller than male, (particularly on belly on lower flanks) and marked with numerous black bars, except on centre of belly, where more or less unmarked. Bill blackish-brown.

Immature male Centre of forehead and crown to nape buffy-rufous with dark feather-bases that show through, giving the appearance of streaking; sides of forehead and superciliary feather-tract unmarked buff. Blue-green feathers of adult male first appear at the rear of the crown. Lores brownish; ear-coverts and area below eye black with a few rufous feathers forming indistinct narrow streaks. Mantle buffy-brown mixed with dark brown (feather-bases); back pale brown mixed with green. Upperwing-coverts brown with paler, slightly buffy notches in tips, forming indistinct paler line in wing; flight feathers brown, the secondaries tinged with green in outer webs. Tail blue, duller than in adult. Throat white; breast mixed green and pale brown; rest of underparts similar to adult. Iris dark brown; bill black, basally pinky-yellow; legs pinky-flesh.

Juvenile Unknown.

MEASUREMENTS Wing 104-110; tail 43-49; bill 29-31; tarsus 36-39; weight 85-97.

10 GURNEY'S PITTA
Pitta gurneyi   Plates 6 & 16

Pitta gurneyi Hume, 1875b, Stray Feathers, 3, p. 296, pl.3: southern Tenasserim [Burma].

Alternative name: Black-breasted Pitta

FIELD IDENTIFICATION Length 18.5-20.5. Gurney's Pitta is probably the rarest species of pitta, formerly well distributed in Tenasserim and Peninsular Thailand, but presently known to survive at only one or two small sites in Thailand. The black mask, vivid dark ultramarine-blue hindcrown and nape, and yellow underparts with black belly and black bars on the flanks distinguish males from all other pittas. Females share the rufous-brown wings and mantle and blue tail of the male, but have ochraceous-yellow crowns, a blackish-brown eye-patch and finely barred, black and yellow-buff underparts. Immatures are brown with various amounts of buffy speckling on the crown and underparts, but largely unmarked dark brown mantle and back, and turquoise-blue tail.

Similar species Within its range, adult Gurney's Pitta is unlikely to be confused with other species. Immatures, which are very similar to the immatures of sympatric Banded Pitta, can best be distinguished from the latter by the lack of white in the wing, and by streaking, rather than scaling, on the breast and by the lack of orange on the sides of the nape. Juveniles are superficially similar to juvenile Blue Pitta, but the latter has less extensive dark brown on the ear-coverts, streaked mantle and back, and broader, bolder buffy markings on underparts. Blue Pittas develop black-and-white bars on the flanks at an early age. Female Gurney's Pitta is similar to female Banded Pitta from Java, but the latter has a narrow black band across the upper breast, narrower bars below, white wing markings, violet-blue rather than turquoise-blue tail and dark chestnut crown.

VOICE Males give a distinctive territorial call, a short but explosive *lilip*. The female also gives a truncated, less emphatic version of this call, transcribed as *llup* (P. Round *in litt.*). Both sexes have a falling-note, slightly tremulous *skyeew* alarm and contact call that is similar to some calls of Blue-winged Pitta and Hooded Pitta, but less squeaky. A sonogram of the *lilip* call is provided by Collar *et al.* (1986).

Gretton (1988) describes three other calls. One is a low-pitched mammal-like *hoo* that carries only some 10-15m. This call is given by both sexes of closely feeding pairs and appears to be a contact call or is perhaps occasionally given when mildly alarmed. It is sometimes given by an adult before it approaches the nest. The second call is described as a cross between the *lilip* and *skyeew* calls; only the female has been observed making this call, the function of which is unknown. Gretton (1988) reported hearing a third sound on one occasion: a male giving the *lilip* call also produced a faint *tchup* that appeared to be associated with slow, half-wing flaps.

Calling is highly seasonal, corresponding to the wet season. Calling therefore starts sometime in the period from late March to mid-April, depending on the onset of wet weather. The incidence of calling remains at a high level until mid-May but then falls rapidly to a low level in mid-June (Gretton *et al.* 1993). The *lilip* call is frequently given by the male from a perch in a tree: Gretton (1988) recorded males calling at 2-10m above the ground. Davison collected a male that was calling from high in a tree in Tenasserim (Hume and Davison 1878).

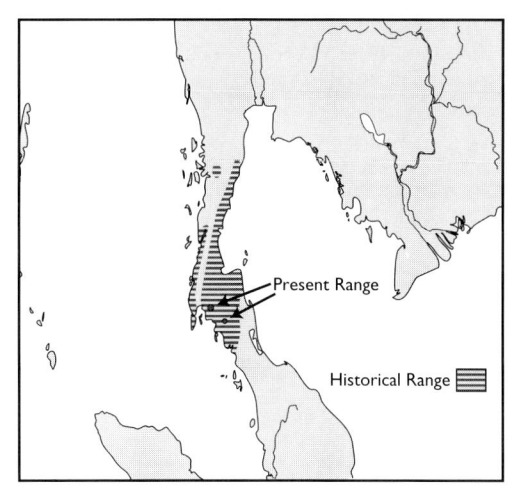

Map shows present known range and historical range. Continued presence of the species in Tenasserim requires confirmation.

DISTRIBUTION Gurney's Pitta is unique in that it is the only species of bird that is endemic to Peninsular Thailand and Tenasserim, Burma. Within this very restricted distribution, it is known only from forests between 11°50'N and approximately 7°25'N. Early collectors also found it on the offshore island of Phuket (Robinson and Kloss 1918). Evidence suggests that within this range the species is rather localised: fieldworkers in Thailand have consistently failed to find the species in many forested areas, many of which would appear to contain suitable habitat. In Burma, all but one record, from Lenya (11°28'N, 99°36'E), are from the extreme southern tip of Tenasserim, and all certain records are from the period from December to June, or July (Collar et al. 1986). Recent surveys in Thailand suggest that Gurney's Pitta now survives only in the lowlands in the vicinity of Khao Pra-Bang Khram (7°55N 99°16'E) and Khao Phanom Bencha in Krabi Province (Gretton 1988), although the status of the species in Burma is unknown. As of 1994, the only area where a population was known to persist was in the Khao Pra-Bang Khram Wildlife Sanctuary.

GEOGRAPHICAL VARIATION None recorded.

HABITAT Within its restricted range, Gurney's Pitta inhabits 'semi-evergreen rainforest'. Forests to the north, at about 12°30'N, are classified as 'moist deciduous forest'; those to the south of the range of Gurney's Pitta are 'evergreen rainforests' (Whitmore 1984). Although Robinson and Kloss (1924b) contended that the species was strictly associated with limestone hills, there is no evidence to support this, and indeed forest areas on limestone in Peninsular Thailand do not appear to have populations of this species.

Research in Peninsular Thailand has shown that Gurney's Pitta is able to occupy a variety of habitats from very degraded forest patches to tall near-primary forest. Since there is no truly primary lowland forest remaining in Peninsular Thailand, however, it is no longer possible to assess whether the remaining population of Gurney's Pitta occupies suboptimal habitat or whether the species has always occupied or even preferred secondary forest habitats. Robinson (1915) implies that he found it commonly in an area dominated by secondary growth in Bandon Province in 1913.

Several territories discovered at Bang Tieo (Khao Pra-Bang Khram) and elsewhere during 1987-1992 were in very degraded forest fragments, often very close to forest edge. The site at which most comprehensive observations have been made is a mere 2ha forest fragment surrounded by rice paddy and containing trees no larger than 20m tall. Indeed, during the course of studying Gurney's Pittas since 1986, P. Round (*in litt.*) has come to the conclusion that it may be preferentially associated with secondary forest.

Studies in Peninsular Thailand suggest that within lowland forest, Gurney's Pitta occupy territories with particular characteristics. It seems likely that the overriding factor is probably the need for year-round water in small streams and gullies within their territory (P. Round *in litt.*): 85% of territories found in 1988 contained small streams or gullies and spiny palms (of the type used for nesting sites) over one metre tall. Some parts of the territories identified in 1987 were dominated by palms. The gullies frequented by Gurney's Pitta usually have concentrations of species of *Licuala* palm, whilst *Salacca rumphii* and numerous rattans, such as species of *Calamus*, were common in the drier parts of their territories. Measurements of tree size around three nests, however, suggested that the size distribution of trees was unimportant, with nests being located in areas of forest in different stages of recovery from disturbance. Unpaired males have also been recorded in swamp forest, although this habitat is judged unsuitable for breeding (Gretton 1988).

Virtually all breeding territories that have been documented in the vicinity of Bang Tieo in recent years have been below the 100m contour (Gretton et al. 1993), with the exception of one territory at 140m on the eastern flank of Khao Pra-Bang, Trang Province, and all other Gurney's Pittas found since 1986 have been below 160m. Although most records and specimens point to Gurney's Pitta as being a lowland forest species, specimens, including a nestling, were recorded as collected on Khao Phanom Bencha (8°17'N, 98°56'E) at altitudes of 610-1,070m (Meyer de Schauensee 1946). However, Round and Treesucon (1986) and Round (in press) provide evidence that these derive from an altitudinal recording error. Recent surveys (1986-1987) for Gurney's Pittas in the region of Khao Phanom Bencha have failed to locate any Gurney's Pittas at higher altitudes (Gretton 1988), although they were present in the adjacent plains (P. Round *in litt.*).

STATUS Early collectors reported that the species was relatively numerous at many of the locations where it was encountered. Robinson (1915) for example stated that it was very common in the vicinity of Khao Nong in 1913, although it did not extend far up the mountain, whilst Robinson and Chasen (1939) noted that it was the commonest pitta species in Trang. Today, however, Gurney's Pitta is apparently extremely rare, as little of its lowland forest habitat persists within its known former range. Until 1986, when Gurney's Pitta was rediscovered in Peninsular Thailand, this species had not been recorded in the wild for 34 years, although small numbers were occasionally recorded in trade.

The only known population is very small, and despite protection and active conservation efforts, a number of serious threats persist and it is consequently considered to be a Critically Endangered species (Collar et al. 1994), and is one of only two species of pitta included in CITES Appendix I (WCMC 1995), making international trade in live birds or specimens illegal. In 1988, only 4.7% of the former area of forest occupied by Gurney's Pitta remained in Thailand (Round 1988), whilst in 1987 the total forest area below the 100m altitude contour that is occupied by Gurney's Pitta was estimated at only 20-50km^2, and this area has undoubtedly been further reduced since this time (Gretton et al. 1993). In the lowlands of Burma, the situation may now be similar, since the 1988 logging ban in neighbouring Thailand has caused an intensification of logging activities in Burma (Collins et al. 1991). The apparent adaptability of the species in its habitat requirements, in particular its ability to survive and breed in small islands of disturbed forest, provides hope that current conservation activities will prevent the disappearance of this attractive and uniquely distributed species.

Surveys aimed at establishing population size at and around Bang Tieo have been conducted in most years since 1987 (Gretton et al. 1993). In 1988, when coverage was most comprehensive, between 23 confirmed territories and six unconfirmed territories were located. Further calling males were heard in sites that were only visited once. In an intensively studied area of 250ha at Bang Tieo, there were 9-15 territories of Gurney's Pitta in 1988 (Gretton et al. 1993), suggesting a maximum population density of 7.2-12.0 breeding individuals per km^2. In prior and subsequent years fewer confirmed territories were discovered in this study area, but observer coverage was poorer.

Based on survey work carried out in 1987-1989, Gretton et al. (1993) estimated that there were some 24-48 Gurney's Pitta territories in Peninsular Thailand. Since 1989, extensive forest clearance has occurred at Khlong Phraya-Khao Phanom Bencha, and it is feared that Gurney's Pitta may have now become extinct at this site, where as many as six pairs were thought to survive in 1988 (Round 1992). Furthermore, most territories outside of the Khao Pra-Bang Khram Wildlife Sanctuary (shared between Krabi and Trang Provinces), in which the remaining population of Gurney's Pitta is found, have now been lost through habitat clearance. This has caused a corresponding population decline, and the population at the end of 1992 was probably in the order of only 20-30 pairs (Round and Treesucon 1990-1992, Gretton et al. 1993), and more recent data suggest that the population size has not increased despite conservation efforts (P. Round verbally).

The future of this tiny population is linked inextricably to the fate of the forest in which it survives, in the Khao Pra-Bang Khram Wildlife Sanctuary, which is the focus of a major BirdLife International-Centre for Conservation Biology (Mahidol University, Bangkok) project. Unfortunately, however, the boundary of the Wildlife Sanctuary (in 1994) excluded most of the breeding territories of Gurney's Pitta, and during 1987-1990 habitat holding approximately eight pairs was cleared (Round 1992). Hence the future is clearly bleak unless reserve boundaries are redefined, or the species is found in Tenasserim, Burma.

MOVEMENTS Hume and Davison (1878) believed that breeding did not occur in Tenasserim, and provided anecdotal evidence that Gurney's Pitta undertakes seasonal movements. Davison reported that, in south Tenasserim, a few birds began to appear in early to mid-February, after which the species remained scarce until mid-April. Oates (1883) collected some at Maliwan in the 'early part of the year'. The species was then more numerous, but most appeared to disappear with the onset of the monsoon, in late May or June. A few birds, however, were present as late as July.

Dates on specimens that have been collected, however, do not support this pattern, since Collar et al. (1986) found that some skins from south Tenasserim were collected in December and January. Furthermore, year-round fieldwork at Khao Nor Chuchi (in Khao Pra-Bang Khram Wildlife Reserve) has shown that Gurney's Pitta is present all year, but that it becomes very difficult to detect during the post-fledging period (Round in press). Serious doubt is now cast on other evidence that has been used to support a theory that Gurney's Pitta is a migratory species (Round in press), such as the alleged collection of specimens at altitudes of 600-1,060m in August (Meyer de Schauensee 1946). In the absence of more conclusive data, Gurney's Pitta should be considered as a resident species.

FOOD Early reports from Tenasserim record snails, worms, slugs and insects in the diet of this species (Elliot 1893).

Food items brought to a nest studied in Thailand in 1986 (Round and Treesucon 1986) and to two nests in 1987 (Gretton 1988) were primarily annelid worms, which formed 73% of identified prey in 1986 and 78.8% in 1987. These worms varied in length from 2-8cm, with the size of prey generally increasing during the fledging period. Up to eight annelids were sometimes brought to the nest simultaneously.

However, since annelids are more easy to identify than other prey items, the actual proportion may have been lower. Round and Treesucon (1986) believe that the male, which, unlike the female, did not hesitate at the nest long enough to allow good views of the food items, may have brought more small items to the nest than were recorded. Gretton (1988) also notes that prey items less than 1cm long were certainly under-recorded. Apart from annelids, items identified in 1986 were large-winged insects, large and small insect grubs and a large grub or slug, and in 1987 unidentified insect larvae, de-shelled snails, slugs, a locust, cicada, beetle, cockroach, a large white butterfly and, on four occasion, a small frog. Although de-shelled snails were included in the nestling diet, no anvils have been found, and no shell-removing activity was noted during these studies.

Other potential food items that were shown to be common in the foraging area of Gurney's Pitta were small spiders, and ants (Gretton 1988). M. Chong (in litt.) observed an adult catch a spider.

HABITS The extreme rarity of Gurney's Pitta has meant that the surviving small population in Peninsular Thailand has been the subject of intense study since 1986, and more is known about it than any other pitta species. In particular, all aspects of breeding biology have been investigated, and these are discussed under Breeding. Most of the information on the behaviour of Gurney's Pitta derives from the studies of Round and Treesucon (1986) and Gretton (1988).

Whilst aspects of breeding behaviour have been meticulously studied, fewer observations have been made of

Gurney's Pitta foraging behaviour, although adults, in particular the males, frequently fed close to the nest and observation hides of researchers. Away from the immediate vicinity of nests under study, few data have been collected. In general, the members of a pair appeared to feed in different areas rather than together.

Gurney's Pitta typically forages by tossing leaf-litter aside with sideflicks of the beak, and by making short probes into loose, damp topsoil. Hence adults are frequently observed with mud stuck to the beak. Between these probing bouts the birds rapidly hop across the forest floor until spotting potential prey or suitable probing substrate. M. Chong (*in litt.*) observed an adult male that had been foraging by flicking leaves aside in a dry streambed suddenly jump c.10cm into the air to take a spider from its web. Occasionally, adults pause to cock their heads to one side, presumably listening for prey. Prey items are often battered on a nearby log or branch and, whilst feeding young, such items are frequently tossed to one side whilst further prey is caught before gathering up four to five prey items to deliver to the nest simultaneously. Both sexes often pause for a minute or more on a vantage point close to the nest, such as a termite mound, before going to the nest to feed the young. Robinson and Chasen (1939) noted that this species has the habit of jerking up its tail and dropping its wings slightly as it hops along (this perhaps in response to danger).

Only very heavy rain appeared to prevent foraging during the breeding period: adults continued to feed during light to moderate rain. After heavy rain, bedraggled adults have been observed to fluff out their belly and flank feathers and, occasionally, to whir the wings to help dry the plumage.

Ranging behaviour is yet to be studied, but evidence suggests that unpaired males may rove widely and that pairs may move territory during the breeding season following an unsuccessful breeding attempt. Males may also mate with more than one female during the same breeding season: the male of a nest that fledged young on 26 July was found at another nest, with four eggs, only 18m from the first site on 8 August. The female may well have been different because data on feeding visits suggested a large difference in foraging pattern (Gretton 1988).

Calling is largely from the ground, but occasionally from low perches, such as fallen logs and less frequently in low trees or on horizontal vines (P. Round *in litt.*). Males may call from several sites within their territories. During this calling they may make a wing-flapping display that has been observed by Hume and Davison (1878) and Gretton (1988). Between giving *lilip* calls the male half-opens its wings, rather slowly, and then more rapidly closes them. This half-wing flapping was recorded 12-15 times in about three minutes by Gretton (1988), accompanied by a faint *tchup* noise that have been made by the closing of the wing.

BREEDING There are no breeding records from Burma. Hume and Davison (1878) noted that dissected specimens collected in Tenasserim in April, May and June showed no sign of breeding. Prior to 1986, when the species was rediscovered after a period of 34 years, there were only three breeding records for Gurney's Pitta: a female containing a shelled egg ready for laying was collected in Thailand in October 1915, near a nest containing four eggs (Herbert 1924); a juvenile was collected in Surat Thani province on 25 July 1929; and a nestling was collected on 19 September 1936. Since the discovery of the first nest in Peninsular Thailand on 15 June 1986, thirteen Gurney's Pittas nests have been discovered at Khao Pra-Bang Khram between 23 May and 8 August, although ten of the 14 nests were found, with eggs, in June (P. Round *in litt*). In Thailand, nests have been discovered in 3-7m tall palms with long spines: those identified were *Salacca rumphii* (Round and Treesucon 1986), rattans *Calamus longisetus* (P. Round *in litt.*) and perhaps *Daemonorops* sp. (Gretton 1988). One of these nests was further supported by a creeper that passed under it. Whilst Thai nests have been situated at heights of 1-2.4m off the ground, Herbert (1924) reported that the nest found in 1915 was situated on the ground at the foot of a bamboo clump (Baker 1934). According to Round and Treesucon (1986), however, Herbert never saw the nest *in situ* himself, so there is the possibility that the fact that it was on the ground, and in bamboo, was recorded in error.

Nests of Gurney's Pitta resemble those of the similar-sized, sympatric Banded Pitta. The nest is a slightly flattened dome of large, dead leaves of broadleaved plants and small sticks with a few bamboo and palm leaves, placed on a base of slightly larger sticks. The roof of the nest may include the large, elongated, thick (i.e., fleshy in life) dead leaves of unidentified plants, some up to at least 30cm in length (and probably of understorey plants), as well as dried pieces of palm leaf. The shallow cup of the nest cavity is lined with fine black rootlets (Round and Treesucon 1986, Gretton 1988, pers. obs.). Nest dimensions are 18-20cm deep, 19-23cm wide and 18-22cm high, with a side entrance about 9-10.5cm wide by 6-7.5cm high. Inside, the diameter is about 11-13cm deep, 9-12cm wide and 8-11cm high (Gretton 1988).

Clutch-size is usually 3-4, but occasionally as many as five: Herbert (1924) collected a nest in 1915 with four eggs and subsequently discovered that the female, shot from the same nest, contained a shelled egg (Baker 1926). Baker (1926) gave egg dimensions as 25.3 x 22.0mm to 27.0 x 22.4mm, and noted that the eggs were very similar to those of sympatric Hooded Pitta *P. s. cucullata*. The eggs were white and boldly spotted with dark purplish-black and brown, with underlying grey spots and markings. Two eggs collected in Thailand by R. Lansdown (BMNH) were slightly different, being more elongated than those in Baker's collection, and more sparsely marked. These two eggs measure c.25.5 x 20.0mm and 27.0 x 20.0mm and are white marked with sparse small brown spots and squiggles and underlying larger, grey, irregular-shaped markings and spots, the markings concentrated at the broader end.

Both sexes brood and feed the young. Whilst the female spent more time incubating than the male at the nest studied by Round and Treesucon (1986), observations by Gretton (1988) suggest that during the twelve hours of daylight each sex would incubate for almost exactly six hours, in shifts lasting from two hours 45 minutes to three hours four minutes, though the female would always incubate overnight. The incubation period is unknown, although it is more than ten days. At a nest studied by Gretton (1988) a clutch of four eggs hatched within a 7.5hr period of daylight and incubating adults ate the entire eggshells when the eggs hatched. The young are brooded almost continuously by the female, and occasionally by the male, for the first three days and then approximately half the daylight hours for the next one to three days. The female also broods the chicks at night for the first seven days after hatching.

Males generally make more feeding visits to the nestlings than do females, although this is probably partly a consequence of the female incubating the young during the first 4-6 days after hatching (Round and Treesucon 1986, Gretton 1988). Males probably also feed the incubating female in the nest. Of a total of 355 visits to a nest studied by Round and Treesucon (1986) over a four-day period, the male made 290 (81.6%) of visits as compared to the female's 65 visits. However, almost half the visits by the latter were during the last two days and it is possible that she was slow in becoming habituated to the observation hide (Round and Treesucon 1986). At a nest studied by Gretton (1988) 58% of feeding visits were made by the female during the second week that the young were in the nest. During 60 accurately timed visits to the nest, males spent on average 26.5 s and females 22.5 s at the nest (Gretton 1988).

Feeding visits to the nest start immediately after sunrise and continue to a few minutes before sunset, with the highest frequency of visits by both sexes being in the afternoon. Most feeding visits at the nest studied by Round and Treesucon (1986) were made at intervals of 5-15 minutes, even during rain, although on one occasion there was a 52-minute interval between consecutive feeds. It was extremely rare for both sexes to visit the nest together whilst feeding the nestlings, though not unusual for both to be momentarily present during incubation and brooding. Usually the entire food load brought to the nest is fed to a single chick. Gretton (1988) provided evidence to suggest that the frequency of feeding visits increases with brood size. The number of feeding visits per hour were calculated as 6.73, 7.32 and 8.53 at nests with one, two and three chicks respectively.

In order to present a faecal sac, chicks turn to face away from the parent immediately after feeding. During the first few days after hatching these were often swallowed by the parent, but thereafter they were carried at least 10m from the nest. Gretton (1988) found two dumps 12-15m from the nest with a dozen or more sacs. Any sac that broke during removal was eaten. Between nine and eleven faecal sacs were removed from the nest each day.

Nestlings fledge at 14-15 days old (Round and Treesucon 1986, Gretton 1988) after spending the last day removing the waxy sheaths off the feather pins of the wings.

As with many tropical bird species, the predation of young Gurney's Pittas is apparently frequent and nesting success must be rather low. Gretton (1988) documented the finding of three nests in 1987. The first, containing four eggs, was predated shortly after discovery on 27 May; only one chick fledged from the second nest, although it had contained two chicks and one egg when found on 12 July; and, finally, two chicks fledged from the nest discovered on 8 August that had contained four eggs. On one day Gretton (1988) replaced a chick in a nest under study on two occasions after it had fallen out, but the next day it had disappeared, presumably after having fallen a third time.

Hence, on average, only one chick was fledged per nest from the three nests observed in 1987, and mortality rate was at least 72.7% when all known eggs and chicks are considered. Although not observed, two young probably also fledged from the nest studied in 1986 by Round and Treesucon (1986), which would mean that overall nest success is higher if all four known nests are considered, with up to 1.25 young fledged per nest. Of the 14 nests documented by P. Round (*in litt.*), young fledged from only 4-5. One of the predators of Gurney's Pitta chicks was identified as a 2m-long Dog-nosed Cat Snake *Boiga cynodon* which was killed near a nest and found to contain a semi-digested 11-day-old Gurney's Pitta chick (Gretton 1988). Intense alarm calling has been noted in response to Black Cobras *Naja naja* and Leopard Cats *Felis bengalensis*, suggesting that these are perceived as potential nest predators.

DESCRIPTION

Adult male Head black with white throat and vivid, fluorescent dark ultramarine-blue patch on crown and nape. Top of mantle black, concolorous with head and occasionally fringed with deep violet feathers; rest of upperparts warm rufous-brown (similar to Banded Pitta); tail bright ultramarine-blue, paler than blue of head; upper throat white, lower throat bright yellow; centre of breast and belly to undertail-coverts black, sides of breast and flanks yellow, narrowly barred with black. Undertail-coverts more or less tipped dark blue. Primaries dark brown with narrow greyish fringe to distal third of outer web of primaries 4-7; secondaries dark brown with outer edge of distal half of outer 2-3 feathers white, bordered by rufous-brown (concolorous with rest of upperparts), these edges becoming progressively broader on inner secondaries; tertials entirely rich rufous-brown. Outer 4-5 primaries with narrow white band across base of inner web forming bar on underwing. Underwing-coverts brown with some white in axillaries. Eye blackish; bill black; legs pink.

Adult female Forehead to nape ochraceous-buff, becoming progressively brighter and more golden towards nape and often having some very narrow black fringes to feathers of forehead; lores mixed ochraceous-buff, broad black patch across and rear side of head black with buff streaks on ear-coverts; upperparts as adult male, but marginally paler; throat white, more or less washed with buff; rest of underparts barred black and buffy-white, the intensity of buff varying considerably and, in most birds this is washed with yellow (especially on breast), and/or with pinky-buff. Some individuals have virtually unbarred buffy-white belly to undertail-coverts.

Subadult Females have the entire underparts suffused with pinky-buff, including the throat, and the yellow bands on the breast mostly replaced by pinky-buff bands.

Immature Forehead to nape dark brown, marked with long buff streaks that become broader and more spot-like on nape; sides of nape more or less buffy-white; sides of head blackish-brown with a few fine buff streaks on ear-coverts; throat white becoming buffy-white at edge. Mantle to rump and wing-coverts dull rufescent-brown with a few indistinct narrow buff lines on mantle; uppertail-coverts fringed with ultramarine-blue; tail blue. Breast, belly and flanks as upperparts, with sparse, rather narrow buffy-white streaks on breast and upper belly; undertail-coverts unmarked. Base to lower mandible and bill tip pale or brightly coloured: bright fleshy-orange in late nestlings (P. Round *in litt.*).

MEASUREMENTS Wing 101-109 male; 99-106 female; tail 45.5-52; tarsus 36-40; bill 26.5-28.5; weight 57-86 (Hume 1875b: 2-3oz).

11 BLUE-HEADED PITTA
Pitta baudii   Plate 6

Pitta baudii Müller and Schlegel, 1845, in Temminck, Verh. Nat. Gesch. Nederland Overz. Bezit., Pitta, pp. 10, 15, pl.2: south Borneo.

FIELD IDENTIFICATION Length 16-17.5. Blue-headed Pitta is endemic to Borneo, where it inhabits forests in the lowlands and hills. Both sexes share the rich reddish-brown colour of the mantle and wing-coverts, the bold white bar in the wing and the blue tail. Males are brighter above, and have bright azure-blue crown to nape, a black mask and white throat contrasting with the dark purple-blue underparts. The head and underparts of females are rich fulvous-buff. Immature males develop adult plumage in patches, so that they appear rather like a female above, often with a few scattered blue feathers on crown, but with extensive patches of orange-buff mixed with black breast and violet belly. Other presumed males may resemble females except for blue feathers on the forehead.

Similar species Adults are unlikely to be confused. Differences between recently fledged Banded Pitta and similar-aged Blue-headed Pittas are not fully documented, but immature birds can probably be distinguished by the pattern on the crown, with the latter species having almost uniform chestnut crown, and Banded Pitta having a blackish crown densely streaked or spotted with rufous.

VOICE The most frequently heard call is a soft descending trisyllabic whistle, *ppor-wi-iil* or *por-y-or*, which is occasionally shortened to a disyllabic *ppor-or*, with emphasis on the first part of the note and a very brief break between this note and the following two. The first note is louder and more abrupt compared to the second part, which is quiet and tails off. It is not clear if both sexes give this call, but when males at close quarters make this or a similar call, there is a distinctive resonant quality to the central note. The alarm note of the female is a nasal, drawn-out *hwee-ouu* (pers. obs.). In Sabah, birds call sporadically, with most calling heard in February-May (pers. obs.).

Distribution based on approximate extent of forest remaining within the altitudinal range of the species. Former range more widespread.

DISTRIBUTION Blue-headed Pitta is endemic to Borneo, where it is widely scattered in Sabah and Sarawak, Malaysia. Its distribution in Kalimantan, Indonesia, is poorly documented, and it is only known from one locality in Brunei, the Rampayoh River (Counsell 1986, Mann 1987).

GEOGRAPHICAL VARIATION None described.

HABITAT AND RANGE An inhabitant of forests in the lowlands and hills of Borneo. Blue-headed Pittas are found in primary and regenerating selectively logged forests (Lambert 1992), particularly near rivers. Unlike many other pitta species, the Blue-headed Pitta has not been recorded in other wooded habitats such as plantations (see e.g. Mitra and Sheldon 1993), or in areas of scrub, and Hose (1898) noted that this species does not visit clearings as do some sympatric species of pitta. During surveys in Gunung Mulu National Park, Wells *et al.* (1979) found Blue-headed Pitta only in floodplain alluvial forest: it was not present in freshwater peatswamp forest.

Whilst confined mainly to lowland areas, this species has been recorded to 610m on Mt Dulit and to 305m on Mt Mulu, Sarawak (Hose 1898, Sharpe 1894); and to 250m in Central Kalimantan (Wilkinson *et al.* 1991b). In Sabah, it is usually restricted to altitudes below 500m (Sheldon in prep.), but A. Johns (*in litt.*) reported finding it at 1,200m in the Maliau Basin. At higher elevations, such as these, it may prefer river terraces. In East Kalimantan, B. van Balen (*in litt.*) found this species in old secondary forest in Kutai National Park and, further inland, at c.300m between Long Punjungan and Long Ketaman. Although sympatric with the morphologically similar Banded Pitta, the latter species tends to be found in areas with steeper slopes and usually at higher altitudes: over a period of 18 months at Danum Valley, Sabah, where both species occur, they were never encountered in the same pieces of forest, strongly suggesting different habitat preferences (pers. obs.).

STATUS Blue-headed Pitta is a locally common species in Sabah, and evidence suggests that it is not an uncommon species in central Borneo, where Wilkinson *et al.* (1991a) recorded up to three calling males in a 1km^2 plot. Thompson (1966) described its status as abundant in primary forest at Kalabakan in eastern Sabah, though he recorded it only once in secondary habitats. Considering the extensive amount of forest left in Borneo, and its tolerance of selectively logged forests, Blue-headed Pitta cannot be considered threatened at the present time. Lambert (1990, 1992) recorded this species in forest that had been selectively logged nine years previously, with a density of at least two pairs per km^2 (pers. obs.). However, sightings suggested that this species is probably dependent on the areas within logging concessions that either are untouched by logging operations, such as those left along the margins of rivers, or else have received little damage. Such areas probably serve as centres for the colonisation of surrounding forest as regeneration proceeds (pers. obs.).

FOOD Food items that have been identified in specimens that were collected in Sarawak and Sabah include ants, caterpillars up to 5cm long, orthopterans (grasshoppers and crickets) and 'soft-bodied arthropods'. This species has been observed eating worms c.4cm in length in Sarawak.

HABITS Blue-headed Pittas usually occur in dispersed pairs, but are rarely seen close together. Males, in particular, often forage along broad leafy trails, when relatively easy to approach. Females seem to be more shy (pers. obs.). This species rarely flies, even when disturbed, but hops away rapidly. Calls are usually made from the ground, males often hopping to a new place before calling again. As well as turning over leaves in search of food, Blue-headed Pittas feed in a manner similar to Gurney's Pitta, making a 'nose-dive' or 'head-butting' action immediately after bounding forward. At Danum Valley, a foraging Blue-headed Pitta was frequently observed to stop feeding to wipe its bill clean on a mossy log (J. Howes *in litt.*).

In Sabah, Blue-headed Pittas have been observed roosting in undergrowth shrubs and on horizontal branches of saplings within 1m of the ground. One bird was found asleep at the tip of a narrow branch, whilst another was on a loop of a vine. When roosting their body is bent over the perch, with the tail pointing down and forward and the entire plumage fluffed out so that they appear as a round ball (L. Emmons *in litt.*).

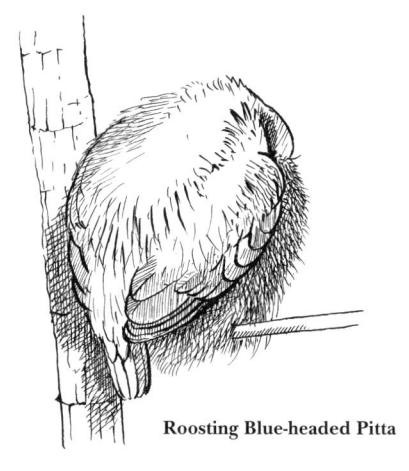

Roosting Blue-headed Pitta

BREEDING Breeding records are scant, but those that exist suggest that breeding occurs principally in the middle of the year. Most evidence of breeding relate to collections or observations of immatures: an immature male, with only a few blue feathers on the head, was collected at the base of Mt Poi, Sarawak, on 20 November; immatures were collected in east Borneo on 2 April and 22 September (USNM), and a juvenile was collected in north-east Borneo between July and mid-October (Stresemann 1938). Specimens with enlarged testes have been collected in Sabah on 5 March, 26 May and 2 June. At Danum Valley, Sabah, a Blue-headed Pitta was observed picking up a leaf on 6 May, although whether this was for nest material or part of foraging activity was not determined.

The only documented nest was found by Davison (1980), on a limestone knoll in Mulu National Park, Sarawak, on 6 May. The nest, containing two eggs, was a dome of matted dead leaves surrounded by herbs, and situated 50cm above ground level on top of a mound of soil that had been thrown up by the root system of a fallen tree. The eggs, measuring c.25 x 17.5mm and 25.5 x 18mm, were glossy white, marked with irregular speckles of purple-brown which were concentrated in a band around the widest point. The male was observed incubating the eggs on two consecutive days. A juvenile male Blue-headed Pitta was trapped in the same area on 3 May.

DESCRIPTION
Adult male Crown feathers, from forehead to nape, long and pointed, basally black with apical half pale iridescent azure-blue, often tinged with violet; sides of head, above and below eye, black, meeting to form a band across back of neck; eye encircled by narrow greenish-blue eye-ring. Mantle, back and rump maroon-red; uppertail-coverts and upper tail iridescent blue to violet-blue. Tail black below. Chin and throat white, breast black, belly to undertail-coverts and flanks glossy purple-blue to dark purple, this colour often extending onto sides of breast. Tertials rufous; lesser wing-coverts dull reddish, less bright than back and mantle and mixed with brownish tones from base of feathers showing through; rest of upperwing-coverts black with distinctive white line formed by broad white edges to the leading edge to distal half of median coverts; inner web of rearmost median coverts brown. Primaries blackish, with narrow band across inner webs at base of outer primaries, this being white on outer four primaries, but becoming grey on next two primaries; secondaries blackish with very narrow white fringe to distal third of outer web of outermost three feathers forming narrow white line in closed wing. Bill black; iris black, legs and feet slate-blue.
Adult female Head dull orange-rufous with dull maroon-red tinge on crown to nape and merging into colour of mantle. Mantle and back maroon-red, slightly duller than male; rump and uppertail-coverts and upper tail pale iridescent blue to violet blue. Throat whitish-buff, merging into orangy colour of ear-coverts and breast; underparts dull orange-buff to buffy-brown, darkest on breast and flanks. Wing as male except for more extensive white bar formed by white lower edge and tips to feathers of median coverts and by more extensive white, sometimes buff, edge to outer secondaries on leading edge. Whitish bar across base of primaries less prominent than that of male, usually confined to inner web of outer three feathers. Underwing-coverts brown with white bar across centre. Bill black; legs and feet pale horn; iris black.
Immature (male) Forehead dark buff; crown to nape darker, rufous-brown, with feathers edged blackish. Area below eye buff with narrow black edges to feathers, giving scaled appearance; ear-coverts black mixed with buff nearer eye; lores mixed black and buff, with narrow black line extending from eye to sides of forehead at base of bill. A few iridescent azure feathers may be present on the sides of crown, the number increasing with age. Mantle dark rufous, mixed with a few crimson-red feathers; back and rump similar but with more crimson-red; uppertail-coverts very dull purplish. Throat white, tinged buff at edges and on lower part. Breast black, flanks and centre of belly purple, with large irregular patches of orange-buff; centre of belly pale buff; undertail-coverts greyish. Wings as adult.
Juvenile Undescribed. However, one specimen in BMNH, although labelled as a female, is perhaps a juvenile male since it has female plumage except that one side of the head has black lores and scattered azure-blue feathers on the forehead and side of crown. An alternative explanation would be that this is an aberrant female showing some male plumage characteristics.

MEASUREMENTS Wing of male 87-96, of female 92-95; tail 33.2-46; tarsus 38.5-42.8; bill 25-29; weight of male 56.0-69.1, of female 65.2-76.0.

12 HOODED PITTA
Pitta sordida   Plates 7 & 16

Turdus sordidus P.L.S. Müller, 1776, Natursyst., Suppl. p.143: Philippines.

Alternative names: Green-breasted Pitta, Black-headed Pitta

TAXONOMY Hooded Pitta presently comprises twelve subspecies. The distribution of these subspecies is rather unusual, and could perhaps be considered as disjunct, since, oddly, there are no representatives on any of the Moluccan Islands. Populations to the east and west of these islands share a number of characteristics with each other but differ in these characteristics from populations in the other half of the global range (see under Geographical Variation). There may therefore be good reasons to consider these two populations as representing two species, although further taxonomic studies are needed to establish whether this might be justified.

FIELD IDENTIFICATION Length 16.5-19. Hooded Pitta is a relatively small pitta that has a widespread distribution, occurring from the Himalayas to New Guinea, although it is absent from most islands in Wallacea. Sexes similar. Head black or black with a chestnut-brown crown; body green, with small azure-blue wing-covert patch and red lower belly and undertail-coverts. Some subspecies have blue or violet suffusion on the underparts. Many subspecies have a large white wing-patch that is visible in flight. Subadults are much duller than adults, differing in having brownish tinge to the upperparts, brown chest and more or less pinkish on the belly and undertail-coverts. Depending on subspecies and age, immatures are dark olive-green to green above, with pale blue rumps, buffy belly and pink undertail-coverts. Breast colour may be dark at first, but rapidly becomes buff. Heads are blackish, with the exception of subspecies with chestnut-brown crowns, which obtain this coloration from an early age. In some subspecies (e.g. *sordida*, *cucullata*) immatures and juveniles have a prominent white bar across the wing-coverts, but in others (e.g. *novaeguineae*) this is absent.

Similar species The black head and green body distinguish this species from adults of all sympatric species. In the Indian subcontinent, immature Hooded Pitta can readily be distinguished from immature Indian Pitta by its chestnut crown and white band across the upperwing-coverts. In areas where both Hooded Pitta and Blue-winged or Mangrove Pitta breed, juvenile Hooded can best be distinguished from these by its white wing-patch and by lack of any trace of blue in the wing. Immature Red-bellied Pittas in the Philippines are much paler, particularly on the head, than resident immature Hooded Pittas, and can also be distinguished from the latter by the absence of white in the upperwing-coverts, mottled (rather than white) throat, and scaled breast. On New Guinea, where both Red-bellied and Hooded Pittas also breed, juveniles and immature Hooded Pittas lack white in the upperwing-coverts, but are much darker than Red-bellied Pitta and lack the mottled underparts characteristic of immatures of the latter species.

VOICE The main call is a fluty double whistle, often given in couplets or produced continuously (in New Guinea at intervals of 3s). Calls of some subspecies, such as *cucullata* and *muelleri*, sound similar, and are best described as *fih-fih* or *whew-whew*, each note being monosyllabic. This call is probably given by both sexes. A three-note variant, often heard in Sundaic forests, has the last note slightly more abrupt and shorter, *fih-fih-fi*. In Palawan, the call is notably different, with each note having a more fluty and disyllabic quality, *hwee-hwee*, somewhat reminiscent of the call of Blue-winged Pitta. Similarly, birds on Biak (*rosenbergii*) sound quite different to those in Malaysia, having a much coarser sounding call (pers. obs.). The call of Hooded Pitta in New Guinea (*novaeguineae*) is described as *wu'wI-wu'wI* (Coates 1990). The call of *abbotti* (Nicobar Islands) is described as a three note (rarely four note), quick, sharp whistle that neither ascends nor descends in tone, and is similar to the yelp of a small pup (R. Sankaran *in litt.*). In the Nicobar Islands this call, which may go on for minutes at a time, is mimicked by Greater Racket-tailed Drongo *Dicrurus paradiseus*.

Sonograms of Hooded Pittas from Peninsular Thailand and of sympatric Blue-winged Pitta show that the double whistle of the latter has two frequency peaks, unlike that of Hooded Pitta. Differences in the calls of these two species are documented in detail by Lansdown *et al.* (1991).

Hooded Pittas in South-East Asia give a brief, harsh alarm note, *skyew*, similar to that given by sympatric Blue-winged Pitta, as well as to that of Gurney's Pitta, although it is shorter and squeakier than that of the latter (P. Round *in litt.*). In New Guinea, the alarm note, *kiaw*, is sometimes followed by one or two short low rasping notes and is similar to the call of sympatric Rufous Babbler *Pomatostomus isidorei* (Coates 1990). In Borneo one call has been described as being rather like a woodpecker. Captive birds performing stretching and crouching at the onset of the breeding season make a vocalisation that sounds like the growling of a small, hoarse dog, during the display (Lee *et al.* 1989).

In parts of the range, calling is seasonal. In southern New Guinea, for example, calling is frequent at the beginning of the rainy season but rare in the dry season. Calling of Hooded Pitta is also frequently heard at night.

Wing patterns of different individuals of Hooded Pitta

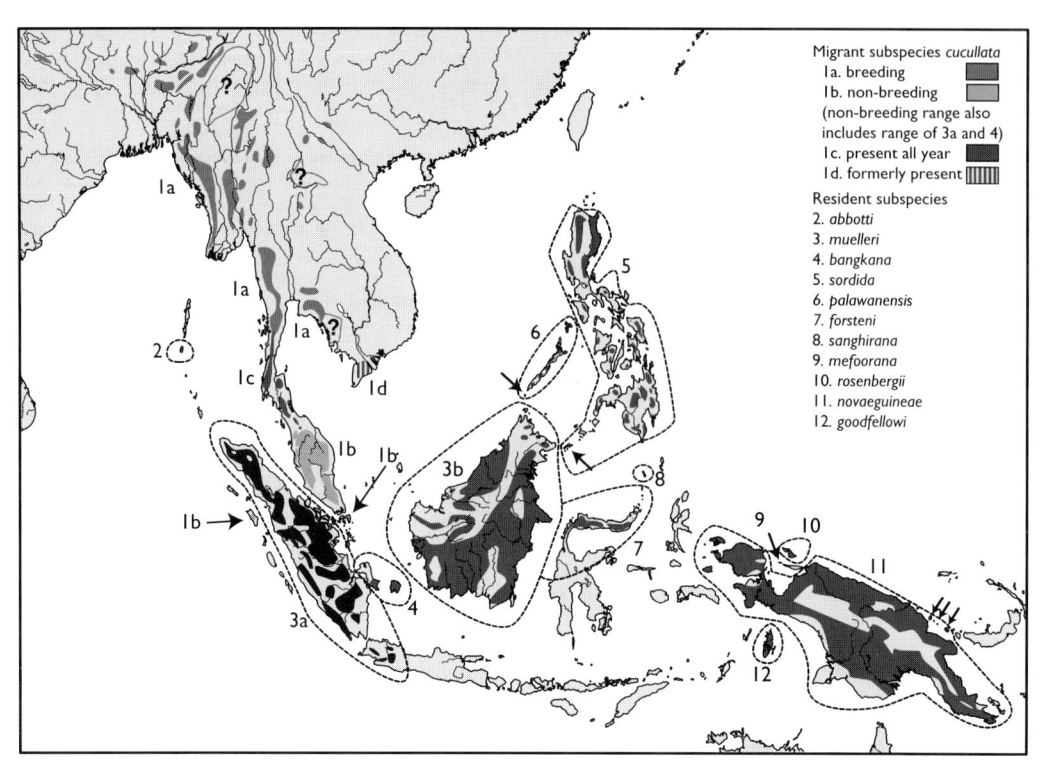

Map shows breeding and non-breeding range of migratory *cucullata* and the distribution of other resident subspecies based on extent of suitable habitat. Former ranges of all subspecies more widespread. Non-breeding *cucullata* occurs within range of *muelleri* and *bangkana*. Distribution of breeding *cucullata* in northern Burma and parts of Indochina needs confirmation.

DISTRIBUTION This is one of the most widely distributed pittas, with twelve subspecies, breeding from the foothills of the Himalayas through South-East Asia to New Guinea, including various islands in the Indonesian and Philippine archipelagos. Hooded Pitta is, however, absent from the majority of the Indonesian islands of Wallacea and, the majority of the eastern satellite islands of New Guinea and the Bismarck Archipelago, in contrast to the similar-sized Red-bellied Pitta. Some northern populations are migratory.

The most northerly distributed population breeds in the southern foothills and adjacent plains of the Himalayas. There are few records west of central Nepal, the most westerly being from Simla in Himachal Pradesh (Jones 1943), but since this refers to a bird found dead at 1,830m on 23 June the possibility that this was an overshooting migrant cannot be ruled out. If it was, the vicinity of Dehra Dun, in eastern Uttar Pradesh, would then be the most western part of the species's known breeding range (Bolster 1921, Betham 1922, Mohan and Chellam 1990).

From central Nepal, the known range to the east is also somewhat disjunct, with records from Sikkim and Bhutan and the north-eastern states of India, including Arunachal Pradesh, Assam, Meghalaya, Nagaland, Manipur, Mizoram and Tripura; in the hill tracts of Bangladesh (Ali and Ripley 1983); Burma, including Arakan, the Chin Hills, central Burma (north to the Bhamo District on the Irrawaddy River), Pegu Hills, Karen Hills, Shan States and Tenasserim (Smythies 1986); Yunnan Province in China, where confined to the south-eastern part of Mengzi (Mengtsz) (Yang Lan 1983, Cheng 1987); Thailand, where present in the hills of the north-west and west, south through the peninsula to at least Phatthalung (c.7°35'N) and in the northern Malaysian states of Perlis and Kedah (D. Wells *in litt.*). In South-East Thailand, it occurs in the hills and mountain ranges that extend into bordering Cambodia as the Cardamom Mountains (Legakul and Round 1991). The distribution of Hooded Pitta in Indochina is poorly known. In Laos, it is only known from the northern province of Phongsali, where collected at Bountai (Bun Tai) (Bangs and van Tyne 1931); in Cambodia from Kampot in the Elephant Mountains (Engelbach 1936, Thomas 1964); and in Vietnam from Cochinchina and Lai Chau Province, Tonkin (Delacour and Jabouille 1929, 1931; Rozendaal 1991). However, the species must be more widely distributed than these records suggest, and it seems inevitable that Hooded Pitta occurs throughout the forests of southwest Cambodia (since populations are known to occur over the border in Thailand) and probably in parts of Laos along the upper Mekong River. Hooded Pittas from the northern part of their range migrate southwards to winter in the Sunda Region.

In the Greater Sunda Islands, populations breed throughout Sumatra (van Marle and Voous 1988), in West Java (Kloss 1931a) and on Borneo. Populations on Sumatra and Java are supplemented by winter visitors from the north-west. In the Philippines, Hooded Pitta is widely distributed, being known from the islands of Balabac, Bantayan, Basilan, Bohol, Bongao, Boracay, Caluya, Calauit, Cebu, Culion, Dinagat, Guimaras, Jolo, Leyte, Luzon, Marinduque, Mindanao, Mindoro, Negros, Palawan, Romblon, Samar, Semirara, Siargao, Sibuyan, Siquijor, Tablas, Tawitawi, Ticao (Dickinson *et al.* 1991) and Panay (G. Dutson *in litt.*). Tablas, Catanduanes and Polillo are the only major Philippine islands where

Hooded Pitta has not been found, and it is probable that the species occurs on the first two of these, if not all.

In Indonesia, endemic subspecies occur on the islands of Bangka and Belitung off south-east Sumatra; Sangihe, to the north of Sulawesi (White and Bruce 1986 note that it may also occur on Talaud); northern Sulawesi (where known only from four localities in the Minahassa peninsula: Tondano, Temboan, Kumarsot and the Dumoge-Bone National Park); the Aru Islands (including Wokan, Dobo, Kobroor and Trangan); Biak Island; and Numfor Island, both to the north of Irian Jaya (Indonesian New Guinea). On New Guinea, this is a widespread species on the mainland, and also occurs on the West Papuan Islands (Misool, Batanta, Salawati, Kofiau, Gag, Waigeo and Gebe), several islands to the north of the Huon Peninsula (Long, Crown and Tolokiwa) and on Karkar Island, off the northern coast of Papua New Guinea.

GEOGRAPHICAL VARIATION Most subspecies of Hooded Pitta have restricted distributions, but four have large ranges: *cucullata, muelleri, sordida* and *novaeguineae*. The other subspecies are either restricted to a few islands (*palawanensis, bangkana, abbotti*) or are single-island endemics (*sanghirana, forsteni, rosenbergii, mefoorana*). Examination of specimens suggests that the population on Karkar Island, formerly treated as a separate subspecies, *hebetior*, should be included in *novaeguineae*, as proposed by Diamond and LeCroy (1979) and followed by Coates (1990). Populations on the Aru Islands may also belong with *novaeguineae* (Rothschild and Hartert 1901), but differ slightly and are kept separate in the following account.

As mentioned under Taxonomy, the various subspecies can be divided into two different groups: those to the west of the Moluccas (mainland Asia to the Philippines, Borneo and Sulawesi) and those to the east (in the Aru Islands and New Guinea region). Populations to the east are quite different in appearance from those to the west, all subspecies being a much darker green than western populations, having a strong dark blue to violet wash on the belly and/or flanks, and having less pink, more orange tones in the colour of the lower belly and undertail-coverts. Furthermore, eastern populations have very little or no white in the wing, in contrast to all populations to the west, which have large white wing-patches. Finally, juveniles may also differ: in at least one eastern subspecies (*novaeguineae*), the white in the wing characteristic of western birds is missing.

For purposes of identification, the various subspecies can also be divided into those with rufous on the crown, and those with black heads, including the crown. In the following description of subspecies, this distinction is made.

Black-crowned subspecies

P. s. sordida (The Philippine archipelago, except in Palawan Province. Recorded from all major islands except Tablas, Catanduanes and Polillo) The upperparts are green, tinged blue, with a variable amount of azure-blue on the upperwing-coverts and uppertail-coverts. The tail is black, narrowly tipped with blue. The underparts are pale greenish with a blue suffusion, the upper belly black and the lower belly and undertail-coverts bright red. The underparts of *sordida* are quite variable, with some birds having a strong blue wash (such an individual, collected on Marinduque, Philippines, was so striking that it was originally described as a separate subspecies, *rothschildi*).

P. s. palawanensis (Philippine islands of Palawan, Culion, Balabac and Calauit, and probably Busuanga) This subspecies is hardly separable from *sordida* and may best be considered synonymous. Although Parkes (1960) reported that the coloration of the wing-covert patch and rump of *palawanensis* is a darker cobalt than that of *sordida*, that of specimens in BMNH from Palawan and Balabac are only marginally different. Parkes (1960) also reported that the bill of *palawanensis* is longer: although this is true of some individuals, there is overlap in length (26-30mm). Wing 99-110; weight 53-63g.

P. s. muelleri (Greater Sunda Islands – Borneo, Sumatra and West Java – Labuan and Sibutu Island, between Borneo and Palawan. May also be present on other small islands in the region. The presence of this subspecies in southern peninsular Thailand and the northern part of Peninsular Malaysia, south to Kedah, needs confirmation) Very similar to *sordida*, but has much less extensive black on the belly. Specimens from Labuan, off north Borneo, are mostly washed with blue on the underparts, whilst several also have blue fringes to feathers on the mantle and back. Wing 102-116; weight 60.

P. s. forsteni (N Sulawesi, where known only from the Minahassa Peninsula, westwards to Mt Tentolo-Matinan at 121°48'E, 0°57'N) This subspecies is large (wing 113-121), lacks the white bar across the primaries and has a green tail.

P. s. sanghirana (Sangihe Island, Indonesia) Darker green above than *forsteni*, with a darker blue rump, more bluish-tinged on the breast and sides and an extensive white bar across the primaries. Wing 106-110; bill 25.2-28.8; weight 65.5-80.0.

P. s. novaeguineae (New Guinea, including the West Papuan Islands, islands to the north of the Huon Peninsula and Karkar Island) The distinctive features are the bluish-green flanks; dark indigo-blue surrounding the small black belly-patch, and usually extending onto lower breast; the glittering pale green or greenish-yellow patch across the upper breast (just below the black throat); and absence of visible white in the wing at rest (white in the wing confined to a few small spots in central primaries). The azure rump-patch is much reduced in size or completely absent. Tail green; uppertail-coverts brighter green than rump. This subspecies, and the following three, from the New Guinea region, are darker below than are those to the west. Wing 102-109; weight 74-77. Immatures are very dark with paler undertail-coverts and blackish head, but lack the white wing-bar.

P. s. goodfellowi (Aru Islands, Indonesia) Hardly distinguishable from *novaeguineae*, but slightly smaller (wing 96-105) and darker, with generally a less extensive area of glittering pale green on the upper breast. Doubtfully a valid taxon.

P. s. mefoorana (Numfor Island, in Geelvink Bay off the northern coast of New Guinea, Indonesia) Intermediate between *novaeguineae* and *rosenbergii*. Differs from the former in having azure-blue lower rump and uppertail-coverts, mostly black tail (with less than 1mm of green at tip), deeper blue on belly, narrow silvery-blue glossy line separating the black throat from green breast, and in lacking any white in the wing.

P. s. rosenbergii (Biak Island, Geelvink Bay, Indonesia) Like *novaeguineae*, but lacks the glittering breast-band, has more extensive, deep violet-blue flanks and lacks black on the lower breast. Centre of belly and undertail-coverts scarlet. Differs from *mefoorana* in being devoid of any metallic gloss on breast, while the scarlet of the belly extends to breast, which has no black patch. Uppertail-coverts and lower rump azure-blue; tail black. Males have strong blue wash to breast. No white in wing. Wing 99-107.

Chestnut-crowned subspecies

P. s. cucullata (Breeds in the Himalayan foothills from Uttar Pradesh – but perhaps also Himachal Pradesh – to Arunachal Pradesh and NE India south and east to Burma, Yunnan, Thailand, including the peninsula, and Indochina. The subspecies resident in northern Peninsular Malaysia south to Kedah is either *cucullata* or *muelleri*. Some parts of the population move south to Peninsular Thailand, Malaysia, Sumatra and West Java in the northern winter. Migrant or wintering Hooded Pittas of this subspecies have regularly been recorded on islands in the Malacca Straits, and occasionally on other smaller offshore islands, such as Nias off west Sumatra. A vagrant has been recorded in the Yaeyama Islands, Japan [Brazil 1991]) This subspecies is large and long-winged (wing 108-119, weight 57-71), and a notably paler, more washed-out green than the black-crowned subspecies. Similar to nominate but crown to nape is rich dark chestnut-brown, with the rest of the head, including a collar below the nape, black. *Cucullata* also has more extensive red to pinky-red on the belly than *sordida*, with the black confined to the upper belly. The extent of white in the wing varies considerably, with some individuals (probably subadults, but perhaps individuals from a different population) having only half the white of others as well as having extensive blue on the tips and sides of the tail and a red gape.

A male collected in Narathiwat Province, Peninsular Thailand on 18 July 1926 with an all-black head was identified as belonging to the subspecies *muelleri* (Riley 1938), and it is therefore possible that *muelleri* replaces *cucullata* in this region (at a latitude of around 6°30'N). A female collected in Perlis in 1911 had broad black bases to its crown feathers and may represent a *muelleri-cucullata* intergrade (Medway and Wells 1976). However, identification of the subspecies breeding in the north of Peninsular Malaysia (Perlis and Kedah) has not been confirmed (Medway and Wells 1976).

P. s. bangkana (Bangka and Belitung Islands, off SE Sumatra) Intermediate between *muelleri* and *cucullata* (Chasen 1937, Mees 1986), with a variable amount of dark chocolate-brown on the crown. Wing of male 105-109, female 100.

P. s. abbotti (Great and Little Nicobar Islands. Calls that were either this species or mimicry by a drongo have been heard on the small uninhabited island of Pilo Milo off Little Nicobar – R. Sankaran *in litt.*) This subspecies resembles *cucullata*, but is slightly shorter-winged (wing 105-109). The plumage is darker, both above and below, and the upperparts suffused with olive rather than the clear green of *cucullata*, with a paler green collar across the upper mantle, bordering the black of nape. The underparts are usually distinctively washed with blue from breast to belly. The uppertail-coverts and upperwing-coverts are a lighter, more azure-blue than those of *cucullata*, and these patches are reduced in size (the rump-patch of *cucullata* being twice as large). There is a dark median line on the crown. The white primary patch is narrower than that of *cucullata*, and is confined to six feathers.

HABITAT Hooded Pitta occurs in a wide range of forested and wooded habitats, sometimes including overgrown plantations, orchards and scrub. In some parts of its range it also occupies a relatively wide altitudinal zone, although in other parts it is apparently largely confined to the lowlands and hills. On migration, Hooded Pittas have been occasionally observed in unusually open habitats, and may be encountered at unusual altitudes.

Ali and Ripley (1983) give an upper altitudinal limit of 2,000m for *cucullata* in the Indian subcontinent, but since this is a migratory subspecies, many records from higher altitudes probably relate to birds on migration. Baker (1926), however, found it breeding as high as 1,525m. In Nepal, where very locally distributed, *cucullata* has been recorded only below 305m (Inskipp and Inskipp 1991), whilst in Sikkim Stevens (1925) found this species to be largely confined to the *terai* lowlands and foothills, although breeding birds have been found at 365m. In Burma, it is also primarily a bird of the foothills (Smythies 1986), whilst in Thailand, breeding birds occur from the plains to 700m (Round 1988).

In the Indian subcontinent, *cucullata* occupies a wide range of habitats. Baker (1913, 1926, 1934) reported that it frequents forest of all kinds, the deepest and wettest as well as the driest and thinnest, while it may also be found within bamboo jungle, scrub jungle, patches of forest within grasslands and, occasionally, in crops such as mustard and rice, or other cultivation outside of forest. Law (1939) caught one of a pair in an overgrown orchard surrounded by bamboo and scrub in West Bengal.

In Thailand, Hooded Pittas are found in mixed deciduous, semi-evergreen and rainforests, principally in the lowlands, as well as in moist secondary growth and scrub and overgrown rubber plantations (Round 1988). Some populations in the peninsula occur in pure bamboo jungle (D. Wells *in litt.*). Wildash (1968) noted that in southern Vietnam *cucullata* frequents isolated forest pockets on grassy plains. In Yunnan, Hooded Pittas have been observed in rather dry forest with abundant bamboo (D. Scott *in litt.*).

Non-breeding visitors to Peninsular Malaysia and Sumatra winter in lowland forests below the steepland boundary, including areas that are occupied by resident conspecifics. In Peninsular Malaysia, although the subspecies *cucullata* has been netted in the hills to elevations of 800m during the migratory period, individuals proven to be wintering by repeated retraps have only been found in the forests of the flat lowlands (Medway and Wells 1976). In West Java, wintering *cucullata* has been collected to at least 460m.

In Sumatra, *muelleri* is an inhabitant of primary and dense secondary forests in lowlands and hills, where occurring up to 500m (van Marle and Voous 1988). It is not clear if it inhabits the extensive areas of freshwater and brackish swamp forests that occur in the eastern lowlands, although its presence in peatswamp forest in Borneo suggests that it probably does. In Borneo, *muelleri* occurs from sea level to at least 760m (SarM), although in Sabah Hooded Pittas have not been recorded above 400m

(Sheldon in prep.). Hooded Pittas apparently favour mainly secondary growth in Borneo, but have been recorded in a diversity of habitats including primary and logged forests, peatswamp forest, mangrove, nipah (*Nypa fructicans*) thickets, bamboo, scrub and old gardens, and is locally common in plantations of *Albizia*, *Gmelina*, overgrown rubber and coffee (Brüggemann 1878, Batchelor 1959, Pfeffer 1960, Mitra and Sheldon 1993). Old collections include *muelleri* from small offshore islands such as Pulau Tiga and Labuan (Sharpe 1889), although these may not all refer to resident birds.

There is little information on the subspecies *abbotti* which is endemic to the Nicobar Islands. On Great and Little Nicobar, it has been found close to the sea in flat primary forests with sparse undergrowth, and observations by R. Sankaran (*in litt.*) suggest that *abbotti* prefers primary forest, although R. Dekker (*in litt.*) observed two in disturbed habitat at sea level on Great Nicobar.

Rand and Rabor (1960) found the nominate subspecies mostly in secondary growth in the Philippines, and other observers confirm this. On Siquijor, for example, Hooded Pitta survives in very degraded habitats, such as in 3m wide 'hedges' between fields in recently cleared forests on limestone (G. Dutson *in litt.*), whilst in the Sierra Madre Mountains of northern Luzon, Danielsen et al. (1994) found it in degraded forests on limestone. However, nominate *sordida* evidently also occurs in good forest: on Dinagat and Siargao, it was reported to occur in dense patches of mixed remnant original dipterocarp and secondary forests (duPont and Rabor 1973a). On Mindoro, the species has been observed in bamboo and dense secondary forest at the edge of relatively undisturbed lowland forest (pers. obs.).

In Sulawesi, subspecies *forsteni* occurs at 250-660m in the Minahassa peninsula (Rozendaal and Dekker 1989), and has been found at 600m in the Tentolo-Matinan Mountains (Stresemann 1940). Stresemann (1940) found *forsteni* in dense habitat at the forest edge, whilst Rozendaal and Dekker (1989) found it along ridge-tops in primary forest where there had been treefalls. On Sangihe, *sanghirana* occurs in a variety of habitats including secondary growth, mosaics of coconut, nutmeg, breadfruit and other plantations, particularly where there is an understorey of cocoa or other shrubs, in sago thickets, and in tree-fern dominated woodland on Mount Awu, up to 300m (Bishop and Coates in prep.), whilst on Mt Sahendaruman it is found in primary forest to at least 750m (RMNH). The subspecies on New Guinea and its satellites, such as *novaeguineae*, *rosenbergii* and *mefoorana* inhabit a wide variety of habitats including primary rainforest, monsoon forests, gallery forests in savanna zones, secondary forests and scrub. Although it is tolerant of habitat disturbance, Bell (1982c) only trapped Hooded Pitta in primary forests during a study in which the avifauna of primary forests was compared with that of secondary rainforest in Central Province, Papua New Guinea.

On mainland New Guinea, *novaeguineae* is apparently very local in its distribution (Diamond 1972), but it is often common in the areas in which it occurs. In the eastern highlands, for example, it has only been recorded from the Jimmi Valley of the northern watershed (Diamond 1972). Rothschild and Hartert (1901) found Hooded Pitta to be primarily a bird of the plains in New Guinea, whilst Diamond and LeCroy (1979) reported that *novaeguineae* was confined to the lowlands in New Guinea. However, Coates (1990) noted that it occurs locally to 1,200m, and in south-east New Guinea Beehler (1981) gives the altitudinal distribution as sea level to at least 1,000m. On Karkar Island, Hooded Pitta has been found only at altitudes of 700-1,200m. On Batanta Island, B. Gee (*in litt.*) found *novaeguineae* in damp, low-lying rainforest near sea level, where he observed a bird calling from dense lush vegetation in a tree-fall clearing. In the Trans-Fly region of southern New Guinea, Hooded Pitta occurs in dense scrubby wooded areas close to water and gallery forest, but is probably replaced in the more open savanna areas by Noisy Pitta.

STATUS In parts of its range, Hooded Pitta has always been considered a relatively common species, but in other areas it may be rare or localised. Thus, in the Philippines, whilst apparently very common on Siquijor, even in degraded habitats, and common in degraded forest on limestone in parts of northern Luzon (Danielsen et al. 1994) it is perhaps rare on other islands, apparently even where favourable habitat occurs: duPont and Rabor (1973a) found it rare on both Dinagat and Siargao, and Parkes (1973) could only trace a single record from the large island of Leyte (Rabor 1938).

Baker (1895) described the Hooded Pitta as uncommon and localised in the Cachar Hills of Assam, but, in contrast, Law (1939) reported that it was common in the hills of northern West Bengal, whilst Cripps declared it very common in the rainy season in the Dibrugarh district of north-east Assam (Hume 1888). Harvey (1990) reported that this is the commonest pitta in Bangladesh. Oates (1883) found *cucullata* to be plentiful in the Pegu Hills of Burma. In the Indian subcontinent, and in Burma, there are still significant areas of habitat where Hooded Pittas are presumably not uncommon, although much of the forest that must have once been occupied in the lowlands of the southern Himalayas and north-east India have long since been cleared. Populations in India are presumably mostly migratory, and as such may also be prone to threats both on migration and in their wintering grounds, which are unknown.

The status of Hooded Pitta in Indochina is unknown, but in parts of Vietnam its habitat has suffered serious degradation or loss. In Cochinchina, Vietnam, for example, where Hooded Pitta was already considered to be rather rare in the first part of the twentieth century (Delacour and Jabouille 1929), little suitable habitat now remains. Other parts of its known Indochinese range still have significant areas of suitable habitat, but documentation of its status, and in many cases even of its presence, remains very poor.

Brüggemann (1878) stated that Hooded Pitta was very common in central Borneo last century, although in eastern Borneo Sclater (1863) reported that it was rather rare. More recently, Batchelor (1959) found Hooded Pitta abundant in overgrown old rubber, scrub jungle and nipah thickets along the Kimanis River, Sarawak, and it is presumably still a relatively common species in Borneo where suitable habitat remains. Sharpe (1889) reported that Hooded Pitta was fairly common on the north Bornean island of Labuan in January, although the present status of this population, which differs slightly from mainland birds (see Geographical Variation), is unknown.

The status of *muelleri* on the other Greater Sunda Islands remains unclear, although suitable habitat is still widespread on Sumatra. Populations in West Java must, however, now be rather fragmented and rare due to high levels of habitat loss. R. Sankaran (*in litt.*) regularly heard this species on Great and Little Nicobar Islands at a number of localities, suggesting that *abbotti* is widespread and relatively common.

Hooded Pitta is reported to be common on Sangihe where it is found in a variety of habitats, but on nearby Sulawesi the lack of records suggests that it is a very localised or rare species, perhaps with more specialised habitat requirements: Wallace (1860) and Meyer (1879) both reported that *forsteni*, of north Sulawesi, was a scarce species last century, whilst Rozendaal and Dekker (1989) describe it as uncommon. Further east, *rosenbergii* is fairly common in suitable habitat on Biak (pers. obs.). Based on the amount of forest remaining in New Guinea and its satellite islands, *novaeguineae* must still have a considerable population. In lowland rainforest near the Brown River, southern Papua New Guinea, the population density is apparently very high, having been estimated to be six birds per 10ha (Bell 1982a), equivalent to 60 individuals per km^2 of suitable habitat.

Gretton (1988) estimated that the population density of resident *cucullata* in patches of secondary forest in Peninsular Thailand was around 10 pairs per km^2. Netting studies at Pasoh, Peninsular Malaysia, suggested a wintering density of two Hooded Pittas (*cucullata*) in an area of around 15ha (Wells 1990), or 13 to 14 individuals per km^2. This suggests that territory size in the wintering range is not dissimilar to that in the breeding range for this subspecies.

MOVEMENTS Several subspecies of Hooded Pitta make migratory or dispersive movements. Northern populations of subspecies *cucullata* are, however, the only long-distance migrants, travelling from the northern parts of their range in South-East Asia to Peninsular Malaysia and Sumatra. The population in northern Peninsular Malaysia is resident, but those in Thailand, southern Burma and Tenasserim are migratory (Smythies 1986, Lekagul and Round 1991). Hooded Pittas are only present in Peninsular Thailand when breeding or during migration (P. Round *in litt.*).

Ali and Ripley (1983) gave the status of *cucullata* in the Indian subcontinent as 'chiefly resident, but with some seasonal altitudinal and dispersal movements'. There can be little doubt that many Hooded Pittas from India undertake migratory movements, since large numbers are reported to be caught for food at lights in the Barail Range (pers. obs., R. Grimmett *in litt.*). In Nepal, Inskipp and Inskipp (1991) reported that Hooded Pitta has only been recorded in the summer, between April and October. It is possible that some individuals are resident but overlooked when not calling, although some of the areas in which they breed, such as the Royal Chitwan National Park, have been well watched for many years, and it is fairly certain that most, if not all, move elsewhere (T. Inskipp verbally). It is possible that birds from foothills of the Himalayas migrate in an east-west direction, passing through the north-east Indian states and spending the non-breeding season somewhere in South-East Asia.

Studies of night-flying migrants at Fraser's Hill, Peninsular Malaysia, have shown that Hooded Pittas pass through the Peninsula from 15 October to 22 December in the autumn, and 7 April to 8 May. The main period of southern migration of the species starts in early to mid-November and ends in the latter half of December (Medway and Wells 1976). Between 1965 and 1973, 1266 Hooded Pittas were trapped at Fraser's Hill during these periods, compared to only 564 Blue-winged Pittas during the same period (Wells 1992). Netting at night at Fraser's Hill has also demonstrated that this species makes midwinter movements, since they have been caught in late January and early February (Wells 1983). Twenty Hooded Pittas attracted to lights at Fraser's Hill during peak passage in 1973 were examined for flight-muscle bulk and all found to have little or no subcutaneous fat, indicating that these low-flying migrants were perhaps near exhaustion (Wells 1975). Ringing has shown that wintering birds may return to the same territory in different years (Medway and Wells 1976). In Thailand, evidence of northward passage is provided by observations of Verbelen (1993), who witnessed a large fall of mainly exhausted Hooded Pittas in Kaeng Krachen National Park, in the northern part of Peninsular Thailand, on 13-14 April. The earliest spring record for Thailand is 30 March at Ban Bang Tieo (P. Round *in litt.*).

Birds from Sumatra also evidently make some relatively long-distance movements, since a specimen of the subspecies *muelleri* was collected on the small island of Anak Krakatau, some 50km from the Sumatran mainland, on 22 June 1955, and there is another specimen from Bangka, c.20km from the Sumatran coast (Mees 1986). Everett collected numbers of *muelleri* on Pulau Tiga, Sabah, in April 1886, but Whitehead did not find it there and these birds may therefore have been individuals from the mainland. These records suggest that *muelleri* might undertake regular long-distance movements in Borneo. Thompson (1966) speculated that this was a migratory subspecies based on the fact that he only observed it in an area of cocoa plantation during May and July, whilst Brüggemann (1878) did not find any in central Borneo between October and April even though it was very common in other months. Six Hooded Pittas were observed among the stream of night-flying migrants crossing a ridge in Sabah in October 1974 (Smythies 1981). Although the subspecies was not determined, they seem most likely to have been *muelleri*, but the possibility that northern *cucullata* sometimes reaches Borneo should not be ruled out.

In the Philippines, *sordida* also apparently makes migratory movements, although these are poorly understood. During a five-year study in the mountains of northern Luzon, 1,265 Hooded Pittas were trapped and ringed at night at Dalton Pass. Of these, one adult ringed in late October 1968 was recovered some 38 months later, in mid-December 1971, 20 miles to the north, at Bambang, in Nueva Vizcaya Province (McClure and Leelavit 1972, McClure 1974). Evidence for local migration is also provided by the sighting of two birds on the fairway of a golf course (not far from gallery forest) in Manila, Philippines (H. Buck *in litt.*). Movements of *sordida* appear to be over rather short distances, in contrast to those of *cucullata*, which is a true long-distance migrant.

In New Guinea, observations suggest that Hooded Pittas are nomadic in areas that are seasonally dry, since moist ground is apparently required for foraging. A minor irruption into the Port Moresby city area was recorded at the end of January 1973, during a period of unseasonal drought (Bell 1982b, Coates 1990). Bell (1982b) found that Hooded Pittas were scarcer at his Brown River study site, southern Papua New Guinea, during the harsh dry season of August to November 1976 than during the more typical dry season of 1977.

FOOD Migratory *cucullata* have been found to contain black ants and various other insects, the eggs of insects and small snail shells, whilst Baker (1929) observed them eating termites. One photograph depicts *cucullata* carrying four small prey items that appear to be millipedes, presumably to feed to young. Specimens of *sordida* have contained insects including beetles and their larvae, and centipedes. On Java, the diet of Hooded Pitta is reported to include a diversity of insects including grasshoppers (Acrididae) and cockroaches (Blattidae), various beetles, including glow worms (Lampyridae) and their larvae, hemipteran and homopteran bugs, Coreidae, hymenoptera, ants, caterpillars and the larvae of flies (Diptera) as well as snails. Specimens from Sarawak (SarM) and Sabah (Sheldon in prep.) have contained ants, beetles, beetle larvae, snails, worms and a leaf, whilst a bird kept in captivity for a month caught soft caterpillars, cockroaches and small grasshoppers, although it spent a large proportion of its time digging up earthworms (Harrison 1964). In New Guinea, earthworms, crickets, cockroaches, grubs and snails have been recorded in the diet of Hooded Pitta. One bird was also observed probably eating a tadpole taken from a drying pool. Nestlings in New Guinea have been observed to be fed primarily with earthworms, as well as grubs and crickets (Coates 1990).

HABITS Hooded Pitta is a highly territorial species, (*cucullata* on both its breeding and wintering grounds), and is most frequently found singly. During the breeding season, pairs may associate more closely, whilst there is some evidence to suggest that migrating *cucullata* may form loose flocks (see Introduction). If disturbed, they usually hop away rapidly rather than flying up into the vegetation in the manner of many other pittas.

Hooded Pittas call most frequently from the ground, although birds have been observed calling from perches up to 7m off the ground (Hachisuka 1934-35). On Mindoro, a bird was observed calling spontaneously and continuously in mid-morning from a perch in the canopy of a tree at the forest edge, some 6m off the ground, and later from perches in bamboo 3-4m off the ground (pers. obs.). At night, Hooded Pittas roost above the ground. In Borneo and Peninsular Malaysia roosting birds have been found on horizontal portions of liane loops up to 4m above the ground (pers. obs.). Hooded Pittas in seasonal forests of New Guinea readily visit pools to drink during dry periods.

Baker (1926) observed Hooded Pittas in India feeding on flying termites by leaping into the air after them or pursuing them with remorseless energy: those that crept under leaves or moss were promptly exposed by the bird kicking away the vegetation before seizing the prey in the bill.

The behaviour at the nest has been studied in New Guinea by Coates (1990). He found that adults feeding nestlings flew directly to the nest rather than approaching on the ground, and would also leave the nest by flying. Adults flush off nests containing eggs relatively easily, but sit tight once the eggs have hatched and brooding of the more vulnerable young becomes necessary. Baker (1926) noted that, like other incubating pittas, adults seldom leave the nest when an observer approaches until almost trodden upon.

Several displays of Hooded Pitta were described by Lansdown *et al.* (1991). During 'bowing', made in response to territorial calling of other birds, an adult would draw its body up rapidly and then, several times in succession, rapidly lower the front part of the body to a horizontal position whilst simultaneously hunching its back. Bows were performed rapidly, lasting only a second or so. Bowing was noted to be associated with alarm calls, as in the case of Rainbow Pitta (Zimmermann in press). On several occasions, Lansdown *et al.* (1991) witnessed presumed pairs of Hooded Pittas that had been feeding 10-15m apart respond to imitation of the double whistle call. Both birds would simultaneously fly to near the observer, where they both reacted to repeated imitation of the call by bowing and producing both the double whistle and the alarm note *skyew*. Another behaviour that was noted was a reaction to the alarm calls of other birds, or to disturbance whilst feeding, in which the bird would stand with the body at 45° whilst bobbing its head up and down. This 'bobbing' behaviour is described as being reminiscent of a raptor gauging distance or position of an object. A third behaviour described by Lansdown *et al.* (1991) was 'wing-flicking', in which the white patches on the primaries are exposed by momentary extension of the wings. Wing-flicking was observed on many contacts with the species, particularly by adults accompanied by young, and is therefore likely to be some kind of alarm reaction or distraction display. Sometimes wing-flicking is combined with bowing and alarm calling. The final display that was witnessed was also in response to alarm by an observer: on one occasion a bird flushed to a perch where it puffed out its crown and rump feathers and fanned the wings and tail, exposing the azure feathers of the wings and rump. The position was held for 30-45s before the adult moved away.

Captive birds have also been observed performing various displays related to courtship and breeding. Birds in captivity are usually very aggressive towards each other until the breeding season starts. Initiation of breeding starts with the adults chasing each other and intermittently performing a stretching display in which both sexes stretch up to their extreme height and then suddenly lapse back to their normal posture whilst rapidly bobbing the tail up and down. During this stretching, growl-like vocalisations are usually given. Occasionally, one bird would also stand in front of the other and flick its wings to display its white wing-patch, and the male was observed crouching whilst making the growling noise. During this period both adults were usually seen together. After about one week of such behaviour, nest-building begins (Delacour 1934, Lee *et al.* 1989).

Recently hatched nestlings have a defensive posture that has been observed in captivity (Lee *et al.* 1989). When disturbed, the nestlings (which are black with long quills, and appear rather hedgehog-like) backed into the nest as far as possible and assumed a posture in which their rumps were exposed and head tucked down under the shoulders whilst the quills bristled out towards the nest entrance.

BREEDING In parts of its range, seasonality of breeding, and breeding biology of Hooded Pitta, are relatively well known, whilst in other parts, virtually nothing is recorded about the breeding of this species. In the following account, aspects of breeding are described under subsections in an attempt to make this long account more easy to use.

Seasonality Ali and Ripley (1983) reported that the breeding season of *cucullata* in the Indian subcontinent usually lasted from April to June, although it may last longer in some years or for a small proportion of birds, since eggs have been found in nests as late as 23 June and 26 July in north-east Assam (Hume 1888), and Baker (1934) reported finding eggs on 8 August. Furthermore, adult Hooded Pittas have been collected in breeding condition in Nepal in June, and young have been taken in August, whilst in Sikkim fledglings have been collected between April and September. Eggs from 17 clutches in BMNH from India were collected between 24 April and 27 July. In Bangladesh, Hooded Pitta breeds locally in May and June, in north-east, south-east and central forests (Harvey 1990). In Burma, nests have been found in the Chin Hills in June and at Amherst on 12 July (Baker 1934, Smythies 1986), and fledglings have been collected in mid-July (Oates 1883).

In Thailand nests with eggs have been found in the south-east on 27 July and at Khao Yai in June, whilst in the south-west (Kanchanaburi), nest building has been observed on 6 June and a nest with eggs found on 3 July (P. Round *in litt.*); in the peninsula, nest building has been observed on 9 and 17 May, and eggs on 9 October and 29 May, 9 June, late June and 12 August (Herbert 1924, P. Round *in litt.*, C. Robson *in litt.*, pers. obs.) and nestlings have been seen in two nests on 30-31 July (Anon. 1988b). An adult with a small fledgling was observed in the peninsula on 15 June (P. Round *in litt.*). In Peninsular Malaysia, eggs and juveniles have been seen in Kedah State in June (Madoc 1956, Medway and Wells 1976). In the east of its range, *cucullata* is poorly known, and the only documented breeding record from Indochina is of specimen of a half-grown bird collected in north-west Tonkin on 20 June (IEBR).

There are few data on breeding of *muelleri* in Borneo, although those available suggest that breeding occurs mainly around the turn of the year and during the first few months: in Sabah, a nest was found on 12 April; several specimens with enlarged sexual organs have been collected in March (Sheldon in prep.), whilst eggs have been collected on Labuan in December (BMNH). In Sarawak, eggs have been collected on 14 May (BMNH), a juvenile was collected in February, and an immature male was collected on 27 October (SarM). A well-grown immature was collected on the Mahakam River in Kalimantan on 7 April (USNM). Well-grown immatures have been collected in central Borneo (Kalimantan) in early May and in August (RMNH). In West Java, eggs of *muelleri* have been collected from December to May (Hoogerwerf 1949, Hellebrekers and Hoogerwerf 1967).

There is no information on breeding of *muelleri* from Sumatra, but records exist from populations on the offshore islands. In Billiton (Belitung), nests with eggs of *bangkana* have been found in March, April and May (van Marle and Voous 1988), whilst two fledglings were collected from a nest on 28 May (MZB), and recently fledged immatures have been collected on 9 August and 15 August (MZB, USNM). A juvenile was collected on Bangka on 16 June (Mees 1986).

In the Philippines, evidence suggests that breeding of *sordida* is later than that of *muelleri* on nearby Borneo. A nest with two eggs was found on Samar on 30 June (McGregor 1909-10) and on Tawitawi on 21 July (BMNH), whilst a juvenile was observed on Tawitawi on 10 August (G. Dutson *in litt.*) and an immature was collected on Mindanao on 8 October.

On Sangihe Island, to the south of the Philippines, F. Rozendaal collected a female (*sanghirana*) containing an embryonic egg on 29 May, and a recently fledged juvenile on 8 May (RMNH), and also observed two adults feeding three recently fledged but older juveniles on 12 May. Another juvenile was collected there on 11 June (USNM). In Sulawesi, a female *forsteni* was collected with an egg in the oviduct on 25 February (Stresemann 1940), whilst a nest with two nestlings was found on 4 June and a juvenile collected on 1 April (Meyer and Wigglesworth 1895). Hence here the breeding season seems to be more similar to that in Borneo, to the west, than to that in the Philippines, to the north.

On Aru, three eggs were collected in February and a fledgling *goodfellowi* was collected in March (BMNH). In seasonally dry parts of south New Guinea, breeding (*novaeguineae*) apparently occurs only in the latter part of the wet season, with eggs from February to early April, a period coinciding with ideal ground conditions for foraging and an abundance of invertebrates to feed the young (Coates 1990). In the less seasonal north of the island, evidence suggests that Hooded Pittas lay eggs throughout the wet season, with eggs having been found in September near Sorong, in the middle Sepik on 22 December (Gilliard and LeCroy 1966) and in the Rawlinson Mountains on 5 July (BMNH). On Karkar Island, a nestling was found in May-June, indicating that in some populations breeding can occur at the very end of the wet season. In western Irian Jaya, *novaeguineae* has been observed feeding recently fledged young in early August (D. Gibbs *in litt.*).

Nests Both adults participate in building the nest (Baker 1934: *cucullata*). In the Indian subcontinent, the nest of *cucullata* is described as being very similar, as are the eggs, to that of partially sympatric Indian Pitta. The nest is described as a loosely put-together oval ball of bamboo leaves, mixed with a variable number of roots, leaves and grass (Baker 1926, Ali and Ripley 1983), although the two species' nests can apparently be distinguished from each other by the presence (Hooded Pitta) or absence (Indian Pitta) of a platform, some 10cm wide, in front of the nest entrance. Nests examined by Baker (1934) measured some 20-25.5cm long by 15.0-23cm broad, with some being almost spherical. A nest found by Hume (1880) measured some 25.4cm in diameter and 24cm in height, whilst the diameter of the entrance hole was about 9cm, and one found by Davison was 24cm in height and 25.0cm in length with an entrance hole 7.6cm wide (Baker 1934).

Baker (1926) noted that, in the Indian subcontinent and in Burma, bamboo was the favourite nest material, and in areas where bamboo is common the nest may be made entirely of bamboo and a few twigs. In other areas, different nest materials may predominate: for example, a nest from Assam was dome-shaped and made of fine twigs and roots with a few dead leaves sticking to the bottom, whilst one from Tenasserim was composed of dry twigs and leaves, lined with fibres, and was resting on a thick foundation of dead leaves (Hume 1880, 1888). A nest on Samar, Philippines, was outwardly of twigs but lined with moss (McGregor 1909-10).

In the Indian subcontinent, the nest is normally built on the ground in thick cover, such as a bamboo clump, and has a rounded side entrance. One nest found in the Dibrugarh district of Assam was placed on the stump of a tree amongst a dense mass of leafy twigs, whilst another was placed among the roots of a fallen tree. Fallen debris may sometimes be piled up to form a slightly raised platform at the nest entrance (Hume 1880, Baker 1926), so that the birds can hop up from the ground to the nest entrance. Such platforms may be 10-20cm wide and, at the highest end, 7.5cm thick (Baker 1934). The platform for the nest and entrance ladder may be added after completion of the nest, or occasionally, before nest-building starts (Baker 1934). Roofing material may also sometimes overhang the entrance to give the suggestion of a porch. Madoc (1956) noted that, in the north of Peninsular Malaysia, nests are often placed in a thick bed of dead leaves and fallen twigs, not infrequently in bamboo refuse on the side of an anthill.

A nest found in the early stages of construction in Thailand on 21 July had a freshly excavated bare earth depression in a bank (where leaf debris had been scraped out), a doorstep, and the beginnings of a roof, woven of bamboo leaves (P. Round *in litt.*).

In Sabah, a nest of *muelleri* was a football-sized dome of dead leaves, grass stems, root fibres and pieces of bark. The nest was placed in the base of a thick clump of a spiny palm, where it had the appearance of a mass of debris rather than that of a nest. A nest discovered in Kalimantan was also built on the ground, where hidden by bamboo leaves. Brüggemann (1878) found a nest made primarily of bamboo leaves in Kalimantan.

Nests from the New Guinea region differ from those described above, in that they typically have mossy roofs and earth in the structure. A nest of *novaeguineae* described in detail by Coates (1990) was situated on gently sloping ground among sprouting regrowth and fallen branches in monsoon forest. The nest was well camouflaged, with a layer of living moss forming the roof and piles of sticks pulled up around the nest. The base was constructed of sticks and leaves bound together in mud. Fine black fibres of 15-17cm length and a few fine twigs formed the lining and provided some support for the roof of leaf skeletons, fine twigs and rootlets. The nest was 20-22cm deep, with an inner diameter of about 13cm and a side entrance of about 8-9cm wide by 4-5cm high. A platform of sticks bedded with earth was built against the entrance. A nest from New Guinea in BMNH was also covered by a deep layer of green moss, interwoven with a few pieces of dead dicotyledonous leaf; the inner nest was made of fine roots interwoven with fine, flexible twigs. A nest of Hooded Pitta in New Guinea described by Mayr and Rand (1937) was also well camouflaged, with grass and herbs overhanging the site, on the ground below a dead bushy branch.

Baker (1934) believed that the nest is built very quickly, since he once observed a pair (*cucullata*) busy gathering nest material, though no nest was visible, yet three days later the nest was practically finished. In captivity, nests have been constructed in 5-8 days (Delacour 1934, Lee *et al.* 1989).

Eggs In India, clutch-size is usually 4-5 (Baker 1934, Ali and Ripley 1983), although nests found in north-east India and Burma have contained 3-4 eggs (Hume 1888, Stevens 1915, Smythies 1986). In South-East Asia, however, such as in Peninsular Thailand, clutch-sizes of three are perhaps more frequent, although nests on Java have contained up to five eggs (Hoogerwerf 1949), whilst one in Thailand and one on Samar, Philippines, contained only two (Hachisuka 1934-35, P. Round *in litt.*). In New Guinea clutch-size is reported to be 3-4 (Coates 1990), although nests in BMNH from north New Guinea and Aru contained 2-3 eggs. A nest found on Karkar Island contained one nestling.

Eggs of *cucullata* from the Indian subcontinent measure on average c.27.1 x 21.0mm, with dimensions varying between 23.0-28.0 x 19.6-22.5mm (based on a series of 50 measured by Baker 1934), and are coloured white and marked with sparse dull or dark purple spots and specks, often with a few lines and scribbles of purple-black. They are more profusely marked than the eggs of Indian Pitta (Baker 1926). Hachisuka (1934-35) gives the size of eggs of the nominate subspecies as 26 x 21mm, noting that they are pure white, rounded ovals, thickly speckled all over with brown and larger underlying spots of grey, the latter being most numerous at the blunt end. Eggs of *sordida* from Samar, Philippines, were described as being pure white and thickly speckled all over with brown and larger underlying spots of grey, the latter most numerous at the broadest end. One egg measured 26 x 21mm (McGregor 1909-10). Thirty-nine eggs from Java (*muelleri*) measured 23.4-27mm x 19-21.4mm, and were creamy-white or white covered in medium-sized speckles and scribbles which were sepia to black in colour. The underlying, often predominant, markings were well-defined greyish-lavender and vinaceous-grey to mouse-grey. Most markings are clustered at the blunt end (Hoogerwerf 1949, Hellebrekers and Hoogerwerf 1967). In New Guinea, eggs are whitish, covered with grey-brown, grey and a few red-brown spots, and are probably typically slightly larger than those from the Indian subcontinent, measuring 25.5-31.1 x 20.2-24.4mm (Rand and Gilliard 1967, Coates 1990).

In captivity, eggs are laid at the rate of one a day, and incubation starts after the last egg is laid. The incubation period is reported to be 15-16 days, and, c.12 days after fledging, chicks are able to feed themselves. In captive situations, where food is plentiful, adults may start building a new nest as soon as chicks are fledged, and a female has laid as many as 10 eggs in one nest (Delacour 1934, Lee *et al.* 1989). In New Guinea, only two young fledged successfully from a nest with four eggs. The young left the nest when c.16 days old (Coates 1990).

Both adults participate in incubating eggs in the Indian subcontinent (Baker 1934), although in New Guinea anecdotal evidence suggests that the female performs most of the incubation and brooding (Coates 1990). Both adults feed the nestlings, and the male may provide more than its fair share of food. In New Guinea, a male brought food to the nest on average every 14.6 minutes, in comparison to an average 18-minute interval by the female. The shortest interval between visits, of two minutes, was made by the male. Usually only one adult visited the nest at a time, and Coates (1990), who studied the nesting ecology in detail, only once saw both adults at the nest simultaneously. The male was responsible for the removal of most faecal sacs from the nest, and, during the period of incubating and brooding, may also have fed the female on the nest.

DESCRIPTION *P. s. sordida*

Sexes alike, although females may be marginally duller. **Adult** Entire head black, extending as hood onto nape and upper margin of breast. Narrow eye-ring grey with red tinge above and below eye. Bare skin around eye grey, broadest behind. Mantle and back dark, bright green; rump and uppertail-coverts iridescent azure-blue; tail black with very narrow (1-2mm) dark blue tip. Breast, flanks and sides of belly green, paler than that of upperparts and glossed with a bluish tinge. Centre and occasionally sides of belly black; undertail-coverts red. Tertials green; large iridescent azure-blue patch on inner wing, extending from the bend across the whole wing, and formed by long fringes to feathers of the lesser coverts, occasionally the blue extending onto the inner basal part of the greater coverts. Secondaries blackish with the outer edge of outer web of all feathers green (less glossy than mantle and coverts), the fringe becoming broader towards inner wing and extending to the whole tip of innermost secondary, the green on the outermost secondaries usually tinged with blue. Primaries blackish, with broad white band across all feathers, near tip of wing and being narrowest on the outer primary and becoming progressively broad so that inner feathers may be two-thirds white. Underwing-coverts and axillaries black. Iris blackish; bill black, legs and feet slate. The nominate subspecies has a plumage variation, which is found in c.40% of individuals (Parkes 1960), in which the green upperparts and underparts are streaked black. This variation also occurs in some of the other subspecies, such as *palawanensis*, but is less common. These markings are not known to be correlated with age or sex.

Juvenile Head blackish-brown with some rufous to pale chestnut feathering on forehead and sides of crown in front of eye; chin and upper throat dusky, lower throat white. Mantle blackish-brown with scattered dark green feathers (progressively more with age); rump pale iridescent azure-blue; tail blackish. Tertials and wing-coverts blackish with dark greenish tinge and distinctive white band across closed wing formed by white distal half to feathers of median coverts and white patches in some outer greater coverts. Secondaries blackish with dull green fringe to outer web; primaries blackish with conspicuous (but often hidden) white patch formed by bands across all but outer 2-3 feathers, narrow on outer edge and broader on inner primary. Breast mostly buffy-orange, mixed with some pale green feathers; undertail-coverts pink. Bill orange with black band across distal half, becoming black with bright orange tip and gape. Bill becomes darker and redder with age.

MEASUREMENTS Sexes (of *sordida* and *palawanensis*) are alike in size. *Sordida* wing 95-107; tail 32-39; bill 24-28; tarsus 36-39; weight 42-70.

13 IVORY-BREASTED PITTA
Pitta maxima   Plate 8

Pitta maxima Müller and Schlegel (ex Forster MS) 1845, in Temminck, Verh. Nat. Gesch. Nederland Overz. Bezit., Pitta, p. 14: Gilolo [Halmahera].

Alternative name: Moluccan Pitta, Great Pitta

FIELD IDENTIFICATION Length 24-27. Ivory-breasted Pitta is a very large, frequently arboreal, unmistakable species that inhabits islands in the north Moluccas of Indonesia. The head, throat and upperparts are black with a glittering light-blue upperwing-covert patch; the underparts are silvery-white, with a bold, bright crimson patch on the centre of the belly. In flight, a broad white band across the primaries is distinctive, and the wings make a loud whirring noise. Juveniles have buff on the underparts.
Similar species None in its range.

VOICE Ivory-breasted Pitta has a very loud, far-carrying double melancholy note of about 1.5s duration, variously described as *wok-wow, wi-whoouw, hwew-wooo* or *werrhh-wuuhhh*, the first note being short and ending rather abruptly, the second note longer and trailing off (Lambert and Yong 1989, A. Greensmith *in litt.*, Bishop and Coates in prep.). When agitated, the call is reported to change slightly; *werrhh-whahlll* (A. Greensmith *in litt.*). Pitch and length become deeper and more drawn out when two birds counter-sing continuously for long periods (Bishop and Coates in prep.). The calls are usually repeated at about 6-8s intervals, and bouts of calling generally last for 2-3 minutes. It usually perches off the ground when calling, sometimes at greater than 10m, in the lower canopy. Calling probably occurs throughout the year.

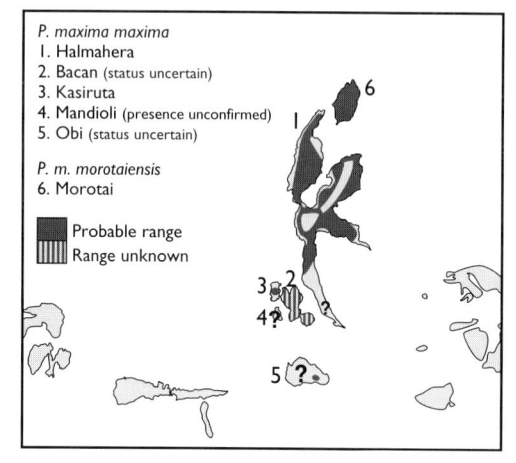

P. maxima maxima
1. Halmahera
2. Bacan (status uncertain)
3. Kasiruta
4. Mandioli (presence unconfirmed)
5. Obi (status uncertain)

P. m. morotaiensis
6. Morotai

▪ Probable range
▦ Range unknown

Distribution based on approximate extent of forest remaining within the altitudinal range of the species. Distribution on Bacan and southern Halmahera is unknown. Confirmation of presence on Obi is required.

DISTRIBUTION An Indonesian endemic confined to the north Moluccan islands. Here it is known from Halmahera, Kasiruta (adjacent to Bacan) and Morotai, and probably from Bacan, although there is no clear evidence that it definitely occurs on this island. It could also occur on the island of Mandioli, close to Bacan. White and Bruce (1986) report that a bird collected on Obi may have been an escaped bird, but Lambert (1994) heard what was most likely this species in the upper catchment of the Widi Besar River. The call sounded slightly different to the calls that Ivory-breasted Pittas give on Halmahera, and it is therefore not certain whether the bird heard was this, or another, species.

The status of Ivory-breasted Pitta on Bacan is unclear, although recent authors (e.g. White & Bruce 1986) have listed it for this island. It was not collected on Bacan by Wallace (Gray 1860) or Guillemard (1885), and was not listed for the island by either Nehrkorn (1894), Hartert (1903) or Heinrich (1956). Indeed, Hartert (1903) stated that it was his belief that the 'Bacan' specimen obtained by Vorderman originated in Halmahera. Harert also mentioned that various collectors had informed him that, although common on Halmahera, it did not occur on Bacan. Furthermore, Lambert (1994) did not encounter Ivory-breasted Pitta, or hear its distinct far-carrying call, during one month's field surveys on Bacan in October-November 1991, even though it was found to be frequently seen and heard on adjacent Kasiruta Island in November. Two specimens in BMNH are said to have come from Bacan. One, however, bears the label of a trader, rather than the collector, whilst the other was procured by A.L. Butler. Although Butler was a well-travelled collector, there is no record of him having visited the Moluccas (White & Bruce 1986) and he may therefore have purchased the skin, in which case it cannot be certain that it came from Bacan. There are also two specimens in AMNH labelled Batchian (=Bacan), but there are no details associated with the birds, and their origin is questionable (M. Lecroy *in litt.*). Whether Ivory-breasted Pitta really occurs on Bacan therefore remains to be resolved. It should be noted, however, that Kasiruta, where the species is not uncommon, is separated from Bacan by a narrow strait of no more than 1-2km and is on the other side of Bacan from Halmahera; the occurrence of this distinctive species on Bacan therefore seems highly probable.

GEOGRAPHICAL VARIATION Two subspecies are recognised.

P. m. maxima (Halmahera, Kasiruta, ?Bacan, ?Mandioli, ?Obi) Wings marked with extensive iridescent azure-blue; leading edge of secondaries green.

P. m. morotaiensis (Morotai) Slightly larger than *maxima* (wing 154-156), with a slightly longer and narrower bill (length 41-42). The blue in the wings is less extensive than in the nominate subspecies, and of a much darker colour, whilst dull blue replaces the green in the leading edge of the secondaries. The white wing-patch is larger and more prominent.

HABITAT Ivory-breasted Pitta occurs in primary, selectively logged and disturbed forests, and occasionally in overgrown coconut plantations (MZB), from sea level to at least 500m on Halmahera. Although it appears to be particularly common in areas with limestone or karst (ultrabasic) rock formations, especially in areas with a rich understorey of spiny palms (pers. obs.), variable-circular plot data suggest that densities may be just as high or higher in forest on sedimentary or volcanic rocks (M. Poulsen *in litt.*: see Status). Heinrich (1956) reported finding this species on Halmahera in low-lying areas that were flooded in the wet season, and in sago swamp. On Morotai, van Bemmel (1939) found this species at about 500-800m altitude.

Ivory-breasted Pitta appears to be rather local in its distribution. No birds were heard or seen, for instance, during five weeks of fieldwork on Bacan in 1991. On the adjacent island of Kasiruta, however, this species was commonly heard and encountered in an area on karst soils at 200-500m altitude at the same time of year (Lambert 1994).

STATUS During the last century, this was considered a common bird on Halmahera (Meyer 1879). Heinrich (1956) reported that it was particularly common in the south of the island. On Halmahera and Kasiruta it is still a locally common species, being locally abundant in hilly areas on karst. In contrast, it is apparently uncommon in flat tall primary forest on the east coast of the north-eastern peninsula of Halmahera. On Bacan, if present it must be very scarce, and if the species occurs on Obi it must also be very rare there (Lambert and Yong 1989, Lambert 1994, Bishop and Coates in prep.).

Although observers have found this species locally abundant in areas of forest on ultrabasic rocks, data collected during surveys on Halmahera by BirdLife Indonesia suggest that the population density of Ivory-breasted Pitta may reach higher values on other rock formations.

In the north-eastern peninsula of Halmahera, density estimates were 8.5-40.4 birds per km^2 on ultrabasic soils, 15.5-59.5 birds per km^2 on sedimentary rocks, and 15.1-54.4 birds per km^2 on volcanic rock (M. Poulsen *in litt.*). However, the value of density on ultrabasic soil was estimated from very few data, whilst no information of the altitudes, state of the forest or time of year have been provided for the sites sampled, and it is therefore not clear if these density estimates are directly comparable.

FOOD Virtually nothing has been recorded about the diet of Ivory-breasted Pitta. However, broken snail shells found at an anvil on Halmahera are thought to have derived from the feeding activities of this species. Locals report that it has been observed eating earthworms and caterpillars (M. Poulsen *in litt.*).

HABITS This species tends to take flight easily if disturbed, often flying off some distance with whirring wings. It regularly perches in trees, particularly when calling, when individuals may call from bare branches in the lower canopy of trees at up to 20m off the ground (pers. obs.). Birds frequently ascend the lower canopy during territorial calling disputes, during which adults may call continuously from adjacent trees. Such birds may fly more than 100m to investigate the calls of other individuals (pers. obs.). Heinrich (1956) saw males displaying in trees, although there is no description of the behaviour he observed. He also heard this species calling at night, and found a pair roosting at night about 4m off the ground on a dry twig.

BREEDING On Halmahera, a female with a nest containing one egg was collected on 2 May 1981 (USNM), whilst local villagers reported finding a nest near Sidangoli in early July (pers. obs.) and a nest with one egg was discovered in June (M. Poulsen *in litt.*). A juvenile was collected on Halmahera on 1 February (von Berlepsch 1901). Males with enlarged testes have been collected on 22 May, 7 June and 17 November (MZB).

Nests are built on the ground, invariably within the protection of the buttress roots of large trees, where pressed against one of the buttresses as it lowers towards the ground: one such nest was situated above a large rock (M. Poulsen *in litt.*). The nest is rather long and narrow, reported by villagers to be c.30cm long, and usually containing two eggs.

A nest photographed in Halmahera (K.D. Bishop), positioned between two buttresses of a tree, was a very ragged, well-camouflaged dome with the appearance of a mass of fallen plant debris. The bulk of the nest appears to be a latticework of interwoven blackish sticks surrounded by a great variety of plant matter, including rotting leaves of broad-leaved plants and their leaf skeletons, various pieces of fine root, and a few medium-size sticks. The nest was virtually covered by the very large palmate leaf of an epiphytic stagshorn moss (*Platycerium* sp.) whilst the strands of green but drying pieces of moss hung down over the nest entrance. An assortment of large sticks and a large piece of dead, fibrous palm leaf were propped up against the side of the nest. The base appears to have been made of large sticks with a few large, entire, dead leaves of broad-leaved trees, these perhaps forming a small platform at the entrance. Locals on Halmahera report that dry palm leaves are used to build the framework of the nest, whilst other palm leaves, which have rotted to produce a skeleton of veins, and other leaves and moss fill the cavities: the whole nest may be concealed by covering it with moss.

Eggs are reported to be white with small chocolate-brown spots, and the approximate size reported by local people is 30 x 25mm (M. Poulsen *in litt.*).

DESCRIPTION *P. m. maxima*
Sexes similar.
Adult Head, including chin and throat, black, concolorous with upperparts, including tail. Most birds have a few iridescent metallic blue feathers on the rump. Breast and entire flanks white; centre of belly dark scarlet, extending to undertail-coverts, where slightly paler in colour. Lesser coverts iridescent azure, with some green at tips of most distal feathers; median coverts green with black basal half and green edges; greater coverts black. Secondaries black with green edge to outer web extending three-quarters down feathers and onto tip of outer web; primaries black with white band midway down primaries formed by spot in inner web of primaries 1-2 and across whole web of primaries 3-7. Underwing-coverts black. Feathers of tarsus white. Iris brown; legs and feet pink; bill black.
Juvenile Whitehead (1893) described the juvenile as having a white throat, metallic blue rump-band and yellowish breast, although it is likely that his 'yellowish' is actually some shade of 'buff'.

MEASUREMENTS Wing 143-154; tail 68-80; bill 37-39, width 18.5-20; tarsus 56.5-62.5; weight of female 166-172, of male 172-206 (MZB).

14 SUPERB PITTA
Pitta superba   Plate 8

Pitta superba Rothschild and Hartert 1914, Bull. Brit. Orn. Club, 33, p.106: Manus Island, Admiralty Islands.

Alternative name: Black-backed Pitta. Locally known as Kuku.

FIELD IDENTIFICATION Length 17.5-20. Superb Pitta is confined to the island of Manus in the Admiralty Islands. It is a relatively large, glossy jet-black pitta with upperwing-coverts largely iridescent pale turquoise-green and a red patch on the centre of the belly, lower flanks and undertail-coverts. The leading edge and tips to secondaries are dark green. The female is slightly smaller and duller than the male. During take-off the red on belly and flanks are visible from behind. Immatures are similar to adults but lack the glossy appearance and have dull pink abdomen and undertail-coverts. The bill tip of immatures is orange-red.
Similar species None in range. In flight the wings whirr, but on Manus only the Metallic Starling *Aplonis metallica* is reported to have a similar wing-whirring, a species usually found in the upper levels of forest.

VOICE The commonly heard call of Superb Pitta is a clear whistle consisting of two almost identical, rather deep melodious notes, *hwouw-whouw*, somewhat reminiscent of the call of Hooded Pitta. At close range, the call may sound shorter and faster. Each note rises and then falls in pitch whilst declining in volume (Dutson and Newman 1991). Two birds recorded calling to each other by D. Gibbs repeated the double whistle every 4-5s. This call resembles the *coo-coo* call of Bronze Ground Dove *Gallicolumba beccarii*.

At close range quieter calls may be heard: these have been described as a chicken-like *gwark*, usually repeated at intervals of more than 10s. Whilst the two-note whistle may be audible over distances of over 0.5km, the *gwark* call can only be heard from around 20m. This latter is perhaps an alarm call, and is usually uttered from the ground. In contrast, the loud call is usually uttered from a perch at least 2-7m above the ground. Sonograms of the call are provided by Dutson and Newman (1991).

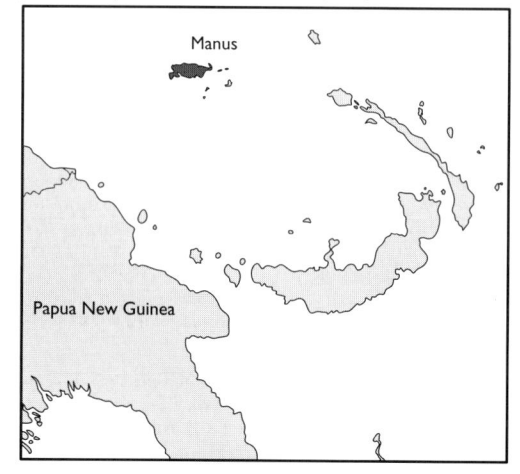

Distribution based on approximate extent of forest remaining within the altitudinal range of the species. However, evidence suggests that the species may be very locally distributed.

DISTRIBUTION The Superb Pitta is endemic to the island of Manus in the Admiralty Islands, Bismarck Archipelago (c.2°N, 147°E).

GEOGRAPHICAL VARIATION None described.

HABITAT Superb Pitta may prefer hilly areas. Dutson and Newman (1991) reported that they found this species only in the more open primary forest of hilltops, but lured calling birds into recent second growth with tapes, and that the apparent absence of Superb Pittas in other areas away from primary forest suggests that proximity to primary forest may be an essential habitat component for this species. However, D. Gibbs (*in litt.*) found it in a mixture of secondary scrub, young secondary vegetation, old gardens that had probably been regenerating for around 10 years, and bamboo scrub. He also saw the species in primary forest close to a road where some secondary vegetation such as bamboo and dense patches of fern were growing. Dutson and Newman (1991) searched for this species in eastern Manus between sea level and 200m, but located birds only at 100m.

STATUS The Superb Pitta has a very restricted range, being confined to the island of Manus, with an area of 1,943km². Mayr (1955) believed that this was a fairly common and widespread species since Meek obtained a large series and it was found at all collecting stations on Manus during the Whitney South Sea Expedition (Petayia, Drabui and Malai Bay). More recently, whilst some observers have failed to find it on Manus, K.D. Bishop (cited in Collar and Andrew 1988) found it to be moderately common in forest near Lorengau, as did M. LeCroy (*in litt.*). However, Dutson and Newman (1991) estimated that the population comprised only one thousand calling birds, based on their detection of no more than three calling birds in an area of around 3.5km^2. However, the proportion of birds calling during their period of fieldwork, in June and July 1990, is unknown. In view of the limited area on Manus in which this species has been found, however, it is suggested that the population size may in fact be lower. D. Gibbs (*in litt.*) was able to locate only two pairs during a ten-day visit to the island in early 1994, and believes that the species is uncommon but probably widespread. Superb Pitta is listed as Vulnerable in BirdLife's 1994 global assessment of the worlds threatened bird species (Collar *et al.* 1994).

Whilst this is a very restricted species which may now be very rare, the people of Manus are said to have a dislike for logging on their land, leading to the conclusion by Dutson and Newman (1991) that the habitat of Superb Pittas should be safe in the short term. Whilst future forest loss through commercial timber exploitation or land clearance for agriculture could endanger the Superb Pitta, its presence in secondary growth and degraded forest suggests that it would survive in selectively logged forests and agricultural/forest mosaics, as do many other species of pittas that inhabit islands.

FOOD Villagers on Manus Island claim that Superb Pittas feed on snails which are smashed on stones. A probable anvil found by Dutson and Newman (1991) was a flat stone of c.10cm in diameter. Fragments of three snail shells nearby were about 1cm in diameter.

HABITS Superb Pitta is poorly known, with little recorded about its habits. However, observations suggest that it is similar to other pittas in that it is shy and retiring and occurs as dispersed pairs. This species can be attracted to an observer by imitation of its call.

BREEDING Rothschild and Hartert (1914) found a nest with two eggs on 11 October and two juvenile Superb Pittas in September/October 1913. The nest (BMNH) was a large, compact, oval structure c.26cm wide, 22cm high and 14cm from front to back, with a 7cm diameter side entrance. The inner nest was made of densely interwoven fine plant fibres, some of them roots, others probably twigs and leaf petioles, but also including fibres from skeletonised palm leaves. The outer part of the nest was more diverse, being a jumbled mass of skeletonised and dried dicotyledonous leaves and small twigs, a few small pieces of rotting wood, pieces of hooked creeper and a few pieces of broken palm or bamboo leaf. These components were intermingled with moss, some of it still green, in particular at the back and on the top of the structure, where mosses and leaves were tightly matted together. The inner nest-lining was constructed entirely from fibres and rootlets. Local people apparently report that the entrance hole is sited to face away from the direction of the prevailing wind, which varies seasonally (E. Lindgren, in Coates, 1990). An old nest was found on the ground, abutting a small vertical drop, in light secondary forest on the slope of a small stream.

Clutch-size is apparently two. Eggs were 34.0-34.4 x 25.6-26.7mm in size. The eggs are white, with purple-grey spots overlaid by smaller purplish-brown spots and a few larger ones of the same colour. These markings are denser at the broader end (Rothschild and Hartert 1914).

DESCRIPTION
Sexes similar.

Adult male Plumage jet-black and glossy, with scarlet-orange patch from centre of belly to undertail-coverts and thighs. Very large, iridescent wing-patch formed by long azure-blue tips to most upperwing-coverts, becoming turquoise-green on some of the innermost lesser coverts. Greater coverts nearest bend of wing are black. Secondaries black with green, sometimes glossed blue, edge to distal half of outer web, broadest on innermost secondary. Primaries and tail feathers black. Underwing-coverts black. Some individuals have a few feathers of the uppertail-coverts marked with dark blue near the tip. Iris dark brown; bill black; feet pale flesh.
Adult female Slightly smaller and duller than the male, with the red of the abdomen and blue of the wings less bright.
Juvenile Dull black without any gloss, the red of the belly and undertail-coverts being replaced with dull pink. Tip of bill orange-red.

MEASUREMENTS Wing 124-131; tail 44-50; bill 32-38.5; tarsus 47-53.5; weight unrecorded (probably in range 100-130).

15 AZURE-BREASTED PITTA
Pitta steerii   Plate 8

Brachyurus Steerii Sharpe, 1876, *Nature* 14, 297: Dumalon, Mindanao.

Alternative name: Steere's Pitta

FIELD IDENTIFICATION Length 18-19.5. Azure-breasted Pitta is a medium-large pitta that is endemic to the Philippine Islands of Mindanao, Samar, Leyte and Bohol. The adult is unmistakable, with a black head and white throat, green upperparts with a conspicuous shining azure-blue wing-patch and pale blue underparts with black centre of belly and bright scarlet undertail-coverts. The rump is azure-blue. Immatures have upperparts similar to those of the adult, but the breast and flanks are a dirty greyish-olive and the entire belly is reported to be pale scarlet. Slightly older birds develop tracts of pale blue feathers along the sides and on the middle of the breast.
Similar species Adults are uniquely coloured, and impossible to confuse with other species in their limited Philippine range.

VOICE This species may be silent for much of the year. However, during the presumed breeding season, which spans April to early June, calls are frequent. The most commonly heard call is a series of five, occasionally four, explosive short whistles *kwo-kwo-kwo-kwo-kwo*, *kva-kva-kva-kva* or *kwo-ho-ho-ho-ho*, repeated every 2-3s. These short whistles are reminiscent of the two-note call of Hooded Pitta. Birds on Bohol also give an explosive whistle, *kWEIOo*, repeated every 2-3s (J. Hornskov in prep.).

DISTRIBUTION Azure-breasted Pitta is one of two endemic Philippine pitta species. It is known from the southern island of Mindanao and the central islands of Samar, Leyte and Bohol. On Leyte, it has been collected in the northern half of the island at Patok, Buri, and Bulog; on Bohol at Anislagan, Guindulman, Carmen, Cantaub (Sierra Bullones), Mayan and Jagna; and on Samar at Paranas, Catbalogan, Matuguinao and Mt Kapoto-an (Rand and

Rabor 1960, BMNH, USNM, L. Kiff *in litt.*). The distribution on Mindanao is poorly-documented, with recent records coming only from southern Surigao del Sur Province, in the eastern coastal region. The type specimen was collected in the Zamboanga Peninsula, on the other (western) side of Mindanao, whilst specimens have also been collected in Agusan del Norte and South Cotabato Provinces (USNM): therefore Azure-breasted Pitta was probably once a widespread species on the island.

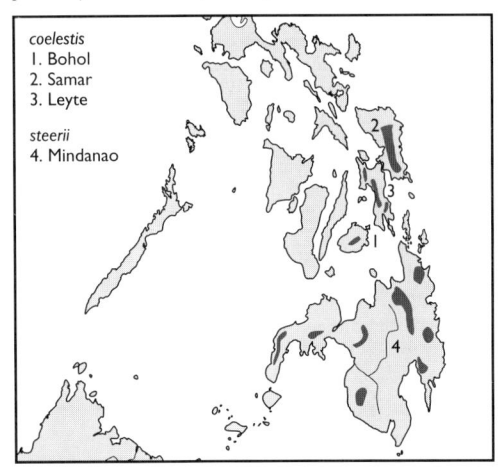

Distribution based on approximate extent of forest remaining within the altitudinal range of the species. Former range more widespread.

GEOGRAPHICAL VARIATION Two subspecies have been described, although *coelestis* (Parkes 1972) is doubtfully distinct.
P. s. steerii (Mindanao, Philippines)
P. s. coelestis (Samar, Leyte and Bohol, Philippines) Differs little from nominate subspecies, but has the blue of the underparts, wing-coverts and rump paler, less turquoise.

HABITAT Azure-breasted Pittas inhabit lowland tropical forests. Observations and collection localities all indicate that the species may favour forests on karst, as suggested by Whitehead (1899). In the mountains of central Samar it is said to prefer fairly thick undergrowth and moss-covered coral limestone boulders (Hachisuka 1934-35). On Bohol, Azure-breasted Pittas were collected by Rand and Rabor (1960) only in areas of dark, dense primary forest in rocky limestone areas. The collecting sites visited by Rand and Rabor (1960) on Samar were also primarily in areas with jagged limestone outcrops.

Recent observations of this species on Bohol have been confined to the Rajah Sikatuna National Park, which is an area of rather stunted forest growing on jagged limestone outcrops and bedrock (Lambert 1993). Recent sightings from Mindanao are also from limestone hills (H. Buck *in litt.*). Although most records appear to have come from primary forest, Evans *et al.* (1993) heard Azure-breasted Pitta in highly degraded forest near Bislig, Mindanao. On Samar, it has been found from 100-600m altitude, whilst on Bohol it occurs at altitudes between 350 and 750m.

STATUS Although McGregor (1909-10) and Hachisuka (1934-35) noted that this species was uncommon throughout its range, it is locally common on Bohol, where Hornskov (in prep.) heard six birds calling around a single clearing on one day. However, the fact that there have been few recent records of this species, and that forest clearance in the Philippines continues unabated, gives cause for concern for the long-term survival of Azure-breasted Pitta, and it has consequently been listed as Vulnerable in BirdLife's recent list of threatened species (Collar *et al.* 1994).

FOOD Specimens collected on Bohol that have been examined contained insects and worms. A bird observed on Bohol was holding a large long-legged (c.3cm) insect.

HABITS Usually solitary, feeding on the ground and on boulders and fallen logs. Whilst primarily terrestrial, Azure-breasted Pittas have been observed in forest trees at heights of up to 13m. A pair of Azure-breasted Pittas observed on Bohol were feeding alongside each other, and within 5m of a Red-breasted Pitta, by flicking over dry leaves (J. Howes *in litt.*). Occasionally, adults may raise the crown feathers to form an impressive helmet-shaped crest (pers. obs.), and one bird that had been trapped raised its crown whilst it was being handled, and simultaneously erected the inner wing-coverts to form a gorget of feathers with the azure facing forward (J. Hornbuckle *in litt.*). This unusual display could perhaps be a threat posture.

On Bohol, Azure-breasted Pittas usually call from perches above the ground, often in relatively dense cover. Birds observed calling in early April were perched on horizontal portions of vines and on horizontal branches some 2-4m off the ground (pers. obs.).

Raised crown and extended blue gorget of Azure-breasted Pitta

BREEDING Little is known about the breeding biology of this species. Fully fledged young have been collected on Samar in June (Whitehead 1899), a juvenile was observed on Bohol on 25 July (G. Dutson *in litt.*), and young birds have been collected on Mindanao in September or October (Dickinson *et al.* 1991). A male with enlarged gonads was collected on Samar in May (Rand and Rabor 1960). These records suggest that breeding starts sometime in the middle of the year, probably spanning May and early June, when adults have been regularly heard calling (Lambert 1993), and perhaps as early as April, although few were heard calling spontaneously on Bohol in mid-April (pers. obs.). This timing would be consistent with the breeding of most Philippine birds, which coincides with the beginning of the wet season, when the southwest monsoon brings rains (Dickinson *et al.* 1991).

DESCRIPTION
Sexes similar.

Adult Head black with white chin and throat and a small greyish-white patch of bare skin behind the back of the eye; mantle to upper rump, tertials and lesser coverts bright dark green; lower rump iridescent azure-blue; uppertail-coverts and tail black. Large iridescent azure-blue patch in wings extending back from bend of wing and formed by majority of coverts (which are unusually long): only coverts at rear of inner wing have a green tinge or are green and concolorous with mantle. Primaries short (covered by secondaries at rest), black with narrow white spot or bar in outer web of primaries 3-6 or 3-7 in centre of feathers; secondaries black with greenish-blue edge to distal half of outer webs, this becoming broader on inner secondaries and covering the entire outer web of innermost feather and becoming progressively greener towards inner secondary until concolorous with mantle; underwing-coverts black. Some birds have black streaks on green parts of mantle and wing-coverts and on blue rump; occasionally also on lower breast. Breast and flanks bright azure-blue (less bright than blue of wing and rump), sometimes with a few greenish feathers on lower flanks; centre of belly black, undertail-coverts bright red. Iris black; bill very robust, and black; legs blue-grey.

Juvenile A juvenile observed on Bohol (G. Dutson *in litt.*) had a dull head with mottled white throat and dull yellow gape. Juveniles have white in the wing (unlike adults).

Immature Grant (in McGregor 1909-10) describes a young male as having upperparts and wings like the adult, but breast, sides and flanks dirty greyish-olive, with only 1-2 of the silver-blue feathers being visible; the entire middle of the breast and belly is pale scarlet. Slightly older birds develop tracts of pale blue feathers along the sides and middle of the breast.

MEASUREMENTS Wing 114.5-125; tail 38.5-43.5; tarsus 40-45; bill 31.5-35; weight (Mindanao) 84-113.

16 WHISKERED PITTA
Pitta kochi   Plates 9 & 16

Pitta kochi Brüggemann, 1876, Abh. Naturwiss. Ver. Bremen 5, 65: Luzon.

Alternative name: Koch's Pitta

FIELD IDENTIFICATION Length 20.5-21.5. This is a medium-large, mostly montane pitta, confined to Luzon in the Philippines. Sexes are similar. The upperparts are dull dark olive-green with an orange-rufous rear of crown and nape. The sides of the head are dark olive-brown with a greyish, pinkish or ivory-yellow malar stripe that broadens under the cheeks, and a darker sub-malar area. The breast, bend of wing, edges to greater-coverts, tail and uppertail-coverts are slaty-blue. The sides of the upper breast and upper flanks are bright olive (latter usually hidden), and the belly to undertail-coverts scarlet. Young birds are mainly blackish-brown, with white spots on the breast and flanks, but obtain the distinctive ivory moustachial streak at an early age.

Similar species The nominate subspecies of Blue-breasted Pitta is very similar to Whiskered Pitta, and may be locally sympatric. In comparison with the former, Whiskered Pitta is much larger, lacks the blue collar and blue on the lower mantle and rump, lacks black on the upper breast, has a darker crown to nape and a pale moustachial streak. In addition, the white in the wing is less conspicuous.

VOICE The call of Whiskered Pitta is unusual for a pitta. It consists of a series of 2-9, but usually five, mournful, rather pigeon-like, descending notes, *whooh-whoh-who-whoo*: the first note is longest and there is a brief pause between this and the following notes, which are uttered at slightly increasing speed. The call lasts a maximum of 3s, and intervals between phrases are 5-30s, most usually 20s (O. Jakobsen *in litt.*, P. Morris *in litt.*). The call is very similar to that of sympatric White-eared Brown Dove *Phapitreron leucotis*, but the notes do not descend like those of the latter (H. Buck *in litt.*). Shorter calls are also reminiscent of Amethyst Brown Dove *Phapitreron amethystina*, but the dove has a deeper song and usually consists of 3-4 elements which last for 2s, rather than the typical five or more notes of Whiskered Pitta (Jakobsen and Andersen in prep.). Some observers have also noted a similarity to the call of Philippine Coucal *Centropus viridis* (M. Poulsen *in litt.* 1995). Calling usually starts at dawn and may be persistent until mid-morning. Birds call primarily from February to May.

× Confirmed Sites
1. Sierra Madre (northern)
2. Dalton Pass
3. Mt. Pulog
4. Mt. Data
5. Mt. Isarog
6. Balian
7. Ilicos Norte

■ Known Range
⬚ Potential Range

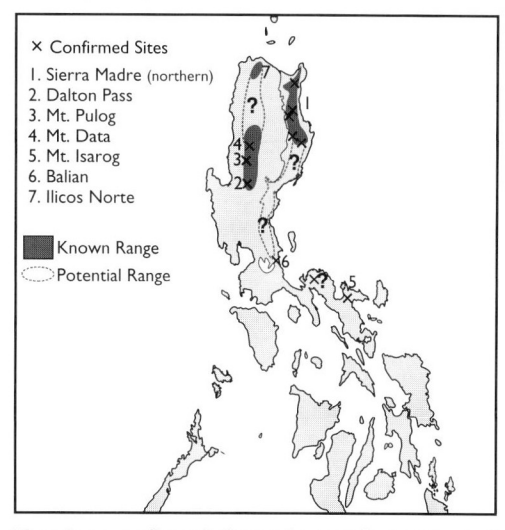

Map shows confirmed sites and approximate extent of known range. The species probably also occurs in other areas where suitable montane forest habitat exists: these areas are also mapped.

DISTRIBUTION The Whiskered Pitta is endemic to the mountains of Luzon, Philippines, where until recently it had been recorded from very few localities. It is known from the Sierra Madre Mountains (Mt Cagua, Mt Dipalayag, Mt Polis, Mt Cetaceo, Los Dos Cuernos, Mt Hamut and Minuma); in the Cordillera Central ranges of northern Luzon (Dalton Pass, Mt Data, Mt Pulog, Mt Sablan and the south-west of Mt Adams peak in coastal Ilicos Norte Province, north-west Luzon); Balian in central Luzon (Laguna Province), and Mount Isarog (Camarines Sur Province) in the south of the island (McClure and Leelavit 1972, Goodman and Gonzales 1990, Dickinson *et al.* 1991, Danielsen *et al.* 1994, Poulsen 1995, Jakobsen and Andersen 1995, D. Allen *in litt.*, K. Mitchell *in litt.*). In the northern part of its range, it is probably more widely distributed than records might suggest, particularly in the Sierra Madre Mountains. Likewise, in the south of Luzon, the apparent disjunct distribution seems likely to be related to inadequate documentation and ornithological exploration rather than to a real absence from intervening mountains.

GEOGRAPHICAL VARIATION None known.

HABITAT In the Sierra Madre Mountains, Whiskered Pitta has been recorded in forest from 350m to at least 1,650m although this species clearly prefers higher elevations, and highest densities have been recorded at 900-1,400m (O. Jakobsen *in litt.*). Altitudinal limits in the Cordillera Central ranges are not well known, although there are no records at low elevation, from forest below the submontane zone. In Ilicos Norte Province, it has been observed only at 1,000m (D. Allen *in litt.*).

Whitehead (1899) first found this species in patches of evergreen oaks within deep ravines in an area dominated by pine forest, noting that the ground was covered with moss, with begonias and other plants forming a dense undergrowth. Most subsequent records have been from primary submontane and montane oak forest, where the forest canopy may be anything between 5-12m high. In mossy forests at higher altitudes, understorey ferns and rhododendron may be common (Danielsen *et al.* 1994, Jakobsen and Andersen in prep.). The species also occurs locally in evergreen forest on slopes and hills at lower elevations. Such areas may be dominated by extensive areas of rather open, scattered understorey and soft forest floor covered with leaf-litter, and with occasional, scattered boulders, as in primary forest at 650m on Mt Diplayag. In degraded montane forest where Whiskered Pitta has been observed, where canopy cover may be reduced to less than 70%, such as at Los Dos Cuernos, young trees, ferns and 2-3m tall grasses may form a dense understorey with few areas of soft, open forest floor (Jakobsen and Andersen 1995). Whiskered Pitta has also been recorded in seriously degraded, rather scrubby secondary forest with thick understorey and patches of bamboo on the lower slopes of Mt Hamut at 350m (N. Redman *in litt.*), and in selectively logged forest on Mt Pulog and at Minuma Creek (Poulsen 1995, Jakobsen and Andersen 1995).

Wild Boar *Sus barbatus* were relatively common at all sites where Jakobsen and Andersen (1995) found Whiskered Pitta, and in many areas these animals had uprooted soil. Such areas may form an important component of Whiskered Pitta habitat.

STATUS In a recent authoritative publication on the threat status of the world's birds, Whiskered Pitta was listed as Vulnerable (Collar *et al.* 1994). However, surveys in various parts of the Sierra Madre Mountains of Luzon, in Isabella and Nueva Vizcaya Provinces, suggest that this species seems to be common at higher altitudes. On Mt Cetaceo for example, Danielsen *et al.* (1994) recorded up to 13 Whiskered Pittas calling in one day between 1,200m and 1,650m, whilst O. Jakobsen (*in litt.*) reports that at least 20 individuals were heard calling in forests at 1,050-1,250m on Los Dos Cuernos in 1991.

As with many other Philippine birds, habitat loss is a serious threat to the long-term survival of Whiskered Pitta. Fortunately, there is still extensive forest in the Sierra Madre Mountains, but this situation could change in view of the increasing scarcity of timber resources in the country. Nevertheless, this species survives in selectively logged forests, and, although its status in such forests is described as uncommon (Jakobsen and Andersen 1995), this ability may help ensure that populations survive in areas where forest cover remains following logging. Hunting with snares for terrestrial bird species is reported to be common within the range of Whiskered Pitta and is also considered a threat, since local people sometimes catch it in such traps (Poulsen 1995). However, legal international trade in this species is effectively banned by its inclusion of Appendix I of CITES (WCMC 1995).

MOVEMENTS None recorded.

FOOD A dead bird contained small beetles, and the remains of beetles were also found in the faeces of other individuals (Poulsen 1995, Jakobsen and Andersen 1995). Nothing else is recorded about the diet.

HABITS Whiskered Pitta is a very shy, skulking species, that is usually found singly. Birds feed by turning aside dry leaves with a sidewards flick of the head, and digging into the wetter soil below, occasionally cocking the head to one side to listen or look (O. Jakobsen *in litt.*). Whiskered Pittas apparently prefer to forage in areas where the soil is moist, and Danielsen *et al.* (1994) and Poulsen (1995) regularly observed them foraging in places where wild pigs have rooted over the soil. Observations on Mt Pulog suggest that foraging pigs expose earthworms and other prey items that Whiskered Pittas perhaps feed on (Jensen *et al.* in prep.). When disturbed, Whiskered Pittas bound away rapidly, or fly close to the ground into dense vegetation nearby, rather than flying up into the branches of trees, and Jakobsen and Andersen (1995) never observed this species more than 1.5m above the ground.

Peak calling apparently coincides with the beginning of the wet season, occurring from the end of April to mid-May, although calling starts as early as the end of February (Jakobsen and Andersen 1995). Birds have been seen calling from trees and from the ground, and one individual was observed using the site of an old campfire, on a logging road, as a song post (Poulsen 1995, O. Jakobsen *in litt.*, P. Morris *in litt.*).

BREEDING Jakobsen and Andersen (1995) observed one juvenile on Mt Dipalayag on 18-19 April in forest dominated by oak at 950m, and another in montane forest on Mt Cetaceo in early to mid-May, whilst an adult carrying food was observed at 650m at the former site.

Calling is apparently very seasonal, from February to May (Jakobsen and Andersen 1995, P. Morris *in litt.*), coinciding with the period that is reported to be the breeding season by local people. However, the fact that two fully grown immature birds, with the breast mostly brown and distinctly spotted white, were collected in Ifugao Province at the end of January 1895 (BMNH) suggests that the breeding season may be longer: although the age of these specimens is unknown, they probably fledged several months previously.

The nest and eggs are unknown. However, local people claim that Whiskered Pitta nests on the ground or in a bush within one metre of the ground, and that eggs are laid from early February (O. Jakobsen *in litt.*).

DESCRIPTION

Sexes similar.

Adult male Forecrown dark brown; lores, ocular area and ear-coverts dark olive-brown; rear of crown and nape orange-rufous; malar stripe bold, greyish to ivory, occasionally appearing pinkish, broadening under the cheeks; submalar area blackish-brown. Mantle, back and rump dull dark olive-green; uppertail-coverts and tail slaty-blue. Tertials and lesser coverts concolorous with dull dark olive mantle; medium and greater coverts slaty blue-grey, concolorous with breast; secondaries dark brown fading to greyish-brown near tips and edged on outer web with blue-grey. Primaries dark brown fading to greyish-brown at tip, and with small white spot in middle of inner web of primaries 3-4. Underwing-coverts grey-brown. Throat, chin and upper breast brown with vinaceous-brown wash to throat and pale purplish-lilac sheen to upper breast; lower breast slaty-blue to blue-grey, clearly but raggedly demarcated from scarlet belly to undertail-coverts. Sides of upper breast and upper flanks bright olive, although Jakobsen and Andersen (1995) report birds with blue extending across whole breast. Bill black, iris brownish to brick-red, legs and feet dull blue.

Adult female Very similar to male, but the scarlet on the underparts is slightly duller, the throat is slightly paler, and there is slightly less blue in the greater wing-coverts than in the male.

Immature female (from McGregor 1909-10) Top of head rather dark brown, shading gradually into a more rufous tint on the nape; all the feathers have rather darker margins, giving a slightly scaled appearance to crown; a few of the orange feathers of the adult are usually present. Rest of upperparts dark olive with a brown shade, though with a few scattered greenish adult feathers. The greyish-blue of the outer wing-coverts and outer webs of the secondaries is replaced by dull olive, and the slate-blue of the uppertail-coverts and tail is not so bright. The ear-coverts are brown with buff centres, the moustachial stripe dirty white. Feathers of chin and throat are white with black margins and bases, those on the foreneck being conspicuously white, but washed with reddish. Chest whitish-buff with scattered slaty-blue feathers; rest of underparts dirty whitish-buff, with most feathers, especially on flanks, margined with brownish-buff; a few pale scarlet feathers indicate the colours of the adult.

More advanced birds have upperparts like the adult except for the wings, which lack any slaty-blue; the foreneck and chest intermixed with white-and-buff-centred feathers whilst the rest of the underparts are mixed scarlet and whitish-buff.

Juvenile (Jakobsen and Andersen 1995) Forehead chestnut-brown with light ochraceous-buff centres. Lores and ear-coverts chestnut-brown. Crown, nape and sides of neck brown with light ochraceous-buff feather-centres. Malar stripe bold and whitish. Chin whitish; feathers of throat and breast cinnamon with broad whitish centres forming droplet-like spots, with a few indistinct blue feathers. Rest of underparts brown with narrow whitish centres forming streaks and pale scarlet or salmon-orange tips. Upperparts, including upperwing-coverts, olive-brown. Primaries and secondaries dark greyish-brown with small white patch across base of both webs of primaries 3-4. Tail dull blue. Bill black with ivory-yellow tip and basal half to lower mandible; gape bright yellow. Iris dark brown. Legs, feet and eye-ring bluish.

MEASUREMENTS Wing 115-123; tail 47-59; bill 34-35.5; tarsus 49-53; weight c.113.

17 RED-BELLIED PITTA
Pitta erythrogaster   Plates 9 & 10

Pitta erythrogaster Temminck, 1823, Pl.Col. Livr. 36, Pl.212: Manila, Luzon, Philippines.

Alternative name: Blue-breasted Pitta, Red-breasted Pitta

FIELD IDENTIFICATION Length 15-17.5. This is a very variable, relatively small species that is distributed on islands from the Philippines to the New Guinea region and Cape York, Australia. The blue breast and red belly are common to all subspecies. Subspecies differ primarily in head colour (which may, for example, include blue on the crown or ear-coverts), the presence or absence of a black breast-band and the colour of the upperparts, which varies between green and blue, both between subspecies and, to a lesser extent, within subspecies. The nominate subspecies has an orange-brown head, with a darker crown that becomes brighter chestnut-orange at the rear, particularly on the nape. The upperwing, mantle and tail are blue-green, and separated from the bright chestnut-orange nape by a conspicuous bright pale blue band. The throat is blackish and the breast blue-green. The rest of the underparts are bright red. In flight, the small white wing-patch is obvious. Immatures of most subspecies are dull brown above with scattered green and blue on the upperparts, and whitish below with dark brown or black scaling on breast and pale scarlet markings on belly. Juveniles lack red or blue in plumage, except for blue on tail and rump.

Similar species Within its range, adults are unlikely to be confused with any other species, with the possible exception of Whiskered Pitta, which is sympatric with the nominate subspecies on Luzon, Philippines. Other pittas in its range lack blue on the breast and extensive scarlet on the belly. Whiskered Pitta is much larger than nominate Red-bellied Pitta and has a prominent pale moustachial streak. Adults of all subspecies differ from Sula Pitta in lacking the broad black band across nape and in having dark, matt-blue or matt-green upperparts. Immature Hooded Pittas have sooty to dark buff underparts, without any scaling, and blackish heads, and are therefore quite different in appearance to immature Red-bellied Pittas, which are pale but scaled below and have brownish, usually mottled, heads.

VOICE The best-known call of Red-bellied Pitta, apparently made by most subspecies, is a low tremulous whistle, first ascending slightly and then descending, with a brief hiccup between the two parts of the call. The calls of different subspecies vary in inflection, and in length (see sonograms below). On Luzon, the first note of this call, *oaoaaaAAH-whoo*, wavers considerably, whilst in birds from Sangihe (*caerulitorques*) the call is reported to be a broken series of notes that resembles the call of other subspecies, so that the bird sounds like it is stuttering (D. Holmes verbally).

The call of *erythrogaster* on Mindanao differs from the tremulous whistle of *celebensis* on Sulawesi in that the 'hiccup' between the two parts of the whistle is shorter. Bishop and Coates (in prep.) describe the call of birds in Sulawesi as a drawn-out, hoarse, slightly tremulous medium-pitch two-note whistle lasting about 2.1s. The first note is upslurred and the second is longer and downslurred. D. Holmes (*in litt.*) noted that in hills near Ujung Pandang, south Sulawesi, the call rises in pitch at the end, but that in south-east Sulawesi the second note consistently rises and then falls. These calls are given at irregularly spaced intervals of one per 15-44s. Birds on Kai Kecil (*kuehni*) give a similar but more musical call to birds on Sulawesi, the second note being almost double.

The call of *macklotii*, from New Guinea, is similar to that of nominate *erythrogaster*, being described as two low-pitched, slow, tremulous whistles, the first rising in pitch, the second falling (Diamond 1972). Coates (1990) provides a similar description of the call, *crooooooi-crooooouw*, of Red-bellied Pittas in the New Guinea region, noting that the two notes take c.4.5s to complete and may be repeated many times at 15-30s intervals. The second, more plaintive note is sometimes repeated to give a three-note call, and occasionally this three-note call is consistently given with an occasional further repetition of the last note to provide a four-note call.

Meyer (1879) described a second type of calls from Sulawesi, made between members of pairs, as a melancholy and protracted *oppo*. On Obi (*rufiventris*), the most commonly heard call, made from the ground, is a brief *foh-foh* (pers. obs.), which is quite different to the calls usually reported as given by other subspecies.

In Australia, Beruldsen and Uhlenhut (1995) witnessed a female apparently calling to a male that had been in the canopy making the usual call. The call made by the female, which almost immediately caused the male to descend, was similar to a single mournful whistle (usually given three times in sequence) sometimes made by female Noisy Pitta.

In Wallacea, calling is seasonal, primarily occurring during the wet season. In southern New Guinea, calling is also frequent during the wet season but rare in the dry season. Calls are delivered from perches at lower and mid-levels of forest as well as from the ground.

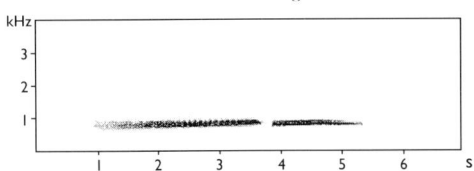

P. erythrogaster ?macklotii
[from Lae, PNG, recorded by H. Crouch, November 1978]

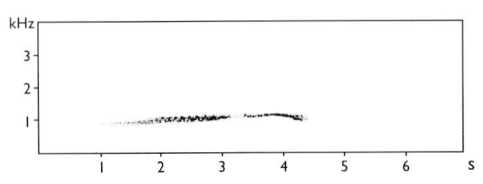

P. erythrogaster celebensis
[from Dumoga Bone, Sulawesi, recorded by A. Greensmith, January 1991]

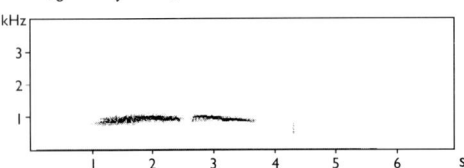

P. erythrogaster erythrogaster
[from Mindanao, Philippines, recorded by S. Harrap, February 1993]

DISTRIBUTION The Red-bellied Pitta ranges from the Philippines through Sulawesi and the Moluccan Islands to New Guinea, the Bismarck Archipelago and northern Australia. The distribution is complex, and for this reason is described under the relevant subspecies account below.

GEOGRAPHICAL VARIATION Twenty-four subspecies are recognised in the following account, many of which are island endemics. Some subspecies, such as the poorly known *P. e. splendida*, differ considerably from neighbouring subspecies. Further study of the taxa that are presently treated as belonging to *P. erythrogaster* would be of great taxonomic value: plumage, morphology (see Table 1) and vocalisations appear to vary, and it is likely that this variation is sufficient in some instances for recognition of separate species. Although it was not possible to undertake a comprehensive taxonomic review of *P. erythrogaster* during the course of preparing this volume, it is clear that the Sula Pitta stands out as being distinct, and this is treated as a separate species (see discussion of taxonomy under its species account).

Subspecies of the Philippines and Wallacea:

P. e. erythrogaster (Widely distributed in the Philippine archipelago, where known from the following islands: Basilan, Bohol, Bongao, Cagayancillo, Camiguin Sur, Camiguin Norte, Cebu, Guimaras, Jolo, Lubang, Luzon, Marinduque, Masbate, Mindanao, Mindoro, Negros, Panay, Romblon, Samar, Sibutu, Sibuyan, Siquijor, Tablas, Tawitawi, Ticao, Catanduanes, Leyte: Dickinson *et al.*, 1991) Crown to nape and sides of head bright pale rufous, throat darker brown becoming black on upper breast. Centre of breast, collar, upperwing-coverts, rump and tail blue; rest of mantle and sides of the breast green. Bright red belly to undertail-coverts, sharply defined from the green and blue breast. Primaries black with a small white bar across the central primary feathers, not usually visible at rest. The amount of green and blue in the plumage varies between individuals, with some birds having a narrow blue collar, little blue on the breast and more green in the inner part of the upperwing-coverts.

P. e. propinqua (Palawan and Balabac, Philippines) Very similar to the nominate subspecies, but with a tendency to be brighter and to have more (cobalt) blue on the upperparts, sometimes reaching from the rump to the upper mantle. Wing 93-98; weight 49-53.4.

P. e. thompsoni (Culion and Calaut Islands, Philippines. Not known from Busuanga, but likely to occur) Similar to *propinqua* but back, rump, tail and scapulars pale blue, not cobalt. Differs from nominate in much paler blue of back, rump, tail and scapulars, and in much narrower, dull green band on mantle, as in *propinqua*. Most similar to *erythrogaster* from Negros in intensity of blue on upperparts, but *thompsoni* is still paler (Ripley and Rabor 1962). Wing 94-94.5.

P. e. celebensis (Sulawesi, Manterawu Island off the Minahassa Peninsula, and the Togian Islands, Indonesia) Differs from *erythrogaster* in having darker, redder crown and nape, with a variable amount of pale blue on the central crown, forming a narrow coronal band, and more extensive blue on the breast, separated from the red belly by a broad black breast-band. Eyes dark brown; wing 102-110; bill length 24.5-27.4, depth 7.3-8.2.

P. e. inspeculata (Karakelong, Salebabu and Kabruang in the Talaud Islands, NE of Sulawesi, Indonesia) Dis-

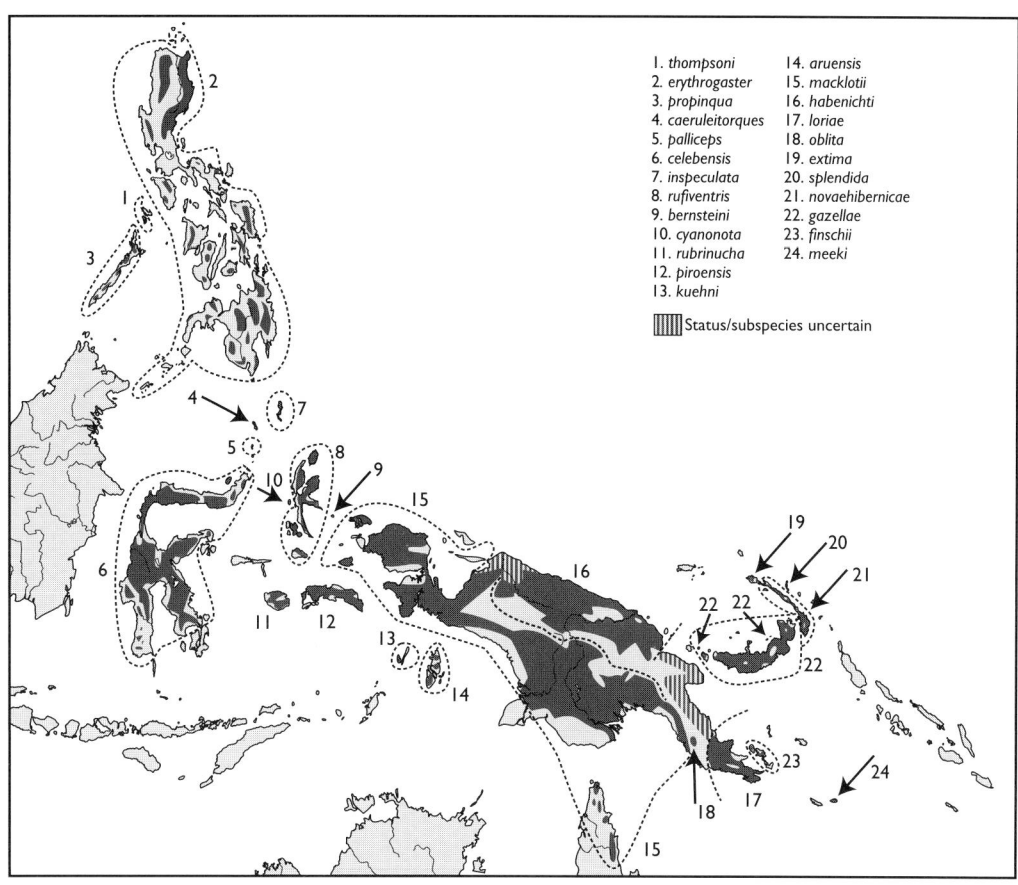

```
 1. thompsoni        14. aruensis
 2. erythrogaster    15. macklotii
 3. propinqua        16. habenichti
 4. caeruleitorques  17. loriae
 5. palliceps        18. oblita
 6. celebensis       19. extima
 7. inspeculata      20. splendida
 8. rufiventris      21. novaehibernicae
 9. bernsteini       22. gazellae
10. cyanonota        23. finschii
11. rubrinucha       24. meeki
12. piroensis
13. kuehni
```

|||||| Status/subspecies uncertain

Distribution based on approximate extent of forest remaining within the altitudinal range of the species. Former range more widespread.

tinctively different from other subspecies, and perhaps a good species. Differs from nominate in darker head, entire breast and upperparts blue. The blue breast-band is narrower than that of *erythrogaster*, with the red of the underparts extending onto the lower breast. The blue of the breast is notably darker than in any other subspecies, which are uniformly coloured. Small patch of pale blue at back of eye. The white wing speculum is poorly developed. Eyes grey brown, wing 95-97; bill length 24.9-27.3, depth 7.4-8.0; weight 50-57.

P. e. caeruleitorques (Sangihe Island, Indonesia) Like *celebensis* with the blue hind-neck strongly marked, but with no blue on the crown. Eyes grey; weight 67.5.

P. e. palliceps (Siau and Tahulandang Islands, Indonesia, between north Sulawesi and Sangihe Island) Similar to *celebensis* but has less blue on the middle of the crown, which is lighter and becomes rusty yellow on the nape. The black breast-band is missing or hardly discernible. Wing 103-104; bill length 26.4-27.2, depth 8.9.

P. e. rufiventris (includes *obiensis*) (The north Moluccan islands of Morotai, Halmahera, Moti, Bacan, Damar, Mandioli, and Obi. Probably also on Kasiruta) Differs from *erythrogaster* in having brighter, reddish tinge to nape and no blue collar; a paler throat; no black but more extensive blue on the breast and a very narrow dark band (of blackish and greenish feathers) separating the blue breast from red belly;

and only the tail and uppertail-coverts are blue. Eyes grey-brown to dark brown; bill heavy. Wing 93-96; bill length 27.0-29.8, depth 8.3-9.6; weight 57-66.

P. e. cyanonota (Ternate, west of Halmahera, Indonesia) Lacks green in plumage, like *inspeculata*, but rear crown a redder, brighter colour, and extending further onto the nape. The throat is pale buffy-brown and the whole breast is paler blue. Some birds have a narrow blackish band separating the red and blue. Differs from *finschii*, which is larger, in having foreneck, head and nape lighter and redder, and a much narrower black line (if present) separating the blue chest and red belly. Wing 93.5-98.5; bill length 26.8-28, depth 8.4-9.7.

P. e. bernsteini (Gebe island, between Halmahera and Waigeo, Indonesia) Similar to *cyanonota*, but slightly paler blue (more silvery-blue) on the upperparts and breast, and slightly larger, with longer wings (100-102) and bill (culmen 21-22) (Junge 1958).

P. e. rubrinucha (Buru Island in the Moluccas, Indonesia) Similar to *erythrogaster*, but lacks the blue collar and has a duller brown head marked with pale blue on the centre of the rear crown and the rear of the ear-coverts, and a distinctive reddish-orange nape-patch. Wing 93-96; bill length 29, depth 8.0; weight 55-65.

P. e. piroensis (Seram in the Moluccas, Indonesia) Similar to *rubrinucha* but darker green dorsally and has a larger scarlet nuchal patch. Wing 96-101, bill length 27.2.

P. e. kuehni (Indonesia on Tual, Kai Kecil and Kai Besar in the Kai Islands, and the small islands of Kur, Kilsuin, Kasiui and Tiur in the Watubela Islands to the northwest, between the Kai Islands and Seram) Intermediate in appearance between *celebensis* and *macklotii* (with a few pale blue feathers on the crown and nape, like *celebensis*, but colour of rufous on the head similar to *macklotii*), but tends to have a narrower black breast-band. Some specimens of *kuehni* are reported to have the entire upperparts blue (Hartert 1901). White and Bruce (1986) treat *kuehni* as a synonym of *macklotii*, and Mees (1982) notes that it is at most a poorly marked subspecies. Wing 98-102; bill length 24.2-27.1, depth 9-9.2.

New Guinea, Australian and South Pacific Island subspecies: the majority have a broad black breast-band separating the blue breast and red belly. The exact distributional limits of subspecies on New Guinea are not known.

P. e. macklotii (New Guinea and Cape York Peninsula, Australia. In New Guinea recorded from western and southern New Guinea – in north to Geelvink Bay, in south to Port Moresby – and the islands of Misool, Salawati, Batanta and Waigeo as well as on Yapen Island. On the Cape York Peninsula it is confined to the eastern coast from the tip south to the McIlwraith Range, and inland to Ducie and the Wenlock: Storr 1973, Blakers *et al.* 1984) Most similar to *celebensis*, but throat darker, crown to nape bright orange-rufous and lacks pale blue on head and upper mantle. The width of the black breast-band varies considerably (usually 1-10mm). Wing 104-112; bill length 27-31, depth 9.5-10.5; weight 70-98 (Diamond 1972, Mees 1982).

P. e. habenichti (N New Guinea from Weyland Mountains at head of Geelvink Bay east to Astrolabe Bay) Hardly differs from *macklotii*, and doubtfully distinct, but has slightly brighter nape.

P. e. loriae (SE New Guinea, west to the Kumusi River in the north and Cloudy Bay in the south) Differs from *macklotii* in having dark chestnut top of the head and hind-neck. Tips of hind-neck and forehead feathers more or less tipped reddish, but the former are not so pale reddish as in *macklotii*. Feathers of the crown darker, almost black, sometimes with faint blue edges. Often has a narrow blue line between the chestnut colour of the hind-neck and the green back. Wing 103-107; bill length 28.5-30, depth 10-11.5.

P. e. oblita (Known only from the mountains of the upper Aroa River, Central Province, Papua New Guinea) Reported to differ from *macklotii* in having a paler, less bright nape and a bluer, less green back (Mayr and Rand 1937). Most similar to *kuehni*, but blue on the neck of the latter is more extensive and it has a brighter, more reddish hind-neck and is smaller (wing 100-105: Rothschild and Hartert 1912).

P. e. meeki (Rossel Island, Louisiade Archipelago) Throat dusky grey. Similar to *macklotii*, but paler, duller rufous crown and nape and greyish throat. Wing 98-99; bill length 30.5-33, depth 10.0-10.8.

P. e. finschii (Fergusson and Goodenough Islands, D'Entrecasteaux Archipelago) Intermediate between *cyanonota* and *macklotii*, with upperparts entirely blue like the former, but underparts like the latter (except for blue, rather than green, sides of breast). Nape dull chestnut, concolorous with the crown. Differs from *loriae* in having blue upperparts, and from *cyanonota*

in larger size and head colour. Iris grey (BMNH). Wing male 109-110.5 (n=2), female 101-105.5 (n=2); bill length 28.5-30.5, depth 10-11; weight 89.4.

P. e. aruensis (Aru Islands, Indonesia) Doubtfully distinguishable from *macklotii*, but slightly smaller. Some specimens have a strong blue wash to the mantle, unlike *macklotii*. Wing 97-102.

P. e. gazellae (New Britain and Umboi [Rooke] Island, Bismarck Archipelago. Birds on Tolokiwa, Lolobau, Watom and Duke of York Islands also presumably this taxon) Differs from *macklotii* in having brighter orange-red sides of the rear head and nape, and in the possession of a few blue feathers on the crown, like *celebensis*. Wing 102-109; bill length 27.5-29.5, depth 8.5-10.

P. e. novaehibernicae (New Ireland, Bismarck Archipelago. Birds on Djual Island are presumably this subspecies or *extima*) Similar to *macklotii*, but throat brownish-red and black breast-band usually missing. Wing 92-103.

P. e. extima (New Hanover Island, Bismarck Archipelago) Like *macklotii*, but nape brighter orange-red and lacks the black breast-band. Differs from *novaehibernicae* in larger size, paler, more orange-rufous nape and in having the blue stripe on the crown less reduced and the back more bluish-green. Wing 101-107; bill length 29, depth 9.

P. e. splendida (Tabar Island, east of New Ireland) Upperparts rich deep blue, nape scarlet, forehead and crown blackish with a reddish wash, occasionally with a few bluish feathers in the centre of the crown, nape separated from back by narrow blackish band which is a continuation of the black throat. Underparts similar to *gazellae* but deeper and richer. Wing 107-111.

HABITAT Throughout most of its range, the Red-bellied Pitta inhabits rain and monsoon forests and secondary growth, usually below 1,000m. It has been found in a number of types of forested habitats, including forests on limestone.

Rand and Rabor (1960) found that on the small Philippine island of Siquijor, the nominate subspecies occupied broken forest and brush but, on nearby Negros, was confined to forest. During a collecting expedition to the southern Sulu Islands and Sibutu, du Pont and Rabor (1973) rarely encountered Red-bellied Pittas, and found that they were virtually confined to dark areas in original forest, although some birds were observed in dense secondary forests. On these islands, and in the Bataan peninsula, Luzon, Red-bellied Pittas apparently prefer dense undergrowth below tall trees bordering rivers, or on the banks of small creeks with shallow water (Gilliard 1950, du Pont and Rabor 1973).

In the Philippines, the nominate subspecies has been recorded from lowland sites, such as in the Bataan peninsula, Mindanao and on Mindoro, as well as in hills and mountains, to at least 1,400m, on Negros (Gilliard 1950, G. Dutson *in litt.*). On Bohol, it has been collected at 350-750m, and on Samar at 300-600m (Rand and Rabor 1960). On Luzon, the species occurs to at least 610m (Goodman and Gonzales 1990).

In Indonesia, this species is found in forest from the lowlands to mountains. In Sulawesi, *celebensis* occurs from the lowlands (Hartert 1897a) to at least 1,220m (Mt Masarang: BMNH), whilst *caeruleitorques* was collected in primary forest at 750m on Sangihe (F. Rozendaal *in litt.*). It has also been observed in mixed plantation agriculture on the island (J. Riley *in litt.*). Bishop and Coates (in prep.)

note that on Sulawesi this species is particularly common on volcanic soils, and is often found in forest with dense understorey of rattan palm, whilst B. Gee (*in litt.*) found it in areas of dense secondary growth. Stresemann (1940) also noted its preference for dense understorey. In the Talaud islands, off Sulawesi, Bishop (1992) found *inspeculata* close to a village on Salebabu in remnant forest patches within a mosaic of coastal cultivation, scrub and secondary woodland, whilst F. Rozendaal (*in litt.*) collected this subspecies on Karakelong in forest edge and primary forest at 10-80m.

Further east, in the Moluccas, *piroensis* has been observed from 100m to at least 900m on Seram (Taylor 1991), *rubrinucha* occurs to at least 840m on Buru, and Lambert (1994) found *rufiventris* to be widely distributed in the north Moluccan islands of Halmahera, Bacan and Obi, from sea level to 550m, where it appeared to favour scrub, patches of bamboo and logged forest. Although it was not recorded in primary forest on Bacan there are specimens said to come from this habitat, and on Obi birds were heard calling in primary forest above 500m. A specimen label from Bacan indicates that *rufiventris* occurs to 2,135m on that island. Stresemann (1914) noted that on Seram *piroensis* was usually associated with bamboo thickets. On the rather isolated Kai Islands, *kuenhi* occurs in severely degraded monsoon and secondary forests and remnant patches of forest within cultivation on Kai Kecil, but only in forest on the adjacent cultivation of Kai Besar (Bishop and Coates in prep.).

In eastern New Guinea, *macklotii* occurs in primary forests, secondary growth and occasionally in scrub in the flat lowlands and in hilly areas to at least 1,220m, whilst *habenichti* has been recorded as high as 1,680m (Coates 1990). Rothschild and Hartert (1901) document specimens taken at 1,525-1,830m in the Eafa District between Mt Alexander and Bellamy. Gibbs (*in litt.*) notes that, in northern New Guinea, both this species and the Hooded Pitta *P. sordida* appear to be absent from the extensive alluvial lowlands that have a tendency to flood. The Australian population of *macklotii* is confined to rainforest and closed gallery forest on the east coast of Cape York peninsula (Storr 1973, Blakers *et al.* 1984). On Goodenough Island, *finschii* has been collected in the mountains at c.600m (Rothschild and Hartert 1914a, USNM). Coates (1990) noted that, in the New Guinea region, Red-bellied Pitta often co-exists with the ecologically similar Hooded Pitta on the mainland but infrequently on islands.

STATUS Although nominate *erythrogaster* is common in parts of its range, such as in parts of Luzon, in other areas it is apparently rare, or overlooked. Parkes (1973), for example, could only trace two records from Leyte, one of which was from a boat anchored offshore (Prescott 1973), whilst Gilliard (1950) stated that it was very uncommon in the Bataan peninsula, Luzon. Watling (1983) and Rozendaal and Dekker (1989) frequently heard *celebensis* in central Sulawesi and Minahassa, suggesting that it is not rare in suitable forest habitat. Lambert (1994) found *rufiventris* to be common and widely distributed on Bacan and Obi in the north Moluccas. However, on Halmahera *rufiventris* may be uncommon or rare since there are few records. Elsewhere in Wallacea, K.D. Bishop found the species to be fairly common on Salebabu (*inspeculata*) and on Kai Kecil (*kuehni*), although it is reported to be uncommon on the nearby island of Kai Besar (Bishop and Coates in prep.). Stresemann (1914) stated that *piroensis*

was probably not rare on Seram, where local people usually knew the birds in the areas he visited.

Although the status in New Guinea and its satellite islands is not documented, much forest remains on this island, and *macklotii* must therefore still have large populations there: Bell (1982a) estimated that the density of Red-bellied Pittas was around 30 per km^2 at his study site in lowland forest near Brown River, Central Province, Papua New Guinea. Storr (1973) described the status of *macklotii* in Australia as common. The status of *finschii* on Fergusson Island and Goodenough Island was described as common by Rothschild and Hartert (1901) at the beginning of the century, but is presently unknown.

MOVEMENTS The nominate subspecies appears to be a local migrant, making movements within Luzon, and perhaps throughout its range in the Philippines, although the extent of these migratory movements is unknown. In northern Luzon, Igorot tribespeople have traditionally caught migrants, including Red-bellied Pittas, by attracting them to lanterns set in high mountain passes during foggy weather on moonless or overcast nights. During a five-year study of migrants at Dalton Pass, in the mountains of north-east Luzon, nearly 2,000 juvenile and adult Red-bellied Pittas were netted at night between September and early December, ringed and then released. One juvenile that was trapped on 1 November 1967 was retrapped three years later, on 5 November 1970, 30 miles to the north-west, as it crossed another high mountain pass near Baguio. Nine other individuals were recaptured and released at Dalton Pass, at intervals of up to two months (McClure 1974, McClure and Leelavit 1972).

Prescott (1973) collected a specimen of *erythrogaster* on a boat, anchored 3-6km offshore of Leyte in Leyte Gulf, on 4 July. Other birds have been collected on boats after flying into searchlights at night in Manila Bay, Luzon, during November (Parkes 1973). There is therefore good evidence that birds in the Philippines make regular movements, although the nature of these movements remains unclear.

Part of the population of *macklotii* from northern Queensland, Australia, may move to southern New Guinea between March and October, although evidence for this is rather inconclusive, with very few records, all from islands in the southern Torres Strait (Draffan *et al.* 1983, Coates 1990). Within New Guinea, there is evidence that the species may be rather nomadic in certain areas, though nomadic movements are poorly understood. Bell (1982b) did not find Red-bellied Pittas in rainforest at his Brown River study site, southern Papua New Guinea, during the harsh dry season of August to November 1976. In contrast, the species was observed regularly during the more typical dry season of 1977. During the wet season, this species was also very rare or absent at Brown River, in contrast to Hooded Pitta, which was commonly observed, suggesting that Red-bellied Pittas had emigrated to other areas.

FOOD Insects and worms have been found in the stomachs of specimens from the Philippines. Meyer (1879) reported that, on Sulawesi, this species's diet included small beetles and caterpillars, whilst small beetles and other insects have been found in specimens from the Talaud Islands (F. Rozendaal *in litt.*). In New Guinea, insects, their larvae and, on one occasion, vegetable matter have been found in the stomachs of Red-bellied Pitta (Coates 1990).

137

At Cape York, earthworms have been recorded in the diet (U. Zimmermann *in litt.*).

HABITS Tribespeople in New Guinea claim that *macklotii* is usually encountered in small groups of about three birds (Diamond 1972), although Coates (1990) notes that, in New Guinea, Red-bellied Pittas are invariably seen alone. However, it is not infrequent for pairs to feed in close proximity to each other in other parts of the range (pers. obs.). Feeding habits are poorly documented, but some food is probably obtained by digging in the ground, since birds are occasionally observed with mud on their bills. Like many other pittas, this species is often observed feeding on forest trails at first and last light. In Luzon, such birds frequently bob their tails downwards, perhaps in response to the observer (pers. obs.).

In Australia, Zimmermann (in press) once observed a Red-bellied Pitta, apparently alone, performing a bowing display similar to that documented for Rainbow Pitta. Beruldsen and Uhlenhut (1995) also observed such a display in Australia involving two birds. In this instance, the female bowed, on one occasion so low that the breast pressed on the ground, but a nearby male was not seen to respond. The male had apparently been called to the ground by the female, prior to the display, immediately after the female began uttering a call that is similar to the mournful whistle of Noisy Pitta.

BREEDING Breeding biology of Red-bellied Pitta is poorly known. In the following section, aspects of breeding are described under seasonality, nests and eggs, in an attempt to make this long account easier to use.

Seasonality In the Philippines, spotted juveniles have been collected between 22 August and 3 April (USNM) whilst breeding has been noted in Luzon in August and in Palawan in September (Dickinson *et al.* 1991). A nest containing two eggs was found in the Sierra Madre Mountains of Luzon on 24 February (D. Allen *in litt.*).

In Sulawesi, half-fledged chicks of *celebensis* have been recorded in early March (White and Bruce 1986), and immatures in September, on 25 November and on 23 January (MZB, BMNH), whilst Rozendaal and Dekker (1989) noted that singing males were holding territories between January and March in Domoga-Bone National Park. To the north of Sulawesi, two immature *palliceps* were collected on Siau on 24 October 1865 (RMNH) and a recently fledged *inspeculata* was collected on Talaud on 16 May 1925 (MZB).

On Buru, Indonesia, nests of *rubrinucha* have been found on 4 May (BMNH), 19 November (with two eggs), and an incubating bird was caught off another nest on Buru on 24 November (Jepson 1993). Wallace (1869) collected an immature on Buru in late June.

F. Rozendaal (*in litt.*) collected a recently fledged specimen of *rufiventris* in primary forest on Halmahera on 31 July, and immatures have been collected on the island on 9 July and 10 August. On Bacan, immature *rufiventris* have been collected between late June and mid-August (Nehrkorn 1894), and Lambert (1994) saw an immature in secondary forest in late October. Juvenile *cyanonota* have been collected on nearby Ternate on 14 January and 10 March (von Berlepsch 1901), with two together on 6 May (RMNH).

In northern New Guinea, nests of *habenichti* containing eggs have been found in April, and nests containing young in mid-March, whilst older immatures have been collected in April, August and September (MZB). On the West Papuan islands, a recently fledged bird (*macklotii*) was collected on Waigeo on 27 March and an older immature on Misool on 30 May.

In the Port Moresby region of southern New Guinea, a nest with eggs was found in late April and observations show that juvenile *macklotii* appear there in the early dry season. Immatures of this subspecies have been collected in Papua New Guinea between June and August and in south-west Irian Jaya on 23 October. Other young immatures from unrecorded localities in Irian Jaya have been collected on 17 November and 6 January, and older immatures on 8 September and 3 May. Adult *macklotii* have been collected in New Guinea with enlarged gonads in late July. This suggests that most breeding in New Guinea is completed in the wet season, although some birds probably breed during the dry season.

In Australia, the breeding season of *macklotii* is reported to be from October or November to January or February (Beruldsen 1980), although breeding may not start until January (Storr 1973) in some areas or in some years.

Eggs of *finschii* have been obtained on Fergusson Island on 16 September and 16 October, whilst a spotted juvenile has been collected on 11 April (Rothschild and Hartert 1914, BMNH). On New Britain, Dahl (1899) found the eggs of *gazellae* between December and March and nestlings in February and early April, indicating that breeding there occurred during the rainy season. However, on the small offshore island of Watom, eggs have been recorded in June, September, December and February (Meyer 1906). A nest of *meeki* was collected on Rossel Island in late January or early February (BMNH).

Nests A nest found in Luzon (*erythrogaster*) was built into roots in the bank of a dry streambed, so that the entrance hole was some 80cm off the ground (D. Allen *in litt.*).

A nest of *rubrinucha* on Buru (Jepson 1993) was located 1.5m off the ground, on the shelf between two buttresses of a massive emergent tree. The nest was a round ball, 25cm in diameter. The base was constructed of twigs up to 20cm long and 0.5cm in diameter supporting a dome of smaller twigs, dead leaves and fine bark and, inside, lined with leaves. The remains of a nest of *rubrinucha* from Buru comprised skeletonised and dried dicotyledonous leaves, with the nest cavity constructed of palm or bamboo leaf fronds and fragments (BMNH). A nest collected on Damar (*rufiventris*) was constructed primarily of flexible fine twigs, up to 25cm in length and perhaps from a vine, with scattered dead broad leaves and leaf skeletons (RMNH).

Villagers on Halmahera claim that *rufiventris* sometimes nests in rattan 30-40cm above the ground, and that nest material also includes dead rattan leaves and mosses, although this has not been verified by ornithologists. Meyer (1879) describes a nest of this species on Sulawesi (*celebensis*) as being found on the slope of a river bank, stating that the adult digs a hole in the bank to accommodate the nest.

In New Guinea, nests of *macklotii* are also usually placed on the ground at the base of a tree (Coates 1990), although Beruldsen (1980) noted that in Australia nests are usually a few metres above ground level and occasionally may be located as high as 10m up. Typical nest-sites in Australia are on logs amongst fallen debris, on tree stumps, in vine thickets or in the fork of a tree amongst fallen leaves and other debris.

Descriptions of nests of Red-bellied Pitta from New Guinea suggest some differences to nests described above. In New Guinea, nests may be lined with a mat of black hairlike fibres (perhaps fungal hyphae), whilst green ferns and many skeletonised large leaves form the roof. Smaller dead leaves, rootlets, grass stems and dead vines form the bulk of the dome, which, like the nest from Buru, is built on a foundation of large sticks. Nests in New Guinea have an outside diameter of 17-22cm with a side opening of 5-6cm diameter and a nest cavity of 11-12cm diameter (Coates 1990). One nest from south-east New Guinea (BMNH) was a bulky mass 15-18cm in diameter, the interior largely of fine twigs or roots of 1mm diameter or less, enclosed by an outer nest that contained skeletonised and entire dicotyledonous leaves, palm leaf pieces, and pieces of fern. A second nest at BMNH from New Guinea was similar, but contained green mosses in the outer nest and a larger proportion of rootlets in the inner nest.

The remains of a nest of *meeki* from Rossel Island (BMNH) was a bulky mass some 20cm in diameter, constructed of long fine twigs and a few rootlets, 1-2mm in diameter, with large leaf skeletons, a few large pieces of dicotyledonous leaf and a few pieces of bamboo and palm leaf, in the outer wall.

Eggs Clutch-size is usually two throughout the range of the species. Eggs from Buru were white, with some spotting (which may have been dirt), towards the pointed end. In New Guinea, eggs are 26.9-34.5 x 20.8-24.3mm in size, and cream or white in colour, with heavy brown and greyish spotting and blotching (Coates 1990). Nehrkorn (1894) described the eggs of *rufiventris* as being creamy-white with regular, clearly marked blackish-brown spots, and measuring 29.5 x 23mm.

Meyer (1879) stated that, in Sulawesi, the female incubated the eggs whilst the male took over when the female foraged for food. Incubating Red-bellied Pittas that were collected from nests in north-east New Britain were all females, suggesting that the proportion of time that males spend helping with incubation may be small. However, a male was collected off a nest in northern New Guinea (Gilliard and LeCroy 1967).

DESCRIPTION *P. e. erythrogaster*
Adult male Centre of forehead rufous-chestnut, sides of forehead and sides of crown blackish, mixed with increasing amount of rufous towards rear; top of crown variable, either deep chestnut mixed with darker tones or deep rufous-chestnut, and extending onto nape where brightest. Lores, ocular area, ear-coverts and sides of throat rufous-orange, becoming brighter towards nape and sides of upper neck, and mixed with variable amount of paler rufous giving a mottled appearance. The bright rufous-chestnut nape is bordered below by a conspicuous bright pale blue band of varying width that abuts the dark green mantle and back; rump and tail bright pale blue (duller than that of the neck), this colour sometimes extending onto the lower mantle. Innermost wing-coverts concolorous with green of mantle, whilst medium and greater coverts are blue. There is a hidden white patch near the bend of the wing, formed by white central spots to several median covert feathers. Tertials may be blue or green or mixed in colour. Primaries black with blue outer edge near tip of outermost feathers, the extent of this blue gradually increasing so that the innermost secondaries are entirely blue on the outer web. Primaries black above, tinged greyish below. Tips of primaries 4-9 are blue, with slight green gloss, this colour extending down the margin of the outer web so that the exposed part of flight feathers at rest is largely blue. Small white band midway down primaries, usually hidden at rest, is formed by white spots on inner web of primary 2, adjacent white band in primaries 3-4 and spot in centre of primary 5. This white spot is variable and can be limited to outer web of primary 3-4. Centre of throat grey, mottled with white formed by central base to feathers, and merging into black bib-shaped patch on upper breast. Rest of breast green, as upperparts, with blue in centre which varies in extent from a small area mixed with green to a large central band that restricts the green to the breast sides. Lower breast to belly bright scarlet-orange. Feathers of tarsus grey. Some individuals have narrow black streaks on the upperparts (formed by black wedges along central shaft of scattered feathers), these usually being confined to the lower mantle, back, rump and occasionally on the uppertail-coverts and inner wing-coverts. Legs pale bluish-grey with a vinaceous wash; iris bluish-slate; bill black, tip bone-coloured, upper mandible with a faint orange wash on the culmen.

Adult female Very similar to male, differing in overall duller coloration, particularly on head and in red tone of underparts. Throat more mottled, and blackish rather than black on upper breast.

Juvenile Crown brown, darkest on sides of crown and palest (buff) on forecrown, sometimes with fine pale streaking towards rear. Sides of head, including ocular area, cheeks, lores and ear-coverts buffish, gradually becoming more rufous with age and mottled with darker tones on ear-coverts. Chin and throat white to buffy-white; breast dull buffy-white with darker sooty scales on upper breast and dark mottling on lower breast and, frequently, a more or less clear white area at top of breast. Mantle and wing-coverts brown, usually mixed with some green feathers; wing-coverts brown to dark olive; uppertail-coverts and tail tip dull dark blue; base of tail brown. Primaries dark brown, with white patch as in adult but lacking any blue on feather tips. Secondaries brown with dark olive leading edge most extensive on inner feathers. Feathers of breast largely buffy to buffy-white with narrow dark brownish edges giving scaled or mottled appearance, gradually darkening with age and with increasing numbers of green or black feathers (depending on position); belly to undertail-coverts white mixed with dirty buff and faint pinky tinge on undertail-coverts.

Immature With increasing age, adult feathers develop in patches so that immatures are rather scruffy and no two birds are alike. Green feathers appear on the upperparts, the tail becomes completely blue, and chestnut feathers develop on the rear of crown and nape. Scaling on the breast usually increases in intensity and a few blue feathers begin to appear, whilst the upper flanks may become cinnamon before this colour is replaced by blue. The belly and undertail-coverts develop into a mixture of pinky, reddish and whitish feathers and the secondaries acquire bluish leading edges. Basal third to half of lower mandible red to orange-red, bill tip reddish-orange.

In New Guinea (*macklotii*), the inside of the nestling's mouth is bright orange-red, and the edges of the gape orange.

MEASUREMENTS Wing 95-101; bill length 24-28, depth 7-8.3; tarsus 33.5-37; tail 36.5-43.5; weight 42-69.

Subspecies	sample size	wing length	bill length	bill depth
inspeculata	4	95-97	24.9-27.3	7.4-8.0
piroensis	1	96-101	27.2	7.8
celebensis	5	102-105	24.5-27.4	7.3-8.2
erythrogaster	18	95-101	24.0-28.0	7.0-8.3
palliceps	2	103-104	26.4-27.2	8.9
kuehni	2	98-102	24.4-27.1	9.0-9.2
rubrinucha	1	93-96	29.0	8.0
cyanonota	8	93.5-98.5	26.8-28.0	8.4-9.7
extima	1	101-107	29.0	9.0
rufiventris	8	93-96	27.0-29.8	8.3-9.6
gazellae	6	102-109	27.5-29.5	8.5-10.0
macklotii	10	104-112	27.0-31.0	9.5-10.5
finschii	4	101-110.5	28.5-31.0	10.0-10.5
loriae	5	103-107	28.5-30.0	10.0-11.5
meeki	3	98-99	30.5-33.0	10.0-10.8
Sula Pitta	1	93-99	31.5	8.5

Table 1. Wing length and bill size (mm) of subspecies of Red-bellied Pitta *P. erythrogaster*, approximately in order of increasing bill size.

18 SULA PITTA
Pitta dohertyi   Plate 10

Pitta dohertyi Rothschild, 1898, Bull. Brit. Orn. Club, 7, p. 33: Sula Mangoli.

TAXONOMY The Sula Pitta has usually been treated as a subspecies of the Red-bellied Pitta, to which it is certainly closely related (e.g. White and Bruce 1986), but it was treated as a full species by Sibley and Monroe (1990). It has a strikingly different head pattern to all other subspecies of Red-bellied Pitta, and uniquely marked upperparts, which appear scaly and iridescent in the field (the upperparts of all subspecies of *P. erythrogaster* are matt); it also has a narrower bill than any subspecies of *P. erythrogaster* for which measurements have been taken (see Table 1), as well, apparently, as a distinctive call (see sonograms). It is also noteworthy, perhaps, that Sula Pittas on Taliabu did not respond at all to playback of Red-bellied Pitta calls from New Guinea (B. van Balen *in litt.*).

FIELD IDENTIFICATION Length 16. Sula Pitta shares a number of features with the allopatric Red-bellied Pitta, including the blue breast and red belly. The sides of the face and entire throat are black, extending as a broad black nuchal collar which separates the rufous-red crown from the rest of the upperparts. The nuchal collar is bordered below by a pale blue band, contrasting with the shining green lower mantle and back, which has a distinctly mottled pattern formed by darker green feather-centres and paler edges. The entire upperwing-coverts and outer edges and tips to the secondaries are pale blue with some green tinge in the tertials. The blue breast is separated from the scarlet-orange belly by a distinct blackish-blue band. There is a narrow white band across the primaries, but this is not usually visible at rest. Females have a white iris (Rothschild 1898).
Similar species None in its limited range. Adults differ from adults of all subspecies of allopatric Red-bellied Pitta in having a broad black band across nape and in pattern and shining texture of green and blue feathers of upperparts.

VOICE The voice of Sula Pitta has not been described, but tape-recordings made on Taliabu by B. van Balen are believed to have been this species (see sonogram). The recorded call starts with an abrupt trisyllabic phrase that is followed, after a very brief pause, by a series of five drawn-out notes that descend in pitch and volume *hk-wha-ha khwa-kwa-wha-wha-wha*, sometimes with an interruption during the latter half of the series. The whole call lasts c.3s. Another call heard, but not tape-recorded, is described as *kwek-kwek-kwek-kwek-kwoOo-(kwoOo)* with the last note (or two) more drawn out (B. van Balen *in litt.*). Although Sula Pitta was not seen to make these calls, the species was observed in the exact area from which calling occurred immediately afterwards, and there seems little doubt that the recording was of this species (B. van Balen verbally).

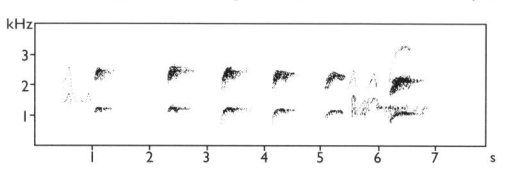

P. dohertyi
[from Taliabu, Sula Islands, recorded by B. van Balen, October 1991]

DISTRIBUTION Endemic to Indonesia, where known only from the small islands of Peleng and Banggai in the Banggai Archipelago, and the Sula Islands of Mangole, Taliabu and Seho (south-west of Taliabu), to the east of central Sulawesi. It is not known whether it also occurs on Sulabesi.

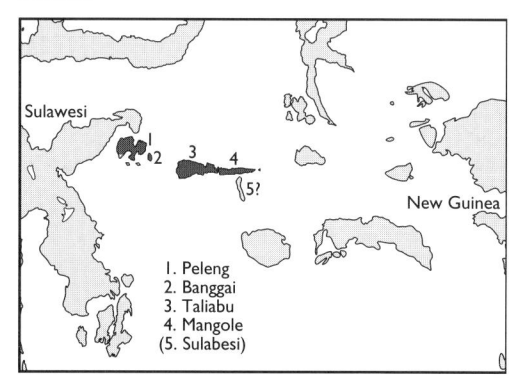

1. Peleng
2. Banggai
3. Taliabu
4. Mangole
(5. Sulabesi)

Distribution based on approximate extent of forest remaining within the altitudinal range of the species. Former range more widespread.

GEOGRAPHICAL VARIATION No subspecies are recognised, but birds from Sula are reported to have darker feather centres on the upperparts, forming a more scaly appearance (Hartert 1898a).

HABITAT Davidson and Stones (1993) and Davidson *et al.* (1994, 1995) reported that Sula Pitta occurs in a variety of habitats on Taliabu, observing it in selectively logged and seriously degraded forests, and particularly in areas where large stands of bamboo dominate the forest understorey. The species appears to be restricted to the lowlands, since it was not recorded above 200m. On

Banggai Island, this species has been observed in tall secondary forest (Bishop and Coates in prep.).

STATUS Davidson *et al.* (1994, 1995) judged this species to be uncommon in forests of Taliabu, finding it at only three of six sites visited. They therefore consider this to be a threatened species, despite the fact that it can evidently survive in seriously degraded and selectively logged forest, and that most of the lowlands still support these habitats. This tolerance of habitat disturbance suggests that it is not seriously threatened at the present time.

FOOD Nothing is recorded about the diet of this species, although it is likely to feed on similar types of invertebrates to those taken by the closely related but allopatric Red-bellied Pitta.

HABITS This is a very poorly known species, with virtually nothing documented about its behaviour. It is usually solitary, hopping away when disturbed. Birds observed on Taliabu were never seen to perch in low vegetation or trees (T. Stones *in litt.*). Stones also reported observing Sula Pitta, Elegant Pitta and Red-backed Thrush *Zoothera erythronota* feeding in close proximity to each other.

BREEDING W. Doherty collected a young male on Mangole Island in October or November 1897 (Hartert 1898a). A probable immature was observed on Seho Island in mid-October 1991 (Davidson and Stones 1993). There are no other records of breeding, the nest and eggs being unknown.

DESCRIPTION
Sexes similar, although eye colour is apparently different.
Adult Forehead, most of forecrown and lores blackish, with a few small patches of pale blue on forecrown. Crown to nape dark orange-red, the feathers basally blackish. Small feathers immediately around and behind eye blackish mixed with very pale blue, giving the appearance of grey, and forming a small pale patch. Ear-coverts, throat and upper breast black, this black extending as a broad collar around the nape and bordering the orange-red crown. Feathers at the lower edge of the black collar have broad pale blue tips that form a second collar at top of mantle. Rest of mantle green with a distinct shining mottled pattern formed by darker green feather-centres and paler edges. These feathers mix with pale blue on the lower back but green feathers are absent from the rump and uppertail-coverts, which are entirely pale, slightly slaty, blue. The black upper breast is bordered below by a band of pale shining blue which is in turn bordered at the bottom of the breast with a black band. Belly to undertail-coverts bright scarlet-orange. Scapulars and innerwing-coverts green, the feathers with darker centres and paler edges; median coverts largely pale blue, the feathers with brighter edges, and mixed with a few green feathers; greater coverts shining blue; primary coverts black. Tertials greenish on inner web, blue on outer web. Primaries brown with pale blue-grey tips to all except outermost two feathers, this blue-grey brighter and most extensive on innermost feathers, and continuing as narrow pale blue margin down distal quarter of outer margin; white spot in base of primaries extends across both webs of primaries 3-4 but is confined to inner web of primary 2. Secondaries brown with narrow bright blue outer margin and tip that broadens towards inner wing so that the margin extends across both webs of innermost secondary. Underwing-coverts pale grey-brown mixed with some pale blue and white markings. Bill blackish-brown with pinky-horn tip and gape; eye brown; legs blue-grey.
Adult female Differs from male in having throat brownish-black, iris white and in slightly smaller size (Rothschild 1898).
Immature Undescribed.

MEASUREMENTS (n=2): Wing 93-99; tail 37-39; bill length 31.5, depth 8.5 (n=1); tarsus 39-40; weight unrecorded (probably in region of 50-70).

19 BLUE-BANDED PITTA
Pitta arquata   Plate 11

Pitta (*Phoenicocichla*) *arquata* [sic] Gould 1871, Ann. Mag. Nat. Hist., ser.4, 7, p.340: Borneo.
Alternative name: Some authors (including Sibley and Monroe 1990) refer to this species, in error, as *P. arcuata*.

FIELD IDENTIFICATION Length 15.5-17.0. Adult unmistakable, with largely orange-red head and scarlet underparts, prominent narrow azure-blue stripe behind eye and band across breast, and dull olive-green upperparts. Immatures are brown with paler belly and red on flanks.
Similar species Adults are unlikely to be confused, both sexed being very distinctive. Juveniles and immatures, however, resemble immature Black-headed Pitta and Garnet Pitta. Those of Blue-banded Pitta differ, in their overall paler appearance with red tones on the flanks, and in lacking blue on their wings, tail and behind the eye. Also note that Black-headed and Garnet Pitta usually occupy a different altitudinal range to Blue-banded Pitta.

VOICE The call is a low monotone whistle that is confusingly similar to Garnet Pitta, Black-headed Pitta and Malaysian Rail-Babbler *Eupetes macrocercus*, all of which occur in Borneo (see sonograms on p.144). The call of Blue-banded Pitta lasts c.1.6s, and is therefore shorter than that typical of Black-headed Pitta, but is more similar to that of Garnet Pitta. The calls of both species are made at virtually the same pitch and only experienced observers are likely to be able to differentiate the calls in the field. B. van Balen (*in litt.*) notes that in East Kalimantan, calls of Garnet and Blue-banded Pittas are about the same pitch, but Blue-banded Pitta never has any upward inflection, and indeed, may have a detectable downward inflection. Wilkinson *et al.* (1991a) report that two birds observed in close proximity gave short, quiet *peep* notes to one another. This species often calls from off the ground, on relatively exposed perches up to at least 3m, particularly if responding to the calls of other individuals.

DISTRIBUTION The Blue-banded Pitta is endemic to Borneo. It has been recorded from scattered localities throughout the central mountain ranges, although there are few records from Kalimantan (Indonesian Borneo), so that its distribution there needs clarification. The lack of Indonesian records, however, may relate to paucity of observer coverage and difficulty in finding the species rather than to a limited distribution, and it seems likely that Blue-banded Pitta occurs throughout the uplands of Borneo.

    In Sabah, it has been recorded from Mt Kinabalu, the Segama River, Bole River, Mt Ensuan, Melangkap, Maliau, Poring, Rayoh, Saliwangan, Togudon and Ulu Samuran

(Sheldon in prep.). In Sarawak, it is known from Mt Poi, Mt Lambir, Mt Dulit, Mt Mulu, whilst Chasen and Kloss (1930) collected it in lowlands from Saribas and Samarahan in southern Sarawak. In Indonesia, it has been found at Ngara in West Kalimantan (Coomans de Ruiter 1938); in the lowlands of the Barito River catchment, central Kalimantan; near Long Pujungan on the Kayan River, bordering Sabah (B. van Balen verbally); on Mt Kenepai, in the upper catchment of the Kapuas River; and on Mt Penrissen, on the border with Sarawak (Chasen and Kloss 1930). There are no records from the hills in the south of Kalimantan, such as from the Meratus Range.

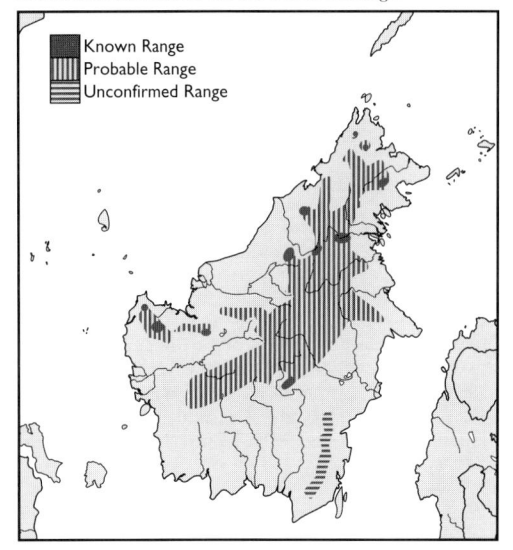

Known areas of occurrence and probable range, based on extent of remaining forest and altitudinal preference. Former range more widespread. Occurrence in mountains of south-west needs confirmation.

GEOGRAPHICAL VARIATION None recorded.

HABITAT The Blue-banded Pitta is a species of mixed dipterocarp forests. In Sabah, it has been recorded in primary and secondary hill forests and lower montane forest, including areas dominated by bamboo. It is most frequently found at higher elevations, but the overall range is from 150m, on Mt Ensuan, to 1,250m (Sheldon in prep.). Whitehead collected it on Mt Kinabalu in 'thick bamboo jungle' from to 610m to 1,220m (Sharpe 1889), and it has been found in the vicinity of very steep slopes in primary forests in the catchment of the Segama River, eastern Sabah, at 300-350m (pers. obs.). Wilkinson et al. (1991b) also noted its apparent preference for steep slopes, and suggested that it is a 'slope specialist'. The species is often observed in areas where fallen tree trunks are found.

Most records from Sarawak are from above 600m, with one record from 1,525m on Mt Poi, although specimens have been collected as low as 305m on Mt Lambir (Baram District, Sarawak) (UMZC). Elsewhere in Sarawak, Hose (1898) found it at 610-1,220m on Mt Dulit and Sharpe (1894) reported it at the same altitudes on Mt Mulu.

In Indonesia, Büttikofer (1900) found it at 550m on Mt Kenepai. Recent records from Central Kalimantan are from 130-250m (Wilkinson et al. 1991b), whilst in East Kalimantan B. van Balen (verbally) reported hearing Blue-banded Pitta at c.280m at Long Pujungan on the Kayan River, in an area where Garnet Pitta was also observed, indicating sympatry. The known altitudinal range would suggest that the species must be widely distributed in Borneo, although there is no evidence that it occurs in the areas of swamp forest or pure heath forest (*kerangas*) that are common at lower altitudes.

STATUS In Sabah, this species appears to be rarer than other pittas that occur within its altitudinal range, although it is probably fairly widespread in areas where suitable habitat is found (pers. obs.). Fogden (1976) found it to be uncommon or rare in Sarawak. There are few records from Kalimantan, and Coomans de Ruiter (1938) noted that it was rather rare in Kalimantan, although it may be more abundant locally. Neither Büttikofer (1897) nor von Plessen (Stresemann 1938) found the species in Kalimantan, although Lumholtz found it on the upper Kayan River (Voous 1961). However, Wilkinson et al. (1991a) commonly heard it in the upper catchment of the Barito River, central Kalimantan, and Shelford (1900) reported hearing its call constantly on Mt Penrissen.

FOOD Three specimen labels (SarM) record the following in the diet: insects and ants, ants, insect parts.

HABITS Generally quiet and unobtrusive, Blue-banded Pitta can be very difficult to observe, usually seeking dense habitat when disturbed. When an adult and juvenile were startled by an observer in Sabah, the adult flew to a log whilst the juvenile flew to a perch 7m off the ground. The adult flicked its wings in an agitated manner and dropped to the ground from where it commenced to call softly to the juvenile, which eventually flew down to rejoin the adult (B. Gee *in litt.*). This species has been observed calling c.3m from the ground, perched on the bough of a large fallen tree trunk (pers. obs.).

BREEDING The little information on breeding that exists, outlined below, suggests that the breeding season is long, probably spanning the period March to August or September.

A nest of this species was discovered at c.490m altitude in primary hill forest in Sabah, on 21 May (pers. obs.). An adult was flushed off the nest, indicating that eggs or young nestlings were present. The nest was situated about 2.5m off the ground on the steep bank of a hill, in a rather exposed position within a small tree. The only other breeding record from Sabah is of a recently fledged juvenile observed with an adult at Poring on 30 April (B. Gee *in litt.*). In Sarawak, a nest with one egg was found in the Hose Mountains on 27 March and another nest on Mt Dulit in September (Hose 1898). In Indonesia, a male was collected from a nest containing two eggs on Mt Penrissen on 25 May (SarM, Shelford 1900), and a nest, also with two eggs, was found on 14 March by Coomans de Ruiter (1938). The latter, procured by a local villager, was reported to have been 1.5-2m above the ground in a *Vitex pubescens* tree, although Coomans de Ruiter was unable to confirm this himself. Two immatures, with some red on the head and the beginning of a blue gorget on the breast, were collected at 1,460m on Mt Poh in July, with another on Mt Poi on 19 November (SarM). Younger birds were collected on Mt Dulit in September and October (at 1,220m) and in the upper Kayan River on 12 April (Sharpe 1892, Hose 1898, Voous 1961).

The nest is a rather ragged, globular dome, built primarily of dead leaves. One nest examined was lined with unidentified plant fibres (pers. obs.). Clutch-size is probably two. Eggs from Mt Penrissen were white, spotted with grey and brown in an irregular band above the middle, and measured 30 x 22mm (Shelford 1900). One egg examined by Coomans de Ruiter (1938) was white with a clear ring of brownish-red at the broadest end and elsewhere brighter violet-brown spotting and lighter-coloured underlying spots, and measured 29.5 x 20.6mm.

DESCRIPTION
Sexes similar, although colour of underparts may be marginally brighter in male.
Adult Forehead dull orange-rufous; sides of head and crown from above front of eye bright reddish-orange, extending across nape to upper mantle. Narrow line of pointed iridescent pale blue feathers extending along sides of crown to sides of nape, starting variable distance behind eye. Some individuals may have scattered dark blue feathers on nape. Sides of head, including lores, ear-coverts and superciliary area above and behind eye, bright orange-buff. Mantle to rump dull dark olive-green; tail dull blue-green. Wing-coverts as mantle, except for greater coverts which tend to be less olive and have conspicuous outer edge of iridescent blue, forming distinct line on wing. Primaries blackish; secondaries blackish with outer web dull dark green and concolorous with upperparts; underwing-coverts blackish. Throat colour variable, often a similar orange-buff colour to the sides of head, but sometimes paler. Orange-buff of throat usually merges into the reddish-scarlet (darker than belly) of breast. The reddish-scarlet is separated from scarlet belly to undertail-coverts by narrow gorget of blue formed by long pointed feathers that are basally brownish but have blue tips that become paler, more azure and shining, towards the end. Iris light grey. Upper mandible black except for tip which has a reddish tinge; lower mandible dirty reddish, with blackish just below tip. Legs blue-grey, feet pale brown to bluish-slate.
Juvenile Upperparts and breast uniform dark brown with olive tinge to mantle, rump and wing-coverts; ear-coverts and throat paler; belly mixed with dark buff; flanks mixed with pink, becoming redder on lower flanks and lower belly, undertail-coverts pale rosy, becoming redder with age. Some scattered red feathers may occur on breast sides. Older birds have a little greenish-blue on the scapulars and wings and along the sides of the hindcrown. Bill grey with orange gape; legs grey (B. Gee *in litt.*).

MEASUREMENTS Wing 85-88; tail 39-40; tarsus 39-41; bill 26.5-27.5; weight 49.5-58.0.

20 GARNET PITTA
Pitta granatina Plate 11

Pitta granatina Temminck, 1830, Planches Coloriées, livr. 85. pl. 506: Pontianak, Kalimantan, Indonesia.

Alternative name: Red-headed Scarlet Pitta

TAXONOMY The two forms of 'garnet' pitta found on Borneo are of taxonomic concern. The north Bornean taxon, *ussheri* (Black-headed Pitta), is very close in appearance and ecology to Garnet Pitta (with two certain subspecies, nominate *granatina* and *coccinea*). *Granatina* and *ussheri* were treated as conspecific by MacKinnon (1993) and Rozendaal (1994), but *ussheri* was considered to be a subspecies of Graceful Pitta *P. venusta* by Mayr (1979), van Marle and Voous (1988) and Sibley and Monroe (1990). Sumatran *venusta* is, however, clearly different to both *granatina* and *ussheri*, and is treated as a separate species, following Rozendaal (1994): the remaining question is therefore whether the Bornean forms belong to the same species or not.

The fact that *granatina* and *ussheri* are (apparently) parapatric but different in appearance, and that the calls differ, is suggestive of specific status. Although they are believed to interbreed in their zone of contact in Borneo (Mayr 1979), there is little evidence to support this.

In west Borneo, the dividing line between Garnet Pitta and Black-headed Pitta is approximately the watershed of the Lawas and Merapok Rivers (Smythies 1957), where the mountains of the central range approach the sea close to the border of Sarawak and Sabah. In the east, the dividing line appears to be in the region of the catchment of the Sesajab (Sesayap) and Mentarang Rivers in the extreme north of East Kalimantan, Indonesia (at c.3°35'N), where mountains of the main range sweep down towards the coast and there is only a narrow belt of lowland forest between the mountains and mangroves. Just to the south of these mountains, B. van Balen (verbally) found Garnet Pittas in the upper catchment of the Kayan River, at Long Pujungan and Long Bia (c.2°35'N, 115°45'E). Both *ussheri* and *granatina* occupy similar altitudinal ranges.

Stresemann (1938) described an intermediate colour pattern of an apparent hybrid from the Kajan (Kayan) River, Bulungan district, but Voous (1961) found no good evidence of hybridisation in a series of nine *granatina* specimens collected in this region by Lumholtz, although one bird showed a trace of black on the crown, with one or two crimson feathers having some black on the outer web and considerably enlarged black bases to all the crimson crown feathers. Specimens of Garnet Pitta in BMNH from the Marabok (Merapok) River and Tutong, Brunei, show no obvious signs of hybridisation. Calls of these two taxa, whilst similar to the human ear, are actually quite different in both length and inflection. Whether hybrids do occur, however, is not the point, since the occurrence of hydridisation is insufficient evidence in itself to lump clearly different forms into one species. Many good species are known to hybridise, and interbreeding between individuals of taxonomically differentiated populations merely indicates the existence of genetic compatibility and not necessarily a sister taxon relationship (Hazevoet 1995). The question that should be asked is why there is no evidence of extensive hybridisation in areas where the two taxa meet if these taxa indeed belong to the same species. One obvious answer is that they are different.

Further evidence that they are probably reproductively isolated can also be found in the voice (see sonograms). Compared to Black-headed Pitta *P.* (*g.*) *ussheri*, the call of Garnet Pitta on Borneo is significantly shorter (2-3s in *granatina*, compared to 3.9-4.2s in *ussheri*), descends in tone more noticeably before rising, and ends rather more abruptly (Rozendaal 1994). Plumage differences are described under Descriptions in the relevant species accounts.

FIELD IDENTIFICATION Length 14-16.5. Garnet Pitta is a small, brightly coloured species that is found in the lowlands of Peninsular Malaysia, Sumatra and parts of

Borneo. The adult is unmistakable, with bright crimson crown to nape and belly to undertail-coverts; iridescent azure-blue bend of wing and stripe behind the eye; black forecrown, throat and upper breast, and dark purple-blue upperparts. Juvenile birds are an almost uniform dark chocolate-brown with a bright orange to red gape and bill tip. Immatures have some crimson on the nape, a blue upper tail and blue in the flight feathers.

Similar species Adults are unlikely to be confused with any other species, the striking crimson crown being distinctive and unique. Juvenile Garnet Pitta is a similar colour to juveniles of Graceful Pitta and Black-headed Pitta, but develops crimson on the nape at a very early age, and immatures should not therefore be confused. However, recently fledged juveniles of these three species are almost identical and not safely distinguished in the field. Juvenile and immature Blue-banded Pitta, of Borneo, are also similar to similar-aged Garnet Pittas, but can be safely distinguished by their paler coloration, the presence of red on the flanks and, in the case of immatures, by the lack of blue behind the eye and on wings and tail.

VOICE The call is a quiet but far-carrying, ventriloquial, drawn-out, quavering, monotone whistle (see sonograms below). The whistle swells in volume and ends very abruptly, and sometimes appears to have a gentle upward inflection. The abrupt end and often quavering qualities are perhaps the best way of separation from the very similar, but usually longer, call of the sympatric Malaysian Rail-Babbler *Eupetes macrocercus*. Separation usually requires some experience with one or both species, and distant birds are sometimes indistinguishable on call, even to observers familiar with both species. In populations of mainland South-East Asia and Sumatra, the call lasts c.1.1-1.5s, whilst those on Borneo have a slightly longer call, lasting c.2-3s (Rozendaal 1994). Differences in the voice of Garnet and Black-and-Crimson Pittas are mentioned under the latter species.

Another vocalisation is a quiet, low, purring call, *prrr, prrr, prrr*, which has a hollow quality (M. Chong *in litt.*). This is reported to be given when two birds meet (see under Habits).

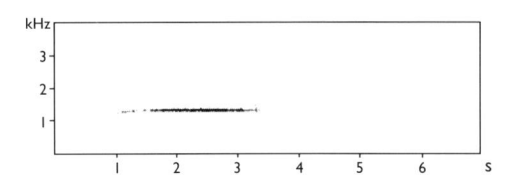

Blue-banded Pitta *P. arcuata*
[from Danum Valley, Sabah, recorded by N. Redman, July 1991]

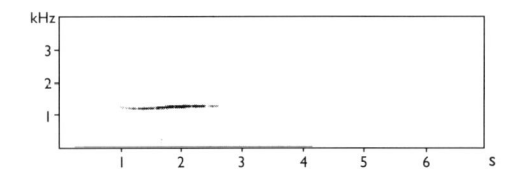

Garnet Pitta *P. granatina*
[from Pasoh, Malaysia, recorded by C. Hails, June 1979]

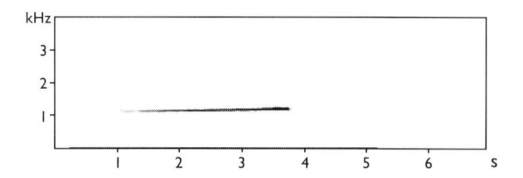

Black-headed Pitta *P.* (*granatina*) *ussheri*
[from Danum Valley, Sabah, recorded by F. Lambert, December 1989]

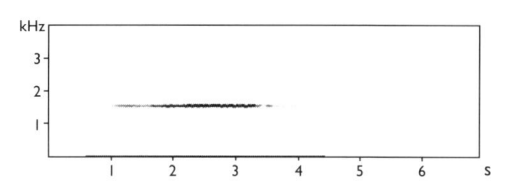

Malaysian Rail-Babbler *Eupetes macrocercus*
[from Negri Sembilan, Malaysia, recorded by J. Scharringa, August 1984]

Comparison of sonograms of Blue-banded, Garnet and Black-headed Pittas with Malaysian Rail-Babbler.

DISTRIBUTION The Garnet Pitta is a bird of the Sunda Region, where it has been recorded from Tenasserim and Peninsular Thailand, throughout the lowlands of Peninsular Malaysia and on the Greater Sunda Islands of Sumatra and Borneo.

Its distribution in the north of its range is poorly documented. In Thailand, it is known only from Prinyor, Narathiwat Province, and from Trang (Holmes 1973), although it was reportedly seen by Davison a considerable distance further north, at the foot of Mt Nwalabo, in Tenasserim (14°2'N, 98°28'E: Oates 1883, Robinson and Chasen 1939). Davison did not collect this individual but had previous experience of the species. This record, however, seems questionable. Further south, in Peninsular Malaysia, Garnet Pitta is widespread in suitable habitats, occurring from Kelantan south to Johore, and formerly in Singapore (Gibson-Hill 1950, Medway and Wells 1976). In Sumatra the Garnet Pitta is known from coastal lowlands of the northern and central provinces, where recorded from Lhokseumawe, Aceh Province; Deli, near Medan, North Sumatra Province (van Marle and Voous 1988); Berbak National Park, Jambi Province (Y. Noor *in litt.*); southern Riau Province (Danielsen and Heegaard 1995) and, probably, from Way Kambas, Lampung Province (J. Hornskov *in litt.*). Hence Garnet Pitta probably occurs throughout the coastal lowlands of eastern Sumatra.

It is widely distributed on Borneo in Kalimantan (Indonesia) and Sarawak (Malaysia), but is not present in north Borneo, where it is replaced by Black-headed Pitta. In west Borneo, the dividing line between Garnet and Black-headed Pitta is approximately the watershed of the Lawas and Merapok Rivers (Brunei Bay), northern Sarawak. In the east, the dividing line is in the region of the catchment of the Sesajab River in the extreme north of East Kalimantan.

The origin of old trade skins labelled 'Java' (BMNH) is unknown, although it seems unlikely that the species ever occurred on Java since there are no published records from that island, even though suitable habitat still remains in a few places.

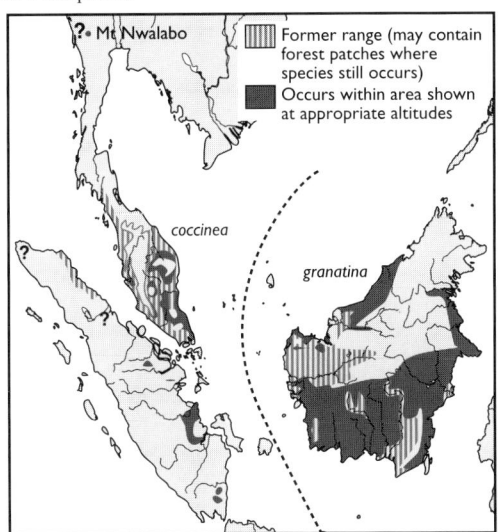

Distribution based on approximate extent of forest remaining within the altitudinal range of the species. Former range more widespread. Areas mapped as former range probably contain pockets of suitable habitat where the species still survives. Probably more widespread than indicated on Sumatra.

GEOGRAPHICAL VARIATION Two subspecies are recognised.

P. g. granatina (Confined to Borneo) The entire forehead and forecrown is black. Crimson scaling on breast indistinct or lacking. Upperparts strongly tinged with purple. Crimson of underparts marginally darker than typical of *coccinea*.

P. g. coccinea (Sumatra, Peninsular Malaysia, southern Peninsular Thailand and Tenasserim, Burma. Formerly in Singapore) Very similar to *granatina*. Individual variation is large, but on average *coccinea* has a greater amount of crimson scaling on the breast and more crimson on the crown than *granatina*. In the majority of individuals, only the forehead is black and almost the entire crown is therefore red. Upperparts are more strongly marked with blue, and lack the strong purple wash of nominate *granatina*. Immatures of the two subspecies are inseparable. Wing 83-94; weight 52-65.8.

HABITAT The Garnet Pitta is a bird of tall lowland rainforest in the Sunda Region, usually restricted to primary forest or the older stages of regenerating, selectively logged forest. Nevertheless, it occasionally occurs in more degraded habitats, and has been heard calling in a thick understorey of gingers (Zingiberaceae) within secondary forest of canopy height 10-15m (pers. obs.). Danielsen and Heegaard (1995) found this species in heavily logged forest in southern Riau Province, Sumatra, as well as in primary forest. Its upper altitudinal limit in Peninsular Malaysia is uncertain, although it is probably below 300m. In Sumatra, Garnet Pitta is known only from the coastal lowlands and lower hills, where it has been recorded up to at least 200m.

In Borneo, Garnet Pitta has been found in primary and secondary forests in Kalimantan (Pearson 1975). In Sarawak it is found throughout floodplain and alluvial forests and also occurs in heath forest (*kerangas*), although it is apparently rarer in the latter (Wells *et al.* 1979, Davison 1980, R. Rajanthan *in litt.*). In northern Borneo, Garnet Pitta is primarily confined to lowlands, although Hose (1898) reported finding it to c.600m on Mt Dulit, Sarawak, and a nestling has been collected at 520m (SarM). It was not, however, present at altitudes of 400-600m in the parts of Sarawak where Fogden (1976) conducted his research, and Wells *et al.* (1979) only found it below 130m in the Gunung Mulu National Park. In East Kalimantan, B. van Balen (verbally) reported finding Garnet Pitta at c.280m at Long Pujungan and Long Bia on the Kayan River, and also made a sound recording of what was apparently this species at c.1,000m in Kayan Mentarang (B. van Balen *in litt.*). In this area, it may be sympatric with Blue-banded Pitta.

STATUS The Garnet Pitta is a relatively common inhabitant of lowland forests throughout its range in Peninsular Malaysia (pers. obs.), but the lack of records from Peninsular Thailand suggests that it must be very rare, if not extinct there, and it is already extinct in Singapore. Little forest remains at low altitude in these parts of its range. The status of the species in Tenasserim is unknown: it has not been recorded there since the last century. Few ornithologists have visited the lowlands of Sumatra in recent times, and hence its status there is unknown: however, the relative paucity of records suggests that it is a scarce and local bird of the eastern lowlands.

The status of Garnet Pitta in Borneo is better documented. In Tanjung Puting National Park, central Kalimantan, in primary and secondary forests of Kutai National Park, east Kalimantan, and in parts of Sarawak, Garnet Pittas are reported to be common (Hose 1898, Pearson 1975, Wilkinson *et al.* 1991a, R. Rajanthan *in litt.*). Davison (1980) found Garnet Pitta to be fairly common throughout floodplain and alluvial forest in Sarawak. Fogden (1976) estimated that the population density of this species was in the order of 10 individuals per km^2 in an isolated fragment of virgin rainforest near Kuching.

FOOD In Sarawak, food items identified in specimens (SarM) include a large cricket, black beetles, a 9cm-long wood grub, a large ant, a bright bronze beetle, medium black ants and hard round seeds, suggesting that fallen fruits are sometimes consumed. A nestling had eaten a black scarab beetle, and a small snail with a whirled shell (SarM). Ants were found in the stomach of a bird collected in Peninsular Malaysia (Wong 1986).

HABITS Garnet Pitta, despite its brilliant colours, can be difficult to observe in its forest habitat, and is usually best located by its call. Once located, however, it is often relatively tame. Individuals, and occasionally pairs, feed in the damper, darker parts of forest, primarily keeping to quite dense cover in the understorey. They feed by probing into damp soil and within leaf-litter, often around fallen logs and branches, and will occasionally perch or forage on fallen logs. However, most time is spent on the ground. During territorial calling disputes, individuals may perch 1-3m off the ground. Most calls, however, seem to be made from the ground, the birds often moving between calls (pers. obs.). Unlike many pittas, Garnet Pittas rarely feed on paths through the forest.

In Peninsular Malaysia, M. Chong (*in litt.*) witnessed an interesting display: two birds that met after a bout of calling faced each other and bowed their heads whilst giving a low purring call. Although Chong interpreted this as a 'greeting' behaviour, perhaps associated with pair-bonding, its similarity to the bowing behaviour of Rainbow Pitta, apparently used as a means of maintaining territorial boundaries (Zimmermann in press: see species account), should be noted.

A bird ringed in Peninsular Malaysia was retrapped eight years later (C. Francis *in litt.*), confirming the suspicion that this, and other pittas, are probably long-lived species.

BREEDING Available data indicate that nesting is seasonal, with breeding occurring during the period that is typical for insectivorous species in the region (see Medway and Wells 1986). This period, from April to August, coincides with the driest period of the year.

In Peninsular Malaysia, nests containing eggs have been found between 2 April and early August and juveniles have been recorded on 6 and 10 June (Medway and Wells 1976, Anon. 1988, UM). In Borneo, a nestling was collected at 520m on 26 April from a nest situated on a log, just above ground level, and a recently fledged juvenile was collected in Sarawak on 4 May (SarM). Wells *et al.* (1979) saw juveniles in April or early May in Gunung Mulu National Park, and an immature was collected there in October (UMZC). Immatures, with some crimson on the crown, have also been collected elsewhere in Sarawak in August and October (SarM), with a subadult in March (BMNH). In Kalimantan, Coomans de Ruiter (1938) found a nest with one egg on 18 April. There are only three breeding records from Sumatra, of an immature collected in Aceh on 13 September (Kloss 1921) and of juveniles netted in heavily logged and in primary forest in southern Riau Province on 16 August and 19 September respectively (M. Heegaard *in litt.*).

A nest found in primary lowland forest in Malaysia on 2 April (Wells 1970) was described in detail. The nest, containing two eggs, was situated on the ground, at the edge of a path through the forest, but sheltered by a small dead fallen, leafy, branch. The nest was a domed cup in a pile of dead leaves, constructed and lined with decomposing leaves, leaf ribs and other plant fibre. The entrance was c.7cm across, dorso-ventrally flattened, and facing down the path, providing a clear flight line for the adults.

Two Garnet Pitta nests have been found in lowland rainforest in the Krau Game Reserve, Peninsular Malaysia. One nest, containing two eggs, was found on 8 May, situated on the ground against a tree root. It was described as a domed leaf pile, with a side entrance. The second, also containing two eggs, was found on 22 March. The eggs were still being incubated on 31 March, but on 18 April the nest was empty. This nest was built on the ground, on an exposed mound (UM).

The nest found by Coomans de Ruiter (1938) in East Kalimantan was made of loosely woven twines and dry leaves, and was c.15cm in diameter, and thus considerably smaller than the nest of Hooded Pitta.

Eggs from Peninsular Malaysia are 26-27 x 20-21mm in size, and weigh 4.5-6.0g. One egg from Kalimantan was 28.3 x 20.2cm in size (Coomans de Ruiter 1938), and therefore somewhat longer than typical eggs from Peninsular Malaysia. Egg colour is glossy-white, with large vinaceous-brown splashes and dots at the broader end, and, superimposed on this pattern, numerous rich brown dots, also concentrated at the blunt end (UM).

DESCRIPTION *P. g. granatina*
Sexes similar.
Adult Entire head black, with bright scarlet crown-patch from above eye to nape, and narrow shining pale blue post-ocular stripe (usually starting immediately behind eye and extending to sides of nape), consisting of rather stiff, long pointed feathers which are black with broad blue tips; throat tinged brown. Feathers of mantle and back basally black with broad dark purple bands across the outer part of feathers, these bands becoming bluer on the lower mantle and blue on the rump. Tail dark blue. Breast colour variable, feathers being black at base and usually with iridescent purple gloss to outer half of feathers giving breast purple coloration. In some individuals, the purple on breast is mixed with more bluish-purple, or, more rarely, the purple is replaced with a more maroon colour. Lower breast, and belly to undertail-coverts, bright, deep crimson. Primaries blackish above, dark brown below, with dull blue leading edge to innermost 5-6 primaries; secondaries similar, with the blue leading edge broader and becoming wider to encompass whole outer web of innermost two secondaries and tip of innermost three; tertials blue with black basal part of inner web; primary coverts black; conspicuous shining bright pale blue wing-patch formed by blue outer web and tip to greater coverts, almost wholly blue median coverts, as well as blue in leading edge of lesser covert feathers. Rest of lesser coverts blackish with purple-blue tinge to tips and outer edges of all except those feathers nearest alula. Underwing-coverts blackish-brown. Bill black, gape and spot at base of both mandibles orange-vermilion; legs and feet blue-grey to pale lavender.
Juvenile Entirely brown, darker above than below, with dull blue suffusion of feathers along post-ocular line, and blue leading edge to secondaries and innermost primaries, broadest on innermost secondaries, narrowest on primaries. Tail blue above. Throat slightly mottled due to contrast between darker centre of feathers and paler, browner, edges and tips. Bill tip orange (Whitehead 1893). The nestling has a coral-red gape and bill tip, the rest of the bill being black.

MEASUREMENTS Wing 85-96; tail 20; bill 25.5-30, width 13.5-15; tarsus 36.5-40; weight 58-67.

21 BLACK-HEADED PITTA
Pitta (granatina) ussheri   Plate 11

Pitta ussheri Gould, 1877, Birds Asia, pt. 29. pl. 75 (and text): Lawas River, North Borneo.

Alternative names: Black-and-Crimson Pitta, Black-crowned Pitta, Black-crowned Garnet Pitta, Black-and-Scarlet Pitta.

TAXONOMY Whilst this taxon and *P. granatina* are treated by Rozendaal (1994) as a single species, they have parapatric distributions and there is no good evidence of hybridisation, suggesting that they are allospecies. For a fuller discussion of taxonomy, see under Garnet Pitta.

FIELD IDENTIFICATION Length 13-15.5. This is a small, striking species endemic to the lowlands of north Borneo. Adults are unmistakable, with black head and breast, crimson belly and prominent but narrow pale blue post-ocular stripe. The upperparts are purple-blue with a conspicuous iridescent azure-blue patch at the bend of the wing. Juveniles are uniform dark chocolate-brown, but immatures are slightly paler below and have a dull blue suffusion to feathers of the post-ocular stripe and blue upper tail and leading edges to the secondaries and innermost primaries. The bill tip is bright orange or red.

Similar species In Borneo, Garnet Pitta is the only pitta that is likely to be confused with Black-headed Pitta, but adults of the former have bright crimson crowns. Juveniles of these two species are identical during early stages, but immature Garnet Pitta develops crimson on the nape before obtaining the blue superciliary stripe, and should not therefore be confused. Younger immatures of both species are not safely distinguished in the field, except on range and altitude. Immatures also resemble immature Blue-banded Pitta, but are generally darker, have blue on the wings, tail and behind the eye, and lack scarlet on the flanks.

VOICE The call of Black-headed Pitta is similar to that of Garnet Pitta, Malaysian Rail-Babbler *Eupetes macrocercus* and Blue-banded Pitta (see sonograms on p.144), although there is little altitudinal overlap between Black-headed and Blue-banded Pitta. Black-headed Pitta makes a prolonged, relatively quiet whistle that gradually rises in power and pitch and then suddenly stops, although it does not appear to stop as abruptly as that of Garnet Pitta, and may occasionally appear to waver, very slightly, in the middle. The call sounds more similar to the longer call of Malaysian Rail-Babbler than to Garnet Pitta, lasting noticeably longer than the latter (typically 3.9-4.2s, compared to 2-3s, average 2.8s in *P. g. granatina*: Rozendaal 1994), and being more even-toned.

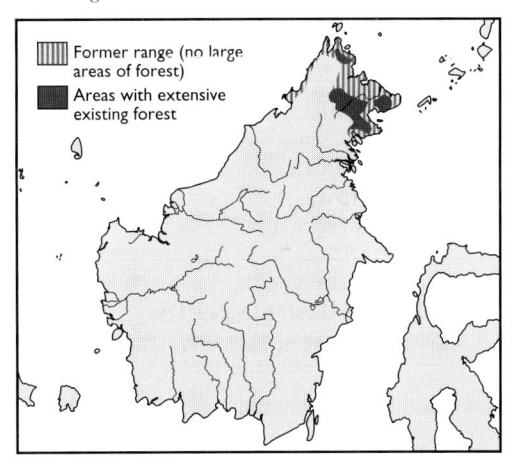

Distribution based on approximate extent of forest remaining within the altitudinal range of the species. Former range more widespread.

DISTRIBUTION The Black-headed Pitta is endemic to north Borneo, where recorded in the lowlands south to the Lawas River (BMNH) in the west (Sharpe 1879, Kloss 1930) and to the border of Sabah and Kalimantan (Chasen and Kloss 1930). Its known range therefore coincides approximately with the borders of Sabah, Malaysia. Details of the approximate limits of range of this species and of parapatric Garnet Pitta are provided in the species account on that species.

GEOGRAPHICAL VARIATION None recorded.

HABITAT Black-headed Pitta is primarily a bird of rainforests, where it frequents dark, damp areas, in particular ravines under dense cover. It is tolerant of disturbance, and can be found in areas which have been selectively logged (Lambert 1990, 1992), as well as in overgrown rubber estates and *Albizia* plantations (Mitra and Sheldon 1993). Exceptionally, this species has been encountered in thick spiny vegetation in an area where the forest had been logged and burnt and few tall trees remained (pers. obs.). This taxon is primarily a lowland forest bird, occurring from near sea level up to c.300m.

STATUS Black-headed Pitta is a common resident in the lowland forests of Sabah (pers. obs.). Studies of ranging behaviour suggest that population densities at Danum Valley, Sabah, are in the order of 21-22 pairs per km² in primary lowland forest (Lambert and Howes in prep.). In logging concessions in Sabah, Black-headed Pittas survive in relict stands of trees that persist in logged forests, but do not appear to favour areas where the majority of large trees have been removed. Since the mosaic of vegetation that forms a typical logging concession contains relatively few patches where damage has been limited or has not occurred, the overall density of the species is anticipated to decline significantly. Capture rates of Black-headed Pittas in regenerating forest that had been logged some nine years previously suggested that this was so, although based on very limited data (Lambert 1992). Nevertheless, the fact that the species can survive in logging concessions suggests that, unless changes in forestry management practices occur, Black-headed Pitta is not severely threatened by logging, despite its relatively small range.

FOOD In Sabah, adults were observed bringing spiders to the fledglings on two occasions (pers. obs.), whilst ants, cockroaches, beetles and snails have been found in specimens.

HABITS Although often silent and difficult to find, this species may be quite tame and approachable, in contrast to many other species of pitta. Black-headed Pittas are usually observed singly. They are often in damp places, particularly around logs and fallen debris, where they frequently feed. Birds radiotracked in Sabah spent a considerable proportion of their time in dark ravines with thick understorey, where they fed in small areas for prolonged periods. These two individuals, followed for 1.5 months and 34 days, ranged over an area of c.c.4.5-4.7ha (Lambert and Howes in prep.). Both birds favoured particular ravines within their territory, spending considerable periods of time in such ravines, but little time between them. The two individuals that were tracked were almost certainly in adjacent territories, but neither bird was ever encountered within the area used by the other bird.

When agitated, Black-headed Pitta has the habit of erecting its elongated bright blue superciliary feathers. Sightings at night suggest that this bird roosts very close to the ground. Two found roosting at Danum Valley were at 50cm and 75cm above the ground, both on horizontal branches of undergrowth shrubs (L. Emmons *in litt.*). Incubating adults will sit on eggs until virtually touched (Sheldon in prep.).

BREEDING Breeding is seasonal, extending from at least early February to late July, and hence coinciding with the driest period of the year. Adults have been collected with enlarged sexual organs as late as 1 August (Sheldon in prep.), suggesting that some of the population must breed later. At Danum Valley, an adult was observed carrying a large dead leaf, perhaps for nest construction, on 24 October.

Breeding records from Sabah are as follows: a nest with two well-incubated eggs was collected near Sandakan on 28 April (Sheldon in prep.); a clutch of two eggs was collected on 29 April (Gibson-Hill 1950a); a single egg was collected on 24 July at Kinabalu (BMNH); an entirely yellow, recently hatched chick was collected from a nest in primary forest along the Meliau River on 2 May (BMNH); and two nests, one containing one egg and one fledgling, the other with two eggs, were found near Danum Valley on 23 February and 17 May, respectively (pers. obs.), whilst a recently fledged juvenile was observed in the company of an adult there on 25 February (J. Howes *in litt.*).

The nest found at Danum Valley on 17 May was situated c.20cm above a small stream gully, nestled into the side of a low muddy bank, in primary lowland forest. It was a dome, built on a pile of coarse sticks and pieces of bark up to 3-4cm long and 2cm broad. The roof was made primarily of leaves, in particular pieces of palm leaves, whilst the nest cup was constructed of fine roots and leaves, with finer roots forming the inner lining. The nest still contained eggs on 30 May, but on 31 May they had hatched, and probably fledged on 15 or 16 June (pers. obs.).

Both adults incubated the egg and nestling in the other nest from Danum Valley (pers. obs.). The egg had not hatched two days later, and on 2 February the nestling had disappeared, presumably eaten by a predator. The nest was situated in a shallow gully, abutting a large root close to the large buttress from which it derived. The nest had a distinctive platform of large twigs at the entrance hole.

The nest from Sandakan was 23cm in diameter, and hidden, in rotting vegetation and dead leaves, among twisted roots on the ground. The nest chamber was lined with strips of palm leaves and fibres and enclosed by dead and decaying leaves. Another nest, in primary forest at 90m altitude, was attached to the base of a clump of saplings and herbs (Sheldon in prep.).

Gibson-Hill (1952) reported that clutch-size of Black-headed Pitta was two, and recent observations from Danum Valley support this. Eggs are variable in appearance, usually being pure white but sometimes off-white or even pinkish. Normally the egg is spotted and blotched with dark red and black spots which are concentrated in a ring around the broader end. One egg, however, was described as having fine pale grey and fuscous spotting (Gibson-Hill 1950a). Egg size varies from 26.7-29.5mm long and 20-20.5mm broad (Gibson-Hill 1950a, Sheldon in prep.).

DESCRIPTION
Sexes similar.
Adult Lores, forehead and crown to nape blackish; sides of head and throat dark blackish-brown. Prominent, narrow post-ocular stripe formed by shining, azure-blue (sky-blue) distal three-quarters to long-pointed feathers along a line starting just behind and above the eye and extending to nape, where the feathers protrude backwards like short horns. Upperparts like Garnet Pitta, but with purple-maroon suffusion to feathers at the base of nape and top of mantle, and mantle and back tends to be more purple (like nominate *granatina*) rather than blue (like allopatric *P. g. coccinea*), and less shining. Wings and underparts like Garnet Pitta, although some individuals lack blue in the leading edge of their primaries. Underparts like nominate Garnet Pitta, usually lacking crimson or reddish tinge to purple breast feathers, although this is evident in some individuals. Bill black, feet blue-grey to grey, iris dark brown.
Juvenile Very similar to juvenile Garnet Pitta. Upperparts uniform dark brown; underparts marginally paler. Head with a dull blue suffusion to feathers along line of post-ocular stripe; blue leading edge to secondaries and innermost primaries, broadest on innermost secondaries and narrowest on primaries. Throat slightly mottled due to contrast between darker centre of feathers and pale brown edges. Bill tip and gape probably red or orange (yellow in dried skins).
Chick Chicks are entirely yellow with pink iris and bill and pale horn feet .

MEASUREMENTS Wing 81-93; tail 33-35; bill length 26-28, width 14-18; tarsus 37-40; weight male 53-64, female 50.8-60.

22 GRACEFUL PITTA
Pitta venusta Plate 11

Pitta venusta S. Müller, 1835, Tijdschr. Natuur. Gesch. Phys., 2, p. 348, pl. 5 (=8). fig 4: western Sumatra.

Alternative names: Black-and-Scarlet Pitta, Black-crowned Pitta.

TAXONOMY The taxonomy of the complex of small, closely related Sundaic 'garnet' pittas has, until recently, been poorly studied. Mayr (1979), van Marle and Voous (1988) and Sibley and Monroe (1990), whilst recognising that Garnet Pitta *P. granatina* was distinct from *venusta*, all also recognised two subspecies of *P. venusta*, the nominate subspecies on Sumatra and *P. v. ussheri* in north Borneo. Rozendaal (1994), however, has provided convincing evidence for treating Sumatran *venusta* as a distinct, monotypic species, as originally suggested by Sharpe (1877), who noted that 'the differences between *ussheri* and *venusta* consist in the much richer colouration of the former. Thus where *venusta* is brown with a reddish gloss, *ussheri* is black with a purple lustre; the top of the head, sides of face and throat are entirely black. In *venusta*, the edge of the wing-coverts are cobalt, but are not very distinct, whereas in *P. ussheri* they are extremely broad and very brilliant, while the secondaries are also shaded with blue.' Rozendaal's (1994) critical appraisal of the taxonomy of garnet pittas shows that is Graceful Pitta clearly different to other garnet pittas not only in plumage, but also in voice, altitudinal preferences and structure, having a proportionally longer tail. The tail/wing ratio of adult Graceful Pitta is 0.55, compared to 0.34-0.35 for all other garnet pitta taxa (*granatina*, *coccinea*, *ussheri*) (Rozendaal 1994). For further discussion of taxonomy of the garnet pitta group, see under Garnet Pitta.

FIELD IDENTIFICATION Length 14.5-18.5. Graceful Pitta is a small species endemic to the submontane and montane forests of Sumatra. Adults are maroon above (though they may appear dark brown in the field), with a

long, fine sky-blue line from above and behind the eye that extends to the nape and protrudes as a short 'horn'. The breast is maroon, tinged with red and merging into the bright scarlet of the belly to undertail-coverts. The wings are dull purple with an indistinct dark azure-blue line in the outer edge of the closed wing. Juvenile birds are uniform dark chocolate brown, often with a post-ocular line of buffy feathers and with a dull orange suffusion on the belly and undertail-coverts. Immatures are more similar to adults, but are duller above, lacking a strong maroon tinge. Below they have orange-rufous fringes to the lower breast and more washed-out colour to the belly.
Similar species Garnet Pitta, which also occurs in Sumatra, is the only species that is likely to be confused with Graceful Pitta, but adults of the latter have bright crimson crowns. The breast of Graceful Pitta is maroon and the belly scarlet: those of Garnet Pitta are dark purple and crimson, respectively. Although lacking maroon tones, juvenile and immature Garnet Pittas may appear to be a similar in colour to adult Graceful Pitta in the field, but develop crimson on the nape before obtaining the blue superciliary stripe, so that any birds that already possess a blue supercilium can be distinguished by nape colour. Immature Garnet Pittas also lack extensive scarlet or orange tones to the belly, in contrast to immature Graceful Pitta. Juveniles of both species are more similar, with Garnet Pitta initially lacking crimson on the nape, and they are therefore not usually safely distinguished by plumage in the field, although the altitudinal range is different.

VOICE The territorial call of Graceful Pitta is described as a low, mournful whistle of 1.3-2.0s duration, mostly on an even pitch and sometimes inflected upwards towards the end of the call. The whistle is usually given at 5-14s intervals. It differs from that of Garnet Pitta in being lower-pitched (Rozendaal 1994).

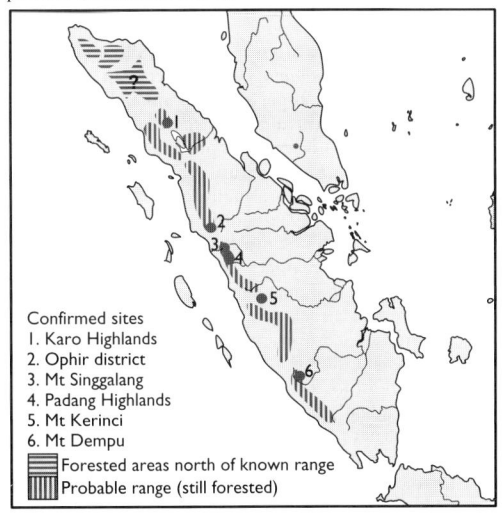

Confirmed sites
1. Karo Highlands
2. Ophir district
3. Mt Singgalang
4. Padang Highlands
5. Mt Kerinci
6. Mt Dempu
Forested areas north of known range
Probable range (still forested)

Distribution map shows known sites and probable range based on extent of remaining forest in suitable altitudinal zone.

DISTRIBUTION Graceful Pitta is endemic to Sumatra, Indonesia, where it is known primarily from the Barisan Range, but also from the Karo highlands, north of Lake Toba (Rozendaal 1994). There is also a specimen of uncertain provenance, labelled as coming from 'Batang Tingalang, south-west Sumatra'.

In the Barisan Range, it is known from Mt Kerinci (Jambi Province), Mt Talamau and the Padang Highlands (West Sumatra Province), Mt Kaba (Bengkulu Province), Mt Dempu (South Sumatra) and near Liwa (Lampung Province) (Nicholson 1883, Robinson and Kloss 1924, 1936, van Marle and Voous 1988, Holmes in prep.). Most records are historical, and the only certain record of this species during the last 60 years is from Mt Kaba in 1989 (Rozendaal 1994). There is also a probable sighting from Rimbo Kulit Manis (0°01'N, 100°22'E) in 1985, and from just south of Mt Dempu (P. Andrew *in litt.*, D. Holmes in prep.).

The range of Graceful Pitta in Sumatra is not known to overlap with that of Garnet Pitta, which is apparently confined to the coastal lowlands.

GEOGRAPHICAL VARIATION None recognised. See under Taxonomy for further discussion.

HABITAT The Graceful Pitta is primarily a bird of rainforests in the hills and submontane zone. It frequents dark, damp areas, in particular ravines under dense cover. Collecting stations where specimens of Graceful Pittas have been collected in Sumatra include areas of virgin forest and secondary forest with much rattan in the undergrowth (Robinson and Kloss 1924).

Unlike Garnet Pitta, which also occurs in Sumatra, Graceful Pitta is apparently restricted to forest at elevations above 400m (Müller and Schlegel 1840, Robinson and Kloss 1924). On Mt Kerinci, Robinson and Kloss (1936) were unable to find this species when collecting from a camp at 1,430m, and its upper altitudinal limit seems to be around 1,400m.

In the Barisan Range it is known from 1,400m on Mt Talamau, from 400m to at least 800m in the Padang Highlands, from 750-915m in the Kerinci valley in Jambi Province, from 915-1,200m on Mt Kaba, from 1,400m on Mt Dempu, from c.900m near Berestagi (Karo Highlands) and a possible record from 400-600m near Mt Dempu (Nicholson 1883, Robinson and Kloss 1924, 1936, van Marle and Voous 1988, Rozendaal 1994, P. Andrew *in litt.*, Holmes in prep.).

STATUS The status of Graceful Pitta is unclear: there have been very few observations of this species in Sumatra in recent years, and it must be considered rare or very local in distribution. Forest in many parts of its range has undoubtedly been cleared, particularly those at the lower end of its altitudinal range, and it should therefore probably be considered as a threatened species.

The taxonomy followed by Collar *et al.* (1994), in BirdLife's review of the world's threatened birds, followed that of Sibley and Monroe (1990), and hence recognised *ussheri* as a Bornean subspecies of *P. venusta*. Since *ussheri* is relatively common, the species was not considered to be threatened. However, given that *ussheri* is not conspecific with *venusta*, a re-evaluation of threat status is needed. As a Sumatran endemic with a range that is probably more restricted than that of sympatric Schneider's Pitta, and with a lower altitudinal distribution than the latter, Graceful Pitta should at least be accorded a threat status of Vulnerable, as is Schneider's Pitta.

FOOD Robinson and Kloss (1924a) found insects and small snails in the stomachs of Graceful Pittas, whilst small molluscs and seeds (Nicholson 1883) and small bugs and worms (Müller and Schlegel 1840) have also been reported in the diet.

HABITS Graceful Pitta is virtually unknown, although its habits are presumably similar to those of other species in the garnet pitta group. The only pertinent information regarding its habits is that it is most often observed on the ground, although occasionally feeding on top of fallen logs (Müller and Schlegel 1840).

BREEDING There are few data on breeding, but nesting appears to be seasonal, with evidence indicating that Graceful Pitta nests from about March to June or July, and perhaps as early as February. Apart from a nest attributed to this species that was found in the Padang Highlands on 26 May (Müller and Schlegel 1840), all breeding records relate to young birds. A well-grown nestling was collected at 1,000m in South Sumatra on 18 June (MZB), whilst a recently fledged juvenile was collected on Mt Kaba on 5 July. Older immatures have been collected there on 13 July, in south-west Sumatra (Batang Tingalang) in May, and at Rimbo Pengadang, Bengkulu Province, on 19 June (Robinson and Kloss 1924, RMNH).

The nest documented by Müller and Schlegel (1840) contained two eggs and was situated among orchids on a fallen tree trunk. It was made of dry leaves, small roots, moss, bamboo leaves and other soft, often rotting, material. Dry leaves of trees lined the inner cavity. Müller and Schlegel (1840) likened the nest to that of Blue-winged Pitta, being a loosely constructed dome some 20cm in diameter and 9.5cm in height. The eggs were white, and unmarked. Although Rozendaal (1994) questions whether this nest was that of Graceful Pitta, he notes that the coloration of the eggs was unlike that of other pittas in Sumatra.

DESCRIPTION
Sexes similar
Adult Forehead and crown blackish; feathers of nape similar but with dull maroon tinge. A line of long, stiff, pointed azure feathers forms a distinct narrow line from above or just behind the eye that broadens towards its end, at the sides of the nape, where these feathers may jut out as a point. Back, mantle, rump and uppertail-coverts maroon, brightest on mantle. Tail dark brown. Lores, ear-coverts and throat dark brown; breast feathers dark brown with long maroon to maroonish-red tips, becoming progressively brighter and redder towards belly. Belly to undertail-coverts scarlet (i.e. a paler red than that of Garnet Pitta, which is better described as crimson). Inner upperwing-coverts mostly dark brown with dull maroon wash; tertials, median and greater coverts dark brown, with narrow blue tips to the leading edge of the greater coverts forming an indistinct blue line in the closed wing. Primaries, secondaries and underwing-coverts dark brown. Iris dark greyish-brown; bill black; feet purplish-black or slaty. The gape and interior of the mouth are also red (Robinson and Kloss 1924).
Immature Differs from adult in duller, browner upperparts with only a tinge of maroon, and in more washed-out underparts, with orange-rufous fringes to the lower breast and paler, more scarlet-orange belly.
Juvenile Forehead, crown, upperparts and breast uniform dark brown with maroon tinge to parts of mantle, back and breast; chin and throat dark brown with indistinct paler streaks; flanks and belly paler brown with distinct dull orangy suffusion, belly centre and undertail-coverts orange-red. The blue post-orbital stripe of the adult is at first absent, but is later indicated by a line of buffy feathers. Iris greyer than that of adult; bill blackish with narrow red tip; legs pinkish-grey.

MEASUREMENTS Wing 88-95; tail 51.4-59.3; bill 27-29.5; tarsus 42.1-45.0; Weight unrecorded.

23 AFRICAN PITTA
Pitta angolensis   Plate 12

Pitta angolensis Vieillot, 1816, Nouv. Dict. Nat., nouv. éd., 4, p-356: Angola.

Alternative name: Angolan Pitta

TAXONOMY Keith *et al.* (1992) considered that African Pitta and Green-breasted Pitta were separate species, but Dowsett and Dowsett-Lemaire (1993) treated them as conspecific, following Louette (1981). However, the basis used for lumping these two forms is very weak, relying on the existence of apparent hybrids (intermediate in size and colour between *P. a. pulih* and *P. reichenowi*) from Cameroon (Germain *et al.* 1973, Decoux and Fotso 1988) and from near Brazzaville in Congo (Dowsett and Dowsett-Lemaire 1989), and anecdotal evidence that the songs, and possibly the display, are similar. Similarity in songs of different species of pitta is well documented from Asia, and is insufficient in itself for considering various taxa as conspecific, particularly in the absence of sonograms. Furthermore, the existence of apparent hybrids between closely related taxa is also of doubtful use in assessing specific status. Louette (1981) does not appear to have considered the possibility that so-called hybrids (between *pulih/angolensis* and *reichenowi*) could in fact have been visitors from the population in Angola, which is intermediate between *pulih* and *reichenowi*. Indeed, there is some evidence that the population from Angola is migratory, and Chapin (1953) identified pittas from Brazzaville as belonging to birds from Angola. Without more reliable evidence of conspecific status, African Pitta and Green-breasted Pitta are best considered as good species.

These two species are not known to overlap in their breeding range, although if hybrids really do exist, then one must conclude that they are sympatric in some parts of their ranges. The nominate subspecies of African Pitta, as well as *P. a. longipennis*, have both been found within the range of Green-breasted Pitta in Central Africa, although apparently as migrants only (Louette 1981). In Cameroon, Green-breasted Pitta has been found at Bitye, whilst African Pitta has been found, apparently breeding (in February), at Efulen, some 160km away (Hall and Moreau 1970): specimens of African (breeding unconfirmed) and Green-breasted Pittas have been collected from localities c.70km apart in the south of the country (Louette 1981). Brosset and Erard (1986) noted that, in northern Gabon, African Pitta is known from Tchibanga (2°49'S, 11°00'E) in the Ivindo basin, whilst Green-breasted has been recorded in other parts of the Ivindo basin as well as from Woleu-N'Tem Province. Clearly, there is still the possibility that areas of sympatry, in which both African and Green-breasted Pittas breed, will be found.

As noted by Sharpe (1903), the East African subspecies (*longipennis*) of African Pitta is quite different (in plumage and morphology) from those in West Africa (*pulih*), and it inhabits much drier habitats. The possibility that this taxon should be considered as a third species of Afrotropical pitta should not be ruled out.

FIELD IDENTIFICATION Length 17-22. African Pitta is a medium-sized pitta that is widely distributed in moist woodlands of eastern Africa and in the lowland coastal forests of West and Central Africa. Subspecies differ markedly in size. Crown and sides of head black separated by a prominent broad golden-buff supercilium. The upperparts are green with an azure-blue rump. The wings are darker with a line of azure-blue feather tips across the innermost coverts, bordered below by feathers marked with azure-blue in West Africa but deep violet in East Africa and Angola. Birds from areas between the Congo River and Dahomey Gap (Benin) may have deep blue in the tips of these feathers, whilst those to the west of the Dahomey often lack any bright tips in the outer coverts. Below, the throat and breast is white, with a distinctive pink wash, the breast and flanks are a cinnamon-buff, whilst the belly to undertail-coverts are deep red. A white wing-patch is conspicuous in flight.

Juveniles are reminiscent of the adult but much darker, particularly on the underparts, which are dark buffish-brown, and usually lack the azure wing-covert tips (depending on age), and have a pink wash rather than red on the undertail-coverts and belly. The bill has a distinct orange-red to red tip and base. Immatures on migration in East Africa have darker buffish-brown supercilium, fewer and smaller, dull azure-blue wing spots, and lack the rosy wash to the throat. The undertail-coverts and belly are a pale pinky-red.

Similar species African Pitta may be found in forests of western Uganda and eastern Zaïre, where Green-breasted Pitta is resident, from May-September (Chapin 1953). It differs primarily from the latter in having a cinnamon-buff rather than green breast. From behind, or when flushed, the two species appear identical. In parts of Cameroon (Sakbayeme, Nkolngem, Bafia, Nkom) birds intermediate in size between Green-breasted Pitta and African Pitta are said to occur (Louette 1981), although it should be noted that there is considerable overlap in size between African Pittas in West Africa and Green-breasted Pittas. The juvenile is very similar to that of Green-breasted Pitta, but usually lacks the distinctive olive wash to the breast and has very few dull azure spots in the wing. In contrast, juvenile Green-breasted Pitta has a prominent line of bright azure-blue spots across the upperwing-coverts.

VOICE On the breeding grounds in East Africa, the main call is a short sharp rising whistle, *wheet* or *hooit*, punctuated by an initial double beat that is mechanical and believed to be made by the wings. Hence the overall noise heard is a *prrt-wheet* or *prrt-hooit*. In Zambia, R. Stjernstedt (*in litt.*) has heard a variation in which the mechanical note was preceded by an extra whistle, *wee-prr-pwheet*, although this call is evidently rare. The mechanical part of the call is apparently sometimes made without the accompanying whistle. Whether West African subspecies differ in their calls from East African birds is unclear. Calling is usually made from a perch above the ground and is most intense during the first hour or so of daylight, and may also occur after sunset.

Harvey (1938) reported that, in East Africa, African Pittas start calling after the first fall of rain towards the end of November or in early December. R. Stjernstedt (*in litt.*) reports that, in the Luangwa Valley and Zambezi Valley of Zambia, territorial calling is intensive when the birds first arrive in mid-November but that the birds become silent by the end of January-February.

When not displaying, African Pittas give an abrupt, low, frog-like croak that rises in pitch (0.5-2.0 KHz) and lasts 0.2s (Keith *et al.* 1992). Captive birds have been heard to utter a low-pitched *skeeow*, a brief whinnying and, whilst standing almost on tiptoe with the belly fanned out, a grunt (Sclater and Moreau 1933). Grunts and low croaks are also made by migrants circling in the mist. A husky *hgg* or *hggg* that has been heard is thought to be an alarm note (McLachlan and Liversidge 1978). Nestlings caught in Liberia regularly gave a soft plaintive note, *pi'-u*, described as being somewhat like a cat's mew, and similar calls, presumed to be this species, were also heard in the forest (Allen 1930).

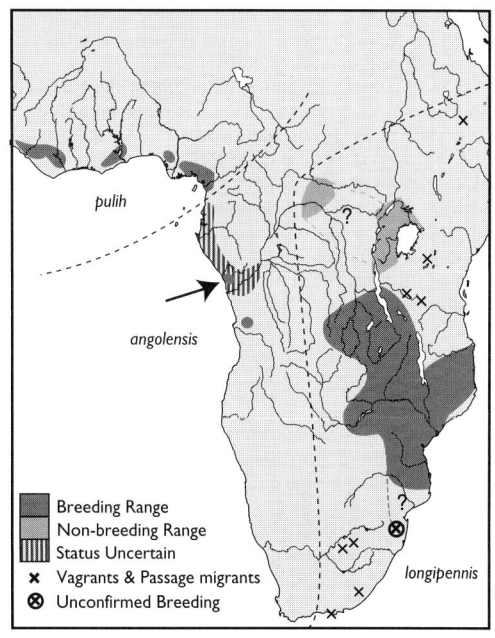

Map shows known distribution of breeding and non-breeding populations of the three subspecies, and sites where vagrant *longipennis* has been recorded. Breeding range in southern Africa and exact limits of breeding and non-breeding range of nominate *angolensis* require confirmation.

DISTRIBUTION The African Pitta is widely distributed in eastern Africa, breeding between c.5°S and 23°S and east of c.25°E, and in equatorial and western Africa from Guinea to north-east Angola in forests bordering the Gulf of Guinea. African Pitta is, however, absent as a breeding bird from the main Congo basin.

In West Africa, African Pitta has a disjunct distribution, and there are apparently either two or three distinct populations. In the most westerly part of its range, it is known to occur in southern Guinea; the lowlands of Sierra Leone (Fraser 1843, Allport *et al.* 1989); Liberia in Lofa County (Schouteden 1970); in Ivory Coast at Lamto, Danané, Maraoué, Taï and Banco National Park (Thiollay 1985, Demey and Fishpool 1991, Gartshore *et al.* 1995);

and southern Ghana, where collected from coastal districts (Grimes 1987). To the east of the Dahomey Gap, there are records from southern Nigeria, where known from the Mbe Mountains, and from near Ibadan (Elgood *et al.* 1994); and in the coastal part of Cameroon, apparently to the west of a line from Ebolowa to Yaoundé (Louette 1981), or west of c.11°30'E.

Birds from Guinea to Cameroon are usually referred to as *P. a. pulih*, but, as described under Geographical Variation, there may be two distinct taxa in this range. Furthermore, L. Fishpool (*in litt.*) notes that the paucity of records within many parts of this range suggests, perhaps, that breeding might occur primarily in the gallery forests north of the forest zone proper, and that birds recorded in the forests to the south may be migrants in their non-breeding range. Within the forest zone a number of areas, such as forests in Ivory Coast, are relatively well-worked and yet there are few records of African Pittas: this is unexpected if the species is a regular breeder at these latitudes. The gallery forests to the north are, in contrast, relatively poorly worked by ornithologists. Liberian garden records (Bouet 1931) certainly give credence to the idea that West African birds make some sort of movement, but present knowledge is insufficient to determine the nature of such movements.

To the south, a third population occurs (nominate *angolensis*). The exact northern limits of its breeding range are unclear, especially since there is evidence that southern populations may undertake northerly migration after breeding (Chapin 1953). Mayr (1979) gave the range as being to the south of the mouth of the Congo and just to the north in Angola (Landana, or Cabinda), based on specimen collection points plotted by Hall and Moreau (1970). There is evidence that birds from this population breed to the north of the Congo, since Chapin (1953) noted that L. Petit had vaguely reported that he had found a nest with four eggs on the Loango coast and that he had often heard it during the wet season. Irrespective of known breeding range, birds identified as *angolensis* have been found from southern Cameroon south through Rio Muni, Gabon, and the Congo Republic to Zaïre (in the lower Zaïre River drainage) and in north-east Angola in Cabinda and the tropical forest region of Cuanza Norte and adjoining Malange and Luanda (Traylor 1960).

The eastern African population breeds locally from about 5°50'S in central Tanzania (Itigi Thicket); Malawi, in the Rumphi District and from the southern end of Lake Malawi to Nsanje in the extreme south (Benson and Benson 1977); south-east Zaïre; eastern Zambia, east of 29°E and including the upper Luangwa Valley, perhaps further west in the Zambezi River valley, as far as Victoria Falls (Berry and Ansell 1978, Clancey 1985); adjacent Zimbabwe below Lake Kariba, south through Mozambique – Clancey (1971) provides a detailed account of the distribution in southern Mozambique – to the Limpopo River, eastern Zimbabwe (east of 28°E); and sporadically in northern and eastern Transvaal and Natal (McLachlan and Liversidge 1978, Clancey 1980, 1985). It possibly also occasionally breeds further south in Mozambique and in northern Natal and Zululand: a bird in near-breeding condition was recorded in November at Pietermaritzburg, Natal (Clancey 1985).

During the non-breeding season, coinciding with the southern dry season (April-September), migrant *longipennis* move north of the breeding zone to spend the austral winter in forests in the vicinity of the equator. Here they have been found in northern Tanzania, Rwanda (Schouteden 1966a), Burundi (Schouteden 1966b), northern and eastern Zaïre (where it occurs to 4°N) (Chapin 1953), south Central African Republic (Lobaye, Mbomou), south and west Uganda north to Budongo (1°45'N) and southern coastal Kenya, where recorded north to Sokoke Forest (3°18'S) and south to Gazi (4°25'S) (Britton and Rathbun 1978, Britton 1980, Keith *et al.* 1992).

Although two female *longipennis* with enlarged ovaries were collected in Sokoke Forest, Kenya, in July, it seems unlikely that breeding occurs this far north in view of the fact that African Pittas in this part of their range are known to be solitary and silent (Britton and Rathbun 1978).

In Kenya, records away from the coast concern exhausted or dead migrants found in the highlands at altitudes of 1,500m or more (Lewis and Pomeroy 1989). Some north-moving non-breeding birds have reached Ethiopia (near Addis Ababa, at 2,440m) and may possibly also reach southern Sudan (Urban and Hakarson 1971, Clancey 1985). During southward migration to the breeding grounds, vagrants have been recorded in southern parts of Zimbabwe (Matabeleland) and eastern South Africa and as far south as Port Elizabeth in the eastern Cape (McLachlan and Liversidge 1978), although this last record may refer to an escape (Clancey 1980).

GEOGRAPHICAL VARIATION Three subspecies have been recognised. Previous authors have stated that the nominate subspecies occurs from southern Cameroon, south to Zaïre and in north-west Angola (e.g. Keith *et al.* 1992) and that populations from Sierra Leone and Guinea to northern and western parts of Cameroon (to the west of a line between Ebolowa and Yaoundé) belong to *pulih*. White (1961) gave the range of *angolensis* as being from northwest Angola to Cabinda, Brazzaville and Leopoldville, recognising that it might be migratory, whilst the range of *pulih* was given as Sierra Leone to South Cameroon. However, examination of skins in BMNH from Sierra Leone, Mt Nimba (Liberia), Gold Coast, Nigeria, Cameroon and Angola suggest that populations to the east and west of the region of the Dahomey Gap (Benin) and from Angola are different, and that therefore there may be three recognisable subspecies in western Africa, rather than two. However, insufficient material was available to investigate this possibility, and in the following account the two western African subspecies that have already been described are recognised, but with a preliminary account of differences between eastern and western populations of *P. a. pulih*. The possibility that clinal variation is responsible for the observed variation in western Africa cannot be ruled out, however, in which case all subspecies would be united under *angolensis*: Mackworth-Praed and Grant (1970) treated *pulih* as a synonym of *angolensis*, as presumably did Serle (1950, 1959).

P. a. angolensis (Breeds in northern Angola, probably migrating northwards to spend the non-breeding season in Gabon, Congo Republic, Zaïre and perhaps southern Cameroon. The exact northern limits of its breeding range are unclear.) Plumage and size are intermediate between the other two subspecies. Rump azure-blue with no violet tinge. Underparts tinged with greenish-golden. Spots in inner wing azure-blue, but those in outer greater coverts are usually distinctly violet-blue. Supercilium relatively uniform. Wing 118-

124; White (1961) gives wing measurements as 116-124.

P. a. pulih (Range usually given as Sierra Leone and Guinea to northern and western parts of Cameroon. The exact limits of range in Cameroon are unclear, but it probably extends to the west of a line between Ebolowa and Yaoundé, and hence at least to the Sanaga and Nyong Rivers. However, as noted above, there may be two distinct populations within this range, with the Dahomey Gap forming the boundary between the two. Furthermore, populations may be migratory, in which case many records may refer to non-breeding birds.) Birds in West Africa are, nonetheless, shorter-winged than *angolensis*, and lack the distinctive violet tips to outer greater coverts, although these feathers may be narrowly tipped cobalt-blue in some individuals from Cameroon to Nigeria. The red on the underparts tends to be more extensive than in *angolensis*, and underwing-coverts and axillaries are blacker. Juveniles have a pink throat and very few or no pale blue spots in the wing. The characteristics of birds to the east and west of the Dahomey Gap apparently differ, as follows.

Sierra Leone to Ghana All spots in wing-covert are usually azure-blue, and most birds have narrower azure-blue tips than found in birds to the west. Some birds have no spots at all in the outer greater coverts (e.g. those from Mt Nimba). The supercilium is clearly bicoloured, being dull golden-ochre above and in front of eye, but pale buffy-ochre to creamy-buff below, extending as a point at rear of crown: sometimes these colours are separated by a narrow black band. The underparts are generally paler than those of birds to the west of the Dahomey Gap. Birds from Mt Nimba also have the border between the red and cinnamon-buff (with greenish sheen) of the breast straighter and more neatly defined than that of other specimens in BMNH. Wing 105-108; bill 22-23. Birds of this population may be only half the weight of *longipennis*: weight (Mt Nimba: Colston and Curry-Lindahl 1986) 45.2-48.2.

Niger to Cameroon. Supercilium uniform or indistinctly bicoloured; spots in outer greater coverts usually azure-blue but occasionally with some cobalt-blue at very tip of feathers. Darker below than birds to the east of Dahomey Gap. Wing 104-115.

P. a. longipennis (An intratropical migrant in eastern Africa, breeding between c.5-23°S and east of c.25°E but moving north during non-breeding season. Non-breeding birds reach northern and eastern Zaïre, western Uganda and Kenya, whilst there are occasional records as far south as eastern Transvaal and vagrants have reached Ethiopia.) Differs from *angolensis* in larger size (wings 122-137, weight 84-92) and more pointed wings. Also differs in having violaceous tinge to the azure-blue of rump and uppertail-coverts and in having violet, rather than azure-blue, tips to the median wing-coverts. Supercilium usually fairly uniform, being darkest on forehead and palest on nape, but not obviously bicoloured. Most individuals have the upperparts slightly more yellowish-green than *angolensis* and the underparts are usually a pure cinnamon-buff without golden-green tinge. Underwing-coverts brown to pale brown. Many birds have dark brownish streaks on the upperparts.

HABITAT In East Africa, this species is most frequent in evergreen, or, occasionally, semi-deciduous bush, and in places they occur in semi-forest thickets along watercourses where the largest trees are evergreen and the thicker bush is semi-deciduous (Harvey 1938, Clancey 1971). Harvey (1938) noted that, in East Africa, he found an abundance of small termites in areas where African Pittas bred.

In the Ulimiri Thicket of Tanzania, African Pitta has been found in 3-4m tall coppice. In Malawi and Zambia, African Pittas breed in deciduous thicket whilst in leaf during the rains, whereas in Mozambique and eastern Zimbabwe it is known to breed in lowland evergreen forest (Benson and Benson 1977, Berry and Ansell 1978). In Zambia, nests have been found in deciduous *Combretum* thicket in areas with scattered tall (20m) trees such as baobab, and birds have been observed displaying in miombo woodland although breeding was not confirmed (D. Aspinwall *in litt.*). J. Colbrook-Robjent (*in litt.*) found them breeding in the southern province of Zambia in low-lying, humid dense thickets of low canopy height with the occasional larger emergent tree (such as baobab). Breeding birds in Zambia usually occupy altitudes at 350-1,000m and possibly as high as 1,100-1,250m (where birds have been observed displaying) although little suitable breeding habitat is present at these altitudes in Zambia (D. Aspinwall *in litt.*). In Chizarira National Park, Zimbabwe, Alexander (1994) found that African Pittas seemed to favour thick scrub in ravines chiefly comprising Sickle Bush *Dichrostachys cinerea* and Buffalo Thorn *Ziziphus mucronata* with a few large specimens of Natal Mahogany *Trichilia emetica* dotted in between. The birds usually frequented the thickest part of scrub in this habitat.

In Kenya the wintering range of *longipennis* is entirely within semi-arid/dry sub-humid areas receiving at least 500mm rainfall per year. At Gedi, Kenya, the habitat occupied by non-breeding African Pitta is tall semi-deciduous lowland forest, dominated by *Combretum schumannii* and *Gyrocarpus americana* trees in the canopy and *Lecaniodiscus fraxinifolius* in the understorey (Britton and Rathbun 1978). A migrant found in Ethiopia was in a forest of young eucalyptus (Urban and Hakarson 1971).

In West Africa, *pulih* has been found in Ivory Coast in the semi-deciduous forests of Maraoué National Park and in evergreen rainforest in Tai National Park (Thiollay 1985). Gartshore *et al.* (1995) found it only in intact forest in Ivory Coasts. Allen (1930) in dense thickets and undergrowth in the forests of Liberia. Altitudinal limits in West Africa are poorly documented, but Tai National Park is lowland, below 400m, whilst rainforests on Mt Nimba extend only to c.850m (Collar and Stuart 1988). Hence it seems likely that West African populations are confined to the lowlands and hills, probably below 1,000m, but perhaps not above 850m.

STATUS Although less often seen than heard, African Pittas are locally common within their breeding range, particularly in parts of eastern Africa. In the Itigi Thicket in central Tanzania they are reported to have been abundant in 1978 (R. Stjernstedt *in litt.*), whilst J. Colbrook-Robjent (*in litt.*) reported that the species is rather common in the area of the Mutulilanganga River, Zambia, during the rains. Keith *et al.* (1992) report that 28ha of dense deciduous thicket in the Zambezi Valley contained six pairs of African Pittas. Hence the population density in this habitat may reach as high as c.43 birds per km^2. In the non-breeding season, marked African Pittas at Gede

153

Ruins, Kenya, occupied very small home ranges of some 0.3-0.36ha during one year, and were observed in the same locality during the subsequent year (Rathbun 1978).

In West Africa, the status of *pulih* is poorly documented. It is apparently a rare resident in Ghana, being known only from old specimens, the last having been collected in 1937 (Grimes 1987). It is also reported to be a rare resident in Nigeria, where it is only known from two lowland forest areas (Elgood *et al.* 1994), and is considered to be uncommon in Ivory Coast (Thiollay 1985). In Sierra Leone, where originally collected at Port Loko in the west (Fraser 1843), there is now little suitable habitat although its continuing presence in the extreme east has recently been confirmed (Allport *et al.* 1989). There is no recent information regarding the status of *angolensis*, although, as in other parts of West Africa, habitat loss and fragmentation is of serious concern and populations must be suffering as a consequence.

MOVEMENTS In eastern Africa, *longipennis* is chiefly a long-distance intratropical migrant. It breeds during the rains between November and February (occasionally as early as September or October) in south-east Africa, and then migrates northwards to non-breeding grounds in central-east and East Africa. Birds that breed in the coastal lowlands of East Africa (central Tanzania) appear to migrate comparatively short distances, whilst those nesting at higher elevations inland probably travel as much as 1,500-2,000km from breeding to non-breeding area (Britton and Rathbun 1978). The non-breeding range of *longipennis* is outlined above, under Distribution.

African Pittas (*longipennis*) are usually present in their non-breeding grounds from April to September, October or November. Extreme dates at Gedi, coastal Kenya are 22 April and 6 November (Britton and Rathbun 1978). Birds returning to breeding grounds in Malawi usually arrive in late November, although occasionally in late October (Benson and Benson 1977). Some birds apparently also move at other times of the year: a bird was found in Nairobi in August, well inland from known non-breeding sites. The presence of African Pittas in Zimbabwe in mid-September, and possibly in early August, suggests that a few birds may remain on the breeding grounds all year (Irwin 1981).

Migration occurs at night and many records of migrants derive from birds attracted to lights. During the migration period, African Pittas are regularly recorded at sites along their migration routes between mid-April and mid-June and mid-November to mid-December. At sites where nocturnal migrants have been studied in detail, such as at Mufindi in Tanzania, African Pitta has been found to be one of the commonest species attracted to lights (Beakbane and Boswell 1984). Many of these sites are at relatively high altitude. In Kenya, for example, dead or exhausted birds have been found at altitudes of 1,500m or more. In Zimbabwe, a main migration route lies in the northern part of the central Mashonaland Plateau, with many birds passing through here in mid-November to mid-December (Irwin 1981).

Migrating birds probably regularly overshoot their destinations. Hence most records of the species south of the Limpopo River in Mozambique probably refer to vagrants since they fall in the periods April-May and November-December.

In West Africa, the occurrence of *angolensis* in towns in the Congo and west Zaïre during May, June and October, and of *pulih* in gardens in Monrovia (on the coast of Liberia) in October, killed at windows on the coast in Sierra Leone in May (Field 1974), and records of birds attracted to light in Nigeria, suggest that local movements, if not migration, are undertaken by both western African subspecies (Bouet 1931). Chapin (1953) noted that records of *angolensis* along the Congo River in Brazzaville and Kinshasa (Leopoldville) on 17 October, 31 May and 17 June and from a garden in Boma in May, suggested that birds that begin breeding in northern Angola towards November migrate northward in May and June to spend the dry season in Gabon.

Such birds could also conceivably reach Cameroon: the record of a window-killed bird from near Korup National Park may perhaps relate to a migrant (Rodewald *et al.* 1994). Although the subspecies of this individual was not recorded, Serle (1950, 1954, 1959) reported that the nominate subspecies had been found on the southern edge of the park. However, the wing measurements of the birds reported by Serle from this area (adults: 107 and 111mm) would suggest that the subspecies was *pulih* rather than *angolensis*. The fact that one bird was on a nest and that an immature was collected (Serle 1954) also strongly point to these birds being *pulih* and it is therefore evident that Serle did not recognise *pulih* as being distinct from *angolensis*.

FOOD Ants, termites, beetles, insect larvae, slugs, snails, millipedes and earthworms up to 10cm in length have been recorded in the diet of African Pitta (Serle 1954, McLachlan and Liversidge 1978, Rathbun 1978, Clancey 1985, Keith *et al.* 1992, BMNH). Stomachs of three specimens examined by Harvey (1938) contained only termites, and Büttikofer is reported to have kept two fledglings alive for several weeks by feeding them with *Termes* nymphs (Allen 1930).

HABITS African Pittas are, like their Asian and Australasian relatives, often very difficult to observe, although they are occasionally very tame once found (Bradley 1994). At rest, African Pittas bob their head and flick their tails. On the breeding grounds in Zimbabwe, they are extremely shy but difficult to flush and will fly only a small distance before dropping back onto the ground, usually relying on ground cover to escape detection (Alexander 1994). When startled, however, they may fly up to a low perch where they sit motionless, often in a crouched stance with the belly resting on the legs: if further disturbed, they will fly away through the forest at about the height of the perch, but if left undisturbed they will drop to the forest floor within a few minutes. Occasionally, if approached slowly they may hop away, keeping some 7m from the observer (Rathbun 1978).

On the breeding grounds, African Pittas are nearly always encountered in pairs, although sometimes as many as four can be observed in the same vicinity (Harvey 1938). Rathbun (1978) noted that this is a very sedentary species on the non-breeding grounds in Kenya, typically staying within a very small area during an entire day: the home ranges at Gede were both small and very close but apparently exclusive, since no territorial interactions were noted.

Most foraging is conducted in deep shade. African Pittas have been observed to feed by standing motionless in deep leaf-litter from 30s up to five minutes whilst watching for prey, and then hopping to a new spot 1-2m away. A single hop may cover as much as 30cm. This behaviour may be combined with flicking leaves aside with the bill, presumably after detecting prey from the motionless position (Sclater and Moreau 1933, Rathbun 1978, Bradley 1994). Leaf-litter in the vicinity of ant and termite mounds appears to be a favoured foraging habitat (Harvey 1938). If a potential prey item is spotted, the bird lunges and captures it in its bill. At other times, birds may forage by sweeping the leaf-litter aside with the bill, and then pausing for a few seconds whilst scanning the ground, with the head held to one side in the manner of a thrush (Rathbun 1978).

At the beginning of the breeding season, African Pittas perform a display during which they stand on a horizontal or more-or-less horizontal branch. Branches chosen are invariably substantial (>10cm diameter), and usually c.5m above the ground. The display is most usually in a tree at the level of the top of the surrounding thicket, and most frequently nearer the centre of a tree rather than the edge (D. Aspinwall *in litt.*). Alexander (1994) reported that the favoured calling perch of an African Pitta in Zimbabwe was in a fairly exposed and prominent position on a large Natal Mahogany branch across a small river.

During the display, the bird jumps 25-45cm above the perch and then parachutes back down to the branch with a few quick shallow wing-beats, although sometimes the wings may apparently be snapped back (Berry and Ansell 1974). The *prrt-wheet* 'call' accompanies this display. As it jumps, the deep red belly is fluffed out (Ginn *et al.* 1989), whilst between jumps and bouts of calling the bird may sway on its hips from horizontal to vertical posture whilst puffing out the breast feathers to expose the deep red belly (Chapin 1953, Keith *et al.* 1992). During and between jumps the bird may also change direction. On occasion, much shorter jumps (of less than 10cm), with wings opened and tail spread, may be made. During such jumps the head appears to remain almost motionless whilst the body moves upwards (D. Aspinwall *in litt.*). According to Chapin (1953) all three subspecies have been observed to make this type of display, although that of *pulih*, mentioned by Reichenow (1903), is poorly documented.

The display is made in bouts. Within each bout, calling and jumping may occur at approximately 10s intervals. Although some reports suggest that calling is made from established display posts (Keith *et al.* 1992), observations in Zambia suggest that this may not always be the case (D. Aspinwall *in litt.*).

Raised posture of African Pitta

Captive birds have been observed to adopt a number of postures in response to an observer. These include crouching low with the breast feathers puffed out momentarily so that the feet are hidden, and stretching up to full height with the breast puffed out and the buff supercilium erect at the rear, forming 'horns'. Aggressive birds adopt a low crouching posture with wings extended sideways and the bill pointed upwards (Sclater and Moreau 1933). Feathers on either side of the upper breast are reported to be erectile, and are used to form a fan of stiffened cinnamon-buff plumage (Clancey 1985), although the situations in which these feathers may be erected is presently unclear.

Nestlings are reported to be very quiet and inactive. They are hard to see since they flatten themselves on the bottom of the nest if alarmed.

BREEDING Records for *pulih* are few and from scattered localities, but all point to breeding in the wet season. At the western end of its range, a fledgling was collected in Sierra Leone on 10 September. In Liberia, nestlings have been found on 15 and 21 September (Allen 1930) and a male in breeding condition was collected in March. Further east, a well-grown immature has been collected in Ghana in July (Keith *et al.* 1992), and, at the eastern end of its range, in Cameroon, nests with eggs have been discovered on 12 September and in October (BMNH) and an immature collected on 15 November (Serle 1954). Hence the breeding season of *pulih* may be rather protracted, and perhaps peaks in the last part of the year.

There is virtually nothing recorded about the breeding season of *angolensis*. The only confirmed breeding record is of a nest with one egg that was found on 6 November in northern Angola (Chapin 1953, BMNH), at the beginning of the wet season. Chapin (1953), however, noted that L. Petit gave the nesting season as December and January and that Petit had reported finding a nest with four eggs along the Loanga Coast of Congo.

In south-east Africa, evidence also indicates that *longipennis* usually lays eggs during the rains, between November and February (Harvey 1938, Keith *et al.* 1992). Eggs have been found in Malawi in December and January (BMNH); Zambia in December and January (D. Aspinwall *in litt.*, J. Colbrook-Robjent *in litt.*) and in February; in Zimbabwe from November to January and in Mozambique in November (Clancey 1971). Fresh, empty nests have been found in Malawi in December and newly fledged juveniles collected in Zimbabwe in mid-January and in Mozambique on 2 March (Clancey 1971, Benson and Benson 1977, Irwin 1981). A bird in near-breeding condition was found in Natal in November (Clancey 1985).

The nest is a large untidy dome of variable size and shape, placed some 2-8m above the ground; it may be within 150m of neighbouring nests. Of ten nests found in Zambia, all but one were less than 3.3m above the ground (J. Colbrook-Robjent *in litt.*). Nests are often situated so as to be protected by thorns. Nests found in *Garcinia* saplings within a thorny tangle had a fringe of projecting twigs at the sides of the roof. In Zimbabwe, one nest was found in a Sickle Bush (Alexander 1994). Nests found on horizontal outer branches of *Ximeria* and *Ziziphus* trees had a distinctive platform in front of the side entrance that is often lacking from other nests. These nests, and a nest in *Combretum* thicket, also had very extensive bases. The base may be 25 x 35cm in area. Normally nests are c.18-20cm high and slightly wider. The side entrance is 7-8cm wide and 6-7cm high.

The base of a nest from Angola (BMNH), which contained a single egg, was made of sticks and twigs up to 6mm in diameter and c.22cm long, and of many types, including spiny acacia and bramble-like twigs, intermingled with a small number of dried and broken pieces of dicotyledonous leaves. The nest dome itself was more compact and made of a mass of fine twigs (1mm or less wide). Largely from one plant species, the twigs were long and flexible with regular short spines along their length, the spines helping to hold the nest structure together. The outer wall of the nest cavity was constructed largely from large dried dicotyledonous leaves, matted together with a few short pieces of twig and hair-like fibres (fine roots or fungal hyphae). A nest from Zambia was built on a base of large dry leaves (c.17cm x 7cm), circled by a rim of the same leaves and intermingled with vine stems. The main nest dome was made of long leaf stems (up to 17cm long plus a few rootlets, enclosing a shallow cup c.11cm and 3.3cm deep (J. Colbrook-Robjent *in litt.*). In some nests, the entrance hole may be primarily composed of leaf petioles. The nest cup, some 10cm wide and 11cm long, is usually lined with fine roots, fine twigs and tendrils, skeletonised leaves and other papery vegetation. The nest cavity is c.8cm high (Harvey 1938, Serle 1954, Keith *et al.* 1992).

Reported clutch-size is most frequently three, but nests may contain four or two eggs and occasionally one. Of ten clutches found in Zambia from the Mutulilangangan River in Gwembe District, Southern Province, one had a clutch of one, four had clutches of two, three had clutches of three and two a clutch size of four (J. Colbrook-Robjent *in litt.*). The eggs of *longipennis* are blunt ovals or almost round, being 27.5-29.0 x 23.2-25.0mm (Harvey 1938, BMNH). Twenty-four eggs from Zambia were of average size 27.8mm x 23.1mm and average fresh weight 7.8g (range 6.6-8.9g) (J. Colbrook-Robjent *in litt.*). Eggs of all races are creamy white, occasionally tinged greenish, and occasionally pink, and marked with irregular small reddish-brown or purplish blotches and a few dark lines on underlying grey, lilac-grey or lavender markings. Markings are usually concentrated in a broad band at the blunter end. An egg of *angolensis* from Angola measured 26 x 23.5mm (Reichenow 1903, Allen 1930, Harvey 1938, Chapin 1953, Benson and Benson 1977, Clancey 1985, Keith *et al.* 1992, D. Aspinwall *in litt.*, J. Colbrook-Robjent *in litt.*, BMNH). Eggs from Cameroon, reported as belonging to *angolensis* (Serle 1954) but probably of *pulih* measured 26 x 22.1mm, 26 x 22.3mm and 24.4 x 22.6mm. These eggs were glossy, creamy-white and elliptical, with markings mainly congregated about one or the other end in the form of a few purplish-black bold clear-cut blotches and spots and similarly shaped underlying ashy-grey marks.

DESCRIPTION *P. a. angolensis*
Sexes similar.
Adult Coronal band (narrow at bill, widening towards nape), lores, ear-coverts and sides of neck black. Long, broad ochre supercilium extending to nape, usually paler towards tip at nape; the rather pointed feathers at the rear supercilium are elongated and capable of erection into a light buffish 'horn'. Mantle to upper rump and uppertail-coverts bright green; lower rump and uppertail-coverts brilliant azure-blue; tail dull black. Lesser coverts dull green with broad azure-blue tips; median-coverts darker green narrowly tipped with violet-blue to cobalt-blue; greater coverts dark brown, the innermost greener with azure-blue spots at tips; primary coverts and alula blackish. Primaries blackish, broadly but diffusely tipped with greyish or greyish-white, particularly on outer web and most prominent on innermost primaries; primaries basally white except on outer web of outer 2-3 primaries (sometimes absent from outer feather); secondaries blackish with some outer feathers tipped with white on the outer web; tertials green with dark shafts. Underwing dark brown. Chin and throat white, often strongly suffused rose-pink; breast, sides and flanks deep cinnamon-buff, often washed with golden-green sheen; belly and undertail-coverts deep red, thighs brownish-buff with red tips to feathers. Bill black with reddish base to lower mandible and red spot on culmen; tongue and inside of mouth red; eyes dark brown to reddish-brown; legs and feet pinkish to greyish-white. Occasionally a few narrow, rather indistinct azure-blue bars across the feathers of the back.
Immature Similar to juvenile, but has an all dark bill tip and basal part to upper mandible, and more extensive pinkish-red on belly and undertail-coverts.
Juvenile Supercilium dark ochre to ochre-buff, becoming almost white along lower margin behind eye and extending to nape as a mixture of white and buff feathers. Crown, lores, cheeks and ear-coverts black. Upperparts dull green, including scapulars. Uppertail-coverts and some feathers on lower rump azure-blue. Upperwing-coverts dark brown, with some dull green in lesser coverts, these feathers also sometimes having indistinct dull azure or blue-green tips forming a broken line of spots, although these markings may be confined to innermost part of wing. Flight feathers similar to those of adult. Throat white to dirty white (juvenile *pulih* has a pink throat: it is not clear if juvenile *angolensis* also shows this). Breast, flanks and upper belly a dark buffish-brown; lower belly and undertail-coverts suffused with pinkish-red. Bill orange-red to red with a broad black central band; feet orange and dark brown. Evidence suggests that migrant African pittas of subspecies *longipennis* moult in the non-breeding range, after migration (Baker and Baker 1992).
Nestling Black with a conspicuous orange stripe along the side of the bill from the gape to tip.

MEASUREMENTS Wing 118-124; tail 39-49; bill 24-29; tarsus 33-35; weight unrecorded.

24 GREEN-BREASTED PITTA
Pitta reichenowi   Plate 12

Pitta reichenowi Madarász, 1901, Ornith. Monatsb., 9, p.133: Middle Congo River, above Stanley Pool.

TAXONOMY For discussion relating to the taxonomic relationships of African pittas, see under African Pitta.

FIELD IDENTIFICATION Length 19-21. Sides of head and crown black, separated by a prominent broad golden-buff supercilium that extends from forehead to nape. The breast and upperparts are green, the throat white, the belly to undertail-coverts deep red and the rump azure-blue. The breast is often glossed with golden-yellowish. Upperwing-coverts are mostly green, with conspicuous azure-blue tips to the lesser, median and outer greater coverts, so that there is an azure-blue patch at bend of the wing and two lines of azure-blue spots in the closed wing. Lower spots may be a deeper blue. The soft, mechanical *prrrt* noise is often the first indication that this shy species is present.

Juveniles have dark brown upperparts with dark green feathers on the mantle and back, blackish head with distinct dark ochre supercilium, and white throat, this often faintly tinged with pink. The wings are similar in colour to the upperparts, but have distinct azure-blue tips to coverts that usually form at least one, and sometimes two, lines of azure spots along length of wings. The breast is a dark ochre-brown, washed with olive, becoming paler towards belly; the belly and undertail-coverts are pink. The bill is distinctly bicoloured, being orange-red to red with a black band near the tip.

Similar species Although quite different to the large, migrant subspecies of African Pitta (*longipennis*), Green-breasted Pitta is similar in size and plumage to western African subspecies, in particular *angolensis*. From behind, or when flushed, the two species appear identical. However, adult Green-breasted Pitta can be distinguished from all subspecies of African Pitta by the coloration of the breast, which is green rather than cinnamon-buff, and in the possession of a black mark across the top of the breast, although this may be difficult to see under field conditions. Whilst *P. a. angolensis* has very similar wing markings to Green-breasted Pitta, the subspecies *P. a. pulih* differs in having fewer azure-blue tips in the upperwing-coverts, and these usually only form one line of spots, compared to two in Green-breasted Pitta and other subspecies of African Pitta. Sometimes Green-breasted Pitta has a strong golden-buff suffusion to the breast but the colour is still clearly different to that of African Pitta. Migrant *P. a. longipennis* differs in much larger size, paler green upperparts, cinnamon-buff breast and a reduced area of red on the underparts, this being confined to the centre of the belly and undertail-coverts. In western Uganda and north and east Zaïre, *longipennis* and *reichenowi* may occur sympatrically during May-September. Juvenile Green-breasted Pitta is very similar to that of African Pitta, but usually has a distinctive olive wash to the breast and the azure-blue upperwing-covert tips are acquired much earlier and are more prominent than in juvenile or immature African Pitta.

VOICE The usual call of Green-breasted Pitta is a pure, bell-like, slightly mournful, whistle that lasts from 0.25-0.5s and is repeated several times in succession at a rate of ten calls in 13-15s. M. Andrews (*in litt.*) observed a bird calling every 3-5s from a branch some 30cm above the ground on Mt Kupe. Another frequently heard noise that is probably mechanical rather than vocal is a short soft purring *prrrt* or *brrrt*. This is made when the bird makes a short flight up to a perch such as a log close to the ground. A loud accelerating *prrrrrt* was also made by a bird flying away from an observer who had disturbed it (M. Andrews *in litt.*). This noise is said to sound like that made by *Smithornis* broadbills during display-flight, but is not so loud or 'wooden' (Brosset and Erard 1986). A captive full-grown immature frequently gave a mournful whistle. The bell-like whistle is reported to be similar to calls of the sympatric Fire-crested Alethe *Alethe diademata* (Keith *et al.* 1992).

DISTRIBUTION Green-breasted Pitta is virtually confined to the Congo basin and parts of southern and western Uganda. Records from the Congo basin are scattered, and the shy, retiring nature of this species means that the exact limits of its range are unknown. Most records from the eastern Congo are from forest along rivers, with intervening areas largely unexplored ornithologically. Hence, although the known distribution of Green-breasted Pitta appears disjunct, it is likely that this may be, to a large extent, due to inadequate data, and the species is probably widespread in the Congo basin. Westernmost records come from north-east Gabon and southern Cameroon. In Cameroon, it has been recorded from Ebolowa and Sangmelima southwards, to the east of c.11°E (Louette 1981), and on Mt Kupe (4°48'N, 9°42'E) (Bowden in press).

In Zaïre, there are records from the extreme southern limit of rainforest (Collins *et al.* 1991), in Kasai-oriental; from the region between Lake Mai-Ndombe and Lake Tumba and the Congo River (north of the confluence with the Kasai River); and in the north-eastern part of the basin north of the equator, in particular along the Ituri and Itimbiri Rivers. Further east, in Uganda, it has been recorded in Bugoma, Budongo and Bwamba (Semliki) forests (along the eastern border of Lake Albert), south to Kigezi and east to Jinja (Chapin 1953, Britton 1980) on the northern shores of Lake Victoria (near Ripon Falls: Mabira Forest).

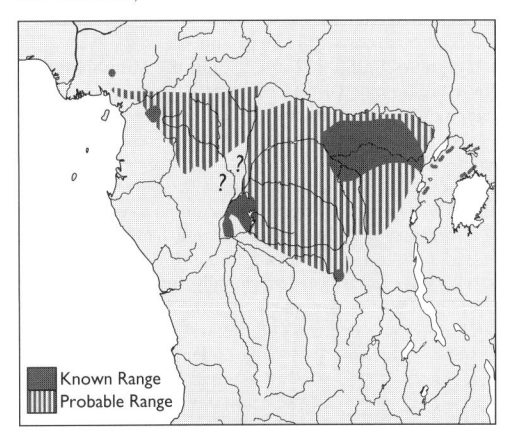

Distribution based on approximate extent of forest remaining in areas where occurrence is confirmed. Probable range, based on extent of suitable forest, is indicated.

GEOGRAPHICAL VARIATION None recorded.

HABITAT Whilst this is sometimes considered a bird of primary and mature secondary rainforest, Chapin (1953) believed that it favoured old second growth and thickets rather than virgin forest. In the Bwamba Forest of Uganda, it has been found in stands of mature ironwood *Cynometra alexandri* (Keith *et al.* 1992), whilst in Gabon it occurs in old plantations where the forest had regrown (Brosset and Erard 1986) and has also been recorded on an island along a river. Chapin (1953) encountered a bird displaying in dense undergrowth at the edge of swampy forest. On Mt Kupe, Cameroon, it has been found at the forest edge, only at 900-950m altitude (Bowden in press). In Uganda, it occurs in forests at altitudes of 1,100-1,400m (Britton 1980).

STATUS In Gabon and parts of Zaïre, the species is reportedly locally common in suitable habitat (Chapin 1953, Keith *et al.* 1992). Four nests have been found in Budongo Forest, Uganda, suggesting that it is overlooked rather than rare (Britton 1980). On Mt Kupe, Cameroon, at the edge of its known range, this appears to be a rare species, with only three reports in as many years (Bowden in press). Habitat loss is unlikely to pose any serious threat at the present time.

FOOD Five of seven specimens collected in Zaïre contained hairless caterpillars, beetles and small millipedes. Beetle larvae were found in two of the specimens and remains of a small centipede, termite, maggot, orthopteran, isopod (pillbug) and tiny snail were found in single specimens (Chapin 1953).

HABITS This is a poorly known species. However, its habits appear to be typical of pittas, being largely solitary, feeding on the ground and occasionally perching in low trees. When alarmed the buff supercilium may be raised to form erect tufts at the rear of the head. Whether Green-breasted Pitta has a display that is similar to that performed by African Pitta is unclear, although the species evidently makes a very similar mechanically produced sound in flight.

BREEDING As a poorly known species, few breeding records exist. Eggs have been found in Budongo Forest, Uganda, on 3 and 10 May, in Gabon in January and Cameroon in May and on 17 September (Chapin 1953, Brosset and Erard 1986, BMNH), whilst half-grown young have been found in September at Medje, Zaïre (Chapin 1953). Fledglings have been collected in Zaïre on 1 May and in Cameroon on 22 June and 15 November (BMNH). A juvenile was collected in Cameroon in 24 August. In Zaïre, birds with enlarged gonads have been collected at Lukolela, just south of the equator, in late October and November (Chapin 1953). These records suggest that Green-breasted Pitta breeds throughout the year, although birds in Cameroon and northern Zaïre apparently breed mainly from May-September, during the rainy months, whilst those south of the equator probably usually start breeding in October (Chapin 1953).

The few nests that have been found have been placed 1.5-2.5m above the ground on horizontal or gently sloping tree trunks and large branches, and once in the tangle of branches and dying leaves of a fallen tree crown (Bates and Ogilvie-Grant 1911). The nest is a fairly large, domed, globular structure with a wide side entrance. The foundation of a nest from Budongo consisted of a few large twigs and tough dried creeper stems intertwined with a large mass of dried and skeletonised leaves. The nest cup was thickly lined with leaves and supported on and mixed with a thick pad of rootlets of various thicknesses (BMNH). Plant stems and petioles or leaves may also be used as nest lining (Brosset and Erard 1986, Keith *et al.* 1992).

Clutch-size is reported to be 2-3 (Keith *et al.* 1992), although a nest discovered in Gabon by Brosset and Erard (1986) contained only one egg, as did a nest found by Chapin (1953), whilst a nest from Budongo Forest apparently contained four eggs (BMNH).

Six eggs from two clutches collected from Budongo Forest varied in size and appearance (BMNH). One set was white, boldly marked with large irregular patches of chocolate-brown, concentrated at the broader ends, and varied in size from 28.4-30.0 x 21.5-22.5mm; eggs from the other clutch were white, with fine chocolate-brown spots and irregular marks and with underlying grey spots. These latter eggs were rounder, measuring c.25.0-25.5 x 21.5 x 22.5mm. Two eggs from Bitye, Cameroon, had fewer markings than either clutch from Uganda, with a few spots of chocolate-brown but more underlying grey markings, all concentrated at the broader end (BMNH). The Bitye eggs were similar in size and shape those of one clutch mentioned above, being c.28.5-30.0 x c.21-22.0mm.

Bates and Ogilvie-Grant (1911) captured a male that was on eggs, suggesting that both parents incubate.

DESCRIPTION
Sexes alike.
Centre of forehead, crown and nape black, bordered below by a long, broad golden ochre-buff supercilium with rather pointed, erectile, tip. Lores, sides of head and ear-coverts to the rear of the head and neck black, joined to black of crown at nape. Upperparts dark green from mantle to upper rump; lower rump and uppertail-coverts shining azure-blue, tinged violet on the latter; tail black, sometimes with narrow azure-blue tips to some feathers. Wing-coverts black with prominent c.4-7mm broad azure-blue tips to lesser coverts (forming a patch at bend of wing), median coverts and inner greater coverts, usually narrowest on the median coverts, where azure-blue is usually replaced by violet-blue tips in centre of wing. Primaries black, tipped grey-buff; secondaries black with inner feathers tinged greenish and two outer feathers narrowly bordered white near tip of outer web; tertials dark greenish-blue. Throat and chin white, occasionally tinged pink separated from the green breast by a narrow but distinct black line; breast feathers pale green with golden-yellowish gloss, or bluish-green with golden-yellowish gloss and buff tips; belly to undertail-coverts deep red, sometimes with greyish feather-bases showing through; thighs brown; underside of tail black. Concealed white patch on the bases of outer primaries (primaries 5-8 or 6-8). Underside of wing glossy black with greyish tips to outer primaries. Eye dark reddish-brown; bill black with reddish base to lower mandible and red spot on culmen; legs and feet pale greyish-pink to greyish-white.

Juvenile Head black except for broad dark ochre supercilium that becomes buffier at the rear, and a white throat; the latter often tinged with pink. Upperparts appear dark brown, although dark olive-green feathers develop on the mantle, back and inner coverts at an early age. Feathers of lower rump and uppertail-coverts tipped with azure-blue, forming a distinct patch. Upperwing-coverts dark brown, often with green sheen to innermost coverts, and with one or two lines of conspicuous azure-blue feather tips to median and lesser coverts. Flight feathers blackish-brown. Breast dark ochre-brown, tinged with olive, and more or less tinged with paler buff tones at lower margin. Belly to undertail-coverts pink. Recently fledged birds have a dull brownish-red to orange bill with black band across the central portion; legs orange-red.

MEASUREMENTS Wing 113-125; tail 39-47; bill 22-27; tarsus 34-40.5; weight unrecorded.

25 INDIAN PITTA
Pitta brachyura   Plate 12

Corvus brachyurus Linnaeus, 1766, Syst. Nat., ed. 12, 1, p.158: 'Moluccas', an error for Sri Lanka.

Alternative name: Blue-winged Pitta, Bengal Pitta

TAXONOMY See under Fairy Pitta for an account of why these two pittas should be considered separate species.

FIELD IDENTIFICATION Length 15-19. This is a small pitta that is unmistakable in its range, which extends from the foothills of the Himalayas to southern India and Sri Lanka. Adult Indian Pittas are best identified by their striking head pattern, with black sides of head and crown and distinct buffy supercilium that becomes white and pointed at the nape, green upperparts, pale cinnamon-buff underparts with scarlet undertail-coverts, and iridescent azure-blue rump and patch on bend of wing. In flight, the large white wing-patch is obvious. Immatures have slightly darker crowns, with black edges to the feathers, dark brown upperparts and paler brown underparts with pink belly and undertail-coverts. Recently fledged birds retain bold supercilium but have generally subdued colours, with the lower parts smoky-black and the scarlet on the belly faint or hardly visible. The lower mandible and bill tip are bright orange to bright red.

Similar species Indian Pitta is unlikely to be confused with any other pitta in its range. However, Mangrove and Blue-winged Pittas both occur in areas where vagrant Indian Pitta could potentially occur. These two species are overall larger and darker species, with much more extensive violet-blue, rather than azure-blue, wing-patches, violet-blue rumps and a broad bold buffy supercilium. They also have considerably more white in the wing and a more chestnut-coloured supercilary band. Both lack the small whitish patch below the eye and white in the supercilium of Indian Pitta. Immature Mangrove and Blue-winged Pittas also have more extensive blue in the wing and are a richer orange-buff below, when compared to similar-aged immature Indian Pitta. Immature Hooded Pittas, which may be found in the same areas as breeding Indian Pitta, have black heads with chestnut crown to nape, and a prominent white bar across the wing-coverts.

VOICE The call is a clear short double whistle, *wheeet-tieu* or *wieet-pyou*, or occasionally, a triple note *hh-wit-wiyu* (see sonogram below). The call is mostly made at dawn or dusk from a bush or tree, sometimes from the ground, and repeated three or four times before moving to a new vantage point from which to call. This call is often accompanied by harsh, high-pitched disyllabic scalding notes. A second, rarer call is a single drawn out *wheeew*. Other notes attributed to this species include several harsh monosyllabic *mews* and *chees*. Calling in the day is more common in overcast weather. In the breeding season, the calls are repeated at the rather extraordinary rate (for a pitta) of three or four every 10s, and this may continue for up to five minutes. A harsh call is emitted by parent birds when the young are threatened with danger.

Calling at dawn and dusk also occurs on the wintering grounds, and the Tamil name translates as the "six-o'clock bird" (Henry 1971). Brown (1931) noted that for the first few days after arrival in Sri Lanka, wintering Indian Pittas give a harsh 'jazz' note or *hizz* that he attributed to warning off other pittas from the calling bird's wintering territory. Calling birds are usually answered by individuals in adjacent territories. In Sri Lanka, drongos *Dicrurus* spp. and Golden-fronted Leafbird *Chloropsis aurifrons* may imitate the call of this species (Brown 1931).

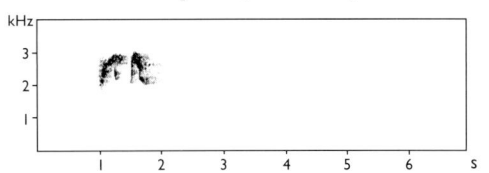

Indian Pitta *P. brachyura*
[from Tamil Nadu, southern India, recorded by P. Holt, January 1994]

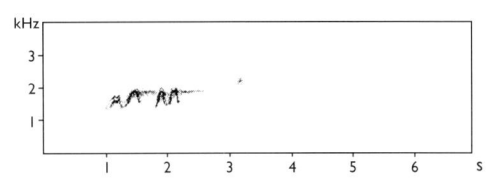

Fairy Pitta *P. nympha*
[from Japan, recorded by T. Kabaya, published in *160 Wild Bird Songs of Japan*]

DISTRIBUTION Indian Pitta is confined to the Indian subcontinent, where widely distributed. It is a widespread breeding bird in the northern part, although it does not usually occur in the desert parts of Rajasthan or, apparently, to the west, in Pakistan (Baker 1934, Roberts 1992). However, there is a small but regular breeding population in the Margalla Hills near Islamabad, Pakistan (Roberts 1992), and an egg in BMNH is said to have been collected in Kashmir in May 1871. To the east, Indian Pitta is known from western Himachal Pradesh, southern Nepal, southern Sikkim, and south to the wetter parts of Rajasthan, Kanara and Bangladesh. It has also been recorded breeding, sporadically, further south, in the plains of central India, such as near Delhi and in the Gir Forest (Gujarat), and in northern Karnataka (Donald 1918, Ali 1962, Ripley 1982, Ali and Ripley 1983, Daniels 1986, Inskipp and Inskipp 1991, Roberts 1992). Abdulali (1976) noted that the southernmost breeding record is near Bombay, and that regular nesting occurs in central India near Mhow (western Madhya Pradesh) (Betham 1922), south and north of the Satpura Range, and also in Dhulia, West Khandesh. After breeding, Indian Pittas migrate south to spend the winter in southern India, south of about 16°N, and in Sri Lanka.

The exact status of Indian Pitta to the east of Nepal and north-east Bihar is poorly known. There are only a few records from Sikkim (Stevens 1925, Ali 1962), and despite mention in major texts of its occurrence in Bhutan, Assam, Arunachal Pradesh, Manipur and the Chittagong Hills of south-eastern Bangladesh (Baker 1926, Ripley 1982, Ali and Ripley 1983, Choudhury 1990), there are no published records in the many detailed papers on birds from this region. Indeed, the only confirmed record of Indian Pitta to the east of Sikkim would appear to be from Bangladesh, where Harvey (1990) reported that it possibly bred in 1989 in Modhupur Forest, central Bangladesh, where it was recorded in June and July. Although villagers from Jatinga, in the Barail Range of Assam, have reported the capture of migrating Indian Pittas during traditional nocturnal bird-catching sessions (R. Grimmett *in litt.*) and the species has apparently been found in this region (A. Choudhury *in litt.*), there are no authenticated records of it being caught there. The exact status and distribution of Indian Pitta in the north-eastern part of India therefore need verification.

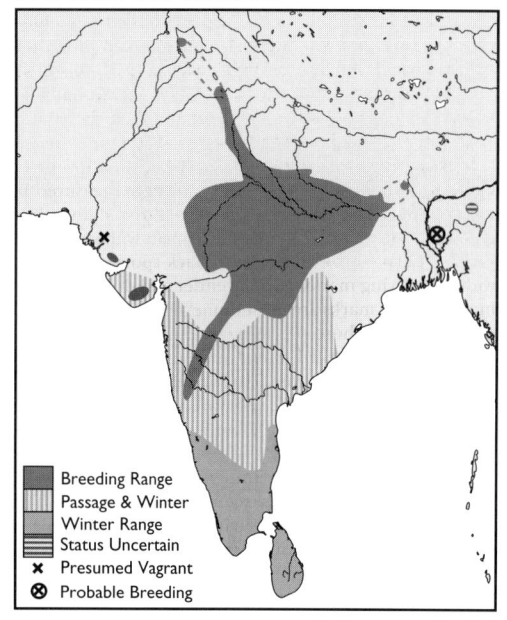

Approximate breeding and wintering ranges shown. Migrants may occur almost anywhere to the south of the breeding range. Range within distribution shown is patchy. Presence of breeding populations in north-east India suspected but needs confirmation.

GEOGRAPHICAL VARIATION None recognised, although Abdulali (1976) suggested that the colour of underparts, lateral crown-stripes and rump might differ between different breeding populations.

HABITAT In its breeding range, Indian Pitta is an inhabitant of deciduous and evergreen forests with dense undergrowth, mostly in the lowlands and hills, breeding in the foothills and plains of the Himalayas, locally up to c.1,200m (e.g. Kangra, in Himachal Pradesh) but is occasionally found up to 1,700m as a migrant or wintering bird in southern India. In northern Pakistan, Indian Pittas are reported to breed during the monsoon in moist ravines, where they are confined to thick scrub at 450-760m. The vegetation in the area of the Margalla Hills where this species breeds is dominated by thorny bushes of *Carissa opaca* and *Ziziphus mauritiana* (Roberts 1992). In the Gir Forest, they apparently prefer cover of *Carissa carandas* bushes and other shrubby undergrowth in the vicinity of ravines, but also occupy teak coppice (Dharmakumarsinhji 1955). Baker (1926) noted that Indian Pittas prefer bamboo and scrub jungle or deciduous forest such as sal forest (dominated by *Shorea robusta*, Dipterocarpaceae) and oak (*Quercus* spp., Fagaceae).

Dharmakumarsinhji (1955) noted that, during the rains, Indian Pittas may be seen anywhere within their breeding range as local migrants. On passage they occur in a variety of habitats, often at unusual altitudes, particularly if migration has been interrupted by mist or fog.

Migrants are attracted to lights in such conditions and may even fly into houses. In Sri Lanka, Indian Pittas have been recorded as high as 1,830m (Henry 1971), although usually spending the winter at considerably lower altitudes. During the winter months, Indian Pittas frequent forested areas as well as small woods and copses, and occasionally well-wooded gardens.

STATUS During the early part of the century, Law (1939) found Indian Pitta to be not uncommon in Bengal, whilst Baker (1926) found that, in its breeding range, it was common in many parts of central India, southern Assam and parts of Bihar but rare in the extreme south. The vast numbers of Indian Pittas that have been recorded as migrants suggest that the population size may have been in the hundreds of thousands.

Although Indian Pittas are not uncommon, changes in the amount of forest in northern India, which has been extensively cleared in the past two decades, are likely to have caused a significant population decline in many parts of its original range. As a consequence, Indian Pittas are now rather patchily distributed. In Nepal, for example, they are only known to be common at one site: Chitwan (Inskipp and Inskipp 1991). Whether the trapping of migrants for food has any significant effect on population size is unknown, but it seems likely that such practices will have an increasing impact as the species becomes rarer due to habitat loss.

Large numbers are traditionally killed during migration at a number of localities in India, particularly on the coast of southern Tamil Nadu in the autumn. Johnson (1972) documents how local people at Point Calimere use snares very effectively to trap birds flying through holes in bushes: 40-50 birds could be caught in a few hours with only 5-8 carefully placed nooses.

Finally, it should also be noted that, as a migrant, Indian Pitta is reliant on the persistence of suitable habitats in its wintering grounds. Whilst it is clearly able to winter in a diversity of habitats, widespread degradation of monsoon and thorn forests by slash-and-burn cultivators has occurred in Sri Lanka (Collins *et al.* 1991), which is an important part of the species's wintering range.

MOVEMENTS Indian Pittas migrate southwards at the end of the rainy season, and return in advance of the south-west monsoon, although in some parts of their range, such as in Saurashtra (Gujarat), they are apparently resident (Dharmakumarsinhji 1955). Migrants reach Sri Lanka in September and October, during the onset of the north-east monsoon, returning north in April and May, and passage migrants are recorded in southern India during these months and early June. Passage in eastern Nepal is recorded as early as May. Records suggest that Indian Pittas often migrate in small flocks, and vast numbers have been observed on regular passage at certain sites, where the first arrivals appear annually on virtually the same date. In Sri Lanka, the date of arrival is also very consistent (Brown 1931).

FOOD Indian Pittas eat a variety of insects, including ants, as well as grubs, earthworms, small snails (including those in the genus *Helix*), and millipedes, and are reported to be particularly fond of termites. According to Ali and Ripley (1970), Indian Pittas extract maggots from human excrement in the vicinity of villages, and they have also been observed eating the larvae of a coffee beetle found in manure. Observations at one nest suggested that the young are fed many orthopterans, especially mole crickets (Gryllotalpidae).

HABITS Indian Pittas feed like thrushes, rummaging in leaf-litter on the forest floor, and digging for insects and worms in wet earth, using their bill. They usually fly off when startled, rather than hopping, often flying to perch in trees, where they wait, motionless, before descending back to the ground. Typical roost sites are also in small trees.

The tail is regularly wagged, slowly, up and down, both whilst feeding and after flushing into the lower branches of a tree, and also when calling (Ingalhalikar 1977). In Sri Lanka, Indian Pittas also have the habit of momentarily expanding their wings and tails to expose the white primary patch, which is otherwise usually only visible in flight. This behaviour has been noted when a rival pitta enters another Indian Pitta's territory, and also when startled. During calling, the head is jerked upwards and backwards during the first note, and then jerked forward again during the second note (Henry 1971). Indian Pittas are territorial, even in their winter range, and in Sri Lanka they have been observed vigorously defending their territory against conspecifics, during which males will often chase each other in flight (Henry 1971).

Territorial calling starts immediately, in May, when birds return to the breeding grounds. Following breeding, a complete post nuptial moult occurs, usually in August and September (there is no moult in the spring).

Daniels (1986) speculated that Indian Pittas wintering in Bangalore were frequently diseased, since hawks, in particular Shikra *Accipiter badius*, were regularly found to catch Indian Pittas despite their shy habits and often dense habitats. On passage, crows may also be responsible for killing some pittas. Neelakantan (1963) recounted how a number of observers had apparently observed crows *Corvus* sp. chasing and killing pittas, as well as eating the flesh of dead birds, although the circumstances were usually such that this behaviour might not be expected to be common, apparently occurring mostly when pittas that had taken shelter in houses were released into the open. One Indian Pitta chased by crows escaped, only to be devoured by a Black Kite *Milvus migrans*.

BREEDING The breeding season is primarily between May and July, although eggs have exceptionally been found as early as 27 April (BMNH) and nests in August and September (Baker 1934, Ali and Ripley 1983). In Gujarat, the principal period for nesting is reported to be from June to August (Dharmakumarsinhji 1955). Of 16 dated clutches in BMNH from various parts of India, one was collected in April, two in May, eight in June, four in July and one in June.

Indian Pittas sometimes nest on the ground under the shelter of dense bushes, but more frequently nest in small trees. Sal trees are often chosen for a nest site (Baker 1934). One nest was found in a bush of *Carissa spinarum* at the edge of Sal forest (A. Osmaston in BMNH). The nest is typically placed on a forked branch, 3-4m but occasionally up to 9m above the ground, and is often conspicuous.

The nest is a large globular structure, usually on a foundation made from a mass of sticks, with the entrance at one end. Foundations of nests in trees are often massive, but on the ground they are smaller, and sometimes not used at all (Baker 1934). The nest itself measures c.20cm long, 15cm wide and 15cm high, with a wide entrance hole of c.10cm diameter. The nest material comprises leaves, strips of grass, twigs and moss, whilst the lining is made of grass and bamboo leaves, and, in some parts of central India, tamarisk twigs and grass roots (Hume 1880, Baker 1926). A nest found in a Sal tree by A. Osmaston (BMNH) was built of small sticks, coarse grass and dead leaves with the nest chamber built of the larger dropped branchlets of *Phyllanthus emblica* (Euphorbiaceae) mixed with a few fine roots and one or two fungal mycelia (rhizomorphs).

The normal clutch-size is 4-5, but occasionally six (BMNH), and both sexes incubate and feed the young. The eggs are broad ovals and some may be almost spherical, measuring around 23.3-28.2 x 20.0-22.4mm, with average size of 24.7 x 21.2 (based on 50 eggs measured by Baker 1934). They are an excessively glossy, china-white and sometimes faintly tinged pink, usually with sparse dull purple, purple-brown or almost black spots and specks, with underlying markings of lavender and dull purple. In most eggs the marks are most numerous at the larger end and sparse elsewhere. Occasionally in a clutch, otherwise quite normally marked, there may be one egg with a big blotch some 4-6mm across. The twisted lines so characteristic of many pitta eggs are very rarely present in the eggs of Indian Pitta (Hume 1880, Betham 1922, Baker 1926, 1934, Ali and Ripley 1983).

Observations of captive Indian Pittas suggest that the female is largely responsible for nest construction. The normal clutch-size of birds breeding in captivity is reported to be five, with incubation starting only after the second egg is laid and with one egg being laid per day. Incubation in captivity took 17 days and the chicks fledged after 15 days. Within five days the juveniles were fully independent (Everett 1974).

DESCRIPTION
Sexes similar.
Adult Broad black crown-stripe, narrowest at forehead and extending to nape where it joins a black band across the sides of the head, which includes cheeks, ear-coverts and lores. There is usually a white patch just below the eye, which may form a short streak behind the eye. A broad golden-buff to buffy-brown supercilium extends from nostril to nape, where narrowing to a largely white point that may be faintly tinged with pale blue. The entire lower border of the supercilium is usually paler, cream or white, though this is occasionally only noticeable behind the eye. The supercilium point may jut out at the rear of the crown. Feathers of the supercilium sometimes have very fine dark tips. Back, mantle, and majority of upperwing-covert feathers dark green; the feathers of the upper back have a variable amount of buff tinge, this sometimes forming an indistinct narrow band of buffish suffusion across the upper back. Most birds have prominent black streaks on feathers of the mantle, back and wing-coverts (including within the blue patch), formed by black along the feather axis, though the amount varies between individuals. Rump and uppertail-coverts bright pale azure-blue, frequently also narrowly streaked with black. Tail black, with broad dark blue band at tip which occasionally extends half-way up tail. Majority of exposed wing-coverts concolorous with mantle, with variable amount of blue tinge to greater coverts, and partially concealed azure-blue wing-patch, formed by broad tips to feathers of bend of wing and lesser coverts. Tertials green, with variable amount of blue towards tip and outer margin of all feathers, and black base to inner web of outer tertial; median coverts primarily

green, but some blue on feathers of inner wing; greater coverts largely blue on inner part of wing, becoming greener towards outer wing. Secondaries black, with blue outer edge to about a third of outer web and usually extending across feather near tip; tip pale greyish to white, usually broader towards the centre of secondaries and sometimes bordered at edge with narrow dark grey. Primaries black with prominent white wing-patch formed by white band across centre of primaries 2-5, inner web of outer primary and outer web of primary 6; primary tips pale brown to grey. Underwing-coverts black, usually with narrow white patch near bend of wing and, occasionally, speckled pale blue along outer leading edge. Chin and throat white, clearly demarcated from cinnamon-buff breast, flanks and upper belly. White of throat extends as a narrow point along the border of the black band across sides of head, becoming buffier towards tip. Lower flanks may be slightly paler buff that rest of underparts. Lower belly and undertail-coverts scarlet. Iris dark brown; bill dark brown or blackish, often with reddish-brown culmen and orange or fleshy-red at gape and on basal half of the lower mandible (perhaps relating to subadult birds); legs and feet pinkish-flesh to purplish-flesh. Males have slightly longer wings, on average, than females.

Immature Head pattern reminiscent of adult: black crown-stripe and sides of head; white chin and throat with variable amount of indistinct dark brown fringes to feathers; feathers of superciliary/coronal area buffy-brown with fine black streaks and narrow dark edges forming a scaled appearance, but becoming whiter towards nape, where a few feathers are usually entirely white. Mantle and exposed wing-coverts dark, unstreaked, olive-green, becoming greener with age. Rest of wing as adult. Rump and uppertail-coverts pale bright blue, less iridescent than that of adult; tail similar to adult. Blue patch on bend of wing smaller and duller than adult. Underparts dark brownish-buff, becoming buffier-white with age; belly and undertail-coverts pink, becoming redder with age. Lower mandible, bill tip and base of upper mandible bright orange to bright red.

Juvenile Recently fledged individuals have even more subdued colours, with the lower parts smoky-black and the pink on the belly faint or hardly visible. Legs of birds just out of the nest are orange (Donald 1918). Everett (1974) noted that development of adult plumage was slowest on the head, and that immatures were indistinguishable in plumage from adults ten weeks after leaving the nest.

MEASUREMENTS Wing 101-111 (labelled as males 104-107; as females 101-105); tail 34-42; bill 26-30; tarsus 33-37; weight 47-66 (average 55.5).

26   FAIRY PITTA
   Pitta nympha   Plate 13

Pitta nympha Temminck and Schlegel, 1850, Faun. Jap., Aves : 135, Suppl.: pl.A and note. Korea.

Alternative name: Chinese Pitta, Swinhoe's Pitta, Lesser Blue-winged Pitta

TAXONOMY Fairy Pitta is closely allied to Indian Pitta *P. brachyura*, and in the past was treated as conspecific. Although there is no published account of why these two taxa should be treated as distinct, most authors now recognise these two pittas as part of a superspecies group that also includes Blue-winged Pitta *P. moluccensis* and Mangrove Pitta *P. megarhyncha* (e.g. Medway and Wells 1976, Mees 1977, Sibley and Monroe 1990). *P. brachyura* and *P. nympha* are clearly different, both with respect to plumage and morphology, and are certainly sufficiently different to be considered as species. Indian Pitta has golden-buff coronal bands, a whiter supercilium that usually becomes faintly blue at the tip, a white patch below the eye, and slightly darker underparts than Fairy Pitta. In addition, the undertail-coverts of Indian Pitta are scarlet, and this colour may extend onto the lower belly, whilst those of Fairy Pitta are a much deeper crimson and the colour extends onto the whole central belly. Fairy Pitta is distinctly larger, being intermediate in size between Indian and Blue-winged Pitta. There is no overlap in wing length (*nympha* 118-128, n=14; *brachyura* 101-111, n=20), tarsus (39-44 *nympha*, n=14; *brachyura* 33-37, n=20) or recorded weight (*nympha* 67.5-155g, with exception of a newly arrived migrant in Borneo of 63g; *brachyura* 47-66g). Tail length is similar and there is overlap in bill length, but bills of Fairy Pitta are heavier and deeper (bill length *brachyura* 26-30, n=20; *nympha* 28-31, n=12, maximum bill depth *brachyura* 7.0-8.3, n=20; *nympha* 9.2-10.8, n=12).

Finally it should be mentioned that the calls of Indian and Fairy Pittas are very different (see sonograms on p.159).

FIELD IDENTIFICATION Length 16-19.5. This medium-sized species breeds in coastal eastern China, Japan, Korea and Taiwan and migrates to Borneo during the northern winter. The sides of its head and a narrow crown-stripe are black, the latter bordered by broad bright chestnut to chestnut-orange lateral coronal bands that contrast strongly with a prominent long creamy-buff supercilium. Upperparts are dark green with conspicuous bright glossy azure-blue wing-covert patch. Throat white to creamy-white; rest of underparts creamy-buff with crimson-red belly and undertail-covert patch. In flight, a round wing-patch, in the centre of the wing, is conspicuous. Immature birds are much duller above and lack the crimson-red of the belly, although this is later indicated by salmon-pink.

Similar species Compared to Blue-winged Pitta, which may be sympatric during the migration and wintering period, Fairy Pitta has much paler, creamy-buff, rather than orangy-buff, underparts; azure-blue, rather than violet-blue rump and wing-patch; and a strong contrast between the pale creamy-buff supercilium and chestnut coronal bands. The extent of red on the belly is greater in Fairy Pitta, and the red is a darker shade. Fairy Pitta is most similar to Indian Pitta, although these species can be separated on range. Indian Pitta has golden-buff coronal bands, a whiter supercilium and richer, slightly darker underparts, and usually, a reddish base to the bill and a white patch just below the eye. For additional differences, see under Taxonomy above.

It is feasible that overshooting migrant Fairy Pittas could reach Wallacea, in which case separation from Elegant Pitta can most easily be on head pattern, the latter species lacking the chestnut coronal bands.

VOICE The call of Fairy Pitta is a clear double whistle, similar to that of Blue-winged Pitta, but is notably longer and slower and comprises two disyllabic notes: *kwah-he kwa-wu*. Mori (1939, cited in Austin 1948) described the call

Heads and bills of (left to right) Mangrove Pitta, Blue-winged Pitta, Fairy Pitta and Indian Pitta.

as *kahei kahei*, noting that at a distance it sounds more like *kai kai*. Austin and Kuroda (1953) note that, in Japan, this species is known locally as *shiropen-kuropen*, which is a rough description of the call. In Japan, peak calling activity occurs between 25 May and 10 June (Brazil 1991). The call is usually uttered from a perch in a tree, at heights of up to 7m.

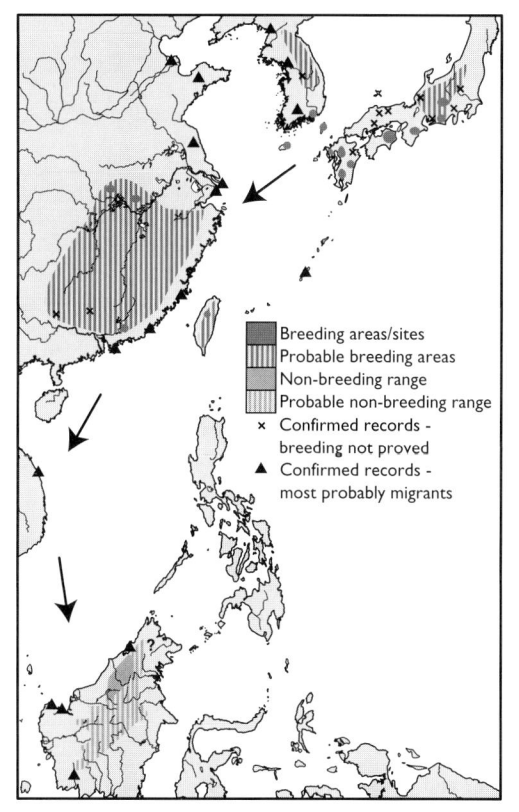

Map shows confirmed breeding sites, probable breeding range and approximate wintering areas on Borneo. Some birds apparently winter in southern China (e.g. Hong Kong) and perhaps in Vietnam. Arrows show hypothetical migration route. Records referable to probable migrants are also shown.

DISTRIBUTION Fairy Pittas are known to breed in southern Japan on Kyushu, southern Shikoku and, perhaps rarely, in central and southern Honshu (Austin and Kuroda 1953, Brazil 1991); the southern slopes of Mt Halla, Cheju-do (Quelpart) Island, and on Koje Island, both off southern Korea, and perhaps on the south Korean mainland (Austin 1948, Won 1969 in Mees 1977, Gore and Won 1971); on Taiwan and in eastern China south to Guangxi and Guangdong. In China, breeding has been confirmed in the provinces of Guangxi (Yao Shan mountains), Anhui and the mountains of Henan, and there are late April and May records from Guangdong and Fujian that could refer to breeding birds or migrants. Records from around Shanghai refer to migrants (Caldwell and Caldwell 1931, Ku Chang-tung 1957, Mees 1977).

The most northerly record from Japan is of one observed in spring in the Akita Prefecture of northern Honshu. Breeding could have been overlooked in this part of Japan, but the record most likely relates to a migrant. There are no records from Hokkaido, and only one record from the Nansei (Ryukyu) Islands, between Kyushu and Taiwan. This was of a bird observed on Yagaji Island, near Okinawa, on 15 July (Brazil 1991). Although this record lies within the period when the species breeds in Japan, there is no evidence that Fairy Pitta breeds in these islands, or even regularly passes through on migration. Although Austin (1972) considered it unlikely that Fairy Pitta breeds on the Korean mainland, noting that the late April and mid-May records from the west coast of Korea probably refer to migrants, there are a number of records from scattered localities (Kwangnung, Kaesong Pagyon, Togyusan and Anju) from June and August, suggesting that some birds may well breed on the mainland, although this has yet to be confirmed (Mees 1977).

During the non-breeding season, Fairy Pittas migrate south through eastern China and Taiwan to winter in Borneo and possibly in southern China and Indochina. Although the evidence for wintering in southern China is poor, birds have been found in Hong Kong during the winter. Sightings in Hong Kong on 22 January and 12 February, perhaps of the same bird (Chalmers 1986), can most easily be explained by an overwintering individual. Mees (1977) found that all nine specimens from Borneo that he examined were smaller than typical birds from Japan and Korea, but similar in size to birds breeding on mainland China. Hence it is possible that birds wintering in Borneo are those from China, whereas the more northerly breeding population from Japan and Korea might winter in southern China or in Indochina. Certainly, measurements of some birds collected in Indochina indicate that they derive from Korea or Japan (Mees 1977).

Although Meyer de Schauensee (1984) believed that populations in Guangdong and Guangxi, China, were resident, there is no clear evidence to support this, although Yen (1933) had stated that Fairy Pitta was 'sedentary' in the Yao Shan Mountains. However, Fairy Pittas from the same latitude in Taiwan are known to be migrants, and all specimens from mainland China reported by Mees (1977) were collected during the breeding or migratory season.

Delacour and Jabouille (1929) encountered this species several times in central Annam, Vietnam, in April. Migrants have also been recorded in Hong Kong in April, July and September (Carey 1994) and on mainland Korea (Gore and Won 1971).

Although it has been established that north-west Borneo is an important wintering area, there are few records and it appears to be localised in its non-breeding range. Mees (1977) located only nine specimens that have been collected in Borneo, with dates between 23 October and March. There are no confirmed records from Sabah, although a specimen collected in mid-November 1959 was later identified as this species by its collector, D. Batchelor (Sheldon in prep.), and there is only one record from Brunei, collected last century on the Tutong River (Mees 1977, Mann 1987). Fogden (1970), however, reported that Fairy Pitta was a relatively common bird in Sarawak. The only record from the Indonesian part of Borneo is of a specimen from Riam, South Kalimantan (Mayr 1938a, Mees 1977).

GEOGRAPHICAL VARIATION Although Stresemann (1923) described the subspecies *P. n. melli* from southern China, based on size, measurements made by Mees (1977) suggest that there is insufficient evidence to separate this subspecies. Data suggest that northern birds tend to be larger than those in the south.

HABITAT In Taiwan, Fairy Pittas inhabit sparsely populated woodlands below 1,300m in central and southern parts of the island. According to Hachisuka and Udagawa (1950), the species is more frequently encountered in the western hills of Taiwan than in the east. In Japan, breeding birds occupy moist deciduous slopes and evergreen forests on low mountain slopes below 500m altitude, although the species has occasionally been observed at altitudes of up to 1,200m on Mt Unzen. Within this habitat, Fairy Pittas prefer areas with a thick understorey of bushes and ferns, in particular near streams. Records of Fairy Pittas in Japan during migration come from a more diverse array of habitats including shady, dry conifer belts. Breeding records from Guangxi and Henan come from 'high altitudes', and in the Yao Shan Mountains they are reported to breed from the foot of the mountains to high altitudes (Yen 1933, Fu 1937 in Mees 1977), where the habitat is described as light deciduous and evergreen forest and scrub jungle (Cheng 1987).

In Borneo, Fairy Pittas apparently winter in primary forest in the lowlands and hills, but are largely found above the steepland boundary (Fogden 1970). Here, Fairy Pitta has been regularly reported from mixed dipterocarp forest on hillsides up to around 1,070m. Banks (1949) reported collecting this species at 305m on Mt Dulit, and Mayr (1938a) reported that a specimen had been collected at Riam in south Borneo, which is in the lowlands.

STATUS Fifty years ago, this was a common species in the Yao Shan Mountains of Guangxi, China (Yen 1933), but present status in China is unknown. However, Mees (1977) reported that Fairy Pitta was a rare species in the northern part of its range, and, in view of the widespread deforestation that has since occurred, its populations must certainly have declined in China. Elsewhere within its breeding range, the species is reported to have become very rare. P. O. Won (in Collar *et al.* 1994) estimated that there were probably less than 20 pairs breeding on the islands off South Korea in 1994, whilst the best-known locality for Fairy Pitta in Japan, Mi-ike, Miyazakiken, holds only around 20 birds in most years (Brazil 1991). In 1953, Austin and Kuroda (1953) reported that this was one of the rarest birds in Japan, and Brazil (1991) more recently described its status there as a very uncommon or rare and local species.

Even last century this was perhaps a rare species in Taiwan, since, in 1894, H. Seebohm wrote on a specimen label (BMNH) that Fairy Pitta was rare there. Hachisuka and Udagawa (1950) also noted that 'numbers are not many' in Taiwan. Recent documentation suggests that large-scale bird-trapping in the spring on Taiwan has depleted the island's breeding population, and may also have affected migrant populations: at Tseng-Wen reservoir, hundreds of Fairy Pittas were caught each spring during the 1980s, with as many as 200 being caught on one day (Severinghaus *et al.* 1991).

In its Bornean non-breeding ground, Fairy Pitta may be a rather local bird. There are no confirmed records from Sabah, but it was reported to be relatively common in parts of Sarawak in the late 1960s (Fogden 1970). There have not been any confirmed records from Borneo in recent years.

All this evidence points to the fact that Fairy Pitta must surely now be a rare and seriously threatened species in much of its range, and in recognition it was included in BirdLife's list of threatened birds as a Vulnerable species (Collar *et al.* 1994). It may, however, be more at risk than this category of threat suggests, and should perhaps be categorised as Endangered, since existing data point to a small population size that may only number in the low hundreds to low thousands. Its status in China clearly needs clarification, since this is presumably the stronghold of this species. Concern for this species has led to its inclusion in Appendix II of CITES (WCMC 1995) as a measure to monitor international trade.

MOVEMENTS Fairy Pitta, like its close relative the Indian Pitta, is a well documented long-distance migrant. However, unlike Indian Pitta, the movements of Fairy Pitta are still poorly understood. Evidence suggests that most if not all populations of Fairy Pitta migrate. Birds from mainland China fly south after breeding to winter in Borneo, but it is not clear, as mentioned above (see Distribution), if some birds, in particular those from Japan and Korea, regularly winter in southern China or Indochina. However, some individuals have been found in southern China (Hong Kong) during the winter, and there is therefore the possibility that some individuals, if not an entire population, spend the winter in this region.

Based on the size of specimens, Mees (1977) pointed out that records from Shanghai strongly suggest that the breeding population of Japan and Korea migrate in a south-west direction, to mainland China, rather than flying south. This fits in with the fact that there is only one record from the Ryukyu Islands, to the south of Japan, and no records from the Philippines (Brazil 1991, Dickinson *et al.* 1991). However, the report of Severinghaus *et al.* (1991) that hundreds have been recorded on migration in Taiwan suggests, perhaps, that part of the Japanese and/or Korean population may take a more direct route to Borneo, perhaps even via northern Luzon and Palawan, where they have so far remained undetected.

Japanese birds arrive on the breeding grounds in mid-May and leave in August and September. Whilst a small breeding population occurs on Taiwan, the majority of birds reported from there probably refer to migrants. All but two October records from Taiwan are from 28 April to 19 July. Hundreds of Fairy Pittas (up to 200 a day) were caught at one site in Taiwan each spring during the 1980s

(Severinghaus *et al.* 1991), perhaps indicating the existence of a distinct flyway. Late April and mid-May records from the west coast of Korea probably also refer to migrants (Austin 1948). Non-breeding birds are in Borneo from mid-October to March.

The April records from central Annam, Vietnam (Delacour 1928, 1929, 1930, Delacour and Jabouille 1929), probably refer to birds returning to their breeding grounds on a northerly route from Borneo, although it is not inconceivable that some birds winter in the hills and mountains of Indochina. Migrants have been recorded in Hong Kong between 9 September and 19 November, with three spring records, in April, July and on 5 May. There are also Hong Kong records that may relate to overwintering birds or to migrants on 22 January and 12 February (Webster 1975, Chalmers 1986, Crosby 1995).

FOOD In Taiwan, Fairy Pittas feed young almost exclusively with earthworms, several of which are sometimes carried to the nest simultaneously (Severinghaus *et al.* 1991). A specimen from Taiwan had fragments of beetles and two small spiral-shaped shells in its stomach (BMNH). Two specimens collected in Sarawak had been eating ants (SarM).

HABITS Very little is recorded of the habits of Fairy Pitta. It spends most of its time in woodland near streams, mostly keeping to the ground, but usually calls from a high branch, flicking its tail as it calls (Austin and Kuroda 1953). On migration, this species sometimes feeds in rather open situations, and is apparently sometimes quite tame.

Nest of Fairy Pitta

BREEDING In Taiwan, a specimen collected on 10 May contained half-developed eggs (BMNH), nests with eggs have been found in late May, and with young in mid-July (Kobayashi and Cho 1977, Severinghaus *et al.* 1991) and an immature was collected on the island on 19 July (USNM). In Japan, the breeding season is from May to July (Brazil 1991). On Mt Hallasan, Quelpart Island, Korea, nesting is reported to occur from May to June (Mori 1939, cited in Austin 1948). A nest with eggs was also observed on Koje Island, Korea, in June (C. Poole and A. Long verbally).

A nest with two large young that were about to fledge was discovered at about 300m altitude in Taiwan on 14 July (Severinghaus *et al.* 1991). The nest was placed about 3m off the ground in the fork of an *Acacia confusa* tree on the edge of a clearing in secondary forest dominated by acacias (80% of trees), with the remaining trees being broadleaved species such as camphor (*Cinnamomum camphora*). The nest, which was about 20cm in diameter and made primarily of twigs and dead leaves, opened to face dense vegetation so that adults could enter the nest without moving into the clearing. Nests in Japan have also been discovered in tree forks, whilst others have been found in clefts in rocks at 1.5-7.5m off the ground.

In Korea, a nest containing at least three eggs was found on Koje Island in June (C. Poole and A. Long verbally). The nest was in very degraded semi-evergreen forest on a very steep rocky slope, and was built between a small branch of a tree that had apparently been cut down and an adjacent boulder, so that it was precariously sited on the boulder side, and very obvious. The nest was very large, measuring c.45cm wide and c.40cm high, with an entrance hole c.14cm wide. The base of the nest was constructed of stout twigs upon which smaller twigs had been added to build the ragged dome. A few dried leaves were placed in the roof. The inside of the nest, lined with stems of a fine-leaved (possibly ericaceous) plant with green and drying leaves, was clearly visible since the entrance hole was so large.

A nest found in Japan in 1937 was in the fork of a cherry tree, c.4.5m off the ground, and contained six eggs. It was made of cherry twigs over a moss lining which in turn was lined with pine needles. This nest measured 15 x 26cm outside, 16cm in inside diameter and the side entrance was 6 x 8cm (Austin and Kuroda 1953). A nest with young close to fledging that was presumably photographed in Japan (on 6 July) was constructed largely of long but narrow twigs with green mosses apparently forming the bulk of the roof over the entrance hole (Takano 1981). A few dead leaves were also evident, and the adult attending the nest was able to perch on a narrow platform of twigs in front of the nest. Of four old nests found in Japan, one was in a cleft in the rocks, and three were in tree forks c.2-7.5m off the ground (Austin and Kuroda 1953).

Clutch-size in Japan is reported to be 4-6 and observations made in Taiwan show that both parents feed the young (Severinghaus *et al.* 1991). The eggs are creamy-white with pale purple-brown spots (Austin and Kuroda 1953).

DESCRIPTION
Sexes similar.
Adult Forehead, crown to nape, lores and ear-coverts black; long pointed creamy-buff supercilium (probably erectile at rear) from nostril to nape that contrasts strongly with broad chestnut coronal band. The supercilium is longer than the coronal bands. Mantle and back bright dark green; rump and uppertail-coverts azure-blue. Tail black tipped dark blue, the extent of blue varying considerably between individuals, from a narrow fringe confined to innermost feathers to broad tips on all feathers, extending down the webs of the innermost pair or two. Upperparts from mantle to rump sometimes with blackish centres to the feathers, forming bold streaks. Throat and chin white, occasionally creamy-white; breast, sides, flanks and thighs creamy-buff; centre of belly to undertail-coverts crimson-red. Innerwing-coverts azure-blue, forming a large conspicuous patch in the closed wing; median coverts green; greater coverts and tertials green with blue tinge. Flight feathers blackish with greyish-white tips to primaries and blue outer webs to the outer half of the outer five secondaries and blue-green outer webs to the

inner secondaries. Primary coverts and alula black. Prominent wing-bar formed by white spots in the outer web of the outer two primaries: across whole web of primaries 3-5 and the outer web of primary 6. Bill black, sometimes lighter towards the tip; iris dark brown; legs and feet pink to greyish-white.

Juvenile Forehead and crown rufous-buff with narrow black edges giving scaled appearance; narrow supercilium creamy-white, broadening on nape, where tinged with buff. Lores black. Black centre of crown, narrow at front, broadening at rear to become a broad black patch that is joined by a narrow band to the black mask across sides of head. Mantle dark brown, tinged olive; lower back greener. Rump azure-blue but much duller than that of adult. Tail blackish-brown with bluish tinge to terminal half. Scapulars a mixture of dull olive and green feathers. Wing coverts dull olive with bold whitish subterminal spots on median coverts, forming broken wing-patch. Tertials dull olive-green. Flight feathers black. Primaries with white spots at base: outermost primary with spot on inner web, primaries 2-5 with spots across both webs and primary 6 with narrow spot on edge of outer web. Secondaries blackish with broad bluish leading edge to terminal half, becoming greener on innermost secondaries and across whole outer web and tip of innermost two secondaries. Throat white becoming buffier on lower throat and with blackish sides. Breast dark ochre, this colour extending along sides of breast and flanks, where tinged with olive. Centre of lower breast paler buff. Belly pale buff with pinkish feathers in centre of belly to undertail-coverts. Bill dark with orange base to lower mandible and extreme tip.

MEASUREMENTS Wing 118-128; tail 34-45; bill 28-31; tarsus 39-44; weight in Taiwan 67.5-155 (a female collected in Kalimantan on 25 November, as a newly arrived migrant, weighed 63g).

27   BLUE-WINGED PITTA
Pitta moluccensis   Plates 13 & 16

Turdus moluccensis P.L.S. Müller, 1776, Natursyst., suppl., p.144; based on "Merle des Moluques" of Buffon (Daubenton), 1765-1780, Planches Enlum., 2, pl. 257: Moluccas; error, emended to Tenasserim, by Baker, 1930, Fauna of Brit. India, 7, p.296; further emended to Malacca by Deignan, 1963, Bull. U. S. Nat. Mus., 226, p.98.

Alternative names: Lesser Blue-winged Pitta, Little Blue-winged Pitta, Moluccan Pitta

FIELD IDENTIFICATION Length 18-20.5. This species is a medium-sized pitta that breeds widely in northern parts of South-East Asia but migrates south to the Sunda Region during the non-breeding season. Crown to nape and sides of head black, strongly contrasting with the broad buffy-brown lateral coronal bands and white throat. Chin black. Upperparts dull greenish with bright violet-blue wing coverts. Underparts cinnamon-buff to orange-buff with crimson belly and undertail-coverts. In flight a large round white patch in the flight feathers is conspicuous. Immature birds are much duller above and lack the red of the belly, although this is later indicated by salmon-pink. Bills of young birds have a bright red or orange-red base and tip.

Similar species Very similar to the slightly more robust Mangrove Pitta, which also has a distinctly larger bill (>38mm in length, compared to <32.5 for Blue-winged). Head patterns also differ, with Mangrove Pitta having less contrast between the lateral coronal bands (which are darker than those of Blue-winged Pitta) and the more uniform, darker brown crown. Blue-winged Pitta also has more black on the top of the crown than Mangrove Pitta, and a narrower black band across the lower nape. The black chin of Blue-winged Pitta, when visible, is also diagnostic. The upperparts of Mangrove Pitta are slightly darker, less bright and more olive in colour than those of Blue-winged Pitta. In parts of their range, these two species may breed in adjacent habitats, and Blue-winged Pittas on migration may occur in mangrove. The other similar species that overlaps in range is Fairy Pitta. Both species may be found in Borneo during the northern winter. Blue-winged Pitta is darker below than Fairy, has violet-blue rather than an azure-blue wing-patch and rump, and a less contrasting head pattern. In flight, Blue-winged Pitta could conceivably be confused with Black-capped Kingfisher *Halcyon pileata*, which has a very similar wing pattern.

VOICE The call given by both sexes is a loud, clear, fluty double whistle *taelaew-taelaew* or *taewu-taewu*, with, at close range, a distinctive disyllabic quality to both parts of the call (see sonogram below). The call lasts less than a second. Calls are sometimes repeated only twice, but at other times may continue for long periods with an interval of about 3-3.5 seconds between subsequent calls: about 14 calls are made per minute during prolonged calling between territorial birds. Calls are often made from a perch in a tree or in bamboo as well as from the ground. A second call, given in alarm, is a harsh *skyeew*. Calling is rather seasonal, occurring mostly during the breeding season. During the non-breeding season, Blue-winged Pittas are usually silent except at dawn and dusk, when they may occasionally call, sometimes from their roosts. In Borneo, this species also has the reputation of calling frequently before it rains and is locally called *burung hujan-hujan*, or 'bird of the rains'.

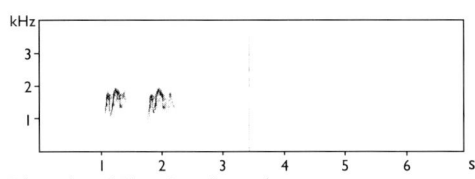

Blue-winged Pitta *P. moluccensis*
[from Krabi Province, Thailand, recorded by J. Scharringa, April 1988]

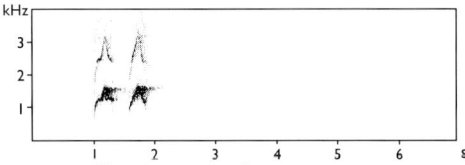

Mangrove Pitta *P. megarhyncha*
[from Kuala Selangor, Malaysia, recorded by J. Scharringa, April 1984]

DISTRIBUTION Blue-winged Pittas are known to breed in southern Yunnan Province (southern Xishuangbanna), China (Cheng 1987); Annam and Cochinchina in Vietnam (Beaulieu 1932, Eames and Ericson in prep., F. Lambert pers. obs.); Cambodia, where known from the Mekong (Oustalet 1903) and the Elephant Mountains (Thomas 1964); throughout Laos (Salter 1993); Burma in the southern Shan States, Pegu Hills, Karen Hills and south Arakan to Tenasserim (Baker 1934, Smythies 1986); Thailand (except the north-east and central plains: Round and Lekagul 1991) and northern Peninsular Malaysia in Perlis, and on Langkawi Island (Medway and Wells 1976, Wells 1990b). Evidence suggests that some individuals breed in Borneo (see below). The occurrence of migrants in the Red River Delta of Vietnam (J. Eames *in litt.*) suggests that Blue-winged Pitta may also breed in Tonkin or in adjacent China, perhaps in Guangxi Province.

During the northern winter, Blue-winged Pittas are found throughout the southern part of the breeding range, but more northerly areas are vacated as the dry season sets in, and birds migrate south to Peninsular Malaysia, Sumatra and Borneo, with birds recorded from many islands in this region (Coomans de Ruiter 1938). It is possible that some individuals in the northern part of the range may stay on their breeding grounds all year: Smythies (1986) saw Blue-winged Pitta in the Pegu Hills of Burma in the winter. Vagrants have reached Tawitawi, Basilan and Palawan in the Philippines; north Sulawesi; Christmas Island and north-west Australia, although records from Australia are questioned by Mees (1971). Although Blue-winged Pittas may reach Java during the winter, as a straggler, there are no acceptable records (Mees 1971).

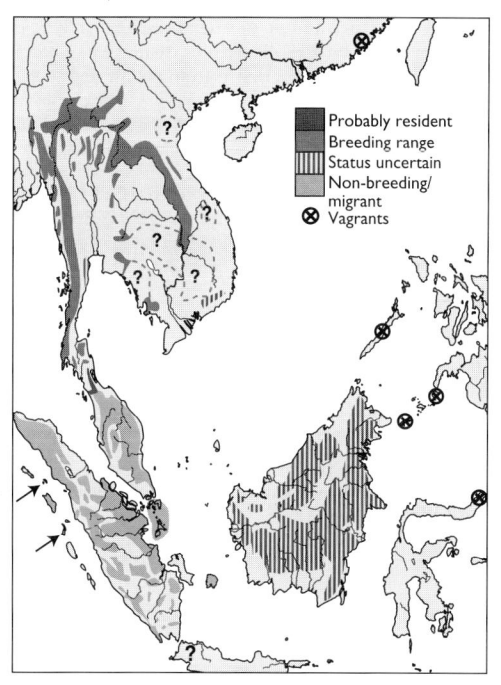

Breeding and non-breeding distribution based on approximate extent of forest remaining within the altitudinal range of the species. Former range more widespread. Presence of breeding birds on Borneo and of populations in parts of Indochina require confirmation. Vagrants have also reached Christmas Island and Australia (not mapped).

Hose (1898) reported that Blue-winged Pittas were present all year round in the Hose Mountains, Sarawak, whilst Pfeffer (1960) reported commonly hearing this species all year round in the lowlands (up to 350m) in east and central Borneo. Wilkinson *et al.* (1991) observed two in central Kalimantan at the end of August, before the usual period that migrants arrive. These observations strongly suggest that this species might be a breeding bird in Borneo. Furthermore, a specimen collected in Sabah at the end of December had a well-developed brood-patch and a slightly enlarged oviduct (Thompson 1966). In Jambi Province, Sumatra, D. Holmes (*in litt.*) regularly heard calling Blue-winged Pittas between November and March, and speculates that the species could also breed in parts of Sumatra, although clearly these could have been unusually vocal wintering birds

GEOGRAPHICAL VARIATION None recorded.

HABITAT Breeding birds occupy a variety of habitats, ranging from 'light forest with sparse undergrowth' in China (Cheng 1987) to mixed deciduous forests, and clumps of bamboo at the edge of secondary forests in Thailand (Junge and Kooiman 1951, P. Round *in litt.*) and tall undisturbed rainforest, as on Langkawi Island, Malaysia (Wells 1983). In Perlis, Peninsular Malaysia, numerous Blue-winged Pittas were heard calling in the latter half of June in teak plantations and the adjacent edge of semi-deciduous forest, and may well have been breeding there (Wells 1990b).

Following migration, Blue-winged Pittas also spend the non-breeding season in a variety of habitats, including the landward edge of mangrove as it abuts onto cleared land, inland scrub, dense secondary growth, gardens with dense cover (Fogden 1965) and coffee plantations (Sharpe 1889). The non-breeding habitat does not usually include primary forest, although passage migrants are sometimes observed in this habitat (pers. obs.). Wintering birds may also venture into very open places, and have been regularly recorded on lawns of suburban gardens.

Throughout its range, Blue-winged Pitta occurs primarily in the lowlands and hills, breeding as high as 400-500m and perhaps at 700-800m in Thailand (Round 1988, P. Round *in litt.*). In Vietnam, Beaulieu (1932) found this species breeding below 170m in Cochinchina, whilst in central Annam a nest was found in limestone hills at c.50m altitude (pers. obs.). Büttikofer (1900) found Blue-winged Pittas in central Borneo at altitudes of 780-1,135m between mid-March and early May, although some or all of these may have been on migration. In Sabah, Blue-winged Pittas generally winter below 200m, but migrants have been found as high as 1,800m (Sheldon in prep.). Wintering birds have been recorded to 1,000m in Sumatra (van Marle and Voous 1988).

STATUS Last century, Oates (1883) described the status in Burma as abundant over the entire country from May to July, and the species was probably common throughout much of its range. In its wintering grounds, Hose (1898) stated that it was common in the lowlands near Mt Dulit, Sarawak. Blue-winged Pitta is still a fairly common species throughout much of its range, although Cheng (1987) reported that it was very rare in Yunnan, China, whilst specimens are rare in collections from northern Thailand (Riley 1938), suggesting that it might be relatively rare there. Hence there is some evidence that Blue-winged Pitta may be rarer in the northern part of its breeding range.

The diversity of habitats in which Blue-winged Pittas can breed means that the large-scale clearance and degradation of forest that is occurring over much of its range is less of a threat to this species, when compared to other pittas, some of which are less tolerant of habitat alteration. In parts of its range, and in particular in Thailand, Blue-winged Pittas are caught for food and trade in large numbers, both whilst breeding and on migration (J. Howes verbally), but the effects that this may have on population dynamics is unknown.

Gretton (1988) estimated that the population density of Blue-winged Pitta in remnant patches of secondary forest in Peninsular Thailand was in the order of 10 pairs per km^2.

MOVEMENTS Northern populations are migratory. This habit enables this species to exploit and breed in drier, more fragmented habitats than most other pitta species (P. Round *in litt.*). Blue-winged Pitta winters in Peninsular Malaysia, Sumatra and Borneo. Elsewhere, the bird is regarded as a vagrant. Vagrants have been recorded in Fukien (south-east China), Hong Kong (Williams 1990), on Tawitawi (G. Dutson *in litt.*), Basilan and Palawan in the Philippines (McGregor 1909-10, Dickinson *et al.* 1991), in north Sulawesi (White and Bruce 1986) and on Christmas Island (Benson 1970, UMZC). The authenticity of two records of stragglers reaching coastal north-west Australia in 1927 and 1930 (Serventy 1968) has been questioned by Mees (1971), although the possibility of ship-assisted passage cannot be ruled out. Migrants have also been recorded from Condore Island (Con Son Islands), Cochinchina (Robinson and Kloss 1930) and on an island in the Red River Delta, northern Vietnam (J. Eames *in litt.*).

The northern limit for resident populations has not been established, although Oates (1883) noted that birds are present all year in extreme south of Tenasserim but populations further north are migratory. Smythies (1986) observed this species in the Pegu Hills in winter, but the great majority of individuals are absent from this area at that time. Evidence suggests that populations in southern Laos are migratory (R. Timmins *in litt.*). Populations breeding in southern Peninsular Thailand and the extreme north of Peninsular Malaysia (Perlis State) are apparently resident (Wells 1990b), although very few of those breeding in Thailand remain throughout the year (Legakul and Round 1991).

Migrants move at night and can be caught under certain weather conditions near lights placed at high points on their migration routes. Trapping at mist-nets behind batteries of powerful lights at Fraser's Hill, Malaysia, has demonstrated that the best weather for this activity is during periods of light or heavy mist, particularly after light rain, and in light winds, during moonless nights. During 1966-1969, 340 Blue-winged Pittas were caught at Fraser's Hill using this method (Medway and Wells 1976). Although Blue-winged Pittas have been caught here between 25 September and 6 December during southern passage, peak southward passage occurs in the second half of October. Interestingly, the migration period of Blue-winged Pitta differs from that of the partially sympatric Hooded Pitta, which tends to migrate south later in the year. The dates of northward migration are not so well known, although at Fraser's Hill Blue-winged Pittas have been caught between 12 April and 9 May (Medway and Wells 1976, Wells 1990b). Whilst in Thailand, P. Round (*in litt.*) has heard this species passing overhead, calling, on various dates between 30 March and 11 April, and has recorded 'falls' of 4-5 birds in Kaeng Krachen and near Krabi on 26 April and 3-4 May respectively. A large number of birds are already in breeding habitats in Peninsular Thailand by mid-April (P. Round *in litt.*). Most Blue-winged Pittas attracted to lights at radio towers and lighthouses in Malaysia have little or no subcutaneous fat, indicating that they are perhaps near exhaustion, and perhaps represent the weak segment of a larger population passing overhead (Wells 1975, D. Wells *in litt.*).

Migrants have also been attracted to lights in the Crocker Range, in north Borneo, during mid-April and early November, whilst Batchelor (1959) reported that a large flock of Blue-winged Pittas landed on his boat, as it approached Miri (Sarawak) on 15 October, and that three flew into his lighted bungalow in Kimanis on 31 October. There are also specimens from Mt Dulit in October, from Labuan in November, and records from many islands in the Malacca Straits (Medway and Wells 1976).

Medway and Wells (1976) give the earliest dates south of the breeding grounds in Peninsular Malaysia as 25 September. The earliest record from Sumatra is 1 August (van Marle and Voous 1988) and a bird was observed even further south, in central Borneo, on 29 August (Wilkinson *et al.* 1991): it has not been established, however, whether these records refer to early migrants or to resident individuals. Wintering birds typically reach Sabah at the end of September and usually leave by mid-May, although individuals have been recorded as late as 5 June (Sheldon in prep.).

FOOD Specimens collected in Borneo have contained a very large hornet, large wood ants, insect larvae, spiders, scarab beetles and the claw of a crab (SarM, Smythies 1981). The stomach of one bird was described as being full of shrimps (Motley and Dillwyn 1855). In Burma, Oates (1883) recorded worms, land shells, ants and other insects in the diet. Four specimens from Sumatra contained ants and bugs (small Coreidae), a large grasshopper (Acrididae), whilst one stomach was filled almost entirely with remains of beetles and larvae of fireflies or glow worms (Lampyridae) (Chasen and Hoogerwerf 1941). An adult photographed near a nest in Thailand was carrying a large earthworm.

HABITS Blue-winged Pittas are territorial on their breeding grounds, and can be netted in successive years. Such netting has shown that individuals can live for more than 5.5 years (67 months; McClure 1974). Deignan (1945) found that this species calls primarily during the breeding season. When breeding, incubating birds sit tight on the nest, allowing very close approach before flying out of the nest. When feeding, however, Blue-winged Pittas are often quite shy, bounding away rapidly or occasionally flying off when disturbed. One wintering bird, however, was unexpectedly observed feeding with Common Mynas *Acridotheres tristis* on a large lawn on a number of consecutive days (Wells 1990b). The bird was unruffled by considerable traffic and human disturbance. Junge and Kooiman (1951) noted that Blue-winged Pittas will often stand motionless in face of danger, although Lansdown *et al.* (1991) reported that wing-flicking may be given in response to anxiety. In the non-breeding range, Blue-winged Pittas frequently roost on small horizontal sections of vines, usually at 3-5m above the ground (pers. obs.). These roost sites are often in rather open situations above paths near the forest edge or clearings.

Birds caught in Malaysia are frequently infested with ticks and larval mites ('chiggers'). Moult is completed on the breeding ground prior to migrating south (Medway and Wells 1976).

In the hand, or immediately after release, Blue-winged Pittas may adopt a posture in which the wings are fully outstretched (so that the tips touch the ground), with the body held forward, head turned back towards the threat, and crown feathers raised.

BREEDING In Burma, Blue-winged Pittas breed during the early part of the south-west monsoon, which usually starts in April or May, although nests with eggs have been discovered as late as 26 July (Evans 1904). Smythies (1986) reported that nests and eggs of this species have been collected in Burma as late as 5 August (although he treated Mangrove Pitta as a subspecies). In northern Thailand, Deignan (1945) reported that the breeding season was between April and July, whilst Baker (1926) noted that nests with fresh eggs had been found as late as 1 August. In Peninsular Thailand, nests with eggs have been found from late May to 11 July (Robinson 1915, Baker 1919, P. Round *in litt.*), whilst in Peninsular Malaysia, nests with eggs have been found from 29 May to 11 July (Medway and Wells 1976).

There is little information on breeding seasonality from Indochina. In central Annam, Vietnam, fledglings taken from a nest by local woodsmen on 9 July were a few days old (pers. obs.). Although Beaulieu (1932) found Blue-winged Pitta breeding in Cochinchina, he did not record any dates. R. Timmins (*in litt.*) found fledglings in southern Laos on 3 July.

Fledgling Blue-winged Pitta

The dome-shaped nest is usually situated on the ground, often against some solid structure such as on the banks of tiny forest streams, with a favourite position being wedged in among the roots of a tree (Baker 1926), but sometimes concealed in dense vegetation such as at the base of dense bamboo. Occasionally, however, nests may be placed above the ground: Oates found a nest of Blue-winged Pitta in the fork of a tree trunk where it separated into three c.60cm off the ground (Hume 1880), whilst Gretton (1988) found such a nest situated in a strangling fig *Ficus* tree, c.4m above the ground, and Eames and Lambert (in prep.) discovered a nest situated in the exposed stilt-like root matrix of a large tree growing at the edge of a bank, and situated c.1m off the ground; and P. Round (*in litt.*) has found nests in a *Salacca* palm and on a tree stump, both c.1m off the ground. The nest, which has a vertical entrance, is usually a rather flimsy, loose structure (J. Scharringa *in litt.*, pers. obs.).

The dimensions of one Burmese nest, situated on a mound and slightly concealed by some blades of grass, was c.43 x 28cm with an entrance hole just above the ground of some 7.6 x 10cm (Evans 1904). A nest from Tenasserim was 20.3cm in diameter and 14cm in height with an internal cavity 14cm in diameter and about 9cm high; the entrance was around 9cm in diameter (Hume 1880). Baker (1934) reported that nest size varied between 15.2-25.5cm in length, and between 12.7-20.3cm in height and width. Davison, writing from Tenasserim, noted that the nest of Blue-winged Pitta is very similar in appearance to that of Hooded Pitta, but is conspicuously smaller, has thinner roof sides and foundation, and often lacks the conspicuous platform in front of the entrance hole that is found in Hooded Pitta (Hume 1880). However, some Blue-winged Pitta nests apparently have platforms in front of the entrance (Baker 1934).

Oates (1883) reported that the nest materials, of sticks, leaves and roots, are bound together with earth, whilst Bingham also noted that a nest of Blue-winged Pitta made of bamboo leaves had been matted together with earth (Hume 1880). Evans (1904) describes a nest from Burma as being constructed of old bamboo and other twigs and a few leaves loosely laid about, but notes that the interior was neatly lined with fibre. Although bamboo leaves are often an important nest material, a nest observed in an area dominated by bamboo in Vietnam contained no bamboo leaves, but was constructed with twigs, dried dicotyledonous leaves and green moss; there was no evidence that earth had been used to bind the materials together (pers. obs.).

Oates (1883) reported clutch-size in Burma of 4-6, whilst in Peninsular Malaysia clutch-size is reported to be 4-5 (Medway and Wells 1976). Baker (1934) reported that clutch-size varied from 4-6 and occasionally seven, although nests have been found (2 May) with as few as two eggs (Hume 1880) (possibly, however, relating to an incomplete clutch). Dated nest records (Evans 1904, Robinson 1915, Baker 1919) suggest that clutch-size increases as the breeding season progresses, although more data are needed to confirm this.

The eggs of Blue-winged Pitta are typical for a pitta but are generally the most richly and profusely marked of all pitta species breeding within its range (Baker 1926). They measure 24.0-28.9 x 20.0-22.9mm (Baker 1934; from a sample of 109), and are fairly glossy. They are white to yellowish-white in colour, thinly spotted and scrawled with purple, usually mainly towards the broader end, though sometimes fairly evenly, with smaller rounded spots of purplish-brown (Oates 1883, Robinson 1915). Eggs of this species in BMNH are very varied, with some eggs largely spotted but without scribbles, and others covered in a fine meshwork of scribbles but having no spots. Markings are usually profuse on both the surface and underlying parts of the shell.

Both parents probably incubate the eggs, since Robinson (1915) collected a male that was incubating in Thailand, whilst other collectors have shot both males and females off nests with eggs (Hume 1880, BMNH).

DESCRIPTION
Sexes similar.
Adult Sides of head, including lores, broad band above and below eye, ear-coverts and back of head and nape black. Crown chestnut, often buffier near lower margin,

and with variable amount of black on centre, forming a broad stripe from above the eyes towards nape, sometimes joining the black band of the sides of head. Eye ringed by narrow band of fleshy greyish skin, this extending as a small pointed patch behind the eye. Back and mantle dark green, rump and uppertail-coverts shining dark violet-blue. Some birds have black streaks on feathers of mantle, back and rump (though this feature is much less frequent than in some other species, such as Indian Pitta). Tail black with narrow dark blue terminal band. Tertials green, concolorous with upperparts; feathers of inner and median coverts blue-green with broad shiny violet-blue edges and tips forming large, bold wing-patch. Outer greater coverts blue-green with dark brown basal half to inner web. Secondaries blackish with bluish fringe to distal half of outer web, becoming broader towards inner secondaries and extending onto feather tip. Primaries blackish with broad white band across central third to half of all feathers, becoming progressively nearer to tip of feathers and reaching the tip on 1-5 innermost primaries. White absent on outer web of outer primary of some birds. Underwing-coverts black. Chin black; throat white, clearly demarcated from cinnamon-orange breast, flanks and upper belly. Lower belly and undertail-coverts red. Iris dark brown; bill black; legs pink to pinkish-white; feet and toes yellowish. Some individuals, perhaps subadults, have small area of reddish at gape.

Juvenile Feathers of crown, from forehead to nape, primarily buffy-chestnut, but with considerable variation in tone and edged narrowly but distinctly with dark brown to give a scaly appearance; feathers of top of crown blackish, forming a broad patch extending towards nape. Upperparts, including wing-coverts, similar to those of adult but much duller, and the blue of the wings is paler. Secondaries as in adult; primaries blackish with white band only apparent on inner 4-5 feathers and present only on inner web of outer primaries. Tail tipped with green rather than blue. Throat white, underparts darker and less buffy than those of adult. Bill tip (up to half of bill) and gape reddish-orange to red.

MEASUREMENTS Wing 114-136; tail 38-54; bill 28-32.5; tarsus 37.5-41 ; weight (night-flying migrants in Borneo, mid-April) 83-87g.

28 MANGROVE PITTA
Pitta megarhyncha Plate 13

Pitta megarhyncha Schlegel, 1863, Vog. Nederl. Ind., Pitta, p.11, pl.4, fig. 2.: Bangka Is., Sumatra.

Alternative name: Larger Blue-winged Pitta.

TAXONOMY Although originally classified as a separate species, Mangrove Pitta was subsequently considered a southern breeding subspecies of the closely similar Blue-winged Pitta, or as a mangrove subspecies of Indian Pitta. It is now recognised as a separate species within the superspecies group that includes these two species (Madoc 1956, Medway and Wells 1976, Mees 1986, Sibley and Monroe 1990). It should be noted, however, that Mangrove and Blue-winged Pitta differ considerably from all other members of this superspecies group in having violet, rather than azure-blue, on the wing-coverts and rump,

and in the extensive white in the wing. It may therefore be more appropriate to separate these two taxa into their own superspecies.

FIELD IDENTIFICATION Length 18-21. This is a mangrove specialist that occurs on coasts, in deltas and along tidal parts of rivers from the Bay of Bengal south along the South-East Asian coastline to Peninsular Malaysia and eastern Sumatra. The head pattern and bill size are the most important features of this species (see under Similar species). The disproportionately large bill, brown coronal bands that do not contrast strongly with the brown crown and the white chin are the most important plumage features. Mangrove Pitta has a black mask that contrasts with the white chin and throat and pale brown lateral coronal bands. The crown is almost uniform brown to the nape, although there is usually a small amount of black. The nape feathers tend to conceal the black band that extends across the lower nape. The upperparts are dull greenish marked with a conspicuous glossy violet-blue wing-covert patch. Underparts are cinnamon-buff with a red patch extending from the undertail-coverts to the belly. In flight the large round white wing-patch is conspicuous. Young birds are much duller above than adults and either lack the red belly-patch or have only a suffusion of red on the belly and undertail-coverts. The bill tip and base is bright red or orange-red.

Juveniles have a rufous-brown crown finely scaled with black and blackish sides of the head, lores and ear-coverts, extending onto sides of nape as a broad streak. The mantle and back are dull brown, with green feathers first appearing on the mantle sides. The wings are much duller than those of adults, being green with some feathers tipped azure. Below, the chin and throat are white, clearly demarcated from the orangy breast, flanks and upper belly. The lower belly is mixed orange-cinnamon and reddish and the undertail-coverts are pinky-red.

Similar species Mangrove Pitta is very similar to the slightly smaller Blue-winged Pitta, but has a distinctly larger bill (>38mm long), a white chin and a different head pattern, with less contrast between the lateral coronal bands, which are darker than the buffy-brown coronal bands of Blue-winged Pitta. Blue-winged Pitta also has more black on the top of the crown than Mangrove Pitta, a broader black band across the lower nape and richer, slightly darker underparts. The upperparts of Mangrove Pitta are slightly darker, being less bright and more olive in colour than those of Blue-winged Pitta, whilst the colour of the wing-covert patch and rump is slightly more violet. These two species may breed in adjacent habitats in parts of their range, as on Langkawi Island, off west Malaysia. Immature Mangrove Pitta is very similar to immature Blue-winged Pitta, but, like adults, also has a notably larger bill, white chin and the entire crown is brown with variable rufous tone (depending on age) and dark edges to the feathers, giving a scaly appearance.

In rapid flight away through the mangroves, Mangrove Pitta can be mistaken for a Black-capped Kingfisher *Halcyon pileata* if views are brief, since both have large white wing patches, but the upperparts of the latter are uniform dark blue and the head black.

VOICE A loud, fluty *hhwa-hwa* lasting just less than one second with the first note rising but the second fairly even, is usually made from the tops or canopy of trees in mangroves, and is repeated at c.4s intervals. Unexpectedly, Allen (1955) noted that an astonishingly loud call, believed

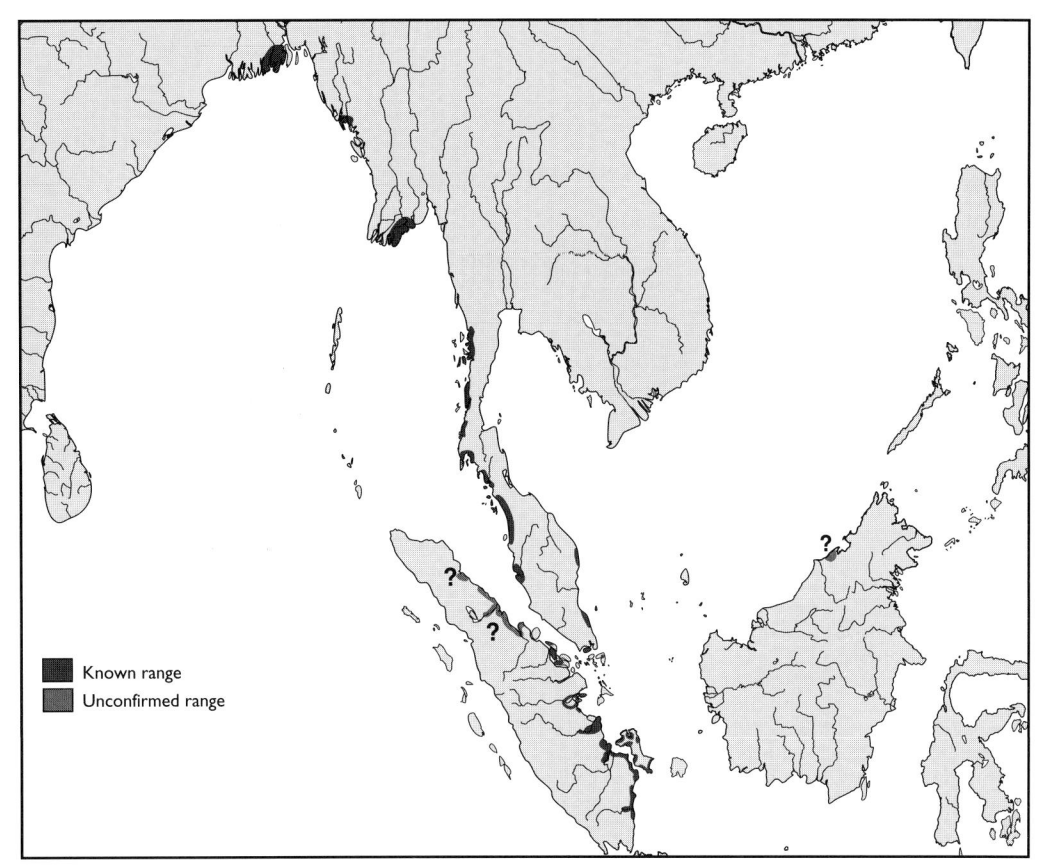

Distribution based on known range and the extent of mangroves within the range. Presence in Borneo requires confirmation. Former range more widespread.

to be this call, was given by birds on the nest on many occasions. At a distance the sound is very similar to the *taelaew* of Blue-winged Pitta, although the latter gives this call in distinct couplets (i.e. as *taelaew-talelaew*, rather than *taelaew* followed by a 3-4s pause before repeating; see sonogram). The individual notes of Mangrove Pitta's call (*hhwa* and *hwa*) are more even than those of Blue-winged Pitta, with no suggestion of any disyllabic quality (the *taelaew-taelaew* notes of Blue-winged are distinctly disyllabic). Although some authors have noted shorter intervals between notes when compared to the calling of Blue-winged Pitta (Lekagul and Round 1991), recordings of a Mangrove Pitta from Malaysia show that this species calls at intervals of about 4s, which is hardly different to the 3-3.5s between couplets typical of Blue-winged Pitta. During prolonged calling between territories, individuals may call about 12 times per minute. Mangrove Pitta also makes a *skyeew* alarm note, like that of Blue-winged Pitta.

DISTRIBUTION Mangrove Pitta is distributed along the coasts, in deltas and along major rivers from Bangladesh to Sumatra and Singapore, and perhaps northern Borneo. In Bangladesh it has been found in the Sundarbans west to c.89°N 30'E (Whistler 1934, Paynter 1970), and although Paynter described it as being a 'conspicuous element of the avifauna', Harvey (1990) asserted that there had been only one record in Bangladesh, in March 1926. Further south, it is known from coastal Burma (Oates 1883, Smythies 1986), coastal Thailand and the west coast of Peninsular Malaysia, south to the islands of Singapore and the islands of the Riau Archipelago (Pulau Karimon i.e. Great Karimun, and Pulau Bintang: Chasen 1931, Riley 1938) between Singapore and Sumatra. From here its known distribution ranges to Bangka Island and parts of eastern Sumatra (Baker 1934, van Marle and Voous 1988).

Populations also occur on the east coast of Peninsular Malaysia, where Mangrove Pitta is known only from the vicinity of Kuantan (Pahang State, Mersing in Johore State) and the south coast of Johore, along the River Lunchu (J. Howes *in litt.*, Anon. 1988). On the west coast of Thailand and Malaysia, it has been recorded from the islands of Phuket, Tarutao (main island), Libong, Mai Ngaam (one of the Lanta Group) and Langkawi (Riley 1938, P. Round *in litt.*). There is one recent sight record from the Gulf of Thailand, where a bird was seen briefly at Samut Sakhon (just west of Bangkok) on 18 November (J. Wolsencroft verbally).

On the Sumatran mainland, Mangrove Pittas occur along the tidal borders of larger rivers, on the coast in Riau and Lampung Provinces (van Marle and Voous 1988, Holmes in prep.) and presumably in Jambi. The most southerly record on mainland Sumatra is from the Seputih River at c.4°40'S (Lambert 1988). An immature collected at 'Deli' presumably came from the vicinity of Medan, in North Sumatra Province, suggesting that Mangrove Pitta occurs along most of the north-east-facing coastline of the island.

Smythies (1957) listed this species from Borneo on the basis of a mounted specimen, originating from the Raffles Museum, Singapore, but now apparently in the Sarawak Museum. This specimen was reported to have been collected in the Baram district of Sarawak in 1891. However, this record can only be treated as tentative since there have been no subsequent records. Although there is ample suitable habitat for the species in Borneo, the absence of other records is rather strange, and the provenance of the specimen is therefore questionable.

GEOGRAPHICAL VARIATION None recorded.

HABITAT As its vernacular name suggests, the Mangrove Pitta is an inhabitant of mangroves throughout its range, although it is also sometimes found in abutting drier areas with good cover. It apparently occurs in a variety of mangrove types, from that dominated by *Bruguiera parviflora* and *Ceriops candel*, and *Avicennia officinalis* (J. Howes *in litt.*), to those dominated by *Rhizophora* spp. (Holmes in prep.) and *Xylocarpus* spp. (P. Round *in litt.*). These species generally occur in the mangrove zone inundated by normal high tides (Aksornkoae 1993). Mangrove Pittas also frequent larger rivers in tidal areas where mangrove or Nipa Palm *Nypa fruticans* groves occur (Lambert 1988). Hence this species can be found well inland along the borders of larger rivers, where the water becomes brackish.

Mangrove Pittas are also found on small offshore islands where habitat is suitable, such as some of the small islands of the Riau Archipelago and of the southern Mergui Archipelago, Tenasserim (Baker 1934). Although Verheugt *et al.* (1993) list this species as occurring in swamp forests of South Sumatra, this is probably an error.

Hussain and Sarker (1973) report collecting this species from a mango orchard to the north-west of Ullapara, Pabna Province, Bangladesh, in May 1970. This is a most unusual record since the site is some 275km from the coast and perhaps 30km from the Jamuna River. The bird had slightly enlarged gonads and the stomach contained a beetle and the remains of other insects. The possibility that this record refers to Blue-winged Pitta rather than Mangrove Pitta cannot be ruled out, and confirmation that the latter occurs so far inland is needed.

STATUS Although this species occurs in secondary and logged mangrove, as well as adjacent habitats, its habitat is being rapidly depleted, especially in Thailand, and in parts of Sumatra and Malaysia, where mangroves are being converted to fish and shrimp ponds, or reclaimed. In a recent study, R. Noske (*in litt.*) reports that although Mangrove Pittas were widespread in mangrove areas of west Peninsular Malaysia, the density was low, with no more than one to two birds per 3ha. In Singapore, Mangrove Pitta is now very rare, with the last few birds inhabiting only two islands, Tekong Besar and Ubin in the straits between Singapore and Malaysia (Lim 1994): birds were formerly present on islands to the south of Singapore, such as Ayer Merbau (Chasen 1939) and on both sides of the Straits of Johore (Bucknill and Chasen 1927).

Holmes (in prep.) notes that Mangrove Pitta was reported as being common in *Rhizophora* sp. mangrove and in nipa swamps along the Rawa and Metas Rivers of Riau Province, Sumatra, in 1991. As mentioned above, the existence of the species in Borneo requires confirmation.

In 1958, Paynter (1970) found this species to be a 'conspicuous element of the avifauna' in the Sundarbans of Bangladesh. A viable population of Mangrove Pitta is presumably therefore present in the Sundarbans, an area of more than a million ha of mangrove (IUCN 1994), although there have been no records from the Indian part of this important wetland.

MIGRATION None recorded.

FOOD Mangrove Pittas probably feed largely on crabs (R. Noske *in litt.*) and the large size of the bill clearly has ecological significance. Specimens shot in mangroves have contained molluscs, ants and other insects in their stomachs (Bucknill and Chasen 1927).

HABITS Mangrove Pittas feed on the exposed mud in mangroves and amongst mangrove roots and adjacent drier land vegetation, sometime hopping around in exposed places at the seaward side of mangroves, and even perching on moored boats at the edge of mangroves (pers. obs.). They are not averse to crossing open areas of mudflat between patches of mangrove. Bucknill and Chasen (1927) noted that they bathe at the landward edge of mangroves. They can be very elusive at certain times of the year, but during the breeding season, adults are often less wary and can be seen feeding, usually alone. At this time, they can also be heard calling from the branches of mangrove trees, although sometimes hidden from view in thick foliage. They frequently fly when disturbed, and in such cases may not alight for considerable distances, although at other times they leap from root to root within the mangroves. Lansdown *et al.* (1991) observed this species performing a wing-flicking display, apparently in response to anxiety. Ringing data from Selangor, Peninsular Malaysia, show that Mangrove Pittas can live for at least 5.5 years (Wells 1982).

Nest of Mangrove Pitta

BREEDING The few breeding records from the north of this species's range point to breeding during the middle part of the year, coinciding with the rainy season. These records are as follows: birds with very enlarged gonads were collected in the Sundarbans (Bangladesh) in April (Paynter 1970), a clutch of four eggs was collected near Rangoon on 1 August (BMNH), and an empty nest and female containing a fully formed egg were found on Phuket on 17 April (Hume 1880, BMNH).

Further south, in Peninsular Thailand, Peninsular Malaysia and Singapore, the breeding season is apparently the same, occurring during the period when the south-west monsoon brings rain, although this is not necessarily the wettest part of the year. In Thailand, nest building has been observed at Krabi on 3 May, a bird found on a nest in the last week of May, whilst a juvenile has been observed on 27-31 July (Anon. 1988a, b, C. Robson *in litt.*). In Malaysia and Singapore, nest building has been recorded between 3 April and 9 July, nests with eggs found from 15 April to 17 July, nestlings found on 7 July and fledglings observed in early June, July and on 25 August (Robinson and Chasen 1939, Allen 1955, Medway and Wells 1976, Anon. 1988, BMNH, UM). In Sumatra, an immature was collected in the vicinity of Medan (at Linngai Rawa, Deli) on 20 August 1888 (RMNH).

Two nests found by Allen (1955) at Pasir Panjang, Singapore, on 3 and 30 April were located between a road verge and mangrove, with the entrance facing the mangrove. The first nest, was placed on the mound of a mud lobster (Thalassinidae), on a solid platform of twigs, some more than 30cm long and 1cm wide. Outside, the nest was composed of fairly slender dead sticks, with a few dead leaves of *Hibiscus tiliaceus* and a seed pod on top of the nest. The nest was damp inside and lined with grass and pieces of coconut fibre: the latter were possibly collected from the road where a number of husks were to be seen, since there were no palms in the area. The base of nest was c.40-41cm across and the nest hole c. 12-13cm in diameter, the egg chamber being c.17cm in width. The second nest found by Allen (1955), on 30 April, resembled the first except that the roof was partly felted over with green seaweed *Chaetomorpha* sp., later found to be common on the mud in the mangrove.

At Selangor, Peninsular Malaysia, a nest discovered on 21 June, still under construction, was built into the side of a bund at the edge of mangrove (Wells 1990b, C. Prentice *in litt.*). It was situated amongst the fronds of *Acrostichum speciosum* ferns, so that it was in a relatively bright, rather open area, protected by low, dense herbage and a *Ceriops candel* tree, and above the area usually flooded by high tides. The adjacent mangrove was dominated by *Bruguiera parviflora*, *Avicennia officinalis* and *Ceriops candel* (J. Howes *in litt.*). The nest was well concealed, however, having mud over the roof, and was c.1.5m above the floor of the adjacent mangroves. It was a large domed structure, c.45cm long and with an internal chamber c.20cm in diameter. Externally the nest was constructed of thick twigs, >4mm in diameter, which formed the base and a platform in front of the entrance, and, on the roof, large twigs were overlaid and mixed with finer twigs. The roof was partially hidden by a number of large dead leaves, still attached to one or two long leaf stalks (each with 6 or more leaves) that had been draped or had fallen over the nest. Inside, the chamber was lined with fine grasses, the leaves of *Bruguiera* and leaf skeletons. When discovered, an adult bird was in the process of binding the lining of the nest interior together with soft mud balls. On 29 June, the nest contained three eggs.

Another nest, found at virtually the same site on 7 July two years later, contained two chicks estimated to be five days old. This nest was c.1m above the water level, on a sloping mud bank on the seaward side of mangrove dominated largely by *Bruguiera parviflora* (C. Prentice *in litt.*). The nestlings had disappeared three days later, assumed predated, but two adults, assumed to be the same pair, were discovered building a new nest on 9 July some 15m from the original site. On 17 July, the nest presumably contained eggs, since an adult was observed on it, but on 24 July it had been abandoned (J. Howes *in litt.*, Anon. 1988).

Nests are not always placed on the ground. A nest discovered near Krabi, Peninsular Thailand, was situated about 1.5m above the ground in some dense scrub at the edge of mangroves. It was in the process of construction, the material being small sticks collected from the mud surface within the mangroves at low tide.

There are few data on clutch-size, but evidence suggests it is 2-4 (BMNH, C. Prentice *in litt.*). Madoc (1956) noted that it was usually three in Peninsular Malaysia, although both nests found by Allen (1955) in Singapore contained only two eggs. Madoc (1956) gave egg size as c.31.7 x 24mm, although two eggs collected in Singapore were less elongated, measuring 29 x 23.5mm and 30 x 24.7mm. Baker (1934) described the eggs as being white with a faint tinge of lilac and richly but not very thickly streaked and mottled everywhere with dull maroon and pale inky-purple. An egg in BMNH is white with fairly evenly distributed pale chocolate-brown spots and irregular markings and a few chocolate scribbles, and underlying grey markings. It is very similar to some examples of eggs of Blue-winged Pitta. Eggs found in Peninsular Malaysia were described as being pinkish-white with large brown and grey-blue blotches all over the shell, but most concentrated at the broader end (J. Howes *in litt.*). Incubation takes at least 15 days. Photos of birds on the nest seem to indicate both birds incubate the eggs (Allen 1955).

DESCRIPTION
Sexes similar.

Adult Forehead to nape brown with variable-size patch or streak of black in centre of crown and usually paler, buffish fringe to crown edges, especially behind eye. Lores, narrow area above eye and broad area below, and ear-coverts black, forming prominent black band across side of head. Mantle and back dark green, occasionally with a bluish gloss to feathers of upper mantle, and rarely (one male specimen from Bangladesh in BMNH), with blue gloss to whole of upperparts: such birds have upperparts that are nearly as bright as those of Blue-winged Pitta. Rump and uppertail-coverts shining dark violet-blue. Tail black with narrow dark blue terminal band. Tertials green, concolorous with upperparts; feathers of inner and median coverts blue-green with broad shining violet-blue edges and tips forming large, bold wing-patch. Outer greater coverts blue-green. Secondaries blackish with bluish fringe to distal half of outer web, becoming broader towards inner secondaries, where extending onto feather tip. Primaries blackish with broad white band across central third to half of all feathers, becoming progressively nearer to tip of feathers and usually reaching the tip of innermost primaries. Underwing-coverts black. Chin and throat white, clearly demarcated from cinnamon-buff to cinnamon-orange breast, flanks and upper belly. Lower belly and undertail-coverts red. Iris dark brown; bill black; legs pinkish-grey (but may appear pinkish-white when covered in dry mud); feet pinker than legs, occasionally yellowish.

Immature Resembles the adult, but plumage is duller above, with indistinct black scaling on the crown, no bright violet colours on the wing-coverts (these being mixed dull blue and green), the tail brown with bluish tip, and the underparts a darker, dirty buff.

Juvenile Feathers of forehead and crown brown with rufous tinge and narrow black edges, giving scaly appearance. Lores and ear-coverts blackish, extending onto sides of nape as a broad streak. Mantle and back dull brown, gradually replaced with green feathers that first appear on sides of mantle. Tips of feathers of rump and uppertail-coverts blue, with a tendency for those of the rump to be azure-blue and those of the uppertail-coverts darker, more violet, and similar in colour to those of the adult. Tail feathers basally black but with broad dull blue terminal band above. Tertials brown on the inner web, glossed with green, and green on the outer web; inner lesser coverts brown with green gloss, outer lesser coverts and medium coverts mostly glossy green, some of which are tipped with azure; primary coverts blackish. Secondaries dark brown with dull blue fringe to distal half of outer web, becoming broader and glossed with green towards inner secondaries, where extending onto feather tip. Primaries dark brown with broad white band on innerwing formed by white third to half of innermost 5-6 feathers and white patch in outer web of primaries 4-5. The amount of white progressively increases towards the innerwing, extending almost to the tip of the innermost 3-4 feathers. In contrast to those of the adult, the feather centres of all primary feathers in the part of the feather that is white remains dark brown, so that there is a narrow dark brown line along the innermost part of both webs of the white parts of the feather. Underwing-coverts blackish. Chin and throat white, clearly demarcated from orangy breast, flanks and upper belly. Lower belly mixed orange-cinnamon and reddish, undertail-coverts pinky-red. Iris dark brown; bill blackish with red gape.

MEASUREMENTS Wing 114-123; tail 37-45; bill 38-42; tarsus 36.5-40; weight 92.1-120.5.

29 ELEGANT PITTA
Pitta elegans   Plate 14

Pitta elegans Temminck, 1836, Planches Coloriées livr. 100, pl. 591: Timor.

Alternative name: Two-striped Pitta

TAXONOMY Mayr (1979) treated *Pitta elegans* as a group of subspecies of Noisy Pitta. However, Elegant Pitta is here treated as a species, following White and Bruce (1986), Sibley and Monroe (1990) and Bishop and Coates (in prep.).

FIELD IDENTIFICATION Length 15.5-18. Elegant Pitta is a small species of pitta that is found on islands of Wallacea, Indonesia. The six subspecies are characterised by green upperparts with shining azure-blue wing-covert patch and rump, and cinnamon-buff underparts with a variable amount of red or scarlet on the belly and undertail-coverts. The head pattern varies according to subspecies, but all have a black crown and broad creamy-buff to rufous-buff or golden-buff supercilium. In three subspecies, the posterior part of the supercilium is faint blue. The throat colour varies from entirely white to entirely black, whilst the belly varies from wholly scarlet to black and red. A concealed white band across the black primaries is conspicuous in flight. Immatures have a chestnut-buff forehead that extends as a broad coronal band along sides of the crown to behind eye, where becoming bluish-white. The rest of the upperparts resemble those of the adult, although the azure-blue wing-patch is broken up by green. Below, the lower flanks and centre of belly are tinged with dull pink. The bill tip and gape of immature birds are orange to orange-red.

Similar species There are no similar species within the range of Elegant Pitta, although it is conceivable that migrant Blue-winged Pitta or even Fairy Pitta might reach islands on which Elegant Pittas occur. Blue-winged Pitta, migrants of which have reached northern Sulawesi, Christmas Island and possibly even Australia, is easily distinguished. It is much larger than Elegant Pitta and has a bold buffy supercilium and brown crown, extensive white in the wings, violet-blue rather than azure-blue wing-patches and a white throat. Fairy Pitta, which could potentially overshoot its Bornean wintering grounds, is more similar in size, and has a similar-coloured, azure-blue, wing-covert patch. However, the head pattern is again very different, with broad chestnut coronal bands. The underparts also differ, with a white throat and much paler, creamy-buff breast and flanks, and in having crimson-red undertail-coverts and central belly, rather than the paler red or scarlet of Elegant Pitta. Noisy Pitta of Australia is also fairly similar, but has extensive chestnut on the crown whilst lacking a buffy supercilium.

VOICE Elegant Pitta usually calls from small horizontal branches 2-7m above the ground. During the breeding season calling is frequent for the first three hours of daylight, and birds also occasionally call at night. On Flores the call is a double note with the first note rising, *kwweee-kwill*, lasting about one second and repeated every three seconds. On Timor it is notably different, and repeated less frequently, every 3-6s: the notes are faster, shriller and harsher than on Flores, and the call sounds trisyllabic *kuw-whaa-whaa* (A. Greensmith *in litt.*). The call of Elegant Pitta from Sumba is also a rapid, abrupt, trisyllabic *ka-wha-kil* or *hwp-hwp-pw*. Sometimes birds on Sumba give a single call, *pwer*, that is abrupt and inflected downwards (pers. obs.). Dutson (1995) noted that the call on Kalaotoa is disyllabic, and sounded identical to calls heard on Sumbawa and Flores, but that those on nearby Tanahjampea sounded trisyllabic. B van Balen (*in litt.*) reports that on Taliabu, in the Sula Islands, the call is also trisyllabic (*kwuwik -kwk*) whilst on the islets of Tayandu and Kai Kecil, in the eastern part of the its range, calls vary from *wrrek-WRREK* to *wrrek-WRREK-WRREK*.

DISTRIBUTION Elegant Pitta is endemic to Indonesia, where found widely scattered on the islands in the southern part of the archipelago. Five subspecies are distributed on the islands of Wallacea, and another is endemic to the small island of Penida, between Bali and Lombok, just to the west of Wallace's Line. Within Wallacea, Elegant Pitta is found throughout the Lesser Sunda arc eastwards to the Tanimbar and Kai Islands, and northwards to small islands to the south of Sulawesi and the Moluccas. In the northern islands, it occurs in Buru, the Sula Islands, and Seram, but may be a migrant rather than a breeding species in this part of its range. With the exception of Ternate, it has not been recorded in the north Moluccas or in Sulawesi, but has been found on two small islands to the north of Sulawesi. Due to the complex distributional pattern of different subspecies, details of distribution are provided in the following section (Geographical Variation).

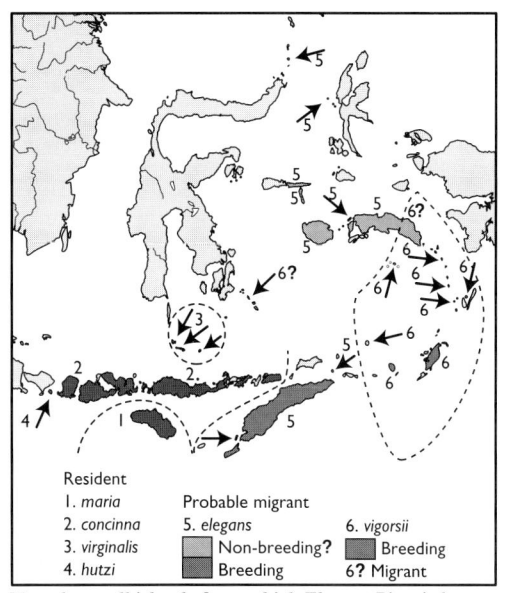

Map shows all islands from which Elegant Pitta is known but does not show range within islands, which is patchy.

Resident
1. *maria*
2. *concinna*
3. *virginalis*
4. *hutzi*

Probable migrant
5. *elegans* — Non-breeding?, Breeding
6. *vigorsii* — Breeding, 6? Migrant

GEOGRAPHICAL VARIATION The following account follows the taxonomy of White and Bruce (1986), who recognised six subspecies, in contrast to Mayr (1979), who listed nine subspecies under the *elegans* group of *P. versicolor*. Sibley and Monroe (1990) noted that *vigorsii* has been regarded as a separate species, but examination of specimens suggests that there is no justification for such treatment.

P. e. elegans (Widely distributed in Wallacea, where known from islands in the central Lesser Sunda arc, i.e. Semau, Timor, Kisar and possibly Roti, where it breeds, northwards to the Moluccan islands of Boano, Buru, Seho, Sanana, Taliabu, Mangole, Ternate, Seram, and two small islands to the north of Sulawesi, Tahulandang and Ruang, where it is probably a non-breeding visitor. Presumably this species could also be expected to occur as a non-breeding visitor on other north Moluccan islands) Head black with a long, narrow pale buff supercilium from bill to upper nape. Centre of belly a mixture of dark red and black, becoming pinky-red on undertail-coverts. Tail narrowly tipped green. Large azure-blue patch on wing-coverts.

P. e. virginalis (Includes *plesseni* and *kalaoenis* of Meise 1929) (Restricted to several very small islands in the Flores Sea, just south of Sulawesi: recorded from Tanahjampea, Kalaotoa, Kalao) Similar to *elegans*, but supercilium broader and cinnamon-buff.

P. e. vigorsii (With the exception of Tanimbar and Seram, where probably a rare migrant, this subspecies is primarily confined to small islands: it is known from the eastern part of the Lesser Sunda arc [Tanimbar, Babar, Damar]; small islands between Seram and the Kai Islands [Kasiui, Tiur, Kur, Taam, Tual]; Banda in the Banda Sea south of Seram; and from the small island of Kaledupa [in the Tukangbesi Islands, east of Butung], south-east Sulawesi) Differs from *elegans* in that the supercilium becomes pale blue behind the eye, the throat is white, centrally buff, and the underparts are slightly darker. There is usually more black on the belly. Tail tip bright green.

P. e. maria (Restricted to Sumba in the Lesser Sunda Islands. Although Hartert [1897b] documents a specimen of *maria* in the collection of Alfred Everett that was apparently collected in the lowlands of south Flores in November 1896, the occurrence of this subspecies there needs confirmation) Similar to *elegans*, but has a large scarlet patch on the belly (occasionally with a few black feathers) and a blue tinge to the posterior part of the supercilium. The distal third of the tail is green. This subspecies has less white in the wing than other subspecies, the white being confined to a small spot on primaries 4-5: other subspecies have white on three to five feathers.

P. e. concinna (includes *everetti* of Hartert 1898b) (Occurs from Lombok to Alor in the northern chain of the Lesser Sunda arc on all the larger [Lombok, Sumbawa, Flores, Adonara, Lomblen and Alor] and some of the smaller islands [such as the 30km^2 Pulau Besar, off east Flores], although not yet recorded on Komodo or Pantar) Supercilium differs markedly from *elegans*, being broader and brighter golden-buff in front of the eye, but narrowing and becoming white behind the eye and strongly tinged with blue near the tip. The throat is black, extending onto the upper breast as a point, and there is a large black patch on the centre of the belly. Sides of the breast are darker than those of *elegans*. The distal half of the tail is green.

P. e. hutzi (Known only from two specimens said to have been collected on Nusa Penida, a small island between Bali and Lombok, at the end of February and in early March 1938 [Meise 1941]. MacKinnon and Phillipps [1993] question the provenance of these specimens, suggesting that they may be *concinna* from Lombok. However, measurements suggest that *hutzi* is smaller than *concinna*: wing of *hutzi* 94-97 [Meise 1941] compared to 97-103 [author]). Meise (1941) stated that *hutzi* also differed from *concinna* in having a lighter-coloured supercilium, more yellowish and brighter breast and sides of body.

HABITAT The Elegant Pitta occurs in a variety of forested habitats, from the dry monsoon woodlands and thickets of the western Lesser Sunda Islands, to the wetter rainforests of the south Moluccas. They regularly frequent forest edge and degraded areas, and, on Lombok and Flores, *concinna* may be found in small patches of trees along watercourses, often close to villages (Rensch 1931, Bishop 1992), as well as in scrub near forest (Jones *et al.* 1994). Rensch (1931) found *concinna* in dense forest in southern Lombok, but on Sumbawa found it in areas where coffee had been planted within light rainforest: he often saw it in open areas with little understorey and scattered coffee. Butchart *et al.* (1993) report that Elegant Pittas appear to be more common on flat ground on Sumbawa and Flores than on steep gradients. They recorded the species in semi-deciduous forests, primary moist forest and in logged moist forest with an abundance of rattans and bamboo in the understorey.

On Timor *elegans* has been observed in swamp forest and forest on coralline rag (A. Greensmith *in litt.*), whilst on Taliabu (Sula Islands) it occurs in selectively logged and severely degraded lowland forest, and in areas dominated by cultivation (Davidson *et al.* 1994, Bishop and Coates in prep.). On Sumba, subspecies *maria* occurs in a variety of habitats, from scrub near forest (Jones et al. 1994), rather open dry forests, particularly in areas where the understorey is dominated by vine tangles, to dense tall primary evergreen forest: many of these habitats are on jagged limestone (B. Gee in litt., pers. obs.). Discriminate function analysis of habitat associations carried out on Sumba suggests that the species may prefer open secondary forest with dense undergrowth on steep slopes, although population densities are apparently higher in forest than in forest edge (Jones et al. 1995). On Taam, in the Kai Islands, Schodde and Mathews (1977) found *vigorsii* to be abundant in low scrub and in the forest around a pool of brackish water. Many of the islands frequented by this subspecies are low-lying, with plateaus of elevated, sharp coralline base rocks.

Elegant Pittas occupy a wide altitudinal range. In the Lesser Sunda arc, they occur from sea level to 1,220m on Lombok (Hartert 1896), to 750m on Flores (B. Gee in litt.), to 1,400m on Sumbawa, to at least 300m on Timor (Bishop and Coates in prep.) and from the lowlands to at least 930m on Sumba (Jones et al. 1994). On the Moluccan island of Buru, Elegant Pitta has been recorded at 1,520m in the Madang Range (Hartert 1924, Rensch 1931), although this record may well refer to a bird on migration. Islands in the east of the species's range, including Tanimbar, are low-lying, so it must be assumed to occur at low altitudes. The only specimens of *hutzi*, from Nusa Penida Island, were collected at 300m (Meise 1941).

STATUS Most accounts suggest that Elegant Pitta is a widespread and relatively common species, and in some localities, such as on the small island of Taam in the Kai Islands, it is abundant (Schodde and Mathews 1977). It has been described as common on Banda, Lombok, and Sumba (Bishop and Coates in prep.), common to uncommon on Sumbawa and uncommon to rare on Flores (Butchart et al. 1993). It was not rare on Alor last century (Hartert 1898b). Encounter rates of Elegant Pitta during surveys in forests of the Lesser Sundas varied from 0.04-0.12 per hour on Flores and 0.42-2.2 per hour on Sumbawa: highest encounter rates on the latter island were recorded in forest that had been logged about three years previously (Butchart et al. 1993). The frequency of calling birds in some forest areas of Sumba suggests that the density in suitable forested habitats must be in the order of one pair per ha, or c.100 pairs per km^2 (pers. obs.), although Jones et al. (1995) reported mean densities of only four birds per km^2, based on Variable Circular Plot data. Jones et al. (1994, 1995) estimated that the population of Elegant Pitta on Sumba (*maria*) was more than 11,000 individuals.

MOVEMENTS The movements of this species are poorly understood, but records suggest that one subspecies, *elegans*, and perhaps *vigorsii*, make migratory movements (White and Bruce 1986). Collection dates of the nominate subspecies suggest that it breeds in the monsoon forests of Timor and satellite islands, but moves northwards to rainforest habitats on the Moluccan islands, such as Buru, the Sulu Islands and probably Seram, during the non-breeding season. However, the exact timing of such movements is unclear.

White and Bruce (1986) note that, on Timor, nominate *elegans* is present between December and February but was not collected there in July-August 1897, despite extensive collecting by A. H. Everett. Furthermore, Noske and Saleh (1993) only encountered the species twice during intensive surveys on Timor from 13 May to 6 July, suggesting, perhaps, that the majority of the population had already departed. Playback of Elegant Pitta calls in areas where they are known to occur during late October failed to elicit any response (pers. obs.). It should be noted, however, that Wallace (1861) also failed to find this species on Timor during January to April, attributing this to the fact that it probably only occurred in the interior of the island. Elsewhere, *elegans* has been observed on Buru in April and on Mangole, in the Sulu Islands, in October (White and Bruce 1976). On Taliabu, also in the Sulu Islands, Davidson et al. (1994) first noted Elegant Pitta on 11 October, even though they had already been on the island for one month. The birds were calling, however, and the exact status of the species on this and other Moluccan islands needs clarification.

White and Bruce (1986) suggested that *vigorsii* may also be a migrant, noting that a female was collected in January on the small island of Kaledupa, in the Tukangbesi Islands off south-east Sulawesi. This is clearly an odd record, although the possibility that it was an escaped bird en route to Butung or Ujung Pandang, which are local centres for much bird trade (pers. obs.) cannot be ruled out. Taylor (1991) failed to find Elegant Pitta on Seram between mid-July and mid-September, but it has been found on the small islands of Babar in March-April (confirmed breeding) and August, on Damar in November and December (Bishop and Coates in prep.), the Kai Islands from 23 April to 19 May (confirmed breeding) (Schodde and Mathews 1977, B. van Balen in litt.), Banda in April (confirmed breeding) and September-October (White and Bruce 1986), and Tanimbar in early November (A. Lewis verbally). A perusal of these dates suggests that if this subspecies is truly migratory, it probably leaves the islands on which it breeds for only a short period.

FOOD Nothing is recorded about the diet of this species.

HABITS In contrast to many species of pitta, Elegant Pittas tend to fly when disturbed rather than to hop away. This behaviour may be related to the rather open nature of many of the woodlands in which they occur. They are usually encountered singly, but occur at high densities in good habitat and several may be visible almost simultaneously in open areas, particularly calling birds, which often perch in rather obvious places.

Two birds observed by B. Gee (*in litt.*) on Flores on 6 August were performing a display that was interpreted as an interaction between two males. The birds faced each other, standing some 50cm apart, with their necks outstretched and the whole body swaying from side to side very slowly and deliberately for some 20-30s.

BREEDING The seasonality of breeding is poorly documented. In the Lesser Sundas, two eggs of *concinna* were collected on Alor on 17 April (Hartert 1898b, BMNH), adults have been collected on Sumbawa with enlarged gonads in May (Rensch 1931), and three birds that had brood patches were netted on Sumbawa from 2-14 August (Butchart et al. 1993). A female collected on Alor in April 1897 (BMNH) had an orange-red gape and bill tip, according to the specimen label, and was presumably an immature bird. Two juvenile *maria* were collected on Sumba in early June, a recently fledged juvenile *elegans* has been collected on Timor in June (RMNH) and an immature in December (Mayr 1944). Finsch (1900) documented a nest with two eggs of *vigorsii*, which was found on Babar sometime before 28 March. Nestlings of *vigorsii* have been recorded on 4 April on Tual, and immatures have been collected on Banda on 16 April, Babar on 21 April and Taam on 25 April, whilst a male collected on Taam on 15 May had enlarged testes (Schodde and Mathews 1977, White and Bruce 1986). On Kalaotoa (south of Sulawesi), nestling *virginalis* have been recorded between May and June (Hartert 1901, White and Bruce 1986).

These records suggest that the breeding season of all subspecies spans the middle of the year, and hence corresponds with the driest season, particularly in the Lesser Sunda Islands. Anecdotal evidence, however, suggests that in some parts of the Lesser Sundas breeding may occur almost throughout the year. The collection of an apparently immature *concinna* in April indicates that the breeding season of this subspecies must be rather long, since this would suggest that the bird had fledged sometime in the period November-February. Since birds with brood patches have been caught in August, the breeding season must last for at least nine months. Furthermore it should be noted that Bishop (1992) found that Elegant Pittas on Flores (also *concinna*) responded aggressively to tape playback in the latter half of September (the middle of the dry season), and suggested that this might indicate breeding. In contrast he found the species to be noticeably less vocal and territorial in June. On Sumba, *maria* is very vocal and responsive to playback at the beginning of November (pers. obs.). Furthermore, B. Gee (*in litt.*) noted that Elegant Pittas were very vocal in the Lesser Sunda Islands (Lombok, Flores and Sumba) during August.

A nest of *vigorsii* from Babar was constructed almost entirely of leaves, at least half of which were leaf skeletons and all of which were from broad-leaved plants. A few large sticks, up to 4mm in diameter and 18cm long, also formed part of the outer dome, whilst the interior was lined with a layer of very fine black rootlets or fungal hyphae (RMNH).

Two eggs of *concinna* collected on Alor were white, covered all over with rufous-brown and purplish-grey patches of irregular form. The latter markings were darker and larger on one egg than on the other. They measured 27.5-28.1 x 22.0-22.5mm (Hartert 1898b, BMNH).

DESCRIPTION *P. e. elegans*
Sexes similar. Within subspecies there may be considerable individual variation in plumages. Schodde and Mathews (1977) report, for example, that on Taam in the Kai Islands, birds in fresh plumage (*vigorsii*) differed in the extent and intensity of red on the abdomen and in the intensity of green on the back.
Adult Crown to nape, lores, ocular area, ear-coverts and sides of upper throat black; bold coronal stripe from bill to sides of back of crown pale buff; upper throat black, extending as broad-ended point onto lower throat, rest of throat pale buff merging into deeper yellowish-buff breast, flanks and sides of belly; centre of belly deep dull red mixed with black (formed by black basal half to red feathers), becoming paler on undertail-coverts. Mantle and back dark green; rump shining azure-blue; tail black with narrow dark green terminal band. Lesser coverts pale shining azure-blue, forming conspicuous bright wingpatch; tertials, greater and median coverts green and concolorous with mantle. Secondaries black with bluish edge to distal half of outer feathers, the bluish edge becoming broader and greener on innermost secondaries, and extending onto whole outer web and tip. Primaries black with white outer band near base formed by white on inner web of primaries 1-3, across both webs of 4-6 and outer web of 7, but occasionally from inner web of primaries 1-2 and whole web of 3-4 and outer web of primary 5 (i.e., the band can vary in position within the wing). Underwing-coverts blackish. Bill black; iris brown to black; feet flesh to purplish-brown. Some individuals (probably of all subspecies) may have narrow black streaks on the upperparts, usually confined to the upper mantle, but occasionally also on the back.
Juvenile (*P. e. concinna*) Forehead chestnut-buff, extending as a broad coronal band along sides of crown to behind eye, where it becomes bluish-white. Rest of head, including lores, ear-coverts and nape, black. Mantle olive-green, the green extending onto the back and upper rump where less olive. Uppertail-coverts bright azure-blue; tail green with dark brown bases. Lesser wing-coverts azure-blue, with green basal halves to feathers breaking up the patch. Rest of upperwing-coverts and tertials green. Primary coverts and primaries blackish-brown. Secondaries blackish-brown with broad green edges to outer web, these becoming broader and extending onto inner web of innermost secondaries, and tinged with blue at base. Primaries 5-6 marked by narrow white band across base. Throat black, bordered by white moustachial streak that broadens towards breast, and mixed with rufous-buff on lower throat. Breast to undertail-coverts dark ochre-buff, becoming slightly paler on lower flanks and centre of belly, the latter tinged with dull pink. Gape and tip of bill orange to orange-red.

Schodde and Mathews (1977) describe a juvenile *vigorsii*, presumably younger than bird described above. Head brown with many feathers tipped ochreish; there is no superciliary stripe; throat washed with grey; most feathers of underparts have light or black tips or bases, giving a scaly appearance; vent extensively washed with red. Rump dark bluish-grey, lacking the iridescent turquoise-blue of the adult; rectrices dark bluish-grey.

MEASUREMENTS Wing 108-119; tail 41-46; bill 28-30; tarsus 35-37.5; weight 48-63 (*concinna*).

30 NOISY PITTA
Pitta versicolor   Plate 15

Pitta versicolor Swainson 1825 (Jan.), Zool. Journ., 1, p. 468: "Australasia" [=New South Wales].

Alternative name: Buff-breasted Pitta

FIELD IDENTIFICATION Length 18-23.0, with northern populations smallest. This is a medium-large Australian species that is found from the coastal region of New South Wales to Cape York and islands in the Torres Strait. Migrants visit southern New Guinea. Compared to most pittas, it is an easy species to observe. Adults have a black head with a chestnut crown that is streaked black in the centre. The upperparts are green with a bold shining azure-blue wing-covert patch and uppertail-coverts. The underparts are yellowish-buff to cinnamon-buff, with a black throat and black patch on the centre of the upper belly, and red centre of lower belly to undertail-coverts. Juveniles are duller than adults and lack black on the throat and chin and azure-blue on the wings, whilst the extent of black on the belly is much reduced and the undertail-coverts are pinky-red. The bill of young birds is black with an orange tip and gape.

Similar species Within its Australian range, Noisy Pitta can easily be separated from other resident species of pitta by its cinnamon-buff underparts. Vagrant Blue-winged Pittas may reach Australia, but are easily distinguished from Noisy Pitta by the white throat, bold buffy supercilium and brown crown, extensive white in the wings and violet-blue rather than azure-blue wing and uppertail-covert patches. Elegant Pitta, the nearest population of which occurs in the Kai Islands of Indonesia, is also superficially similar in appearance to Noisy Pitta, but has a different head pattern, with a black crown and broad pale buff to golden-buff supercilium.

VOICE The calls of Noisy Pitta resemble those of Hooded Pitta and Elegant Pitta. The most frequently heard call, lasting less than one second (Coates 1990), comprises two upslurred notes separated, without pause, by a short low note, *hhWup-hu-hWip* or *hrWeep-hu-whu*, sometimes expressed by birders as 'walk-to-work'. The first note is almost disyllabic. In New Guinea, this call may be given only occasionally or repeated frequently at short intervals. In comparison to calls of Elegant Pitta, the call of Noisy Pitta is more fluty. The high-pitched nasal alarm call, *kiaw*, is usually repeated regularly when given. Zimmermann (in press) has heard this species give a call that is reminiscent of the purring of a cat and identical to a call made by Rainbow Pitta, as well as an alarm call, *keow*, that is also very similar to a call of that species. The latter was made in response to playback of its usual 'walk-to-work' call. Beruldsen and Uhlenhut (1995) report that female Noisy Pittas sometimes make a single mournful whistle, usually given three times in a row, when calling to the male. Taylor and Taylor (1995) noted that the call of the male was more assertive, whilst that of the female was more hurried and higher-pitched. During courtship, males may start calling as early as 3am, prior to dawn, and then throughout the day, although frequency declines during the day.

In New Guinea, during the non-breeding season, Noisy Pitta is largely silent (Coates 1990). MacGillivray (1914) reported that, although some individuals on the coastal plains near Atherton, Queensland, are present all year, they are almost silent between November and March or April. In contrast, birds at higher altitudes (460m) are noisy during this period but are silent at other times of the year.

DISTRIBUTION The Noisy Pitta is found in eastern Australia, in the northern part of the Great Dividing Range and coastal plains to the east, and in the Cape York Peninsula; and in New Guinea, where it occurs mainly as a non-breeding visitor, although breeding is believed to occur in the southern Trans-Fly region (Coates 1990). In the Torres Strait, Noisy Pitta (*simillima*) has been recorded from many islands (including Turtlebacked, Bet, Mt Ernest, Banks, and West), but it occurs on smaller islands only as a migrant. Birds have been recorded in New Guinea between October and December and in April, June and mid-August (Bell 1968, Storr 1973, Coates 1990). In Papua New Guinea, Coates (1990) documents records from Daru Island and the adjacent mainland (Binaturi River), Bensbach River, and near Port Moresby. In Irian Jaya, the Indonesian part of New Guinea, Noisy Pitta has been found near Kurik (Mees 1982).

Australian populations are disjunct, at least during the breeding season: Blakers *et al.* (1984) reported that there were no records of birds from the Atlases that link the eastern Cape York Region, Atherton, Eungella and southern populations. However, two subspecies are migratory, and overlap in their ranges occurs.

In Queensland, one population occurs from the eastern Cape York Peninsula south to the McIlwraith Range (as well as on larger islands in the Torres Strait), and another in the east, where present north to Cooktown and inland to Helenvale, Mt Lewis, Atherton, the Eungella Range, Rockhampton, Biggenden, Bunya Mountains, Ravensbourne and the upper Condamine (Storr 1973). In New South Wales, Noisy Pitta occurs in the far north-east and southwards patchily to Dorrigo, in the eastern New England Mountain Range, and the Camden Haven district, some 320km to the north of Sydney, with occasional reports further south (Hindwood and McGill 1958, McGill 1960). Birds overshooting on migration, or wanderers, have reached as far south as Sydney (Edden and Boles 1976), although it should be noted that the species has been kept in captivity in Australia since the 1850s, and records from this area may have referred to escapes (Hindwood and McGill 1958). The species also occurs on islands off the east coast of Queensland (*versicolor*) during the austral winter, including Dunk, Palm, Whitsunday, Peel and Stradbroke.

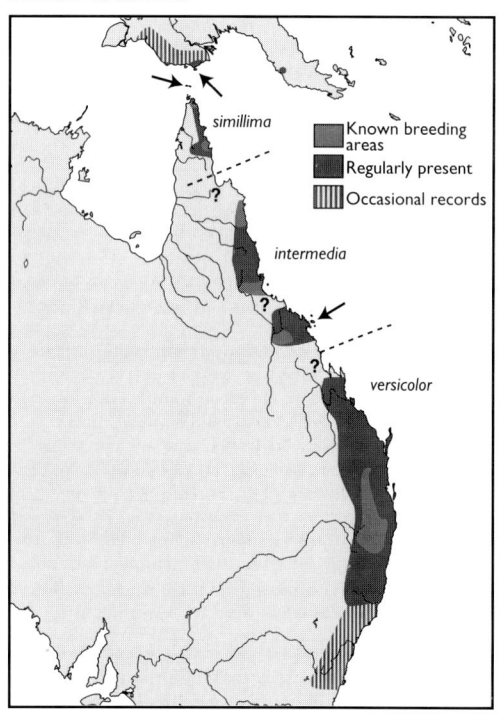

Map shows confirmed breeding range and areas where the species has been regularly encountered, probably during dispersive or migratory movements. Breeding in New Guinea requires confirmation.

GEOGRAPHICAL VARIATION Up to three subspecies have been recognised, but differences between them are rather small, primarily based on size and coloration of the underparts. In particular, *versicolor* and *intermedia* are very similar, and these two were lumped by Storr (1984).

Woodall (1983) studied biometrics throughout the range of Noisy Pitta, finding that there was significant latitudinal increase in size, from Cape York southwards, following Bergmann's Rule: however, he also found significant differences in some variables that were consistent with the existence of three subspecies. Although *versicolor* and *intermedia* might arguably be lumped, there appear to be consistent differences, and they are tentatively treated as separate below. Specimens from the northernmost population clearly stand out as different from series of more southerly populations, and *simillima* therefore appears to be a good subspecies.

P. v. versicolor (S Queensland north to Cooktown and New South Wales, and islands off the east coast. During the austral winter this subspecies is apparently also found in the Cape York Peninsula) Large (wing 120-129, weight c.92-106), with large white spot in wing, usually extending across primaries 4-6 and onto outer web of primary seven.

P. v. intermedia (Cairns District, N Queensland, Australia) Intermediate in size (118-121); less white in wing than *versicolor*, usually restricted to a small spot on the outer web of primaries 4-5, and occasionally on primary six.

P. v. simillima (Breeds from eastern Cape York Peninsula, south to the McIlwraith Range, N Queensland and on larger islands in the Torres Strait, also Haggerstone, off east coast; non-breeding birds migrate to southern New Guinea) Small (wing 113-118, weight c.74), with slightly darker, richer cinnamon-buff underparts than other subspecies. White spot in wing on primaries 4-6 restricted to outer web of primary six but more or less across both webs of primary 4-5.

HABITAT The Noisy Pitta is primarily an inhabitant of rainforest in Australia, where it has been recorded from sea level to 1,500m (Blakers *et al.* 1984). Driscoll and Kikkawa (1989) reported that, in Australia, Noisy Pitta is the only inhabitant of tableland rainforests that is intolerant of selective logging. Nevertheless, in parts of Australia, it has been found in woodland adjacent to rainforest, whilst in the Tooloom Scrub it is in eucalypt forest (Norris 1964, Blakers *et al.* 1984). Norris (1964) found Noisy Pittas in dry sclerophyllous forest at altitudes over 535m in the Tooloom Scrub of New South Wales, where it nested. On migration, it can be found in mangrove and coastal thickets (Storr 1973). In New Guinea, Noisy Pitta is generally found in dense gallery forest, monsoon woodlands and mangroves of the southern plains that form the Trans-Fly; elsewhere it is a vagrant (Coates 1990). Howe (1986) found that Noisy Pitta was rare or absent in rainforest patches of 2.5ha or smaller in north-eastern New South Wales.

STATUS Driscoll and Kikkawa (1989) found that selective logging on tablelands in north Queensland, Australia, led to the disappearance of Noisy Pittas. In the Tooloom Scrub, Norris (1964) recorded an average of one Noisy Pitta every 1.5km whilst surveying along six 3km-long transects. In New Guinea, the Noisy Pitta is generally rare (Coates 1990). Storr (1973) described the status of this species as locally moderately common in Queensland, whilst McGill (1960) stated that it was a rare inhabitant of the rainforests of New South Wales.

MOVEMENTS Noisy Pittas from northern Queensland and islands in the Torres Strait (*simillima*) migrate to southern New Guinea during the austral winter. Birds move northwards in March and April and south in October-December (Storr 1973). Barnard (1911) first recorded migrants on southward passage in scrub at Cape York on 10 October, noting that numbers built up over the next few days.

Birds on passage have been recorded on the islands of the Torres Strait from September to December. Noisy Pittas have not been found on these islands during the early dry season (Draffen *et al.* 1983). Specimens collected on Daru Island in June or July (Rand 1938) were presumably wintering birds. Most records of Noisy Pitta in New Guinea are from October, November and December, although there are also mainland records from April, June and August (Coates 1990). Whilst most records are presumed to refer to non-breeding visitors, some birds are almost certainly resident.

MacGillivray (1914) found the species fairly common in Cape York scrub habitats from July to the end of March, overlapping with the period when *simillima* are known to be in New Guinea. Hence, part of the population is perhaps resident or perhaps winters on the Cape York Peninsula rather than moving further north to southern New Guinea. U. Zimmermann (*in litt.*) also found Noisy Pittas (of unconfirmed subspecies) in the Iron Range (Cape York Peninsula) in July 1994 and January 1995. Considering that the timing of breeding of *simillima* on Cape York is November-February and on islands of the Torres Straits is January-April (see under Breeding), and that passage on small islands in the Straits occurs from September to December, it seems likely that birds in New Guinea in October-December originate from the Torres Straits population rather than from Cape York, whilst birds present in New Guinea at other times may derive from either area, or be residents.

Bell (1968) suggested that northern populations (*simillima*) move to New Guinea but are replaced by wintering birds of the nominate subspecies from Queensland and New South Wales. The size and coloration of some specimens collected on Cape York (e.g. in BMNH) support this idea, as noted by Elliot (1870), who identified some specimens from the vicinity of Moreton Bay, Cape York, as being the large, long-winged nominate subspecies. These specimens are in BMNH and are undoubtedly from southern populations. However, it is by no means certain that *versicolor* does undertake long-distance migration, but if it does, the evidence below would suggest that only part of the population is involved.

Storr (1973) reported that *versicolor* moves coastwards, and presumably northwards, in March-April, returning to the breeding grounds in September to November. Thus he believed that *versicolor* breeds in the highlands and foothills, north to Bloomfield, from October to December, but winters on the coastal plains and on eastern islands. Blakers *et al.* (1984) also stated that birds in the Atherton region and south-east Australia do not undertake large-scale movements. Movements in the Atherton region, for example, are altitudinal rather than true migration, with birds descending from the tableland in winter and moving away from the coastal region and islands around Cairns before the wet (breeding) season starts (Anon. 1987, U. Zimmermann *in litt.*).

Woodall (1993) found a lack of a seasonal component in variance in size of specimens in Australian collections from throughout the species's range. He argued that this indicated that the movement of southern birds is unlikely to extend far up the coast, if at all, and that movements are therefore more likely to be limited to east-west and altitudinal, although limited north-south movements might not have been detected by the analysis. However, as mentioned previously, there are specimens in BMNH said to have been collected on Cape York that, based on size and colour, are clearly of southern origin. Hence it would seem that at least some southern birds do make considerable latitudinal movements. The exact nature of movements by southern populations therefore requires further study.

FOOD Noisy Pittas are reported to eat insects, ants, woodlice, earthworms and, in particular, snails (Anon. 1976, U. Zimmermann *in litt.*). Stomachs of three specimens from Mt Sapphire, Australia, contained insects and snails, whilst the stomach of a specimen from New Guinea contained a small bone that probably came from a young lizard. Shells of large forest snails are cracked open at anvils (Favaloro 1931), these usually being a stone or piece of wood, which may be worn smooth with continuous use. In one case, the discarded pieces of a green bottle were used as an anvil (Edden and Boles 1976). Small berries and other fruit have also been reported in the diet of Noisy Pitta (Anon. 1976, Edden and Boles 1976): U. Zimmermann (*in litt.*) reports that the fruits of the Alexander Palm *Archontophoenix alexandrae*, Blue Quandong *Elaeocarpus angustifolius* and a fig *Ficus* sp. are eaten by this species.

A wide variety of food is fed to nestlings. Taylor and Taylor (1995) observed adults bringing centipedes, grasshoppers, flies, ants, snails, caterpillars and earthworms to a nest. In captivity, adults fed nestlings with mealworms, earthworms and snails (Parrish 1983).

HABITS Noisy Pittas feed alone or in pairs, particularly when breeding, and frequently bob their heads and flick their tails. Calls are most frequently made from the vegetation, and birds have been regularly observed at 10m off the ground. Noisy Pitta also roosts in trees.

During a study of a nesting pair of Noisy Pittas, Taylor and Taylor (1995) noted that, before nesting, both adults foraged in the immediate vicinity of the nest site, and even roosted in nearby trees. However, once nest construction was complete, they foraged further up the same creek, and rarely near the nest. Once feeding young, the parents used their own distinctive flight path and observation perch when approaching the nest.

Mating was observed twice by Taylor and Taylor (1995), occurring on the ground on both occasions. On the first occasion, it occurred after the female picked up a piece of debris and ran along the forest floor with it: the male gave chase and subsequently mated her, following the utterance of a short, single note that was not heard at other times.

Norris (1964) observed a probable threat display, regularly performed by a captive bird when the cage was approached. The bird would turn its back on the source of danger and crouch slightly whilst looking over its shoulder. The tail was then cocked and finally the wings spread. Assumption of this position took only about one second, whilst the final pose was held for 15-20s.

BREEDING The Noisy Pitta is known to breed in Australia and on larger islands of the Torres Strait. In the Torres Strait, breeding is confined to islands with rocks, although Noisy Pitta is present on sand and mud islands immediately before and after breeding. Woodall (1994) investigated, in detail, the breeding season of Noisy Pitta at different latitudes, and discovered that birds breeding on Cape York had a delayed breeding season. Rainfall, or perhaps some food resource correlated with rainfall, appears to be the proximate cue for the start of breeding, with peak egg-laying some 2-3 months before peak rainfall. Egg-laying is reported to occur in the Cape York area from November to February with a peak in January (Woodall 1994), although some birds may breed in March (MacGillivray 1914). Further south, in Queensland and New South Wales between 15-32°S, most birds lay eggs between October and December, with a peak in November, although eggs have been found in every month from June to February. Early egg-laying in the southern region generally occurs in years with above-average rainfall (Woodall 1994). On islands in the Torres Straits, breeding is reported to occur in the first four months of the year (Draffan *et al.* 1983). The presence of calling Noisy Pitta in dense monsoon woodland near the Bensbach River, Papua New Guinea, throughout April suggests that breeding also occurs in New Guinea (Coates 1990).

The nest is built on or close to the ground, usually on a slope and on the lower side of the tree or rock against which they are built. Some nests are protected by placement within the buttresses of trees, although the trees chosen as a site are invariable medium- to small-sized trees rather than larger individuals (Beruldsen 1980). Some nests have also been found on tree stumps. One nest measured 30 x 31 x 20cm (Edden and Boles 1986). The nest is similar in appearance to that of Southern Log-runner *Orthonyx temminckii*, but lacks sticks in the main structure (Beruldsen 1980). Both parents share in nest construction (Taylor and Taylor 1995).

The nest is loosely constructed and placed on a base some 2-7cm in depth that is constructed of short sticks. The nest is a bulky roofed structure that is usually made of soft dry vegetation such as leaves, leaf skeletons, moss, ferns, vines, rootlets, and strips of bark. Sticks are often absent from the main dome (Beruldsen 1980), but used in the adjacent platform and base, although some nests may have a large number of twigs in the upper structure (Taylor and Taylor 1995). The nest cavity is neatly lined with materials such as blackish fern fibres and root fibres, but may also be lined with decayed wood and debris from the forest floor. The top of the nest is usually covered with a layer of green moss that may partially obscure the side entrance, and the moss in turn is usually camouflaged by the addition of a few dry sticks and leaves. One nest found by Beruldsen (1980) was hardly recognisable as a nest since it was built on a 7cm base and had a 12cm layer of moss on the top, making it a very substantial structure. Mammal dung has been recorded as a material used in the matted platform of sticks that abuts the entrance hole, which is at the side of the nest (Beruldsen 1980, Edden and Boles 1986, Coates 1990, Taylor and Taylor 1995).

Clutch-size is usually 3-5, occasionally two, and one clutch collected near Kempsey, New South Wales comprised nine eggs (Woodall 1994). Nests discovered that contained only one egg seem likely to relate to incomplete clutches. Of 118 clutches traced by Woodall (1994), clutches of 3-4 were most common. However, Woodall (1994) discovered that mean clutch-size was significantly correlated with latitude, with a mean of 3.7 in southern Queensland and New South Wales (25-32°S) but one of 2.8 north of 22°S in Northern Queensland and Cape York. The female apparently lays more than one egg per day, since a clutch of four eggs was recorded two days after a female first entered a nest to lay and incubate (Taylor and Taylor 1995). Both sexes incubated, the incoming parent calling from some distance before arriving and the incubating parent leaving the nest before the other bird arrived to continue incubation. Two adults were only observed at the nest once, when the eggs first hatched, and on this occasion the male fed the female. At most changeovers, soft nest material was brought to add to the nest. Both parents also feed the nestlings, which are fully feathered and close to adult size when they fledge: nestlings fledge after a minimum of two weeks (Taylor and Taylor 1995).

Eggs are white or slightly creamy-white ovals. Eggs of *simillima* in BMNH have scattered chocolate-brown, chestnut, dark brown and blackish spots and marks that are generally distributed over the entire surface, and underlying grey markings. Most eggs lack scribbles, but two slightly larger eggs (32.5-33 x 25) of uncertain origin (and therefore not necessarily *simillima*), have more scribbles than spots, with most of these being below the egg surface. Two eggs in BMNH labelled as *versicolor* were more densely marked and had more irregular, slightly paler and more scribbled markings than the two eggs marked as *simillima*. The two eggs marked as *simillima* measured c.29.5-30 x 25mm, whilst two labelled as *versicolor* were more elongated, c.31 x 22.5-23.5mm. Four other eggs of uncertain origin measured c.30-32 x 24.5-25.5mm.

The last egg to be laid is usually much less and more finely marked than the others. Eggs may be stained by animal dung from the adults' feet if such material is used to make a platform in front of the nest. Incubation takes 15-18 days in the wild (Beruldsen 1980, Edden and Boles 1986, Coates 1990, Taylor and Taylor 1995).

This species has been successfully bred in captivity in Australia (Parrish 1983). In captivity incubation took only 14 days and the young fledged after a minimum of 16 days. When the young first left the nest they were aggressively protected by the parents, but the two young did not stay together. When they first fledged, they were only around half the size of an adult, but after 32 days were apparently independent of their parents. Other observations of interest from captive breeding include the observation that the male constructed a nest in three days and also incubated the eggs more frequently (for longer periods) than the female. After hatching, neither sex kept the young warm by sitting on the nest during the day, and the male did most, if not all, of feeding the young whilst they were in the nest: the female only fed young once they had fledged.

Nest predators include snakes: Taylor and Taylor (1995) observed a snake eating a nestling Noisy Pitta that was close to fledging.

DESCRIPTION *P. v. versicolor*
Sexes similar.
Adult Crown from forehead to nape dark chestnut, usually with more or less well defined black coronal band starting above the eye and extending to upper nape. Lores, ocular area and ear-coverts black, joining as broad black band across nape and neck and across throat. Mantle to rump and uppertail-coverts dark green; broad tips to uppertail-coverts shining azure-blue. Upper tail black with green band across tip of variable width, so that occasionally as much as half of tail is green. Tail from below black. Upper malar area and throat black; sides of neck and breast to upper belly and flanks rich cinnamon-buff and extending around black throat towards lower malar region on sides of neck. Centre of belly black, undertail-coverts pale red with a distinct pinky tinge, and occasionally pink. Primaries black with olive bands across tips and white patch half-way along formed by white band across primaries 4-5 and on outer web of primary 6, and frequently extending onto the outer web of primary 7. Secondaries black with green edge to distal half of outer web that becomes longer and broader on innermost feathers, extending across whole tip and entire outer web. Long iridescent azure-blue tips to lesser coverts forming distinct solid wing-panel; in some birds these tips are mixed azure-blue and green. Median and greater coverts green, concolorous with back and brighter than green of secondary edges. Underwing-coverts blackish. Some individuals have broad black streaks on the mantle, formed by dark central streak along feather axis; this feature occasionally extends to azure-blue feathers of uppertail-coverts. Iris dark brown; bill black; legs and feet pink or brownish-pink.
Immature Similar to adult, but duller, best distinguished by colour of throat, which is black mixed with grey. Bill black with pinkish gape.
Juvenile Crown chestnut, scaled with black edges to feathers and usually with black patch or short stripe on top of crown; sides of head, including lores, ocular area, ear-coverts and upper malar area, black, meeting across nape as broad black band which is bordered below at sides by pale buff, merging into white sides of lower throat. Chin and centre of throat greyish, mottled with white; breast dark ochre, becoming paler and buffier on flanks and belly and usually with some dull blackish-brown scaling at the top of breast; undertail-coverts pinky. Upperparts and wing-coverts dark green (wing-coverts lack azure-blue); uppertail-coverts and lower rump shining azure-blue; tail black with narrow green terminal band. Tertials green with blackish basal half of inner web; secondaries blackish with broad green edge to outer web of distal half, broadest on inner feathers. Primaries black with pale greyish-brown tips and hidden white spots towards base on outer web of primaries 4-6. Bill black with an orange tip.

MEASUREMENTS Wing 112-129; tail 54-65; bill 35-37; tarsus 43-47; weight 92-106.

31 BLACK-FACED PITTA
Pitta anerythra   Plate 14

Pitta anerythra Rothschild, 1901, Bull. Brit. Orn. Club, 12, p.22: Isabel.

Alternative name: Masked Pitta

FIELD IDENTIFICATION Length 15-20. This is a very shy medium-small pitta that is confined to three islands in the Solomons, where it is the only species of pitta. The head is largely black with a chestnut crown or a chestnut band across the rear of the crown, depending on subspecies. The upperparts are a dark deep green, marked with a conspicuous shining azure-blue patch on the wing-coverts. In the field, this brilliant wing-patch is the most striking feature. The underparts are buff or almost white, depending on subspecies, with a paler lower throat. The immature is undescribed.

Similar species No other species occurs in its range, but adult Black-faced Pitta is superficially similar to Noisy Pitta of Australia and southern New Guinea. However, Black-faced Pitta lacks black and red on the underparts.

VOICE Mayr (1945) reported that the call was a low-pitched *kree-kreeee-o*. On Santa Isabel, the call is a rather harsh, tuneless, double note with a slight pause near the end of the second, drawn-out note, after which it rises sharply in pitch and terminates abruptly, *hww-hwe(-)hhwe*. Occasionally, only the second phrase of the call is given, *hwe(-)hhwe*. It has the quality of the drawn-out call of Red-bellied Pitta but the duration of Hooded Pitta (D. Gibbs *in litt.*). This is a very far-carrying call, which, at close range, has a tremulous quality that is lost at distance. D. Gibbs noted that local people on Santa Isabel reported hearing this species throughout the year. Calling is from a perch, usually 1-4m above the ground, but birds responding to playback may ascend to the subcanopy at some 10-15m to call.

D. Gibbs (*in litt.*) found that, in late March, some Black-faced Pittas started to call before light at 5am, but that most birds started at first light and called persistently until soon after dawn. The birds then remained silent throughout the day until around 5pm, when more or less constant calling started, lasting until completely dark. Very wet mornings dramatically reduced calling, and Black-faced Pittas were not heard during rain, though they called occasionally between showers. Birds responded to playback throughout the day.

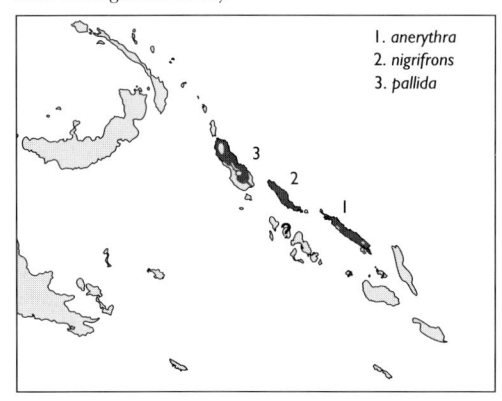

Distribution based on approximate extent of forest remaining within the altitudinal range of the species. Former range more widespread.

1. anerythra
2. nigrifrons
3. pallida

DISTRIBUTION The Black-faced Pitta is endemic to the northern Solomon Islands of Choiseul and Santa Isabel, and to Bougainville (Papua New Guinea). A recent sighting from Kolombangara (Buckingham *et al.* in prep.), in the New Georgia group to the south of Choiseul (8°S, 157°E), requires confirmation.

GEOGRAPHICAL VARIATION Three subspecies are recognised.

P. a. anerythra (Santa Isabel, Solomons) Crown rich chestnut-brown, buffier on the nape; underparts deep ochre.

P. a. nigrifrons (Choiseul, Solomons) Similar to *anerythra*, differing in having much more black on the forehead and forecrown. Wing 96-101 (Mayr 1935).

P. a. pallida (Bougainville, Papua New Guinea) Larger than other subspecies (wing 98-112), and with whitish underparts, the flanks and belly to undertail-coverts being almost white. Crown black with variable band of chestnut on the nape. Some birds lack chestnut on head: one specimen (BMNH) has virtually no trace of chestnut on the head and the throat entirely black. Two specimens in BMNH from Olimai and Buin, both in 'Agoli' (southern Bougainville, 6°52'S, 155°42'E), have wings 106 and 112 and more distinct blue fringes to feathers of uppertail-coverts than other specimens of *pallida*. All other specimens in BMNH have wings of 98-105.

HABITAT The Black-faced Pitta is an inhabitant of forest in the lowlands and hills. Rothschild and Hartert (1905) report that specimens were collected on the coastal plains of Choiseul and Bougainville, whilst D. Gibbs (*in litt.*) found this species at c.400-600m in southern Santa Isabel. Here the species occupies a variety of habitats, including cyclone-damaged primary forest, secondary forest, old gardens where regrowth is well advanced, and secondary scrub, often within hearing distance of habitation.

STATUS D. Gibbs (*in litt.*) found this species to be common on the southern part of Santa Isabel in 1994, but searches on southern Choiseul failed to locate any. Whilst local people on Santa Isabel were very familiar with this highly vocal species, locals interviewed on southern Choiseul were not familiar with it: hence the subspecies *nigrifrons* must be either very local in distribution, or very rare, or both.

Black-faced Pitta is one of seven species of pitta that are considered to be threatened (Vulnerable) by BirdLife (Collar *et al.* 1994): declines in its population are speculated to have occurred due to predation by introduced mammals, such as cats.

FOOD Reported to feed on worms, snails and insects (Haddon 1981).

HABITS This is a very shy and alert species that is difficult to see, even when calling. Nothing is recorded of its behaviour, apart from its habit of calling from above the ground (see under Voice). Local people on Santa Isabel claim that it calls every day, all year round (D. Gibbs *in litt.*). In flight the wings are almost silent, in contrast to those of many species of pitta.

BREEDING The nest is a bulky structure of leaves and moss that is placed on or near the ground (Haddon 1981). An egg measuring 30.8 x 25.0mm was collected on Bougainville on 12 May (BMNH). It was creamy-white, marked all over with short brownish-red lines and scribbles and with underlying greyish-purple spots (Rothschild 1904, Rothschild and Hartert 1905). Two males collected in July 1901 on Santa Isabel were moulting (Rothschild 1901), suggesting that breeding had already been completed at this time of year.

DESCRIPTION *P. a. nigrifons*
Sexes similar.
Adult Crown chestnut, darkest in centre and in front, where mixed with some black feathers in centre of forecrown; becoming brighter towards the rear, with golden tone at sides of rear crown, and fringed with some golden-buff feathers at top of nape. Narrow band of black across forehead, broadest in centre. Chin, upper throat, lores, narrow margin above eye and ear-coverts black, con-

tinuing across sides of neck to form a broad collar extending onto upper mantle and encircling the chestnut crown. Lower mantle, back and uppertail-coverts deep dark green. Lower throat white, merging into deep buff of breast, this buff fading towards belly and palest on undertail-coverts. Tail feathers green above with blackish base, the black extending along the outer web of outer feathers. Tail black from below. Uppertail-coverts tipped with azure-blue, but usually concealed at rest by longer feathers of rump, which may cover entire tail. Primaries blackish with short white band across primaries 4-6, midway along feather, on inner web of primary 4 and both webs of primaries 5-6. Indistinct dull green outer web to tips of all but outermost two primaries. Primary coverts black. Secondaries black with broad dark green edge to leading edge to distal half to two-thirds of all feathers, broadest on innermost secondaries. Tertials green with hidden blackish bases. Median coverts green, partially hidden by long lesser coverts; lesser coverts basally green, with broad terminal fringe of shining azure-blue (nearer base) that becomes mixed with shining pale green near tip. Underwing-coverts black. Iris brown; bill black; legs and feet pale greyish to smoky-brown.

Juvenile Undescribed.

MEASUREMENTS (one female) Wing 98.5; tail 39; bill 29.5, width 16; tarsus 44; weight unrecorded. *Pallida*: wing 99-111; tail 36-43; bill 29-33, width 14-18.5; tarsus 41-48.

32   RAINBOW PITTA
Pitta iris   Plate 15

Pitta iris Gould, 1842, Birds Australia, pt. 6, pl. 9: Cobourg Peninsula, northern Australia.

Alternative name: Black-breasted Pitta

TAXONOMY Rainbow Pitta has been considered to be conspecific with Noisy Pitta *P. versicolor* (e.g. Mayr 1979), but they are very clearly different in plumage, biometrics and call, and are treated here as separate species, following Blakers *et al.* (1984), Sibley and Monroe (1990) and other authors.

FIELD IDENTIFICATION Length 15.5-17.5. An unmistakable north-eastern Australian endemic that is relatively easy to observe. This rather small species of pitta has the head and underparts black except for rich pale chestnut-brown post-ocular coronal bands, these often meeting on nape, and pinky-red undertail-coverts. The upperparts are bright green and marked with a large conspicuous shining azure and violet-blue wing-covert patch. Immatures can be distinguished from adults by duller coloration and white edges to the feathers of chin. Bare-part colours also differ, immatures having an orange gape and reddish-brown legs.
Similar species Within its Australian range, the black underparts and green upperparts of Rainbow Pitta are unique and it is unlikely to be confused with other species. The only other pitta with extensive black in the plumage is Superb Pitta, endemic to Manus Island, but in this species the black extends to the entire upperparts.

VOICE The most common call, heard especially during the breeding season, is a double whistle, although this species also makes a trisyllabic call. Crawford (1972) found that birds near Darwin called most frequently in October-December, usually from within trees and up to 7m off the ground. Another often-heard call is a single note, *ptu* or *pptu*. Additional vocalisations that have been recorded are a warning call that is reminiscent of the call of a buzzard *Buteo*, but weaker, and a call used in aggressive interactions (during bowing displays: see under Habits) that sounds exactly like the purring of a cat (U. Zimmermann *in litt.*, in press). Crawford (1972) heard an agitated adult give a high-pitched, drawn-out call to a juvenile which was probably an alarm, since the juvenile responded by standing motionless. Zimmermann (in press) describes the alarm call as a sharp, loud *keow*.

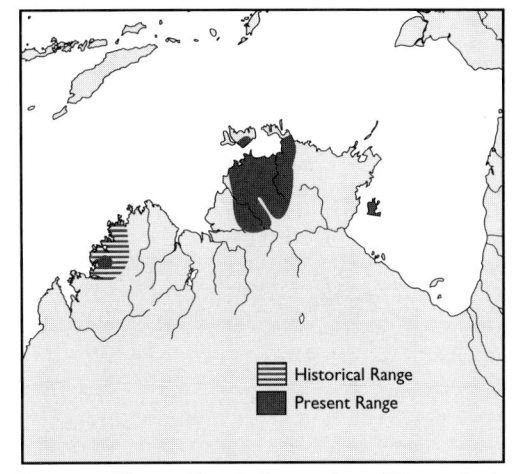

Present and known historical range shown. The species is patchily distributed within this range.

DISTRIBUTION The Rainbow Pitta is endemic to Australia, where it is apparently resident in the Top End region of the Northern Territory, north-west Australia, and the coastal part of the Kimberley region of Western Australia. In Northern Territory, Rainbow Pitta is known from Port Bradshaw west to Port Keats and on Melville Island. In the Kimberley region, there are historical records from Parry Harbour and Napier Broome Bay south and west to the Prince Regent River Reserve including some of the small islands off the coast, and within this area Rainbow Pitta is still present on the Mitchell Plateau (Bransbury 1992). These records from Western Australia suggest that there may be two distinct populations (Blakers *et al.* 1984).

GEOGRAPHICAL VARIATION None recorded.

HABITAT Rainbow Pittas inhabit a diverse array of habitats, including patches of monsoon rainforest, gallery forest along rivers, mangrove, in scrub and forest in sandstone gorges and eucalypt forest adjacent to the shore (White 1917, Crawford 1972, Hall 1974, Blakers *et al.* 1984). In the Kakadu National Park, Woinarski *et al.* (1989) and Woinarski (1993) found that this species was restricted to monsoon rainforest on sandstone dominated by the tree *Allosyncarpa ternata* (Myrtaceae), where more common in the interior of forest patches. Zimmermann (in prep.) found Rainbow Pittas in monsoon rainforest, *Melaleuca* forst and three other types of mixed forest, but not in mangroves. These three mixed forest types were dominated by the following species: *Syzygium nervosum/*

Barringtonia acutangula; *Lophostemon lactifluus*; and *Eucalyptus* sp. In contrast, the monsoon rainforests inhabited by Rainbow Pittas are characterised by having no dominant tree species, and during the dry season, many of the trees are more or less leafless. Zimmermann (in prep.) also found the species breeding in mixed forest dominated by *Lophostemon lactifluus* trees, and, at lower densities, in mixed forest dominated by eucalypt trees. Hall (1974) noted that the species frequented areas with dense vegetation with fallen logs, and reported that specimens were collected from areas where the ground was hard and dry.

STATUS Within its range, in suitable habitat, Rainbow Pittas are not uncommon. On the Cobourg Peninsula there was a pair in every patch of rainforest larger than 1ha, and at least seven pairs were present in a 5ha patch (Frith and Calaby 1974). Zimmermann (in prep.) found that the usual breeding territory was slightly bigger than 1ha. In Kakadu National Park, birds are significantly more abundant in larger forest patches (>2ha), although occasionally occurring in patches as small as 0.5ha in size (Woinarski 1993). Zimmermann (in prep.) found that the density of Rainbow Pitta was similar in monsoon rainforest and mixed forest dominated by *Lophostemon*, but much lower in mixed forest dominated by eucalypts.

MOVEMENTS This species is believed to be sedentary (U. Zimmermann *in litt.*). Crawford (1972) recorded Rainbow Pittas near the Coastal Plains Research Station, near Darwin, in all months except February-April, and believed that single birds found in very small isolated patches of monsoon forest in mid-January and mid-June could have been migrants. However, U. Zimmermann (*in litt.*) asserts that these observations can be explained by dispersal of young birds: Rainbow Pittas are very quiet at this time, when undergoing a post-nuptial moult, and were no doubt overlooked: U. Zimmermann (*in litt.*) found colour-ringed Rainbow Pittas at the same site all year round. White (1917) found a Rainbow Pitta on a small offshore island near King River on the northern coast of Northern Territory in late January. Such records are probably attributable to dispersive movements of juveniles or adults searching for territories.

FOOD During a study of Rainbow Pittas, Zimmermann (in prep.) found that the diet varied seasonally, with earthworms being the major food consumed during the wet season (October to March) and other invertebrates predominating during the dry season. Apart from earthworms, Rainbow Pittas were observed to eat centipedes, caterpillars, ants, bugs (Heteroptera), dipteran larvae, spiders, butterflies, beetle larvae, snails, small frogs, skinks (Scincidae) and the fruits of Carpentaria palms *Carpentaria acuminata*. Snail shells are broken on a branch or root, but in contrast to some other species of pitta, no anvils have been documented (White 1917, Frith and Calaby 1974). White (1917) found beetles and other insects in the stomachs of specimens, whilst Hall (1974) recorded grasshoppers, wood-boring beetles, a moth and ants, as well as small skinks, in the diet of this species.

HABITS Colour-ringing studies show that this species is territorial, staying within the same territory throughout the year (Zimmermann in prep.). Blakers *et al.* (1984) noted that Rainbow Pitta will forage on the breeding mounds of Orange-footed Scrubfowl. Crawford (1972) found that the sound of birds hopping through carpets of dry leaves, with a pattern of 4-6 hops followed by a pause, was distinctive and aided detection.

Zimmermann (in press) documents several displays and postures that this species makes. During the so-called 'bowing display', interpreted by Zimmermann as a social signal used to maintain territorial boundaries, the bird adopts a very stiff posture, standing upright, with the legs straighter than normal, body vertical and neck extended. Usually, two birds adopting this posture then start the bowing display, in which they stand a 1-10m apart and alternate between an upright posture and bowing movement, made in slow motion. During the movement, the breast may almost touch the ground. Bowing was performed along territorial boundaries, each bird displaying for 5-20s at one spot before moving a few metres to repeat the display. Bowing is accompanied by a call reminiscent of a purring cat that was only heard during such displays. During this bowing display, the two birds stand parallel to each other, perhaps in order to maximise display of the conspicuous shoulder patches (U. Zimmermann *in litt.*). The adults performing this display were sexed on ten of c.30 occasions, and were all found to be males.

A second behaviour documented by Zimmermann (in press) is 'wing flicking'. Wing flicking appears to be used as an alarm or to deter or distract the approach of potential egg or nestling predators. It was observed when a human approached a nest, and on one occasion when a Sulphur-crested Cockatoo *Cacatua galerita* was near an occupied nest. During wing flicking, the adult flicks its wings, at c.5s intervals, and regularly gives a sharp, loud *keow* alarm call. Occasionally, when an observer approaches a nest with young, nervous or highly alarmed adults perform a 'wing spreading' display, and simultaneously give the alarm call. In this display, the wings are partly opened and held in a vertical position for c.3s.

A final behaviour recorded by Zimmermann (in press) was a 'ducking posture' that was seen when several Spangled Drongos *Dicrurus bracheatus* dived to capture insect prey flushed by a foraging Rainbow Pitta. The pitta bent forward so that its tail was in the air and its breast virtually touching the ground but the head is stretched backwards with the bill pointing upwards whilst the wings were slightly opened.

U. Zimmermann (in prep.) notes that one sex (thought to be the male), calls significantly more often than the other (one frequently calling bird killed by a car was found to be male). Blood-sucking flies (Hippoboscidae), bird lice (*Picicola* sp.) and mites (Acari) were found to parasitise Rainbow Pitta (Zimmermann in prep.).

Rainbow Pitta - upright posture, wing-spreading and bowing

BREEDING Breeding is seasonal, usually starting in mid-October and ending at the end of February (U. Zimmermann in prep.), or in early March. The nest is most frequently built in trees, including palms, shrubs or vines, up to heights of 20m, but also occasionally in clumps of bamboo, dense mangrove thickets, on tree stumps or on a log amongst fallen debris (Beruldsen 1980). Sometimes however, nests are also built on the ground, where usually situated between the buttress roots of trees: U. Zimmermann (*in litt.*) found only three of 27 nests on the ground. White (1917) found a Rainbow Pitta nest on a ledge near the top of a rock within scrub along buffs at the edge of sandstone hills, whilst nests are reported to have been found in tussocks of grass in the Kimberleys.

Nests are dome-shaped with a side entrance but no hood over the entrance, no entrance platform and no base of sticks, in contrast to the similar nest of Noisy Pitta (Beruldsen 1980). Size and shape varies little, with nests typically c.30cm long, 30cm wide and 21cm high. The entrance is some 9cm wide and 7cm high. U. Zimmermann (*in litt.*) reported that the nest is constructed principally from small twigs and sticks. Other nest materials used in more than 100 nests examined by U. Zimmermann (*in litt.*) during 1992-94 were leaves, bark, ferns, climbers and parts of palm fronds. Beruldsen (1980), however, reported that bamboo leaves and plant sheaths are used more often than other materials. These apparent differences in reports of nesting material used by Rainbow Pitta suggest that it is highly adaptive in choice of materials, and that these may reflect local availability.

Clutch-size is 3-4 (although seven eggs in BMNH may have come from one clutch), incubation takes 14-15 days, and the nestling period lasts 14 days (U. Zimmermann in prep.). The eggs are white or creamy-white, glossy, and have purple-black blotches and spots on the surface, with underlying dull purple-grey markings, the markings being concentrated at the broad end. The pattern of markings is quite variable, with some eggs having well-defined fine spots and others having more messy, ill-defined markings (BMNH). The eggs are variable in shape, some being rather elongated (e.g. 20.5 x 28mm) and others rather round (e.g. 21.5 x 25.0mm) (BMNH). Five eggs in BMNH measured c.20.5-22 x 25-28mm, whilst 32 eggs measured in the field averaged 21.1 x 25.9mm (U. Zimmermann in prep.).

Both sexes participate in nest building (White 1917), incubation and feeding nestlings and fledglings, with male and female feeding nestlings with equal frequency (Zimmermann in prep.).

DESCRIPTION
Sexes similar.
Adult Head, including chin and throat, black, extending as hood onto upper edge of mantle and marked with a broad rich pale chestnut-brown coronal band. This stripe varies from a short streak from behind eye to sides of nape, to a patch that starts on sides of forecrown and broadens on the rear of crown to form a solid patch on nape. Upperparts, including rump, bright green, often with slight iridescent golden gloss on sides of mantle, and occasionally tinged with blue so that the upperparts are paler and brighter than usual (as in one BMNH specimen from Port Essington). Tail colour variable, usually with inner feathers entirely green, outer feathers with brownish inner webs near base. Bright shining blue wing-patch, formed by long azure-blue tips to feathers on inner four rows of outermost lesser coverts and deeper violet-blue tips of outer row of lesser coverts. Rest of upperwing-coverts bright green with distinctive golden sheen; tertials green. Secondaries blackish, with broad outer green edge extending across most of outer web of outermost feathers and across whole of both webs of innermost secondaries. Primaries blackish with paler, greyish tips and small hidden white spot towards base of primary 4 (inner web), primary 5 (both webs) and primary 6 (outer web). Underwing-coverts black. Lower flanks and sides of anterior belly cinnamon-yellow; undertail-coverts reddish-pink. Feathers of tarsus grey. Bill black; iris dark brown; legs light flesh to reddish-grey; feet dark brown or blackish.
Immature General plumage duller than adult. Head black, extending as hood onto upper mantle, and contiguous with black to blackish underparts. Sides of crown behind and above eye marked with a few inconspicuous small warm pale chestnut-brown dots. Feathers of throat dark greyish, fringed white and forming scaled effect. Upperparts as adult, but duller. Feathers of lesser wing-coverts and scapulars basally grey and green, terminally azure-blue but lacking the long fringes found in feathers of adult, so that the wing-patch is an ill-defined dull azure-blue mottled with green. Rest of wing-coverts and flight feathers as adult, but with duller coloration. Green of mantle duller than adult, with strong bronzy wash along margin of black upper mantle. Immatures are very similar to adults, being identical at an age of six months. Gape orange; legs reddish-brown.

MEASUREMENTS Wing 103-108; tail 35-38; bill 29-31; weight 59.5-65.5.

SMITHORNIS

The three species in this genus are confined to Africa. Subspecies are poorly differentiated. Sexually monomorphic but perhaps polygynous. Two species inhabit tropical rainforests, whilst the other occupies a variety of forest types as well as scrub and thickets. Usually encountered alone or in pairs, but evidence suggests that birds may congregate when displaying. The display involves short circular flights accompanied by distinctive, loud mechanically produced trills, and exposure of puffed-out white back feathers. The plumage is otherwise rather dull, and these broadbills have the appearance of robust short-tailed flycatchers, with a very broad flat bill. Insectivorous, usually catching arthropods by sally-gleaning from perches, from the understorey to mid-canopy.

33 AFRICAN BROADBILL
Smithornis capensis   Plate 17

Platyrhynchus capensis A. Smith 1840 Illustr. Zool. S. Afr. Bds, pl.27: Natal; restricted to the coastal forests of northern Zululand, Vincent, Bull. Brit. Orn. Club, 55, p. 96.

Alternative name: Delacour's Broadbill

FIELD IDENTIFICATION Length 12-14. African Broadbill is widely distributed in the southern tropics of central and eastern Africa, although absent from drier areas, and has isolated populations from central Ghana to Sierra Leone, in Cameroon, Gabon, Central African Republic and Angola. It is the dullest broadbill in Africa, and resembles a robust short-tailed flycatcher, with a very broad flat bill. Males have black crown, buffy forehead, buff or grey cheeks that are often streaked with white (depending on subspecies), greyish-brown to rufescent-brown upperparts, more or less distinct buffy or creamy wing-bar, and silky-white underparts streaked black on the breast, flanks and sides of throat. Most subspecies have a buffish wash at the sides of the breast, whilst some subspecies have a buff or grey wash across the breast. The bill is bicoloured, with black upper and white lower mandible. Females are similar to males but generally duller. The crown varies between subspecies, being either black or brown, and sometimes streaked. Juveniles closely resemble adult females, but the crown is brown with indistinct dark streaking and there is little buff on the forehead. African Broadbill is easily overlooked but, like other species in the genus *Smithornis*, often detected from the noise made whilst performing the elliptical display-flight. Both sexes have a white patch on the mantle that is conspicuous during display-flights.

Similar species African Broadbill is smaller than Grey-headed Broadbill, and easily identified by the lack of orange on the underparts. Males also differ in having a black, rather than grey, crown. It differs from partially sympatric Rufous-sided Broadbill in having white or buffy sides of the breast, instead of bright orange, a less uniform head pattern and, usually, in its choice of habitat. African Broadbills prefer secondary forests and more open habitats with trees, whilst Rufous-sided Broadbills are usually confined to primary forests.

VOICE The voice of African Broadbill is generally considered to be weak, in contrast to the noise made by the primary feathers during display-flights. Two quiet calls are documented, one a plaintive whistle *huiii*, the other a mewing noise (Manson 1983), possibly associated with stress. Wells (1985) also describes a quiet contact call *twee-uu*.

During courtship, birds in southern Mozambique have been heard to utter a loud noise like a klaxon (Lawson 1961) whilst hanging from a perch. It seems that this noise was not the same as that made during display-flights documented below, but Lawson (1961) concluded that the origin of the well-documented noise made during display-flights is vocal rather than mechanical. Most observers, however, have attributed the source of the sound made during display-flights to the vibration of stiff, twisted outer primaries.

The sound emitted during display-flights is a mechanical-sounding whistle that is very loud, being audible at distances of at least 60m, and may be made by either sex (Chapin 1953). It is a very fast, staccato trill of c.0.7s duration that sounds like the rattle produced by pinging the free end of a wooden ruler clamped across the edge of a wooden table (Keith *et al.* 1992). This trill has been described as *tttt-rrrrrrrrrrrrrr* (Keith *et al.* 1992), as *kerrrrrrr* in Liberia (Rand 1951), but in southern Africa as *purrup*, suggesting variation between different populations. The trill starts suddenly, slows then accelerates and abruptly stops. Keith *et al.* (1992) reported that the trill comprises some 4-6 notes in 0.25s followed instantly by 15 notes in 0.5s. Whilst this mechanical note sounds like a trill to close observers, the buzzing noise disappears increasingly with distance, and the notes sound more musical than mechanical (D. Aspinwall *in litt.*).

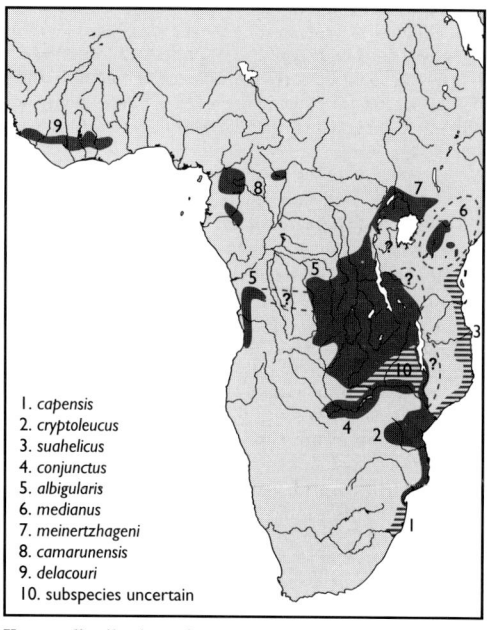

1. capensis
2. cryptoleucus
3. suahelicus
4. conjunctus
5. albigularis
6. medianus
7. meinertzhageni
8. camarunensis
9. delacouri
10. subspecies uncertain

Known distribution of all subspecies.

DISTRIBUTION The African Broadbill is the most widespread species of the endemic African genus *Smithornis*. It is widely distributed in Central and eastern Africa south of the equator, but absent from the dry south-west and central parts of southern Africa. The most southerly populations are found in Natal. In the north, its range extends along the eastern border of Zaïre and western Uganda, through eastern Tanzania to just north of the equator in Kenya. Rather isolated populations are also found in West Africa from central Ghana to Sierra Leone, and in Cameroon, Gabon and the Central African Republic.

Due to the complex distributional pattern of different subspecies, details of distribution are provided in the following section (Geographical Variation).

GEOGRAPHICAL VARIATION Subspecies of African Broadbill vary in size, upperpart coloration and pattern, the intensity of streaking on the underparts, and the crown colour and pattern of the female, which may be streaked or plain. Nine subspecies are recognised here, following Keith *et al.* (1992), but the taxonomy and relationships of populations are still poorly understood, and this account should not be considered definitive. For most subspecies, the exact range limits are poorly documented:

S. c. capensis (South Africa in coastal Natal and S Zululand) Males have more or less buff forehead, greyish ear-coverts, a narrow grey nape-band and grey tinge to rufous-brown mantle. The underparts are silky-white, boldly streaked blackish-brown on sides of throat, breast and flanks; throat and breast usually tinged with buff, flanks tinged greyer. Females have grey crowns with black streaking, and, sometimes, brown ear-coverts.

S. c. camerunensis (Cameroon, Gabon, Central African Republic. In Cameroon, it occurs in lowlands in the south and in Gabon, east of the Sanaga River [Louette 1981]. Dowsett and Forbes-Watson [1993] note that there are unverified records from Equatorial Guinea) Mantle to rump and wings of both sexes richer, more rufescent than in other subspecies; underparts heavily streaked, especially on breast where streaks are particularly broad; strong buff wash across breast. Crown of female varies from black, almost lacking streaks, to heavily streaked black and rufous. Wing 69-73.

S. c. delacouri (Sierra Leone, Liberia, Ivory Coast, north to Comoé National Park [Demey and Fishpool 1991] and Ghana. Probably also in S Guinea, since it has been collected close to the border in Liberia [Rand 1951]) Similar to *camerunensis* but slightly paler, with greyer nape and top of mantle. Wing 67-71.

S. c. albigularis (Central Africa in N Malawi, N Zambia, Zaïre and in East Africa in west and possibly central Tanzania, where it is patchily distributed from Namabengo and Kasanga north to Mt Mahari, Gombe Stream Game Reserve, the Tabora area and Dodoma, extending to the east of Bahi [Fuggles-Couchman 1984]. Populations in Zaïre occur in the Kasai and Katanga regions [Chapin 1953]. A population in the lowlands of W Angola may be isolated. In Angola, occurs in S Cuanza Norte and adjacent Malange, extending south along the escarpment to Gabela and Chingoroi and reaching SW Malange on the Luce River. It also occurs at Cazombo, extreme NE Moxico [Traylor 1960]) Upperparts much greyer than most subspecies, including nominate, and breast washed yellowish-buff; ear-coverts of male grey, streaked white; crown of female blackish with buff edgings.

S. c. meinertzhageni (Highlands of NE Zaïre and adjacent forests in Rwanda, Uganda [Budongo, Kibale, Kibirau, Kanungu, Kalinzu, Maramagambo, Impenetrable] and in W Kenya from south of Mt Elgon to Kavirondo [Lake Victoria], including Kakamega, Lerundo and Nyarondo) Similar to *albigularis*, but underparts less buffy and female has blackish crown. It is darker above, with heavier breast-streaks and shorter wings than *medianus*, and lacks the yellowish wash to the belly and ochraceous breast-patches. Wing 65-70.

S. c. medianus (includes *S. c. chyulu* van Someren 1941) (Highlands of C Kenya [Crater highlands to Nairobi, the Aberdares, Mt Kenya, Chyulus and Ngaia Forest] and NE Tanzania [E and W Usambaras, Mt Kilimanjaro]) Similar to *albigularis*. Both sexes have more uniform, less streaked mantle and rump than other subspecies and grey wash across upper mantle. Crown of female blackish-grey, usually with indistinct paler grey scaling. Wing 73-77.

S. c. suahelicus (includes *S. c. shimba* van Someren [1941]: see Clancey 1970) (Coastal districts from Shimba Hills, Kenya, to Beira, Mozambique, and inland in Tanzania to the Uluguru Mountains and Nguru) Male has dark grey ear-coverts, the female has buff wash to underparts. Male differs from *medianus* in having extensive black on centre of mantle, and from *meinertzhageni* in paler, more ochraceous upperparts and finer streaking below. Differs from *cryptoleucus* in being more rufescent, and richer above, and in having cream-buff tones to the underparts, which also have finer, paler streaking. Female *suahelicus* has a darker, more uniform dorsal surface to the head than *cryptoleucus* (in which this surface is streaked with black) and a browner back. Wing 70-71.

S. c. conjunctus (S Angola, the Caprivi Strip [NE Namibia] and N Botswana [including the Okavango] through the middle Zambezi valley in S and SE Zambia and N and W Zimbabwe to the Tete district of NW Mozambique) Yellowish to buffy-olive wash to breast and flanks; male has buff ear-coverts.

S. c. cryptoleucus (SW Tanzania and S Malawi south through E Zimbabwe and Mozambique to South Africa in NE Zululand, E Swaziland [Lebombo Mountains] and NE Transvaal. Populations south of the Zambezi River in Mozambique occur locally in the Maputo district, the Lebombo Mountains and the littoral of Sul do Save [Clancey 1971]) Smaller than *capensis*, with more sparsely streaked underparts. More buffy-olive above than *conjunctus*. Females have olivaceous slate-grey cap, broadly streaked with dull black. Wing 68-71.5; bill 16.5-18.

HABITAT The African Broadbill is resident in a variety of habitats. Compared to the closely related and partially sympatric Rufous-sided Broadbill, the African Broadbill occupies a variety of habitat types. These include: the lower storeys of evergreen forest; riparian forests; dense savanna woodland, including areas with stands of the bamboo *Oxytenanthera abyssinica*; dense deciduous thickets; miombo woodlands; coastal bush; disturbed areas with a mosaic of forest and scrub; montane areas dominated by tree-ferns;

and, in parts of its range, such as in Gabon, plantations and even cultivation or open country around villages (Sharpe 1907, Chapin 1953, Rand *et al.* 1959, Prigogine 1971, Brosset and Erard 1989, Keith *et al.* 1992, C. Carter *in litt.*, R. Colbrook-Robjent *in litt.*). Van Someren (1941) found the species in patches of small *Catha edulis* trees in the Chyulu Range of Kenya.

Although occasionally found in primary forest in parts of East Africa, it is apparently absent or very rare in this habitat in West Africa. Thiollay (1995) reported that it occurs in the undergrowth of both primary and secondary rainforest as well as in semi-deciduous forests in Ivory Coast, but his records from primary and secondary rainforest seem doubtful: neither Fishpool and Demey (1994) nor Gartshore *et al.* (1995) found it in primary forest in Ivory Coast, and L. Fishpool (*in litt.*) believes that African Broadbill is primarily a bird of the drier zone to the north of the evergreen forest belt in this part of its range. In Gabon, African Broadbill is reported to be strictly a bird of the lower levels of secondary forest and disturbed habitats (Brosset and Erard 1989): Rand *et al.* (1959) found it in heavy growth in plantations in the Massif du Chaillu. Similarly, in Sierra Leone, it occurs in logged forest near the border with Liberia (Allport *et al.* 1989). In Cameroon, G.L. Bates only found the species in bushes in open country (Sharpe 1907), whilst Chapin (1953) reported that it was absent from undisturbed forest but present in bushes in clearings in the south of the country. In Zaïre, a displaying female was found in a valley full of tree ferns (Chapin 1953), plants that are often associated with disturbed areas of forest.

In Zambia, African Broadbills occasionally leave thickets and evergreen forest patches to forage in *Brachystegia* woodland and groves of *Acacia*, and have also been recorded in mature conifer plantations (C. Carter *in litt.*). Madge (1972) found it in Zambia in pine plantations where there were bamboo clumps growing on termite mounds, as well as in miombo woodland. In Zimbabwe habitats of this species include thickets in *gusu* (*Baikiaea*) woodland. In Mozambique, *conjunctus* occurs in riverine thickets along the Zambezi River and its major tributaries, whilst *cryptoleucus* occurs in similar habitats as well as in forest clumps and primary evergreen forest in central and southern parts of the country (Clancey 1971). Angolan populations are found along the edges of evergreen and gallery forests (Traylor 1960).

As well as occupying many types of habitat, African Broadbills have a wide altitudinal range: from sea level to 1,150m on the Mashonaland Plateau, Zimbabwe; to 1,560m in the Vumba Mountains of Zimbabwe (near the border with Mozambique) (Manson 1983); in Kenya from sea level to 2,150m in North Nandi Forest; to c.2,150m in Zambia on the Nyika Plateau and 1,980m in the Mafinga Mountains (Benson and Benson 1977); and to at least 2,440m in the Usambara Mountains, Tanzania (BMNH). In southern Malawi, *cryptoleucus* has not been found above 1,370m (Benson and Benson 1977). In eastern Zaïre, *meinertzhageni* has been found in secondary forests from 1,230-2,550m (Prigogine 1971), whilst in Rwanda it has been collected at 1,600m in Rutshuru Valley (Schouteden 1966a). In Gabon, *camerunensis* has been found at 820m in the Massif du Chaillu (Rand *et al.* 1959). Throughout most of its range, however, it is generally reported to be commoner at altitudes below 700m (Benson and Benson 1977, Manson 1983, Keith *et al.* 1992), although one subspecies, *medianus*, is wholly montane in the highlands of north-east Tanzania and central Kenya, east of the Rift Valley.

Habitat choice may be influenced by the presence of other species of broadbills: in the Impenetrable Forest, south-west Uganda, where African Broadbill is sympatric with African Green Broadbill at high altitude, the former is reported to be a forest-edge bird, even though it occurs in primary forests elsewhere in East Africa (Britton 1980). Furthermore, in West and Central Africa, where its range overlaps with the two other species of *Smithornis* broadbills, African Broadbill tends to be found in secondary habitats, and rarely in primary forest, whereas Grey-headed and Rufous-sided Broadbills appear to favour primary forests and secondary forests at an advanced stage of regeneration.

STATUS This species is reported to be locally common throughout much of its range. However, in some parts of its range this may not be the case: *cryptoleucus* is described as a rare resident in Swaziland (Parker 1992) and is apparently very scarce in adjacent Transvaal (Tarboton *et al.* 1987), *conjunctus* is apparently a sparse resident in riverine forest in the Okavango Delta and along the northern perennial rivers of Botswana (Penry 1994), and Rand *et al.* (1959) considered that *camerunensis* was a rare species in Gabon. Habitat destruction has severely affected populations in some areas, such as in Natal (Ginn *et al.* 1989).

The population density at Kakamega Forest, western Kenya, was found to be very low for a bird of this size, with only one adult in 8ha of forest, corresponding to around 12 birds per km^2 (Zimmerman 1972).

MOVEMENTS None recorded.

FOOD Despite its relatively common status and wide distribution, little has been recorded about the diet of African Broadbill, although it is clearly insectivorous. Beetles (Coleoptera), grasshoppers and crickets (Orthoptera) and their larval stages, bugs (Hemiptera), ants, spiders, caterpillars and butterfly eggs have been recorded in their diets (Prigogine 1971, Keith *et al.* 1992).

HABITS The African Broadbill is a shy, inconspicuous and usually quiet forest species that is only easily detected when calling and displaying. Its behaviour and habits probably differ little from the better-known Rufous-sided Broadbill. Both species are territorial, with pairs frequenting dense vegetation in the understorey, and both commonly flycatch and make display-flights. African Broadbills usually keep to within 6m of the ground. Although they tend to inhabit thick vegetation, Clancey (1970) notes that fading of plumage coloration suggests that they regularly expose themselves to the sun, presumably in glades or clearings.

Invertebrates are caught during sallies from perches in the manner of a wattle-eye *Platysteira* sp., returning to the same perch after pursuing prey, or searching under leaves and branches. Occasionally, African Broadbills drop to the ground whilst feeding, or when collecting nest material. This species is also reported to joins flocks of mixed bird species as they pass through their territory (Prigogine 1971, Keith *et al.* 1992).

Elliptical display-flights, apparently performed by both sexes, are also very similar to those made by Rufous-sided Broadbill, although the flight path is usually of slightly greater diameter, being c.60-80cm. Zimmerman (1972) reported that, in Kakamega forest, Kenya, display-flights are rarely observed outside the breeding season, although

Brosset and Erard (1986) found this species displaying all year round in Gabon. The frequency of such flights tends to be lower than those of Rufous-sided Broadbill, being about one flight per 45s. In Zambia, African Broadbills occasionally jump off their perch vertically into the air, attaining a height of c.25cm, immediately prior to making the horizontal display-flight (C. Carter *in litt.*). The white back-patch is made very conspicuous by puffing out the back feathers during display-flights, and hence exposing the white bases to black feathers (van Someren 1941, Rand 1951).

Whilst most aspects of behaviour are similar to Rufous-sided Broadbill, the reported courtship display of African Broadbill is quite different. At Chimonzo, Mozambique, Lawson (1961) observed what was believed to be a courtship ritual in early September. This involved two birds situated about 45cm apart on a low horizontal branch. The birds flicked their wings in the manner of a chat, then swung down to hang from their perches while continuing to flick their wings and to make loud klaxon notes for about 30s before righting themselves and flying to separate perches some 100m apart.

BREEDING In western Africa, African Broadbills have been collected in breeding condition in Angola from December and March. A juvenile just out of the nest was collected in December at Canhoca, Angola (Traylor 1960), suggesting that breeding may start as early as November in this part of its range. Further east, Brosset and Erard (1986) note that the display is performed throughout the year in Gabon, though more intensively during October and March. However, eggs have been found in Gabon only in January, and Rand *et al.* (1959) collected a male in breeding condition on 6 August in the Massif du Chaillu. G.L. Bates collected females with ova along the River Dja, Cameroon, on 5 May and 26 July (BMNH), specimens that were breeding in March and a female off a nest containing three young on 17 May (Sharpe 1907). Further east, in the Itombwe region of eastern Zaïre, eggs have been found in September, October, April and May. Clearly both *camerunensis* and *albigularis* have long breeding seasons, extending for at least six months, but more data are required to elucidate seasonal patterns.

In East Africa, eggs have been collected in November, December and March (van Someren 1941, Keith *et al.* 1992), including four clutches originating from southern Mozambique that were all taken in December (Clancey 1971). Eggs have been found in every month from October to February in Malawi and Zimbabwe, from September to January in Zambia and from September to January in Natal, although most breeding records are from November (Benson and Benson 1977, Irwin 1981, Ginn *et al.* 1989, R. Colbrook-Robjent *in litt.*). However, breeding occurs in the Shimba Hills of Kenya slightly later, since van Someren (1941) collected birds with enlarged gonads and shot a bird off a nest with an egg towards the end of March. In Zambia, a recently fledged juvenile has been caught as late as 10 April (R. Colbrook-Robjent *in litt.*). Most of these records coincide with the rainy season, with a tendency for eggs to be laid at the beginning of the rains.

The nest is similar but usually larger than that of Rufous-sided Broadbill, being an untidy bag hung from a horizontal branch 1-2.5m above the ground, usually from a bush or small tree or sapling in dense, shady understorey, although nests have occasionally been discovered in fairly open situations, such as one found in bushes on waste ground near a village in Cameroon (Sharpe 1907). Nest size apparently varies considerably, measuring from c.20-100cm in height, including the fibres dangling from the bottom, with a slightly porched entrance on the side near the top. Two small nests in BMNH, one 19cm long and the other 25cm long, were composed almost entirely of rather weak plant fibres: it seems possible that the nest material used in this instance was too weak to construct a larger nest. The nest chamber is about 9cm deep and 6-7cm wide, with an entrance hole of around 4-5cm high and 3.5-5cm wide (BMNH, Keith *et al.* 1992).

The material used in nest construction varies considerably, perhaps regionally. The nest is constructed from plant fibres, dead leaves, green moss and twigs and, like that of Rufous-sided Broadbill, this material is often woven together with the fine black fungal strands of *Marasmius* (Chapin 1953). Bates and Ogilvie-Grant (1911) reported that a nest from Cameroon was made of long stringy fibres, probably the dried bark of the weed *Triumfetta* or of dried plantain stalks, noting that such materials were not derived from inside the forest. Jackson (1938) noted that the fibres used in nest construction were torn from dried stems. Van Someren (1941) found nests in the Shimba Hills of Kenya that, externally, were made of long strips of bark and strips of dead cycad leaves intermixed with hair-like fungi. Bark, leaves, forest debris were also present inside the structure, and it was lined with finer fibre.

Nest of African Broadbill

Long dry leaves may be used in a broad band to attach the nest to the horizontal branch, and these may also be used in weaving the other nest materials together (Keith *et al.* 1992), although two nests in BMNH were made primarily of plant fibres that were either derived from the leaves of palm or bamboo, or fibres from rotting wood; these fibres formed the sole material used in attaching the nest to the twigs from which they hung. Only a few dicotyledonous leaves were present in these nests, and one also contained a few small twigs. Neither had a long tail. In one nest, the cavity was made from pieces of palm or bamboo leaves, and the lining was of very fine strands of plant fibre; in the other no palm or bamboo leaves were present, only fine plant fibres with a few small dicotyledonous leaves, suggesting, perhaps, that nest construction was incomplete. Other nests have apparently been lined with soft bark, dry stems, leaves and grasses, and matted together with spiders' silk (Keith *et al.* 1992). In southern Africa the nest is frequently made entirely of 'old-man's beard' lichen (Ginn *et al.* 1989).

Clutch-size is reported to be typically larger than that of other *Smithornis* broadbills, most frequently being 2-3 (Bates and Ogilvie-Grant 1911, Jackson 1938, Chapin 1953, Clancey 1971). The eggs are rather long, unmarked, glossy white ovals. Keith *et al.* (1992) reported that eggs measured 19-23.6 x 14.5-16mm, although eleven eggs from southern Africa (*capensis/cryptoleucos*) were 21.3-24.9 x 14.5-16.6mm (average 23.3 x 15.9mm) (McLachlan and Liversidge 1978), whilst eggs of *camerunensis* are reported to measure 22.0-24.5 x 15.5-16.0mm (Chapin 1953). Twenty-five eggs of *conjunctus* from southern Zambia measured on average 21.2 x 15.4mm, and one egg of *albigularis* from northern Zambia measured 21.9 x 15.7mm (R. Colbrook-Robjent *in litt.*), whilst an egg from the Shimba Hills (*suahelicus*) was 19 x 15mm (van Someren 1941).

A nest of African Broadbill discovered in Zimbabwe in early December contained the egg of a brood parasite in addition to two broadbill eggs. The former egg, which measured 22.0 x 16.5mm and was unglossed, white with a zone of faint reddish-brown freckles at the obtuse end, was identified as probably belonging to a Long-tailed Cuckoo *Cercococcyx montanus* (Dean *et al.* 1974).

It is not clear whether both sexes participate in nest building and incubation. However, Bates and Ogilvie-Grant (1911) report collecting a female off a nest with eggs and incubation behaviour is likely to be similar to that of Rufous-sided Broadbill, for which only females have been reported to incubate eggs.

DESCRIPTION *S. c. capensis*
Adult male Crown black with creamy-grey to greyish-buff patch on forehead, usually broadest at sides of forehead and extending to lores near bill, where greyer and occasionally with some rufous feathers at centre of forehead; lores creamy-white to buffy-white but appear darker due to dense black rictal bristles, and lores in front of eye usually browner; ear-coverts greyish to grey-brown with distinct white feather-shafts; sides of neck grey-brown with diffuse blackish-brown streaking. Upperparts including rump and tail rufescent-brown to grey-brown with rufescent wash, and marked with bold broad black streaks on mantle and back and (often hidden) silky-white basal half to two-thirds to feathers of mantle and upper back. Greyer birds often have a narrow greyish band, with indistinct darker streaks, across nape. Feathers of rump and uppertail-coverts may be unstreaked or indistinctly streaked, the streaks formed by narrow dark brown centres of feathers; tail brown. Scapulars and upperwing-coverts usually concolorous with mantle and back and with diffuse dark centres to feathers forming indistinct streaks; median covert tips fringed with creamy-white to buff, forming distinct bar across closed wing; greater coverts narrowly fringed or tipped with buff, often forming a second pale bar. Primaries, secondaries and tertials dark brown, with buffy wing-panel formed by narrow buff leading edges to secondaries; inner web of tertials and inner web of secondaries edged with buffish-white, this broadening towards base of feathers; inner underwing-coverts long and silky white, rest grey-brown with paler tips. Outer 4-5 primaries distinctly twisted, with slightly emarginated inner webs. Underparts silky white, boldly streaked blackish-brown on sides of throat, across breast, on flanks and belly sides, and usually tinged with buff or grey-buff (usually greyer on flanks) on sides of breast and on flanks; centre of belly and undertail-coverts unstreaked; underside of tail and underside of flight feathers silvery-grey. Eye dark brown, upper mandible black, lower mandible white to horn; legs and feet olive-green to yellow-green. Bill tip hardly hooked; rictal bristles well developed, up to 13mm.
Adult female Feathers of crown grey with black centres, forming bold, broad streaks; forehead rufous-buff to buffy-white; lores creamy to grey-brown, mixed with brown at rear; ear-coverts grey-brown streaked with white; nape to rump and uppertail-coverts rufescent-brown diffusely streaked blackish (dark feather-centres) and with some hidden silky white in terminal half of mantle feathers. Underparts, wings and tail as male. Bare parts like male.
Juvenile Similar to adult female, but crown rufescent-brown with narrow black streaks; pale area on forehead confined to small patches at sides; lores and ear-coverts rufous-brown with dark streaks, and mantle feathers lack hidden white. Rest of plumage as adult female.

MEASUREMENTS Wing 75-76; tail 55.5-56.5; bill length 18.5-19, width 15-16.5; tarsus 16-18; weight of *capensis* unrecorded, but other subspecies range from 20-31.

34 GREY-HEADED BROADBILL
Smithornis sharpei   Plate 18

Smithornis sharpei Boyd Alexander 1903, Bull. Brit. Orn. Club, 30 Jan, p.34: Mount Ysabel, 4,000 ft, Fernando Po. Col. pl., Ibis, 1903, pl.7.

FIELD IDENTIFICATION Length 15-18. This is an African species that is confined to the Congo basin, where isolated populations occur in both the east and west and on the small island of Bioko. Males usually have a blue-grey crown and ear-coverts with diffuse slaty mottling on the crown, orange sides of forehead, white throat and a prominent broad orange breast-band, divided by white in the centre. The rest of the underparts are more or less white with bold black streaks; the upperparts are rufescent-brown with a few black patches on the back and mantle and hidden silky white feather-bases. Wings are browner than the mantle and back and lack any pale bars or spots. The upper mandible is black, the lower white. Females differ in having less extensive and much duller orange on breast. Some birds, in particular females, may have blue-

grey crowns, but most females have a dark grey crown. Immatures have brown crowns, sometimes streaked with orange, and dark brown breast-bands mixed with orange-buff and overlaid with bold dark streaking. Immatures also differ from adults in having two narrow orange or orange-buff wing-bars. The display-flight and accompanying loud mechanical rattle is a good guide to locating the species.
Similar species Distinguished from African Broadbill by rufous breast, and from Rufous-sided Broadbill by grey head with buff lores, lack of wing-bars and larger size. Grey-headed Broadbill has a lower-pitched trill than either African or Grey-headed Broadbill. Immatures differ from similar-aged African Broadbills in lacking greyish tones and black streaks to the upperparts, and in having a more or less dark breast: the underparts of immature African Broadbill are silky white but lack the dark breast-band. Immatures are very similar to immature Rufous-sided Broadbills. However, the latter have dull orange across the breast, narrower, less prominent breast-streaks, and brighter, more rufous upperparts, and usually have the outermost median covert-spot white: all pale wing markings in immature Grey-headed Broadbill are buff or rufous-buff.

VOICE The call is a quiet, high-pitched short whistle *whee...whee...*, or *huiiii*. During display-flights males emit a loud, rapid, mechanical trill that is very similar to that of Rufous-sided Broadbill but is slightly deeper and perhaps slower (C. Bowden *in litt.*). This trill, lasting c.0.5-0.6s, starts abruptly, slows and then accelerates almost imperceptively before ending abruptly (Keith *et al.* 1992). Both sexes probably perform this display.

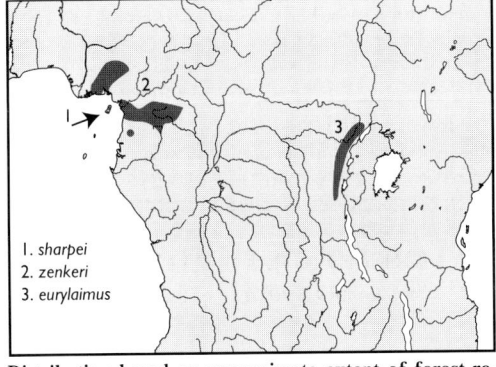

1. *sharpei*
2. *zenkeri*
3. *eurylaimus*

Distribution based on approximate extent of forest remaining in areas where occurrence is confirmed. Probably more widespread than indicated.

DISTRIBUTION The Grey-headed Broadbill is virtually confined to the Congo basin, where it apparently has a disjunct distribution. It has been found only in the extreme east and extreme west, with two subspecies in central-west Africa and another some 1,300km to the east, in eastern Zaïre. Chapin (1953) noted that there was little likelihood of it occurring in the forests of central Zaïre or the Mayombe.

In central-west Africa it occurs in northern Gabon, southern Cameroon, Equatorial Guinea and in eastern Zaïre, with a population on the small (2,017km^2) island of Bioko (formerly Fernando Po), where it has been found on Basilé Peak and in the southern highlands (Pérez del Val *et al.* 1994). Its distribution does not appear to be restricted by altitude, having been recorded at high altitudes as well as in lowlands. In Cameroon, it has been found in montane forests, including those of Mt Kupe, at Kounden (Louette 1981), the southern slopes of Mt Cameroon, Mt Nlonako, the Rumpi Hills and the Bamenda Highlands (Stuart 1986), and in the Bakossi Mountains between Mt Kupe and Korup (Anon. 1994, C. Bowden *in litt.*), and also occurs in the coastal lowlands east to Yokadouma (Louette 1981). From here the distribution extends southwards to the Woleu-N'Tem region of northern Gabon and to Equatorial Guinea, where it is known from Mount Alén (c.1°30'N, 10°30'E) (J. Pérez del Val *in litt.*). In eastern Zaïre, Grey-headed Broadbill has been found in the region between Bondo Mabe, Ituri, and Kanyaa, Itombwe, and also near the Semliki Valley, Ruwenzori (Chapin 1953).

GEOGRAPHICAL VARIATION Three subspecies are recognised, differing primarily in subtle differences in the colour of the crown and pattern of streaks on the underparts:
S. s. sharpei (Confined to the island of Bioko) Crown blue-grey with diffuse streaks; underparts marked with broad orange breast-band and blackish streaking.
S. s. zenkeri (Cameroon and N Gabon) Crown and ear-coverts more sooty than nominate *sharpei* (although some individuals, especially females, may have blue-grey crowns); underparts typically with denser, bolder black streaking.
S. s. eurylaemus (E Zaïre) Very similar to nominate, but smaller (wing 75-79, bill 21-22). Male differs from *sharpei* in more uniform, less mottled, clearer grey crown; females are reported to differ in having sides of breast darker, more olive-brown.

HABITAT Little data have been recorded on the habitat of this species, although it appears to be a rainforest bird that has little tolerance to disturbance. Chapin (1953) noted that, in Zaïre, it is a bird of the undergrowth.

On Bioko, the nominate subspecies has been found in the belt of wet montane rainforest that occurs above 800m. Here it has been recorded in both primary and secondary forest (Pérez del Val *et al.* 1994). In the vicinity of Mt Cameroon and Mt Kupe, Grey-headed Broadbills appear to favour closed-canopy primary and relatively undisturbed secondary forest with relatively open understorey, on both steep slopes and in flatter areas (C. Bowden *in litt.*). Whilst the species survives in old logged forests it may not be able to tolerate seriously degraded areas (M. Andrews *in litt.*).

In Zaïre, *eurylaemus* is primarily found in forests at c.1,000-1,500m (Prigogine 1971), although there are specimens from lower altitudes, such as one from the Semliki Valley, Ruwenzori, from c.915m (BMNH). The nominate subspecies, on Bioko island, some 32km from the mainland in the Gulf of Guinea, has been found only at 800-1,800m (Pérez del Val *et al.* 1994, J. Pérez del Val *in litt.*), but in Cameroon *zenkeri* is found in lowland forest (at c.100m) as well as on mountainsides, where recorded to at least 1,400m at Kounden (Chapin 1953, Louette 1981). On Mt Kupe, the altitudinal range is 850-1,950m (Bowden in press). Stuart (1986) found Grey-headed Broadbill at 500-800m on the southern slopes of Mt Cameroon, at 1,050-1,600m on Mt Nlonako, 1,100-1,300m in the Rumpi Hills, and also found it in the Bamenda Highlands.

STATUS This status of this species is poorly known, although it has been described as frequent to common in certain parts of its range which are relatively well studied, such as in Cameroon. On Mt Kupe, for example, it is not uncommon, and is reported to be evenly distributed throughout the forest (Bowden in press). At several sites in Cameroon visited by Stuart (1986) at least 1-3 observations of the species were made per day, perhaps indicating that it was not uncommon. Its status in other parts of its range is less clear. There is only one confirmed record from Gabon (Makokou) and it is only known from one site in mainland Equatorial Guinea (J. Pérez del Val *in litt.*). It is apparently not common on Fernando Po, none being caught during a prolonged mist-netting programme in montane forests (J. Pérez del Val *in litt.*).

MOVEMENTS None recorded.

FOOD Grey-headed Broadbills feed on a a variety of invertebrates. In Cameroon, they have been seen catching flying insects such as grasshoppers, cicadas and a butterfly, and a female with a bill full of caterpillars has been observed during the breeding season (M. Andrews *in litt.*). Basilio (1963) reported that a specimen from Bioko had eaten orthopterans and beetles with a metallic shine.

HABITS This is the least known of the African genus *Smithornis*. Individuals or pairs usually inhabit the lower and middle storeys of forest, where they are generally quiet, sitting still for long periods. They occasionally join mixed bird flocks (Prigogine 1971). The display-flight, which is a tight circle accompanied by a loud mechanical trill (see under Voice), is made mostly in the evenings. Bates (1905) noted that these display-flights were made around small openings of the undergrowth, although other observations (see below) suggest that they are made in a variety of situations. On Mt Kupe, Cameroon, the display-flight has been heard in most months but more frequently at the end of the wet season, from September to November (C. Bowden *in litt.*).

On Mt Kupe, Grey-headed Broadbills have been observed at a gathering resembling a lek, with at least four birds in attendance: two males, a female and an unsexed individual (M. Andrews *in litt.*). Two males were observed displaying from a low horizontal branch over a gradual incline in rather open primary forest, whilst the female looked on from a slightly higher perch. The males alternated in flying out in a tight circle (making the mechanical trill noise), returning to the branch in quick succession and showing the white lower back continuously. Even whilst perched the males swivelled with puffed-out throats with their wings held low to accentuate the slightly raised white feathers of the lower back. The two males were also observed jumping up and down, occasionally doing this together in an alternate fashion, in a manner described as being somewhat reminiscent of the displays of Neotropical manakins (Piprinae). This behaviour was observed over a period of some 5-10 minutes.

BREEDING In Cameroon, the breeding season is during the dry season, from November to the end of April. Brood patches have been found on birds trapped on Mt Kupe in November (Bowden in press) and eggs found in Cameroon between 2 January and February (Bates 1905, Keith *et al.* 1992). A nest was found in the Bakossi Mts on 17 November (C. Bowden *in litt.*), whilst a nest found on Mt Kupe in late March contained two nestlings, and another nest was observed in late April (M. Andrews *in litt.*). Family parties have been trapped on Mt Kupe in January and March, and moulting birds from May to July (Bowden in press), whilst immatures have been collected on Mt Cameroon on 12 May and on the River Dja on 24 February (BMNH). In Zaïre, a nest containing eggs has been found in April (Keith *et al.* 1992). There are few breeding records from eastern Zaïre: Prigogine reported that eggs were laid from February to April and in September, and a well-grown immature has been collected in the Semliki Valley on 11 August (BMNH). Prigogine (1971) collected another in eastern Zaïre on 28 February. The only breeding record from Bioko is of a male with enlarged gonads on 17 December (Basilio 1963, J. Pérez del Val *in litt.*)

The nest is suspended from a twig so that it hangs 1-16m above the forest floor: one nest found in Cameroon was hanging over a stream. A nest found by C. Bowden (*in litt.*) at 680m altitude in the Bakossi Mts was c.16m up a 32m high tree, but nests found by Prigogine (1971) and Bates (Bates 1905, Chapin 1953) were only 1-2m above the forest floor. The nest is typical of the genus, and resembles a huge sunbird nest, with bits of moss and lichen hanging from the bottom (C. Bowden *in litt.*). Bates (1905) described a nest that he found hanging from a bush some 1.5m above the ground as a pocket-shaped mass of moss, whilst a nest found by Prigogine (1971) was made principally of green moss with some dried leaves and twigs, with dried leaves lining the interior. Approximate dimensions are 22cm high and 9-10cm broad, with an entrance hole of 6 x 3.5cm located on the upper side (Prigogine 1971). One nest was described as having a porch over the entrance (Keith *et al.* 1992, M. Andrews *in litt.*).

Nest materials apparently vary: one nest was built entirely of fine black fungal fibres (probably *Marasmius* sp.) and dried leaves, others have been made of fresh green moss lined with dry leaves, fibres and stems (Bates and Ogilvie-Grant 1911, Chapin 1953, Keith *et al.* 1992) or of mosses, old fern strands and material that appeared to be lichens (M. Andrews *in litt.*).

Normal clutch-size is reported to be two, though occasional nests with one egg have been found. The eggs are long, pointed, pure white ovals, 22-25 x 16-17.5mm (Bates 1905, Bates and Ogilvie-Grant 1911, Chapin 1953, Prigogine 1971).

DESCRIPTION *S. s. sharpei*
Adult male Crown to nape dark slate-grey to blue-grey, mottled with darker sooty centres to feathers on crown. Sides of lower forehead and lores orange-rufous mixed with black. Ear-coverts slate-grey with sparse, indistinct, long narrow buffy-white streaks. Eye large, bordered with very narrow whitish eye-ring. Mantle, back and rump rufous-brown, vaguely tinged with olive tones; uppertail-coverts brighter rufous. Tail dull dark rufescent-brown, being brighter along edges of tail feathers. Bases of feathers of mantle pure silky white, and lower mantle feathers with white bases and black subterminal areas (these white and black areas usually hidden). Upperwing-coverts concolorous with mantle and back or marginally duller, with brighter, rufescent-orange tips and leading margin to outer web of median coverts and narrow brighter rufescent leading edge to greater coverts. These pale edges, however, do not form a distinct wing-bar. Flight feathers brown, the secondaries narrowly edged with rufous-buff to rufous-orange. Hidden outer margins to inner webs of secondaries and inner primaries white to grey-white towards base. Outermost four primaries distinctly twisted and emarginated on inner web. Underwing-coverts pale or-

ange, mixed with silky white in axillaries. Chin white; throat white bordered with dark orange, this colour extending as a broad band across the breast that is usually broken in the centre by white. White in sides of throat and in centre of breast streaked with bold blackish-brown. Outer flanks duller dark orange mixed with diffuse dark brown streaks; centre of belly and undertail-coverts white, the inner flanks boldly marked with long, broad blackish-brown streaks. Rictal bristles on lores and directly above bill well developed and dense, up to 14mm long. A few rictal bristles also present on chin. Legs olive; basal two-thirds of lower mandible yellow, rest of bill black; eye dark brown.

Adult female Similar to male, differing in having slightly paler, more slaty crown; paler, more buffy-white, lores; and less extensive orange-rufous on breast, this generally being overlaid with diffuse brown streaking, and often slightly paler in colour than that in male. Occasionally the crown may be more blue-grey.

Immature *S. s. zenkeri* Forehead, crown, nape, mantle and rump brown with rufous tinge (and some individuals with diffuse orange streaking on crown and forehead), this tinge increasing in intensity on lower back and most intense on rump; ear-coverts sooty mixed with dull rufous-brown; eye large, surrounded by narrow yellowish-white eye ring; lores sooty, mixed with rufous nearer bill, this rufous sometimes extending as a narrow line along side of crown to above eye. Tail dull grey-brown with narrow rufescent edges to outer feathers. Feathers of wing brown, with buffy-rufous terminal wedges or marks in outer webs of median coverts and innermost greater coverts, and narrow rufescent leading edges to secondaries. Central throat, centre of breast and belly silky white. Sides of throat, breast (except narrow centre) and flanks brown mixed with buffy-rufous and overlaid with bold, but often diffuse, dark brown streaks. Underwing as adult. Bill black with yellowy-orange basal third to lower mandible and narrow yellow stripe along entire length of culmen ridge. Eye brown; legs olive with yellowish tinge.

MEASUREMENTS (nominate, n=5) Wing 80-82 (*zenkeri*: 78-84 in Keith *et al.*, Stuart 1986); tail 54-61 (*zenkeri*: male 54-58, female 50-54 in Keith *et al.*); bill length 24-25, width 17.5-19; tarsus 16-18; weight (*zenkeri*) 34-40.3. Based on the biometrics of six birds from Cameroon, Stuart (1986) noted that birds at higher elevations (Mt Kupe and Mt Nlonako) were the longest-winged and the heaviest.

35 RUFOUS-SIDED BROADBILL
Smithornis rufolateralis   Plate 18

Smithornis rufolateralis G.R. Gray 1864, Proc. Zool. Soc. London, p.143: Gold Coast (=West Africa).

Alternative name: Red-sided Broadbill

FIELD IDENTIFICATION Length 11.5-13.5. This species, the smallest of the African *Smithornis* broadbills, has a rather scattered distribution in Central and West Africa. In behaviour and appearance it somewhat resembles a wattle-eye *Platysteira* sp., often flycatching in the understorey. The male has a black head and rufescent-brown upperparts marked with extensive black on the mantle and back. White spots in the wing form a broken bar and there is a hidden white mantle-patch. The underparts are silky white with black streaking on the breast and sides and a conspicuous bright orange patch on the sides of the breast. The white mantle-patch shows in flight. Females differ considerably from males, having dark brown head, tinged rufous and indistinctly streaked darker; rufous brown upperparts with no black, or occasionally a few black marks; and duller orange sides of breast. The wing-coverts are a paler, browner colour than those of the male and the underparts washed with greyish or buffish tones, primarily on flanks. Both sexes have a bicoloured bill, with black upper and white to yellowish-white lower mandible. Adults frequently draw attention to themselves through the loud mechanical rattle made during the display-flight. Juveniles have bright rufous upperparts, often including the crown, although this may be darker than the mantle and back; and buff tips to the median coverts form a broken bar, with the outermost spot usually white or cream. The underparts resemble those of adult females, with dark orange on the sides of breast.

Similar species Best distinguished from partly sympatric Grey-headed Broadbill by smaller size, possession of white wing-spots and having orange restricted to the sides of the breast. Males also differ in having a black head. Both sexes differ from African Broadbill in having bright orange sides of breast. The mechanical trill is reported to be higher-pitched than those of congeners. Juveniles can be distinguished from juvenile African Broadbill by the presence of an orange band across the breast from an early age and unstreaked upperparts, but closely resemble young Grey-headed Broadbill. Immatures of the latter, however, have dirty orange-buff rather than dull orange across the breast, and darker, less rufous upperparts, without any white in the wing (juvenile Rufous-sided Broadbill usually has the outermost wing-spot white).

VOICE The call of Rufous-sided Broadbill is a weak, plaintive, high-pitched short whistle that is variously described as *whee...whee...*, *theew theew* or *huiiii*. A more frequently heard noise is the mechanical trill made during display-flights that may be attributed to a toad or tree-frog when first heard. The latter is a very loud staccato *ttttt-rrrrrrrrrrrrr* of around 0.7s duration that starts abruptly, slows and accelerates and then ends abruptly. It is remarkably similar to that of the closely related African Broadbill, but is higher-pitched. During courtship, the *huiiii* call and other rhythmic calls (untranscribed) run together to form a simple song (Brosset and Erard 1986). Most calling and displaying occurs at daybreak and sunset (Chapin 1953, Demey and Fishpool 1994).

DISTRIBUTION The Rufous-sided Broadbill is known from Central and West Africa, where its distribution is apparently rather dispersed. In West Africa, it has been found from Sierra Leone to Angola. Allport *et al.* (1989) found it at Madina, Wemago and Gola West in eastern Sierra Leone; in Liberia and Ivory Coast there is a population in the vicinity of Mt Nimba (Thiollay 1985, Colston and Curry-Lindahl 1986); whilst elsewhere in Ivory Coast it is known from Yapo Forest, Tai Forest, Oumé, Mopri (5°50'N, 4°55'W) and Irobo (5°50'N, 4°35'W) (Thiollay 1985, Demey and Fishpool 1991, 1994, Gartshore *et al.* 1995); in Ghana it has been found in coastal areas and Bia National Park and forest reserves bordering the Tamo River on the border with Ivory Coast (Davidson 1978, Nash 1990); in Nigeria it is known from the delta of the Niger

(Elgood *et al.* 1994). Its range also extends through southern Cameroon to the south-west Central African Republic and southwards through northern and western Gabon to coastal Congo (Keith *et al.* 1992). In Angola, it has been recorded from Cabinda, near the coast to the north of the Congo River (Traylor 1960). It is also reported to occur in Equatorial Guinea and in Togo by Dowsett and Forbes-Watson (1993), although the range in these two countries is unclear. In Liberia, it may also be more widespread than previously recognised since there was (until recently) extensive forest (Sayer *et al.* 1992) but the whole central and eastern parts of Liberia are being logged by foreign companies (Gatter and Gardner 1993). It has also recently been discovered in adjacent parts of Sierra Leone.

In Central Africa, populations are known to occur in Zaïre: between the Congo River and Lake Mai-Ndombe; in the central Congo, at c.1°S 22-23°E; near Kavanga, Kasai-oriental, on the extreme southward edge of the rainforest; and in the east, north to the equator to the edge of the forest-savanna ecotone, in particular in the region of the Lindi, Ituri, Epulu, lower Uele and Bomokandi Rivers to the north of the Congo River. This last population extends to forest in western Uganda in the region of Lake Albert, including Budongo, Bugoma and Bwamba (Semliki) Forests, and Bunyoro. Chapin (1953) gave the northern limits in eastern Zaïre as the Nava River north of Medje (2°25'N, 27°18'E) and Bondo Mabe near Arebi (2°47'N, 29°35'E).

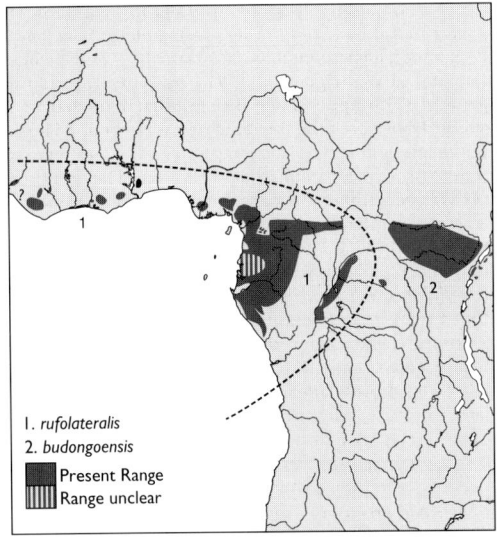

Distribution based on approximate extent of forest remaining in areas where occurrence is confirmed. Probably more widespread than indicated.

GEOGRAPHICAL VARIATION Two subspecies have been described but they are very similar and the species may be best treated as monotypic. Presently recognised subspecies are:

S. r. rufolateralis (West Africa from Liberia and Cameroon to about 20°E in Zaïre)

S. r. budongoensis (Central and north-east Zaïre and western Uganda) The female has a greyer crown than that of female *rufolateralis*, but males show no appreciable differences.

HABITAT This broadbill occurs in primary and secondary forests in the lowlands and mountains to at least 1,300m, and occasionally higher, as on Mt Nyombe, eastern Zaïre, where it occurs to 1,500m (Prigogine 1971). Virtually nothing is known about the habitat preferences of eastern *budongoensis*: the published details outlined below all concern the nominate subspecies.

In Gabon, Rufous-sided Broadbill does not occur in the young secondary growth occupied by its congener, African Broadbill, but it has been found in old regrowth on land that had been under cultivation (Brosset and Erard 1986). It is often found near water, and apparently prefers dense growth, often within tangles of lianes. Rand *et al.* (1959) found it on Massif du Chaillu, Gabon, in primary forest at 730m, whilst in Cameroon Bowden (in press) encountered it at 900-1,050m in good forest on Mt Kupe. In Korup National Park, Cameroon, it occurs in both primary and logged forests (Rodewald *et al.* 1994).

In the western part of its range, Allport *et al.* (1989) recorded Rufous-sided Broadbill in primary and logged forests in Sierra Leone. In Ivory Coast it has been observed in high primary forests in the zone of semi-deciduous forests, where known from Oumé, as well as in evergreen rainforest in Tai National Park (Thiollay 1985). Demey and Fishpool (1994) found it in lowland evergreen forest at Yapo in Ivory Coast. The habitat here has been classified as *Diospyros* spp. and *Mapania* spp. forest, and characterised by an abundance of such species as *Dacryodes klaineana*, *Piptadeniastrum africanum*, *Heritiera utilis*, *Anopyxis klaineana*, and *Scottellia chevalieri* (Demey and Fishpool 1994). In Ghana, the species has been observed in the understorey of primary moist evergreen forest in Bia National Park (Davidson 1978), and it was often heard or seen in forest that had been selectively logged some 17 years earlier in two forest reserves bordering the lower Tano River to the south of Bia (Nash 1990).

STATUS Although reported to be rare and local in Uganda (Jackson 1938), Rufous-sided Broadbill is generally reported to be a frequent to common species, and perhaps has a larger range than current records suggest. Dowsett-Lemaire and Dowsett (1991) considered that it was locally common in forest in the Kouilou basin, coastal Congo. At M'Passa, Gabon, the species is reported to be common, with a population density estimated to be 12 pairs per km^2 (Brosset and Erard 1986). Rodewald *et al.* (1994) found it fairly common in Korup National Park, Cameroon, but on nearby Mt Kupe Bowden (in press) considered this species to be rather rare. To the west, Chapin (1953) found it to be common in Ituri Forest, Zaïre, but scarce in the Semliki Valley.

MOVEMENTS None recorded.

FOOD The Rufous-sided Broadbill is an insectivorous species that often catches flying insects. Small cicadas and small hairless caterpillars are reported to be favourite items in the diet. Other food items that were recorded in the stomachs of 20 specimens examined by Chapin (1953) included coleopterans (including click beetles), orthopterans (crickets, grasshoppers), an earwig, ants, a spider and a millipede. A male has been observed to feed nestlings with a small green cicada (Chapin 1953).

HABITS Despite being confined to one habitat type and having a relatively restricted range, the behaviour and habits of Rufous-sided Broadbill are better known than those of its congeners. It is an inhabitant of thick understorey and mid-level forest, usually at 3-15m above the ground, where it usually perches quietly on horizontal portions of small creepers or thin horizontal boughs (Chapin 1953, Demey and Fishpool 1994). It does, however, occasionally chase insects into the canopy. Occasionally, Rufous-sided Broadbills have been observed in mixed bird flocks, where they are silent (Nash 1990, Thomas 1991).

It is apparently a monogamous species (Keith *et al.* 1992) with well-documented display-flights and courtship behaviour. The latter, performed by the male, may occur between display-flights and is often a prelude to mating. Display-flights are usually performed solitarily, primarily by the male but also by the female. Brosset and Erard (1986) have, nevertheless, observed three individuals displaying in close proximity to each other without any suggestion of aggression, on several occasions. Similarly, Chapin (1953) observed two males that were sitting a few metres apart uttering short whistled sounds whilst a single female sat nearby: the males seemed to be excited since they were wagging their tails up and down and spreading the white patches of the back.

Display-flights are performed all year round and at any time of the day, although most frequently at dawn and dusk. The display is performed from a horizontal branch or vine 5-15m above the ground. One to two abrupt jumps into the air are made, sometimes turning around in mid-jump, before flying in an elliptical flight of around a metre diameter and returning to the same perch. The white bases of the feathers on the back are exposed by puffing out the feathers. The plane of the ellipse varies from horizontal to an inclination of 45° (Brosset and Erard 1986). During the flight a mechanical sound, thought to be caused by the vibration of the stiff, slightly twisted outer primaries, which are unusually narrow towards the base (Chapin 1953), can be heard up to 150m away. The timbre of this sound varies with the plane of flight (Brosset and Erard 1986).

These display-flights serve a territorial function (Brosset and Erard 1986), and hence the display often starts in response to the display-flights of conspecifics. In such instances, successive display-flights are usually made from a succession of many perches as the bird travels up to 1.5km in a period of around 15 minutes. Less often, display-flights are performed consecutively from the same perch, often accompanied by the *huiiii* call between flights, although this call is not always followed by display-flights (Demey and Fishpool 1994). A frequency of one flight per 60s for an hour or so is typical, although if neighbours are also displaying, the frequency may double: one bird has been observed to make an average of two flights per minute for a period of 14 minutes (Keith *et al.* 1992). In Ivory Coast, Demey and Fishpool (1994) found that displays started before dawn, when still dark, continued until around 9am and then started again in earnest in the late afternoon and at dusk. During the day displaying was sporadic.

Courtship is initiated when the male approaches the female, giving *huiiii* calls and with the head turned to one side. The plumage is then puffed out, so that the white mantle and orange breast-patches become conspicuous, as the body is held horizontally with bent legs and the bill held almost vertically. The wings droop and are rhythmically opened and closed, and occasionally flicked up high over the mantle. This is followed by a succession of short, excited jumps up and down in front of the female whilst uttering rhythmic calls that are described as running together to form a simple song (Brosset and Erard 1986).

BREEDING Nests containing eggs of the nominate subspecies have been found in Cameroon and Gabon in every month from September to April. A female collected by G. L. Bates on 23 March along the River Dja, Cameroon, contained an ova, and a juvenile was collected in the same area on 17 October (BMNH), perhaps suggesting that the breeding season starts earlier than September. In Cameroon, displaying males were encountered on Mt Kupe only in February (Bowden in press) and females with eggs in oviduct or brood patches have been netted at Korup in late October and early March (Serle 1950, Rodewald *et al.* 1994). Records from Gabon (Brosset and Erard 1986) suggest a peak in egg-laying activity there in January. Birds have been found with enlarged gonads in Liberia from August to April, and with moderately enlarged testes in September and November (Colston and Curry-Lindahl 1986).

Nests of *budongoensis* containing nests have also been found in most months, between January and October and in December, but nests have only been found between August and January in the forests of the middle Congo River (Chapin 1953, Prigogine 1971). A nest has been found in the Budongo Forest, Uganda, in early June (L. Bennun *in litt.*).

The nest is a slightly ragged globular bag that is hung from a horizontal creeper or branch by fibres draped over c.8-12cm of its length (Brosset and Erard 1986, BMNH). A nest found by Chapin (1953) at Lukolela, Zaïre, on 6 September, was suspended c.2m above the ground, whilst one found by Prigogine (1971) was only 1m off the ground. The appearance is somewhat like that of a sunbird's nest, but the broadbill nest is larger and lacks a porch. Nests are c.15-17cm long, and have an attenuating trail of fibres protruding from the bottom. The nest hole, placed c.3cm from the top, is around 3.5-5cm in diameter.

Nests may be made of coarse dead leaves, plant fibres, mossy roots and dry twigs, thickly woven together with the long black hair-like strands of fungi *Marasmius* sp., and lined with softer dead leaves (Chapin 1953, Brosset and Erard 1986). A nest found by Prigogine (1971) in eastern Zaïre had moss around the entrance hole but the basal half was constructed principally from dried leaves. Dried leaf scraps also lined the nest. One nest in BMNH, collected in West Africa on 17 May, was made almost entirely of very fine, grass-like fibres of palm fronds or bamboo leaves and strips of fibres derived from rotting wood, these perhaps being the main source of the fibres used. The only other material in the outer nest was a single short, 3mm wide, twig. The cavity wall was made of more entire, fresh bamboo or palm leaf pieces, whilst the cavity was lined with fine fibres. In appearance, this nest was remarkably similar to some of African Broadbill.

Clutch-size is usually two, sometimes one. The eggs, of dimensions 22-23 x 15.5-16mm, are pure white and glossy (Bates and Ogilvie-Grant 1911, Bannerman 1936, Prigogine 1971). The female incubates the eggs, but both sexes feed the young in the nest. Males guard young in the nest by clinging to the entrance when the female is absent (Chapin 1953, Brosset and Erard 1986).

DESCRIPTION *S. r. rufolateralis*
Adult male Head black except for white throat and white or buff lores, occasionally also sides of forehead. Mantle to upper rump more or less black with concealed white

feather-bases that irregularly show through, and usually mixed with a few rufous-brown feathers; sides of mantle and scapulars rufous-brown. Uppertail-coverts dark rufous-brown. Tail and flight feathers blackish-brown; tertials and lesser coverts dark brown with rufescent edges, remaining upperwing-coverts blackish or black, with two rows of spots on closed wing formed by triangular white wedges in tips of greater and median covert feathers. A large, rather elongated alula feather is blackish with a rufescent border to the leading edge. Secondaries have narrow rufous leading edges, forming wing-panel. Tail and flight feathers with greyish sheen from below, the flight feathers with narrow whitish borders to inner webs, broadest on secondaries and nearly absent from outer primaries. Outermost primaries (2-5) slightly but distinctly twisted. Innermost underwing-coverts long and silky white, rest of underwing-coverts mottled dark grey and blackish-brown and mixed with white at bend of wing. Underparts white with heavy black streaking on sides of throat, breast, flanks and belly, and sometimes on centre of throat, and large orange patch on sides of breast; breast and belly usually washed with creamy or buffy tones; undertail-coverts unstreaked. Primaries 2-5 are unusually narrow towards base and slightly twisted. Eyes dark brown; upper mandible black with a little blue-grey at sides of base, lower white or yellowish-white with pinkish tinge; legs and feet olive-green. Rictal bristles well developed, up to 11mm. Bill tip hooked.

Adult female Forehead more or less orange-brown, lores whitish to creamy-buff but appear darker due to dense rictal bristles, and often bordered above by a narrow black line or streak. Crown dark rufous-brown, diffusely streaked blackish, the streaks formed by dark feather centres; ear-coverts brown, this colour usually mixed with buff or black; Rest of upperparts rather bright rufescent-brown, with hidden silky white bases to mantle and back feathers and, occasionally, some black markings on back and rump. Tail dark brown with rufescent edges. Median coverts dark brown with broad white tips forming line of spots in closed wing; greater coverts dark brown, often edged with rufous-buff; Primaries and secondaries as adult male. Underparts as adult male except that the sides of breast are a duller orange, and there is usually more creamy or buffy tones.

Juvenile Forehead and crown variable, from bright rufous-brown to dark-brown with rufous tinge, with dark feather-centres forming indistinct streaks. Lores orange-buff to buff, often bordered above by black. Ear-coverts rufous-brown, but this area may be mixed with black or white on feather-shafts. Rest of upperparts bright rufous-brown. White bases to mantle and back feathers may be present, and males develop a few black patches at an early age. Wings similar to those of adult female, but the median covert spots are buffy-orange, with the exception of the outermost (leading) spot, which is usually white. Underparts as female, but the orange patches on breast sides are duller and darker.

MEASUREMENTS Wing of male 60-72, of female 60-65; tail 39-45; bill length 19-21, width 14.5-17; tarsus 12.5-15; weight (Liberia) of male 17.8-23.2, of female 10-23.1 (mean 17.7); *budongoensis* 19-29.

CALYPTOMENA

Three, strongly sexually dimorphic species, confined to the rainforests of South-East Asia. Two species are endemic to Borneo. Plumage largely green, marked with black, and in one case blue, markings. The nostrils are covered by a distinctive tuft of feathers which extends almost to the tip of the bill in males. *Calyptomena* broadbills are highly specialised frugivores that feed on a wide variety of fruits and play an important ecological function as seed-dispersal agents. Unlike other broadbills, these species appear to be relatively intolerant of habitat disturbance. They are often quiet and easily overlooked, but small numbers may congregate in the vicinity of fruiting trees. *C. viridis*, which is wide-ranging and perhaps nomadic, apparently performs elaborate displays, suggesting that it may be a polygamous species. Only females have been observed at nests.

36 GREEN BROADBILL
Calyptomena viridis   Plate 19

Calyptomena viridis Raffles, 1822, Trans. Linn. Soc. London, 13, pt.2

Alternative name: Lesser Green Broadbill

FIELD IDENTIFICATION Length 14-17. *Calyptomena* broadbills are very distinctive, being green and robust with relatively short bills and short tails, and having an unusual tuft of feathers on the forehead that virtually hides the bills of males, but is less well pronounced in females. Green Broadbill is the smallest species, with a widespread distribution in southern South-East Asia, Borneo, Sumatra and various offshore islands. Males are iridescent green, with a prominent black ear-patch and broad black bars and patches on the wings. The female is a uniform paler green, with a brighter lime-green eye-ring, and is best identified by rather plain appearance and shape. In flight the plump body, short tail and broad-based but pointed wings are distinctive.

Similar species Green Broadbill is sympatric with Hose's and Whitehead's Broadbills in parts of Sabah and Sarawak. Hose's differs from Green in having blue on the underparts and black spots rather than bars in the wing. Green lacks the black throat characteristic of its larger, longer-tailed congener, Whitehead's Broadbill.

VOICE Green Broadbills make a variety of distinctive, but sometimes quiet, calls. Most notes have a bubbling, liquid quality. The most frequently heard call is a soft, bubbling trill, starting with a quieter note, that increases in tempo and has an upward inflection *toi, toi-oi-oi-oi-oick*. In captivity, Webster (1991) differentiated two variations of this or a similar call, *goik-goik* and *goik-goik-doyik*, with the last syllable faster and higher pitched. These latter calls are believed to be given in alarm, but they may serve as a contact call in some instances. During the deep head bobbing behaviour that has been observed in captivity (see Habits), a series of seven or so faster *goiks* are uttered, each note being accompanied by a head bob. Webster (1991) also describes a call made by males in captivity, *doy-doy-doy-ee-oh*, which is believed to be a territorial courtship song. Females also occasionally utter this call, but it is very quiet and sounds slightly different: *go-hohohohoho* or *coo-whowhowhowho*. Males also make an assortment of wheezes, whines and cackling notes in captivity.

Soft mournful whistles that recall green pigeons *Treron* spp. and a single loud *oik* are most frequently given near fruit sources. Another vocalisation, thought to function as a contact call, is a frog-like bubbling rattle, *oo-turrr*, sometimes given preceding the longer trill (King *et al.* 1975). What is presumably the same call is described by Medway and Wells (1976) as a soft initial note followed by a brief low-pitched rising trill, *u-trrr*.

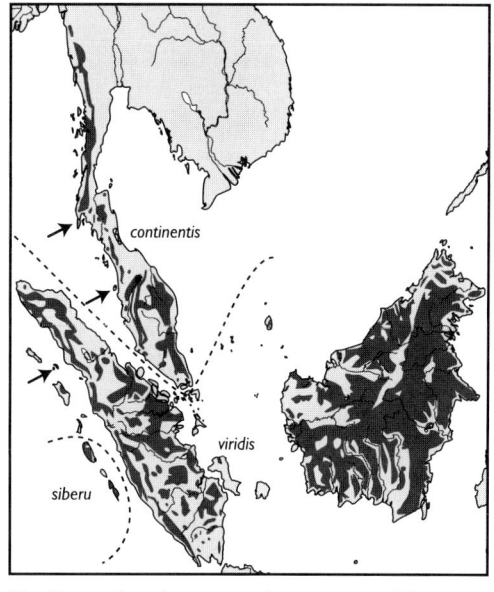

Distribution based on approximate extent of forest remaining within the altitudinal range of the species. Former range more widespread. Former occurrence in Cochinchina unconfirmed.

DISTRIBUTION The Green Broadbill is distributed from Tenasserim, Burma and western Thailand to Peninsular Malaysia and Singapore and the Greater Sunda Islands of Borneo and Sumatra. There are also populations on a number of islands off western Sumatra such as the Batu Islands (Tana Masa), Lingga Archipelago, Nias, the Mentawai Islands (Siberut, South Pagai and North Pagai); on the North Natuna Islands, between Peninsular Malaysia and Borneo (Chasen 1935a); on Ubin Island off Singapore (Bicknill and Chasen 1927); and on Phuket and Penang off the west coast of Peninsular Thailand and northern Peninsular Malaysia.

The most northerly record in Peninsular Thailand is from Hat Sunuk at c.11°50'N (Glydenstolpe 1916) and the species has been found as far north as Amherst in the Gulf of Maraban, extreme north of Tenasserim (Robinson 1927, Riley 1938). Green Broadbill may have formerly occurred in Cochinchina, southern Vietnam, north of the Mekong (Bô Khoa Hoc *et al.* 1992) where Tirant (1879) claimed to have encountered it. However, whilst Delacour (1970) concluded that most of Tirant's records were sound, this particular record has been questioned (J. Eames *in litt.*). Since little, if any, suitable habitat remains in Cochinchina, it is unlikely that the occurrence of Green Broadbill there can ever be verified.

GEOGRAPHICAL VARIATION Three subspecies are recognised, although differences between *viridis* and *continentis* are minor.

C. v. viridis (includes *gloriosa* Deignan 1947) (Borneo, Sumatra, Nias Island, Batu Islands, the Lingga Archipelago and the North Natuna Islands) See description. Wing of male 92.5-100.5; of female 98-106.

C. v. continentis (Tenasserim, Burma, western Thailand and from c.11°50'N in Peninsular Thailand south through Peninsular Malaysia to Singapore) Marginally paler and larger, on average, than nominate. Wing of male 98-106; of female 107-113.

C. v. siberu (Siberut, North Pagai and South Pagai in the Mentawai Islands off north-west Sumatra) Largest and darkest subspecies, the female being darker and duller than female of *continentis*, particularly on the upperparts. Males have a distinctive bright blue wash to plumage, particularly noticeable on throat and belly to undertail-coverts. Females may have some blue feathers on nape. Length 18.5-19; wing of male 106.5-112; of female 107-113.

HABITAT Green Broadbills frequent the understorey and lower levels of rainforest, and occasionally the upper levels and forest edge when feeding. However, they are usually very rare in logged and secondary forests (see also under Status). Studies in Sabah, Borneo, have shown that Green Broadbills also occasionally frequent overgrown *Albizia* plantations (Mitra and Sheldon 1993), and overgrown rubber estates, although such birds may be transients. Thompson (1966) also observed the species in cocoa plantations in eastern Sabah. In Kalimantan, Green Broadbills are common in some areas of heath forest or *kerangas* (Nash and Nash 1987) and have also been found in this habitat, and occasionally in tidal swamp forest in Sarawak (Wells *et al.* 1979, Duckworth and Kelsh 1988). In South Sumatra, Green Broadbill also occurs in swamp forest (Verheugt *et al.* 1993), and in Peninsular Malaysia in selectively logged peat-swamp forest (Kang and Lee 1993). Danielsen and Heegaard (1995) found Green Broadbills in lightly and heavily logged forests in southern Riau Province, Sumatra.

Green Broadbill is found throughout Borneo where mostly recorded below 700m, but occasionally as high as 1,300m; Gore (1968) reported that it was common at 1,220m in the Crocker Range of Sabah, but in Sarawak Wells *et al.* (1979) did not find it above 470m in Gunung Mulu National Park. In Brunei, it is reported to be widespread in riverine and primary forest (Mann 1987). The species also ranges throughout Sumatra, where recorded as high as 1,700m in the Padang Highlands, though more usually below 1,000m (van Marle and Voous 1988); and throughout Peninsular Malaysia to about 800m, south to Singapore. Records of specimens said to have been collected at 1,070m in Thailand (Meyer de Schauensee 1946) need confirmation.

STATUS Smythies (1986) noted that in Tenasserim this was a rare species in the north but relatively common in the south. In Peninsular Thailand, where habitat loss has undoubtedly isolated populations, it is still fairly common where habitat remains (P. Round *in litt.*). In Peninsular Malaysia Green Broadbills are relatively common in primary forests in the lowlands and low hills within their range (pers. obs.). In Borneo, Pearson (1975) reported that it was common in the lowlands of East Kalimantan, and Gore (1968) noted that it was common in the Crocker Range. Chasen and Hoogerwerf (1941) described the status as fairly common in north Sumatra, but there is no information regarding populations on the western Sumatran islands, including those occupied by the distinctive subspecies *siberu*.

In Sarawak, Fogden (1976) estimated that the population density of Green Broadbills was 60 per km^2 in an isolated fragment of primary rainforest near Kuching. However, unlike many other fruit-eating birds they appeared to be much rarer in logged forests. Indeed, during a study of the effects of selective logging on birds at Danum Valley, Sabah, Lambert (1990, 1992) found that most bird species which are important seed dispersers of forest trees survive the effects of logging. However, the Green Broadbill, the only understorey frugivore at the site, declined dramatically. Ringing and transect data suggest that this species was only about a quarter to a tenth as common in forest selectively logged some nine years previously as in nearby primary forest. Wong (1986) also found fewer Green Broadbills in forest that had been selectively logged some 22 years previously when compared to an adjacent and contiguous are of virgin forest in Peninsular Malaysia.

Lambert (1990) pointed out that the decline in numbers of Green Broadbill in logged forests could have serious consequences for the regeneration of understorey plants. Many plant species with large seeds, such as understorey palms, may be largely dependent on Green Broadbills for dispersal, since there are no other avian frugivores with such broad gapes which regularly feed in the understorey. Hence, the decline of Green Broadbills may be one of the most important consequences of selective logging in the Sunda Region with respect to the role of birds in ecosystem recovery.

MOVEMENTS Possibly nomadic, or at least wide-ranging when food resources are scarce. Four individuals were netted shortly after daybreak or just after dark in montane forest at Fraser's Hill between 8 and 24 December 1968, at 1,300m (Wells 1970). Since these birds were unlikely migrants, and this species does not usually occur in montane forest, it seems most likely that these birds were moving in response to food shortages in their normal range, at lower altitude. Fogden (1972) reported that Green Broadbills were rarer at his study site in Sarawak when fruit resources were also rare.

Radiotelemetry was used to document the ranging behaviour of three female and one male Green Broadbills in rainforest at Kuala Lompat, Peninsula Malaysia (Lambert 1989b). Although it was only possible to radiotrack the birds for a short period, during this time Green Broadbills ranged over 2.5-6.0ha per day (these are minimum values since the birds were not followed continuously). A female Green Broadbill covered an estimated area of 13ha in seven days and a male 24ha in six days. The limited data suggested that ranging was to a large extent determined by the availability of fruiting trees.

FOOD Green Broadbills are specialist frugivores, although they occasionally supplement their diet with invertebrates and may feed young with considerable numbers of insects (Fogden 1972). For example, they have been observed catching emerging alate termites whilst in flight on a number of occasions (pers. obs.). The large gape enables broadbills in the genus *Calyptomena* to swallow very large fruit relative to body size. Green Broadbills are reported to feed on a wide variety of fruits, in particular figs and those with large seeds and lipid-rich arils, such as species of *Eugenia* (Büttikofer 1900, Leighton 1982, Wells 1988, Lambert 1989a).

The smallest seeds regurgitated by Green Broadbills caught at Danum Valley, Malaysia, were a mere 1 x 1mm in size, but large seeds reached 20 x 16.7 x 10.3mm and 24.2 x 16.2mm (pers. obs.). The latter were from the bright red arillate fruits of the genus *Knema* or *Horsfieldia*, with ripe fruit dimensions of 27-30.5mm x 18-23mm. Leighton (1982) suggests that the major fruit sources for Green Broadbills in Kalimantan (Indonesian Borneo) were trees in the families Meliaceae and Myristicaceae ('nutmegs') that have fruits of diameter less than 21mm. In Borneo, fruits eaten also include the very hard, oily, purple-black fruits of *Canarium denticulatum* (L. Emmons *in litt.*), and D. Wells (verbally) reports that they apparently eat the large fruits of understorey palms.

However, as noted by Fogden (1970), the lack of a sharp cutting edge to the bill, such as those possessed by sympatric frugivorous barbets *Megalaima* spp., means that *Calyptomena* broadbills are unable easily to manipulate large fruits and therefore tend mostly to eat soft fruits such as figs *Ficus* spp.. Though often large, these fruits are, nevertheless, usually swallowed whole. Lambert (1989a) listed 21 species of strangling or epiphytic figs with fruit eaten by Green Broadbills at a single site in Peninsular Malaysia. These ranged in size from the 5.4 x 5.9mm figs of *Ficus caulocarpa* to the 32.4 x 27.4mm figs of *F. stupenda* (see Table 2).

During a brief radiotracking study (Lambert 1989b), two Green Broadbills that were followed for more than two days were found to spend prolonged periods at or within 30m of fruiting trees, and to visit the same trees for several days. Green Broadbills spent substantial periods of time at fruiting *Ficus*. Apart from brief consecutive visits to an unidentified fruiting tree made by one male during 80 minutes, Green Broadbills were not seen to visit fruit sources other than *Ficus* during the period of data collection.

The activities of the female Green Broadbill that was radiotagged during a seven-day period demonstrated the great importance of figs to this species. At least five different *Ficus* trees were visited during this period: two (*F. sundaica* and *F. bracteata*) on the first day, a different *F. sundaica* on the second day, a *F. obscura* on the four subsequent days, and another *F. obscura* on the last day. In total, 960 minutes were spent in or close to fruiting *Ficus* trees during five days (days 2-6), representing c.50% of total observation time for those days. One of the fruiting *F. obscura* trees was evidently a very important source of fruit since 62.4% of total time observed (during the four days that it was used) were spent at or very near to this tree. Figs were also shown to be important to the male Green

Broadbill that was followed for a six-day period during the same study. Two *Ficus* trees were visited during the period: a fruiting *F. benjamina* on days 2-3, and a fruiting *F. pellucido-punctata* on days 3-6. The total time spent in or in close proximity to fruiting *Ficus* was 587 minutes > representing 30.7% of total observation time for the five days when these trees were visited by the male.

HABITS Cryptic and often motionless for prolonged periods, the Green Broadbill is most frequently observed in pairs and small groups, although these are often dispersed and individuals may appear to be alone. Once found, Green Broadbills often suddenly apparently disappear without trace. Groups occasionally congregate at fruiting trees. Most frequently, Green Broadbills are encountered in the lower and mid-levels of the forest, but they also enter the canopy to feed on fruit. For example, four Green Broadbills have been observed feeding on *Eugenia* fruits at 25-30m in the upper canopy of lowland forest in Malaysia, and others have been seen, albeit very briefly, in the canopy of large fig trees at c.40m above the ground (pers. obs.). Ripe fruit are almost invariably plucked from the tree during short flights, but one observer has seen Green Broadbills picking up fallen fruit from below a tree in which about ten Green Broadbills were feeding (A. Owen *in litt.*). This latter behaviour appears to be very unusual: coming to the ground to bathe or drink at small puddles, ponds or streams in forest is, however, regularly observed (pers. obs.).

Male Green Broadbills have a very realistic wounded-bird display, usually within a few metres of the ground and presumably near the nest. During this display, the bird may appear to be entangled in a spider's web or glued to its perch, whilst continuously screaming with the bill held wide open and vigorously flapping its wings. When approached, birds performing this display fly a short distance and continue the display on another perch (Hopwood 1919, L. Emmons *in litt.*, pers. obs.).

At Kuala Lompat, the roost site used by a radiotagged Green Broadbill on two consecutive days was within a swampy area of forest, where the bird appeared to be in the subcanopy of a small tree c.10m above the ground. At least one other Green Broadbill was at the roost site just before dusk. A second female Green Broadbill also roosted within the swampy area on three successive nights. It was not established that the roost site was shared, but the detached transmitter of another bird was found within 10m of the site (Lambert 1989b).

In Thailand, A. Greensmith (*in litt.*) observed two males performing what he interpreted as a display ritual that lasted around five minutes. The birds were perched c.1-2.5m apart on the same small horizontal branches within 0.5-2.5m of the ground, and gave a continuous low *quit-quit-quit* call. As they called, their heads were flicked up and down quickly and the wings were flapped rapidly in a half-open position. The males changed position during the display, sometimes facing each other, and at other times perched back-to-back.

Green Broadbills in captivity have provided some fascinating observations of behaviours poorly documented in the wild (Webster 1991), including observations similar to those observed by A. Greensmith (above). After a female had built a nest, which took some five days to complete, she then virtually ignored the nest for the next two days. During this period the male was observed to display vigorously to the female by ruffling the head feathers and performing a deep head-bobbing behaviour. The pair would sit crosswise on a branch and make head-bobbing movements under the level of the branch. The pair sat next to each other so that they almost touched and ruffled their back feathers. After three days of this displaying, the male was observed to feed the female after regurgitation, and then proceeded to copulate: three days later the first egg was laid.

The deep head bobbing behaviour, in which the head is bobbed to below the level of the branch on which the bird is sitting, has been observed in more than one zoo collection. At San Antonio Zoo, Texas, this behaviour appears to be performed from set perches by the male of a pair. When approaching these perches, and occasionally when approaching the female, males may adopt an unusual, slow butterfly-like flight in which the body is held in an upright position (Webster 1991).

Another male display observed in captivity involves raising the characteristic tuft of feathers that covers the bill and exposing the tiny patches of bright yellow and black feathers just in front of the eyes. This is performed whilst the bird adopts a very low, hunched, neckless posture, and is accompanied with repeated wing-flashing (Webster 1991). This display is usually immediately preceded by the bird ricocheting around the exhibit in a very regular pattern. Accompanying the wing-flashes is a gaping display in which the male rapidly opens the bill very wide, exposing the orange-pink interior. This is performed most frequently when in close proximity to the female. In one collection the female was observed to respond to gaping by the male by crouching low on her perch with the tail horizontal, the body feathers fluffed and wings partly open and shivering slightly. The male responded by fluttering onto her back and mating for 3-4 seconds with fluttering wings before hopping off and perching nearby (Holyoak 1970).

Courtship feeding by Green Broadbills has also been recorded in at least two collections (Webster 1991). This usually follows the ricocheting flight of the male around the exhibit. He then regurgitates a berry which is offered to the female. Up to four berries have been observed to be offered by the male in succession, and up to three accepted by the female. Sometimes the female apparently solicits such offerings by sitting with the body plumage slightly fluffed-up and vibrating all over, whilst simultaneously making slight head bobs. The male usually immediately responds by offering food, which is invariably accepted.

The most astonishing display performed by male Green Broadbill in captivity, documented by Webster (1991), is usually performed after much ricocheting around the aviary. During this display, the male started on a perch c.0.5m off the ground. After ascertaining that the female was on her 'normal' perch, about a metre above the male and to one side, the male wing-flashed briefly and then launched himself vertically to a 30cm wide (fake) tree limb on the ceiling of the aviary and spun there with his bill agape, occasionally touching the limb with his beak. After a few of these 'beak pirouettes', lasting 5-10 seconds, the male would fly to another perch in the display. Although the female watches the male's acrobatics during this display, she has never been observed to show any interest. Webster (1991) notes that this spinning display is seasonal, being observed only from December to July.

Green Broadbills are known to be long-lived. Ringing records indicate that this species can live at least five years in the wild (Wells 1990b), whilst in captivity one bird reached an age of at least 19 years (Webster 1991).

BREEDING

Seasonality Evidence suggests that populations of Green Broadbill in mainland South-East Asia breed after the heavy rains associated with the early part of the north-east monsoon. Nests with eggs have been discovered in Peninsular Thailand between 28 April and 14 June (P. Round *in litt.*), and Peninsular Malaysia between 26 May and 16 August (UM). Two nestlings have been found in nests on 5 July, 16 and 31 August, and 27 July, whilst a recent fledgling has been observed as early as 13 May and two juveniles were observed being fed by adults in Malaysia on 9 September (M. Chong *in litt.*). In Peninsular Thailand, C. Robson and R. Lansdown observed recently fledged birds at Ban Bang Tieo between 25 March and 7 July and on Phuket Island on 11 June (P.Round *in litt.*). In Tenasserim, breeding may begin earlier: Smith (1943) found nests, both containing three eggs, on 28 February and, at Tavoy, on 3 March; Hopwood (1919) found three nests in Tenasserim on 7 March (one under construction, one empty and one with a single nestling); Darling found two nests containing eggs in Tenasserim, on 3 and 10 April. Two weeks later he found another nest containing three eggs. Baker (1934) gave March-April as the usual breeding season in Tenasserim.

On Borneo, Green Broadbills apparently have an extended breeding season that may last more than six months, corresponding with the driest period of the year. Fogden (1965) found a nest with three eggs at Serian in Sarawak on 31 January, whilst Wells *et al.* (1979) saw a juvenile and caught adults with brood patches in Gunung Mulu National Park in April or early May. Sharpe (1877a) noted that specimens from Borneo collected during January were in breeding condition, whilst in Sabah males with enlarged testes have been collected in mid-February, early May and late June, though females collected at the same time showed no signs of breeding (Sheldon in prep.). A newly vacated nest was found in Central Kalimantan at the end of June (Nash and Nash 1987).

The only breeding record from Sumatra is of two juveniles netted together in southern Riau Province on 18 September (M. Heegaard *in litt.*). A juvenile collected on Nias on 30 July (Salvadori 1887) is the only breeding record from the west Sumatran islands: there are no breeding records for the subspecies *siberu*.

Nests In contrast to most other genera of broadbills in Asia, but like *Smithornis* broadbills of Africa and *Serilophus* broadbills of Asia, *Calyptomena* nests are built across branches rather than being suspended from their tip. The nest is neat, laterally compressed and gourd-shaped, and is usually slung below a thin twig 1-2m off the ground on the edge of a small tree or sapling. One nest was suspended over water (Robinson 1915). Typically, the nest is c.23cm in length and 10cm in diameter, with a tail up to 65cm long, although often much shorter. There is an elongated side entrance situated in the upper half which is some 5-6.4cm wide and 7.5-8.9cm high. The chamber of a nest built in captivity was 14-16.5cm high and diameter 7.5-8.9cm (Webster 1991).

The nest is constructed from very closely and compactly woven coarse fibres, in particular those of palms or strips of palm leaf, bamboo leaves, sometimes rootlets, and usually includes a few dead leaves and other plant materials. However, the nest is not usually camouflaged with mosses, green leaves, spiders' webs etc. that are so characteristic of the nests of some sympatric broadbill species, or lined with green leaves. Nests vary considerably in the materials used to build them: some are reportedly made almost entirely of fine grass (though these are perhaps more likely strips of palm or bamboo leaves that appear like grasses without proper examination), others contain fibres, fine hair-like black roots and, apparently, may have moss, sometimes still green, incorporated into the outer structure (Hume 1880, Hopwood 1919, Baker 1934, Smith 1943). The nest chamber may be lined with fine dried palm or bamboo leaves, grass-like fibres or hair-like black roots. One nest was reported to have been built of 'lallang flower stalks', *lalang* being the Malay name for long, coarse grasses.

In captivity, the female of a pair was alone responsible for nest construction. She was observed to twirl plant fibres in her beak, two or three at a time, and to drape and fasten these to a branch and onto each other whilst hanging head-over-heels (Webster 1991). The nest cavity is presumably rather small in relation to the size of the adult, since the female sits with her head clearly protruding from the side of the nest during incubation (pers. obs.).

Nest of Green Broadbill

Eggs Clutch-size is 1-3: of thirteen nests from Thailand and Peninsular Malaysia, only one contained one egg (Wells 1984), nine contained two eggs, and four, two in Thailand and two in Tenasserim, contained three eggs (UM, P. Round *in litt.*). Adults have also been observed with three small young in Thailand (P. Round *in litt.*). Eggs from the wild in Thailand were elongated ovals, rather pointed towards the smaller end, and measured c.21.2-21.7 x 31-31.5mm (Robinson 1915), whilst 11 eggs examined by Baker (1934), including at least six from Tenasserim, averaged 29.7 x 20.7mm, with a range of 28.4-30.6 x 19.7-22.0mm. Two eggs laid in captivity were about 28.9 x 19.7mm (Webster 1991). The eggs are spotless and somewhat glossy. The colour has been variously described as white, cream and creamy-yellow.

In captivity, copulation was observed after three days of displays by a male (see Habits). No further copulation was observed during the next three days, but the female was then found on the nest, incubating a single egg. Three days later another egg was found in the nest. The eggs were removed after the female failed to incubate them properly and one hatched after a period in an incubator but the nestling later died. The female laid two more eggs, with a two-day interval between them, some three weeks after the original eggs had been removed. After 17 days of incubation one chick hatched, but this and the other egg disappeared the following day (Webster 1991). In all documented cases, both in the wild and captivity, the female was the only adult observed to incubate the eggs. Two eggs that were present in a nest on 16 August hatched, on c.27 August, and the young fledged when 22 to 23 days old (C. Francis *in litt.*).

Observations of colour-ringed birds in Sarawak have shown that replacement clutches are laid if the nest is predated (Fogden 1972). During this study, a fledging success of 0.5 young per pair was recorded, based on observations at six nests.

DESCRIPTION *C. v. viridis*
Adult male Upperparts bright green, palest and brightest on head and rump. Feathers of forehead and lores basally black, elongated and stiff, forming distinctive tuft that obscures all but the tip of the bill. The black bases may be shown during display, but are otherwise not usually seen in the field. Large black spot at rear of ear-coverts and an obscured bright yellow spot just in front and above eye (shown during display). Occasionally some bluish feathers present around edge of lower ear-coverts. Upperwing-coverts basally dark green with black subterminal band and broad bright paler green terminal band, giving the wing a striped black-and-green appearance. Tertials green, sometimes with small black subterminal spot on innermost feather; secondaries black with broad green edge to outer web, broadest on innermost feathers. Innermost primary black, the rest black with bright green leading edge at base, this extending further towards feather tip and becoming broader from outer wing inwards. Some males have very narrow iridescent blue leading edge to underside of outer secondaries and basal inner primaries. Underwing-coverts black. Underparts entirely dark green, often with blue sheen on belly, and palest on undertail-coverts, where feathers are basally white. Uppertail-coverts bright green and very long, extending almost to tip of tail. Feathers of tail green above but black below with faint blue sheen. Iris dark brown to black; bill colour variable, upper mandible black to grey or greenish-grey; lower mandible dull olive to greyish-green, usually with yellowish cutting edges; legs and feet light green to olive-green with paler yellowish soles.
Adult female Entire plumage dull green. Head marked with narrow but distinct brighter lime-green eye-ring. Feathers of lores and sides of forehead form tuft that conceals basal half of bill. Pattern of flight feathers like that of male, but much duller. Underparts green, palest on belly and undertail-coverts. Feathers of sides of neck and breast usually brighter than rest of underparts. Uppertail green, undertail pale blue. Bare parts as male.
Immature male Like adult female, but has bright pale green spot just in front and above eye and develops diffuse blackish spots to centres of median coverts at an early age. Occasionally has some pale blue feathers scattered on head.
Juvenile Green plumage, as adult female, with very pale greenish breast and paler greenish-white belly and undertail-coverts. Flight feathers browner.

MEASUREMENTS Wing of male 92.5-100.5, of female 98-106; tail of male 35-40, of female 42.5-52; bill length 22-28, gape 20.5-23; tarsus of male 19-20.5, of female 19-22; weight 43-72.8 (mean 57).

SPECIES	FRUIT SIZE	FRUIT COLOUR
F. caulocarpa	5.4 x 5.9	Pink/pale yellow
F. virens	7.0 x 7.7	Greenish/ochre
F. obscura *	7.1 x 7.8	Red/deep red
F. heteropleura *	7.6 x 7.7	Red
F. binnendykii	8.5 x 7.8	Pink/purple
F. benjamina	8.9 x 7.6	Pink/purple
F. delosyce	10.1 x 11.0	Pink/yellow-green
F. pisocarpa	11.6 x 12.3	Yellow-orange
F. kerkhovenii	13.3 x 11.9	Orange-red/red
F. pellucido-punctata	17.7 x 11.6	Purple
F. parietalis *	14.8 x 15.4	Red
F. sundaica (type 2a)	14.4 x 16.7	Red/deep red
F. stricta	15.8 x 15.7	Red
F. consociata	14.3 x 17.8	Red/deep red
F. trichocarpa *	15.5 x 17.5	Deep red
F. crassiramea	20.0 x 17.5	Red/deep red
F. sundaica (type 2b)	19.6 x 18.3	Red/deep red
F. bracteata	18.5 x 21.5	Red
F. cucurbitina	30.9 x 20.3	Deep red/black
F. dubia	28.6 x 27.6	Deep red/purple
F. stupenda	32.4 x 27.4	Red/deep red

Table 2. Figs eaten by Green Broadbills in a Malaysian lowland forest, indicating fruit size and colour. Data from Lambert (1987). Species of *Ficus* marked with asterisks have separate male and female fruits; only the male fruits are eaten by birds (Lambert 1992a).

37 HOSE'S BROADBILL
Calyptomena hosii   Plate 19

Calyptomena Hosii Sharpe, Ann. Mag. Nat. Hist. (6), 9, 1892, p.249: Mt Dulit, Borneo.

Alternative name: Magnificent Green Broadbill

FIELD IDENTIFICATION Length 19-21. Endemic to the montane forests of Borneo. Males are unmistakable, being bright iridescent green with prominent round black spots on the wing-coverts, a black spot behind and just in front of the eye, and another across the nape and top of the mantle. The underparts are beautifully marked with indigo-blue on the breast and deep blue on the belly to undertail-coverts. Females are a paler, slightly olive-green above, with black spots confined to the wing-coverts. The head is marked with a prominent lime-green eye-ring, and there is a small black spot in front of the eye. The underparts are lime-green, notably paler and yellower than the upperparts, with sky-blue on the belly to undertail-coverts. Juvenile males are similar to females but have dark feathers on the lower nape. Immature males are more similar to adult males, but have less extensive blue on the underparts and lack most of the black head markings.
Similar species Both Green and Whitehead's Broadbill may be observed in areas where Hose's Broadbill occurs, though only female or immature Green Broadbills are likely to be confused. However, in contrast to these species, Hose's Broadbills have blue on the lower underparts

at all ages. Hose's also differs from Green in having distinct spots, rather than bars, on the wings. Whitehead's differs from Hose's in being considerably larger and having a very distinctive prominent black throat-patch, as well as much more prominent black markings on the upperparts and wing-coverts, and a relatively long tail.

VOICE The call is reported to be a rather beautiful, soft, dove-like cooing (Hose 1898, Fogden 1965).

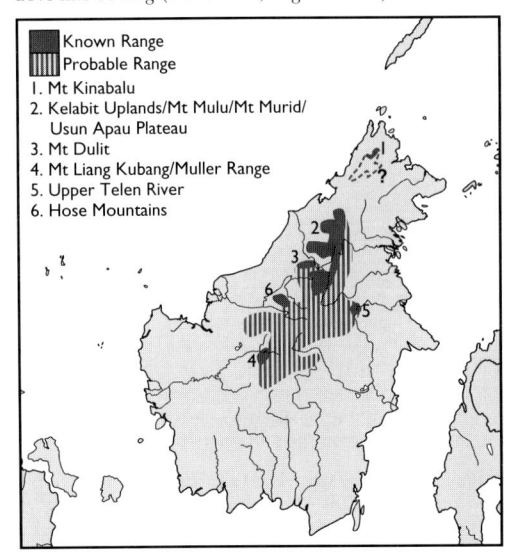

Known Range
Probable Range
1. Mt Kinabalu
2. Kelabit Uplands/Mt Mulu/Mt Murid/ Usun Apau Plateau
3. Mt Dulit
4. Mt Liang Kubang/Muller Range
5. Upper Telen River
6. Hose Mountains

Map shows known distribution and areas where suitable habitat remains. Confirmation of occurrence in these areas is required.

DISTRIBUTION Endemic to Borneo, where patchily distributed and locally common from the lower slopes of Mt Kinabalu, Sabah (where rare), to at least Mt Liang Kubang and the Mandai Valley in the Muller Range, Kalimantan (Büttikofer 1900). In Sabah, Hose's Broadbill has only been observed in south-central Sabah and on the lower slopes of Mt Kinabalu (Sheldon in prep.). Further south, in Sarawak, it is a more widely distributed species. In East Kalimantan, it has been found in the upper Telen River (Smythies 1957 at 1°00'N, 116°58'E (MZB) and in the upper catchment of the Kayan River in the proposed Kayan Mentarang National Park, near the border with Sabah (B. van Balen verbally). The species appears to be absent from east coast forest in north Borneo, even at suitable altitudes.

GEOGRAPHICAL VARIATION None recognised.

HABITAT Hose's Broadbill is an inhabitant of lower and middle levels of submontane forests and occurs locally in lowland rainforest. In Sabah, it has also been found in forest in areas dominated by limestone pinnacles (Sheldon in prep.). On the slopes of tall mountains, it may be found almost at sea level, but it does not occur in the flatter lowlands.

In general, Hose's Broadbill is most common between 610 and 1,220m. In Sabah, it occurs at altitudes of 300-1,150m. Fogden (1965) found it to be widely distributed in the submontane area of the Baram and Tutoh headwaters of Sarawak above 310m. At Tutoh, Fogden found it above 760m but further down the Tutoh River it occured as low as 365m, whilst on the Kubaan River it was common at 305m and above. In Kapit, Sarawak, this species is regularly found as low as 300m in Hill Dipterocarp forest. On Mt Dulit it has been found at 370-915m (Sharpe 1892, Banks 1935), on Mt Mulu at 610-1,220m (Sharpe 1894, Banks 1935, Wells *et al.* 1979) and on the Usan Apau Plateau at c.90-1,070 m (Smythies 1957). Occasionally it is also found at higher elevations in mossy forest: specimens have been collected as high as 1,680m in the Kelabit Uplands.

STATUS Hose's Broadbill appears to be very rare in some parts of its range, such as on Mt Kinabalu, but has been reported to be fairly common in other areas such as the Kelabit Uplands, Mt Murid, Mt Batu Song, Mt Dulit and the Usan Apau Plateau (Büttikofer 1900, Smythies 1981). Fogden (1976) found it to be common at 400-600m in pristine rainforest at the headwaters of the Tutoh River, northern Sarawak. In Sarawak, populations of this species are rarer in forest immediately following logging (Z. Dahaban *in litt.*).

MOVEMENTS None recorded.

FOOD Although primarily a frugivore, Hose's Broadbill also eats some insects, and occasionally, leaf buds. Fruits eaten include figs *Ficus* spp., and soft greyish-yellow berries (SarM). In East Kalimantan, it has been observed to eat small orange figs c.1cm in diameter that were covered in short spiny hairs (B. van Balen verbally).

HABITS This is one of the least-known broadbills, and little is documented regarding its habits and behaviour. It is usually encountered in pairs or small groups, primarily at lower levels of the forest, and occasionally feeding in the same trees as Green Broadbill *C. viridis* (L. Emmons *in litt.*). Like the latter, small groups of Hose's Broadbill may congregate at fruiting fig trees.

Whilst calling, the head and neck are bobbed in a bowing motion (Hose 1898, Fogden 1965). A similar, more jerky, movement is also made when the bird is nervous (Fogden 1965).

BREEDING The little information that exists suggests that breeding occurs during the period typical of other passerines in Borneo, starting in March or April and extending to the middle of the year. Wells *et al.* (1979) caught a male with a brood patch on Mt Mulu, Sarawak, in April or early May. A half-grown juvenile female was collected in the Liang Kubang Range, West Kalimantan, in mid-May (Büttikofer 1900), and an older juvenile male was collected in Sarawak in October (SarM). A male caught in mid-May was moulting flight feathers, suggesting that it had recently completed breeding.

The nest has not been fully described, but Hose (1898) noted that this species builds a nest that hangs from the end of a bough, and that it is similar in appearance to that of Whitehead's Broadbill, but he provided no other details. No nests have been documented since last century.

DESCRIPTION
Adult male Head, mantle, back and upperwing-coverts bright deep green, with slight blue iridescence formed by very narrow blue edges to some feathers of mantle and sides. Green around eye slightly paler, forming narrow indistinct eye-ring. Prominent tuft of stiff, erect feathers from upper lores and forehead protrude over bill, almost obscuring it. Feathers of tuft with obscured black bases;

those on sides of crown above and in front of eye almost entirely black, forming a small black spot (occasionally obscured by tuft). Large prominent black spots also present at rear of ear-coverts, on rear of crown (oval) and at base of nape, where forming a short horizontal fringe rather than a spot. Feathers of greater and median upperwing-coverts marked with prominent round black subterminal spot. Tertials green with hidden black basal half of inner web. Primaries black with green leading edge to basal part of outer web, absent on outermost primary, narrowest and shortest on outer primaries, where confined to base, becoming broader and extending further along primaries from base on inner primaries and almost reaching tip of innermost primaries. This pattern extends to secondaries, with the green reaching the tip of inner secondaries. Underwing-coverts blackish-brown. Feathers of chin and throat green with strong pale blue iridescence and some scattered pale blue-green feathers. Sides of breast and flanks green; centre of breast vivid indigo-blue; centre of belly to undertail-coverts deep blue in colour and diffused on undertail-coverts by broad green bases to feathers (blue of all feathers is formed by broad blue edge to feather-tips, with rest of feather green). Tail feathers mostly green above with broad black terminal band, black below with dark blue sheen to basal two-thirds, most prominent on inner webs. A few hidden uppertail-coverts are basally black with broad, bright violet tips; the others are tipped green. Iris dark brown to black, bill blackish-horn, legs and feet dark olive to green. Bill tip hooked, but not strongly; rictal bristles short and somewhat obscured by the long, stiff feathers of the lores.

Immature male Head and mantle duller than that of adult male, and tinged olive, with more prominent pale green eye-ring; black markings on head absent except for small indistinct black spot at base of rear of ear-coverts. Wings and uppertail-coverts like adult, though lacking any violet in the latter. Compared to adult male, has only a few indigo-blue feathers in centre of breast, and less intense blue on belly and undertail-coverts. Sides of breast and flanks yellowy-green, not dark bright green as in adult male. Like the adult, has a prominent tuft of feathers over the bill.

Adult female Head and mantle to back and rump dark green. Small tuft of feathers from sides of forehead and lores obscure base of bill. Head marked with prominent lime-green eye-ring and very small black spot on side of crown in front and above eye. Upperwing-coverts and wings as adult male. Throat and breast yellowy-green, becoming yellower on flanks. Centre of belly to undertail-coverts pale blue. Uppertail feathers entirely green; undertail greyish with distinct pale blue sheen.

Juvenile Juvenile males are similar in appearance to females, but have distinctive dark feathers on the lower nape. Slightly older males lack the black marks on the nape, in front of the eye and behind the ear-coverts that are typical of full adults, have less extensive blue on the underparts, particularly the breast, and are overall paler in colour than adults.

MEASUREMENTS Wing of male 130-131, of female 125; tail 64-67; bill length of male 26.5-28.0, of female 24; gape of male 22-25, of female 20-21; tarsus of male 25-26, of female 23.5; weight of male 102-115, of female 92.

38 WHITEHEAD'S BROADBILL
Calyptomena whiteheadi   Plate 19

Calyptomena whiteheadi Sharpe, 1887 (1888), Proc. Zool. Soc. London, p.558: Mt Dulit, Borneo.

Alternative name: Black-throated Green Broadbill

FIELD IDENTIFICATION Length 24-27. A very large, vivid green-and-black broadbill of montane forests in Borneo. Adults have a large bold black patch on the centre of the breast and extensive black on the wings and mantle. Males have a prominent black spot behind the eye and on the nape, as well as strikingly marked black-and-green streaked underparts. Females are slightly smaller than males, lack the black head-spots and are a uniform, duller green below, with no black markings. Immatures resemble adults but have fewer black markings, although the black throat-patch and black centres to the upperwing-coverts are present.

Similar species Hose's and Green Broadbill, though notably smaller than Whitehead's Broadbill, are superficially similar, but both sexes of Whitehead's, including young birds, have a large black throat-patch, whilst neither of the smaller congeners has the extensive black markings on the upperparts and wings shown by Whitehead's. Hose's Broadbill has blue on the underparts at all ages.

VOICE In general, Whitehead's Broadbill appears to be more vocal than Hose's or Green Broadbill, although it is still a species that may sit quietly for prolonged periods. Numerous calls made by Whitehead's Broadbill have been described, although several of these descriptions may refer to the same call. The most notable calls are a loud, high-pitched, staccato, woodpecker-like *eek-eek-eek* (Sheldon in prep.), which may in part be used as an alarm note, since a specimen label from a bird shot in Sarawak (SarM) records that, when shot, the bird gave a screeching *e' ek ekek*. A second alarm note that has been recorded is a hissing *ee-ooo*. Other hissing notes apparently form part of the repertoire of calls associated with contact. Two males recorded by J. Scharringa repeatedly gave a short, very harsh, rather woodpecker-like *tzip* note followed immediately by a 2-3s coarse rattle, and in flight, between these calls, a rapid phrase of harsh notes that were very woodpecker-like. A group of 3-4 males sitting together were calling repeatedly, the call being a coarse, harsh, rather abrupt *kerrrrrrr* or *kh-khrrrrrr* (pers. obs.); a similar call heard on Mt Lunjut was described as *teek-waaaaarrr* (B. van Balen *in litt.*). Another contact call is described as a hollow, deep, trogon-like *go-up*, which, if imitated, apparently attracts the species (Sheldon in prep.). Other calls documented on specimen labels are a loud wheezing chatter, and a shrill *saaat*, the latter given by a lone male.

DISTRIBUTION Endemic to Borneo, where patchily distributed from Mt Kinabalu to at least Mt Batu Tibang, on the border of Sarawak and Kalimantan (Pfeffer 1960), roughly in the centre of the mountains (c.1°30'N, 114°30'E). Although the species occurs in Kalimantan, the Indonesian part of Borneo, there are few records, having been recorded from Mt Duk Nan (Pfeffer 1960), and on Mt Lunjut in the upper catchment of the Bahau River, in the proposed Kayan Mentarang National Park on the border with Sabah (Robson 1993, B. van Balen *in litt.*).

In Sabah, the species is known from Mt Kinabalu, Poring, Mt Trus Madi, Rinangisan, Tambuyukon and the Sinsuran Road (Crocker Range) (Sheldon in prep.). In Sarawak, it occurs in the Kelabit Uplands as well as Mt Murid, Mt Mulu and Mt Dulit (Hose 1898, Banks 1937). It is possible that the range of Whitehead's Broadbill extends further south in the central mountains along the border of Sarawak and Kalimatan, and could also occur on some of the other tall mountains in East and Central Kalimantan, although Banks (1952) states that it is certainly absent from the isolated Poi Range. Smythies (1960) noted that it is also apparently absent from Mt Penrissen and Mt Liang Kubung.

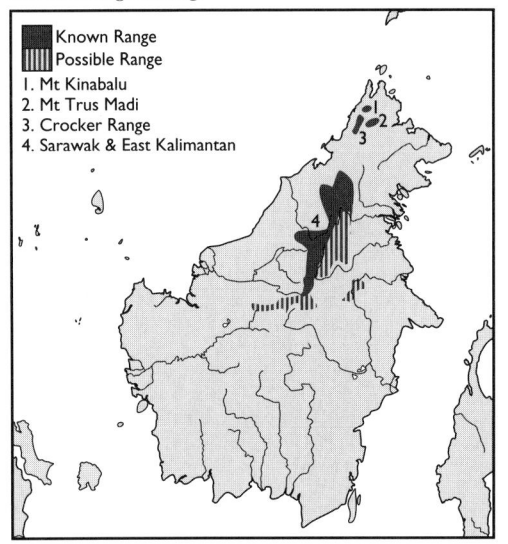

Map shows known distribution and areas where suitable habitat remains. Confirmation of occurrence in these areas is required.

GEOGRAPHICAL VARIATION None known.

HABITAT Whitehead's Broadbill is an inhabitant of montane rainforests, most frequently observed in the middle and upper levels of the forest, or forest edge. It prefers tall growth, and is generally not found in stunted montane forest (Sheldon in prep.). Like congeners, Whitehead's Broadbill is frugivorous, and often observed in proximity to trees bearing large dehiscent fruits.

In general, it occurs at higher altitudes than Hose's and Green Broadbills, and is not usually found below c.900m, although it has been observed as low as 600m on Mt Kinabalu (Sheldon in prep.). Fogden (1970) stated that Whitehead's Broadbill is commonest in montane forest between 1,220m and 1,680m, but an examination of specimen labels suggests that this species is common down to about 915m. Specimens have been collected as high as 1,980m on Mt Murid, Sarawak (Banks 1937), and at 1,850m on Mt Trus Madi in Sabah. Elsewhere in Sarawak, it has been collected at 1,035-1,525m on Mt Dulit, at c.1,220-1,525m on Tamo Abo (Hose 1898, Banks 1937), c.1,050-1,680m in the Kelabit Uplands and at c.1,220m on Mt Batu Tibang (Smythies 1957). In Sabah, it is known at 600-1,700m on Mt Kinabalu, 1,550-1,850m on Mt Trus Madi, 900-1,200m at Tambuyukon and 1,200m in the Crocker Range (Sheldon in prep.). In East Kalimantan, it has been observed at 1,600m on Mt Lunjut and on the border with Sabah (Robson 1993, B. van Balen verbally).

STATUS Although rather patchily distributed, this species is relatively common in suitable habitat. There are still extensive forests in Borneo at the altitudes inhabited by this species.

MOVEMENTS None recorded.

FOOD Like the other two species in the genus *Calyptomena*, Whitehead's Broadbill is a specialised frugivore. Fruits eaten range from small berries to drupes larger than plums, most of which are swallowed whole (Smythies 1981). Fruits as large as 15 x 20mm have been found in the stomachs of Whitehead's Broadbill. An examination of specimen labels found only one reference to insect remains in the stomach contents of specimens of this species, although it is likely that arthropods are eaten occasionally, particularly when feeding young. Sheldon (in prep.) has observed this species feeding on moths, at dawn, around streetlights at Kinabalu National Park.

HABITS Whitehead's Broadbill is often solitary and shy, but occasionally small noisy groups gather close to favoured species of fruiting tree. The behaviour of these loud and obtrusive groups is quite different to that of the other two species of *Calyptomena*, which are generally quiet. Some observers have described the behaviour of Whitehead's Broadbill as reminiscent to that of a woodpecker, partly because some of the calls are woodpecker-like, but also because members of small, active groups occasionally perch on trunks. During the course of moving between trunks and branches, individuals of such groups may repeatedly screech and flap their wings (Sheldon in prep.). Whitehead's Broadbill is also sometimes observed as a member of the often large mixed bird groups that occur in montane forests in Borneo, although whether this is a frequent behaviour is not known.

On Mt Kinabalu, Sabah, at least three male Whitehead's Broadbills were observed performing a display on 22 July (pers. obs.). These males were sitting on small horizontal branches just below the canopy, calling almost continuously. Two of the males were perched close together, and appeared to be competing as they called, whilst at least one other male was perched and calling on another branch close by.

BREEDING Despite being a relatively easy species to find, breeding by Whitehead's Broadbill is very poorly known. Whitehead found a nest containing two nestlings on Mt Kinabalu on 13 March 1888 (BMNH), and discovered another nest a few days later (17 March?) containing two eggs, whilst Ryves found a nest with a single fresh egg on 23 March 1939 at Kiau, Sabah (Gibson-Hill 1952). No other nests have been documented, although Whitehead collected a juvenile just out of a nest on Mt Kinabalu on 13 May (Sharpe 1889). Females with enlarged oviducts have been collected in Sabah on 1 and 2 April, whilst sexually active males were collected on 4 and 13 April (F. Sheldon *in litt.*). Two male specimens collected on Mt Dulit in early August were in primary moult (BMNH). Hence, evidence suggests that breeding occurs principally from around February or March to May or June.

The nest discovered by Whitehead in 1888 was hanging from a slender bough c.15m above the ground. The outside was of fresh green moss, bound over the bough and worked into the sides, whilst the inside was lined with dry bamboo leaves. A degree of camouflage was provided by a long dangling streamer emerging from the bottom of the nest, since many long dripping streamers of moss and lichen were present on nearby branches. Eggs are glossy, creamy-white, pyriform, and c.33-36.8 x 24.5-25mm (Sharpe 1889, Gibson-Hill 1950a, BMNH).

DESCRIPTION
Adult male Plumage entirely iridescent bright green and black. The green of the plumage is glossed with yellow so that it is perhaps best described as lime-green. Head green with large bold black spot behind ear-coverts and on nape, and concealed black bases to feathers of forehead and crown which may show as a black spot above and in front of eye at certain times. Feathers of lores and forehead stiff and erect, forming distinctive tuft that covers most of the bill. Feathers of mantle to rump black with bright green pointed tips, giving a more or less green-and-black streaked pattern that varies from individual to individual, though usually with more diffuse green streaks (and hence more black showing) on the sides of back. This pattern extends onto the upperwing-coverts which are predominantly black with bold green streaks. Green tips to greater coverts confined to outer web. Tertials green with blackish-brown base to inner web confined to base of innermost but extending towards tip of outermost tertial. Secondaries blackish with broad dark green edge to outer web, broadest on innermost feathers, as well as a very narrow bright iridescent blue-green leading edge. Primaries blackish-brown, with short, narrow bluish mark on edge at base to innermost 6-7 feathers. Tail black, with green basal half to feathers of central uppertail and green basal half to outer web of other tail feathers. Undertail black with blue tinge to outer edge of feathers at base. Underwing-coverts black with green tips and edges to feathers near bend of wing. Chin and upper throat bright green, rest of throat black, forming a large black patch that extends onto the upper breast. Rest of breast green, occasionally with some black bases showing through; belly to undertail-coverts as breast but generally black bases more visible, giving streaked appearance. Undertail-coverts often tinged with blue. Iris dark brown; upper mandible black, lower mandible grey-brown or dark horn; feet greenish-olive to greenish-grey.
Adult female Green and black, as male, but green is less iridescent and darker. Head green with paler green eye-ring, partly developed tuft covering basal half of bill, and partially hidden black nape-spot. Mantle to rump largely green, with black feather-bases showing as patches in places, mostly on lower back. Median and lesser wing-coverts largely black, with green wedge at tip, the extent of green greatest on median coverts. Greater wing-coverts black on inner web, green at tips and along leading edge of outer edge. Primary coverts black with green leading edge, except on innermost feather, which is entirely black. Tertials and secondaries as male, but blue tone on leading edge is much duller. Primaries blackish with narrow green border on all but outermost, from basal quarter to half, shortest on outermost feather. Underparts green, with large black lower throat/upper breast-patch, and slight blue tinge to belly. Rump and uppertail-coverts green with concealed black bases. Undertail blackish; uppertail as male. Iris dark brown or bluish; feet grey-green to olive-green; upper mandible black with grey-green cutting edge, lower mandible grey-green.
Immature Similar to adult female, but duller green with little black in plumage, visible black on upperparts being largely confined to centres of greater coverts and a few feathers on lower mantle (base showing through). Throat/breast-patch reduced in size and blackish rather than jet black.

MEASUREMENTS Wing of male 157-160, of female 144-153; tail 86-91; bill 32-36; bill width at base 26.5-29; weight of male 142-171, of female 150-163.

CYMBIRHYNCHUS

A monotypic genus of South-East Asia. Only one subspecies is well differentiated; the others are all very similar and may not be valid. *C. macrorhynchos* is a colourful, sexually dichromatic species that inhabits forests and forest edge in the lowlands and hills, most frequently close to water. It is insectivorous, but occasionally also snatches aquatic prey such as crabs, snails and fish from the water's edge or from shallow pools and streams. Conspicuous, untidy, bulky nests are built over water, where they are protected from predators but vulnerable to change in water levels. Occasionally, nests may also be suspended from telephone lines.

39 BLACK-AND-RED BROADBILL
Cymbirhynchus macrorhynchos Plate 21

Todus macrorhynchos Gmelin, Syst. Nat., 1, 1788, p.446: no locality, but probably Borneo.

FIELD IDENTIFICATION Length 21-24. Black-and-Red Broadbill is an unmistakable species that inhabits forests, often near water, in the Sunda Region and northwards to Indochina and Burma. The striking deep maroon underparts and rump, black breast-band, conspicuous brightly coloured blue-and-yellow bill and black upperparts with a bold long white wing-flash, make this an easy species to identify throughout its range. Juveniles are browner, but have a white wing-patch, black breast-band and patches of maroon on the rump and, usually, on the central belly and undertail-coverts.
Similar species The largely sympatric Banded Broadbill is similar in that the underparts are maroon, but this species has extensive yellow spotting on its wings and upperparts at all ages. The bill of Banded Broadbill is almost entirely blue, whilst that of Black-and-Red Broadbill is bicoloured, with only the upper mandible entirely blue.

VOICE The Black-and-Red Broadbill is often silent, and its calls are generally quieter than those of the other species of insectivorous broadbills in the Indomalayan Region. One call is a rising trill rather similar to that of the Black-and-Yellow Broadbill, but is much slower, softer and briefer. R. Timmins (*in litt.*) described the commonest-heard call in Laos as an accelerating series of notes, *parnk*, reminiscent of the wing-beat of Wreathed Hornbill *Rhyticeros undulatus*. Other calls that have been described include grating cicada-like notes, a monotonous and repetitive *tyook* and, in Peninsular Malaysia, a rasping *wiark*. The alarm call is a rapid series of *pip* notes.

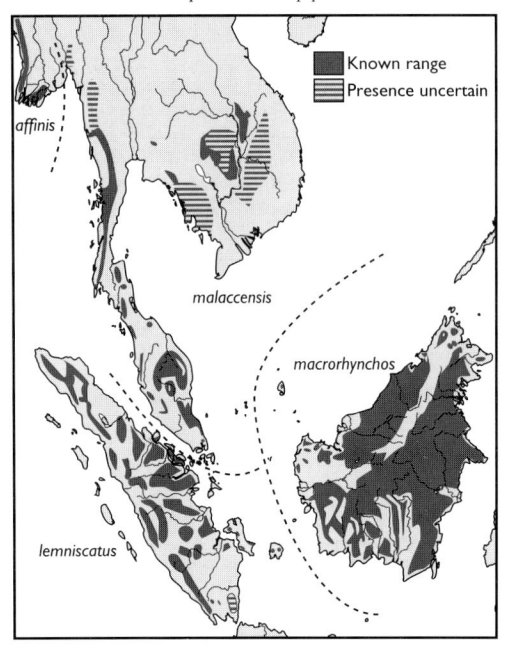

Distribution based on approximate extent of forest remaining within the altitudinal range of the species. Former range more widespread.

DISTRIBUTION The Black-and-Red Broadbill is distributed from Indochina and Burma and through Peninsular Thailand and Malaysia to the Greater Sunda Islands of Sumatra and Borneo. It also occurs on many small offshore islands, including Labuan off north Borneo; Bangka and Belitung off eastern Sumatra; and Penang in the Malacca Straits.

On mainland South-East Asia, one population is distributed throughout Peninsular Malaysia and north through Peninsular Thailand and Tenasserim to the lowlands of much of Thailand and parts of Indochina. It is found in southern Laos, north to at least the Dong Hua Sao Biodiversity Conservation Area (R. Timmins *in litt.*), whilst in Cambodia it is known from forests near Angkor, at Sambor (near Kompong Thom), Siêm Réap and near the Mekong River at Kratie (Delacour 1929, Delacour and Jabouille 1931, Engelbach 1953, Thomas 1964), as well as in the mountains along the border with Cochinchina (Oustalet 1903). In Vietnam, it occurs only in Cochinchina and southern Annam. In Burma, an isolated population (*affinis*) occurs in the Irrawaddy Delta, the Arakan Hills and south Arakan. Elsewhere in Burma, it apparently occurs only in Tenasserim, where it is perhaps confined to the south (Hume and Davison 1878, Smythies 1986).

On Borneo, Black-and-Red Broadbill is well distributed, and has been found throughout the island. Van Marle and Voous (1988) reported that this species was found throughout the mainland of Sumatra, as well as on the islands of Bangka and Belitung. However, its distribution in Sumatra may be rather localised, since there are few recent records and it seems to be absent from some well-watched sites where apparently suitable habitat is present (Holmes in prep.).

GEOGRAPHICAL VARIATION Four subspecies are recognised, although individual variation and the differentiation of these subspecies is poorly understood. The Sumatran subspecies *lemniscatus* has been referred to both *macrorhynchos* and *malaccensis*. Robinson (1915a) believed that variation within subspecies was such that *lemniscatus* and *malaccensis* might best be lumped with nominate *macrorhynchos*, whilst Chasen (1937) thought that birds from east and south Sumatra and Billiton were referable to *malaccensis*. Chasen and Hoogerwerf (1941) treated Sumatran birds as *lemniscatus*, based to a large extent on bill size, but noting that the population was not homogeneous. Mees (1986) discussed the taxonomy of this species in Sumatra in detail. His conclusion was that, in view of the considerable individual variation in the extent of white in the tail, populations on Borneo, Sumatra, Billiton and Bangka all belonged to the nominate subspecies. However, this opinion has not been recognised by subsequent authors, such as van Marle and Voous (1988), who recognised *lemniscatus* as an endemic Sumatran subspecies. Although *lemniscatus* is also treated as a subspecies here, it is a very poorly differentiated form, and should perhaps be included, along with *malaccensis*, in *macrorhynchos*.

Following Medway and Wells (1976) and van Marle and Voous (1988), *tenebrosus* of Lampung Province, southern Sumatra, and *siamensis* of Tenasserim, southern Thailand, northern Peninsular Malaysia, Cambodia and southern Vietnam (both described by Meyer de Schauensee and Ripley 1940), are not recognised here, because there appear to be no consistent differences in plumage or biometrics between these and the subspecies within which they are here included.

C. m. macrorhynchos (Borneo) White on the tail almost obsolete; if present, usually as a trace on innermost pairs of tail feathers. Wing 96-111.

C. m. affinis (W Burma, from S Arakan to Pagoda Point and east to Rangoon) Differs from larger nominate subspecies in having long crimson spots on the innermost secondary and the outer web of the next two, a more conspicuous white wing-spot extending onto both webs, broader white tail-bars (extending to all but the central pair of tail feathers), and narrow black edges to the crimson feathers of the rump. Wing 88-93.

C. m. malaccensis (Tenasserim in Burma, S Thailand, Cambodia, S Laos, S Vietnam and Peninsular Malaysia, including Singapore [formerly], and Penang) Red of plumage marginally paler than that of *macrorhynchos* and belly more or less spotted with orange-yellow. White bar on inner webs of tail feathers usually well defined, although sometimes restricted to outer tail feathers. Slightly smaller than nominate. Wing 94-106.

C. m. lemniscatus (Sumatra. Birds from Bangka and Belitung [Billiton] are probably this subspecies) Intermediate between *malaccensis* and *macrorhynchos*. Differs from *malaccensis* in having orange spotting on belly absent or much reduced, and differs from *macrorhynchos* in usually having more white in the tail, with 2-3 or occasionally the outer four pairs of tail feathers having white. Bill averages longer (29-34) than *malaccensis* (28-31), but wing length similar (102-106).

HABITAT The Black-and-Red Broadbill is a forest species that is rarely found away from water. In most of Thailand, for example, it is restricted to evergreen forest in the immediate vicinity of rivers and streams in forests but, unlike other species of sympatric broadbill, does not exist in areas where only mixed deciduous forests occur (Round 1988). In southern Laos, Thules (1993) reported that Black-and-Red Broadbill inhabits semi-evergreen forest, including seriously degraded areas, along the larger rivers, whilst R. Timmins and J.W. Duckworth (*in litt.*) also found it in mixed dipterocarp forest. In southern Annam and Cochinchina, Vietnam, C. Robson (*in litt.*, Robson *et al.* 1991) only recorded this species in riverine forest.

Elsewhere in its range, such as in Peninsular Malaysia, the Black-and-Red Broadbill also frequents plantations near water, and Baker (1934) noted that it often breeds in mangrove. Verheugt *et al.* (1993) listed both mangrove and swamp forest as habitats of this species in South Sumatra, whilst in Peninsular Malaysia it has been recorded in highly disturbed peat-swamp forest in which logging had left a vegetation of low secondary forest trees and scrub and few, scattered tall trees (Kang and Lee 1993). In Tenasserim it has been found in villages and gardens as well as forested areas (Hume and Davison 1878), whilst in Sumatra Chasen and Hoogerwerf (1941) found it in low secondary vegetation such as small clumps of open forest in pasture land. In Sabah it has been recorded in a diversity of habitats, including rubber plantations, peat-swamp forest, nipa swamp and gardens (Gore 1968, Sheldon in prep.).

Like many South-East Asian forest birds, Black-and-Red Broadbill persists in small numbers even in areas where much of the original habitat has been destroyed, and is sometimes present in areas where relict tall trees remain in otherwise very degraded secondary growth along rivers. In Sumatra, for example, Chasen and Hoogerwerf (1941) found Black-and-Red Broadbills in low secondary vegetation in small clumps of open forest within pastureland.

On the South-East Asian mainland, Black-and-Red Broadbill is primarily a bird of the lowlands and hills. In Peninsular Malaysia it has been found to at least 300m; in Thailand, it was formerly widespread throughout the lowlands to 300m – specimen records from 1,050m (Meyer de Schauensee 1946) are very doubtful – whilst in Cochinchina, southernmost Vietnam, Beaulieu (1932) found it at below 170m, and it has also been found in lowlands (150-300m) of southern Annam in Nam Bai Cat Tien National Park (Robson *et al.* 1991). On Sumatra and Borneo it is frequently found at higher altitudes, and has been recorded to at least 900m on both these islands.

STATUS This species is a fairly common resident in lowlands and hills throughout much of its range, although it is now a rare species in northern parts of its mainland range. In Thailand, where it has been affected by lowland deforestation, although abundant along rivers earlier this century (Robinson 1915) it is now the scarcest species of broadbill. It is only in the peninsula of Thailand that the species is widely present, in part because it shows some ability to survive in deforested country where relict taller trees and moist secondary growth occurs along riverbanks (Round 1988). Further south, in Peninsular Malaysia, this is a relatively common species where habitat exists. In Borneo it is a common species in the lowlands, but very rare in the interior and at higher altitudes (Pfeffer 1960, Gore 1968, Pearson 1975). In Sumatra it was reported to be the commonest species of broadbill (van Marle and Voous 1988), but observations by Holmes (in prep.) suggest that, while widely distributed, the species is now very scarce.

There are few recent data on the status of this species in Indochina. In central Cambodia, and particularly around Angkor, Black-and-Red Broadbill was apparently common (Delacour 1929, Engelbach 1953). Further east, Delacour and Jabouille (1931) reported that it was abundant in the forests of Cochinchina, although little forest remains there today, and Robson *et al.* (1991) found this species to be scarce in Vietnam. In southern Laos it is described as fairly common in the mosaic of degraded forest and taller scrub along rivers (Thules 1993). R. Timmins and J.W. Duckworth (*in litt.*) found it at highest densities in very degraded semi-evergreen forest in southern Laos.

The status of the very distinctive population (*affinis*) from western Burma is unknown.

MOVEMENTS None recorded.

FOOD The Black-and-Red Broadbill is insectivorous, but it has also been recorded eating small riverine animals, including molluscs, freshwater crabs, and on one occasion a small fish (SarM). Invertebrates recorded on specimen labels (SarM) include a longhorn beetle, a shiny green-and-bronze beetle, ants, crickets, large grasshoppers, hemipteran bugs and a caterpillar. Sharpe (1877a) found the gizzards of two young birds to be full of caterpillars, with some beetle remains and, in one, a small *Helix* snail. Robinson (1927) observed this species catching large blue-and-yellow moths fluttering over small streams. Of eight specimens for which stomach contents had been recorded (SarM), three contained fruit or seeds (specific mention being made of green large-stoned berries and small white seeds), suggesting that Black-and-Red Broadbills may regularly consume fruits. A specimen label in BMNH also records a bird as having three seeds and a green leaf in its stomach, whilst small pips were found in a specimen collected by Chasen and Hoogerwerf (1941).

HABITS The Black-and-Red Broadbill is normally found in pairs, but is occasionally observed in small parties (Robinson 1927). It often forages in thick foliage, but is also regularly encountered in more open situations, particularly at the water's edge. It is generally silent, and may be easily overlooked, despite its bright plumage, since it spends considerable periods of time sitting inactively. When hunting, however, it may pursue prey, gleaning from various substrates and even chasing flying insects. In contrast to other broadbills, it occasionally drops to the ground at the edge of water to pick up small riverine animals.

Two pairs that Smythies (1981) disturbed from feeding on the ground flew into nearby trees and greeted each other with loud churring calls and melodious whistles, whilst simultaneously bowing at each other and depressing their tails under their perches.

In Malaysia, a bird ringed as an adult was retrapped 60

months later (McClure 1974), indicating that this species can live at least six years. A bird banded by McClure (1974) in Sarawak was found dead six months later c1.5km distant from where it had been originally caught.

BREEDING The nest of Black-and-Red Broadbill is conspicuous, and numerous nests have been found in Peninsular Malaysia and Borneo.

Seasonality In general, most nest records from Peninsular Malaysia are from the drier months of the year, presumably because the chances of losing nests that hang over water is lower at this time. Building, in which both adults participate, has been recorded in Peninsular Malaysia between 11 February and 28 July, and eggs recorded between 26 March and 19 August (Medway and Wells 1976, M. Chong *in litt.*), although most nests contained eggs during May. Nests with eggs have been found in southeast Thailand from 25 May to 19 June (Riley 1938), and with young in Tenasserim from April to June (Hume 1880). Smith (1943) found a nest of *macrorhynchos* containing four eggs in Burma on 25 February.

Data indicate that the nesting season in the Greater Sunda Islands also spans the driest period of the year. In Sabah, Borneo, nests have been found in every month from December until August, except June, although nests with eggs have been found in Sarawak in June (Fogden 1965, Blaber and Milton 1994, Sheldon in prep.). Nest building has also been observed in June in Kalimantan (Nash and Nash 1987), and a nest was found in Brunei in mid-May (Counsell 1986). In Sumatra, eggs have been collected between March and June (van Marle and Voous 1988), corresponding to a relatively dry time of year.

There are only two records of nesting from Indochina: an adult flushed from a nest in Cochinchina on 19 June (C. Robson *in litt.*) and another occupied nest observed in southern Laos on 28 May (R. Timmins *in litt.*), at the end of the dry season. Immatures have been observed in southern Laos in mid-January (R. Timmins *in litt.*) and young birds collected at Siem Riep and Angkor, Cambodia, on 24-25 July and 3 August (Eames and Ericson in prep.).

Nests The nest of Black-and-Red Broadbill is invariably suspended in a conspicuous position, most frequently over a river, stream or lake, but also occasionally over other waterbodies such as pools in coastal slacks, tidally inundated mangroves, seasonally flooded pools, or man-made drainage ditches. Rather rarely, nests are also constructed over paths or roads, sometimes even suspended from telephone wires, and one was found suspended from a lone bush in an open rice paddy. More frequently, the nest is suspended from the end of a branch of a tree, or bamboo, aerial root or creeper, and hangs 1.5-8m above the surface of water. One nest, built in a rubber tree near to a stream, was in close proximity to a wasp nest (Hume 1880, Baker 1934, Medway and Wells 1976, Blaber and Milton 1994, R. Timmins *in litt.*). In Peninsular Malaysia, nests along the Tahan River are very frequently suspended from lianes or vines climbing on *Dipterocarpus oblongifolius* trees (M. Chong *in litt.*).

A study of the distribution of nests along the Lingga River, Sarawak (Blaber and Milton 1994), provided evidence that Black-and-Red Broadbills here prefer nesting in tidally inundated swamp forest, though nests were also found in tall mangrove trees near the estuary and suspended from *Pandanus* vegetation. M. Chong (*in litt.*) found that nests of Black-and-Red Broadbills along the Tahan River, Peninsular Malaysia, were most commonly encountered over the faster-moving stretches of water.

Occasionally, nests may be built on large poles or sticks that emerge from water, but such nests are very vulnerable and may be swept away in floods, as are those suspended very close to the water surface. Two nests that were built successively by one pair of birds on telephone wires were both destroyed by strong winds (G. Noramly *in litt.*). Observations at these nests suggested that the first may have taken as long as 49 days to complete. Nests are frequently destroyed: of 17 found in Peninsular Malaysia, at least six were destroyed before breeding had been completed (UM nest record cards).

The nest is constructed of a variety of plant material, and usually appears rather ragged and messy. One detailed account from Sabah describes the nest as a pear-shaped structure of interwoven grass, vines, sticks and leaves measuring about 25 x 25 x 65cm, with a nest cavity of 7 x 8 x 10cm deep (UM). The cavity was lined with wide-bladed reed-grass and the bottom with fresh green leaves. The opening to the nest cavity, protected by a roof of grass, was 5cm in diameter and situated 38cm below the top of the nest. 'Streamers' of vines and twigs were dangling below the nest. Another nest was built of creepers and dry bamboo leaves, c.20-30cm deep and 12-13cm wide at the broadest point, with a long untidy dangling tail. Other nests have been built of fine twigs, rootlets and leaves, and green moss is sometimes added to the outside. In one case, a nest in Tenasserim had the upper part composed almost entirely of moss and the lower part primarily of bamboo leaves netted over with moss and plant fibres. The cavity, some 15cm high and 7.5cm in diameter, was lined with fine, stiff grass (Hume 1880).

A nest from Sabah (BMNH) was 30cm deep and 15cm broad at the widest point with a nest hole some 11cm from the top and 4cm in diameter; it was suspended entirely by a 12cm width of very fine black hair-like fibres, probably those of fine rootlets or fungal hyphae, and densely packed and compact, with the black fibres throughout the nest, intermixed with pieces of dried leaf, spiralled pieces of vine, strips of bamboo leaf or palm and short twigs up to 8cm long and 3mm wide. The nest chamber was made of interwoven lengths of broad palm frond. The bottom was adorned with a few pieces of cotton-wool-like material within which were a few small burred seeds.

A nest collected in Indonesia (RMNH) was built primarily of small roots, strips of rope-like bark and fine rootlets or fungal hyphae, with a few scattered dicotyledonous leaves, pieces of moss and the small rubbery leaves and stalk of an epiphytic climbing plant. The inner lining was a neat layer of monocotyledonous leaves (possibly bamboo) with a few entire dicotyledonous leaves that were probably added when green.

During the construction of a nest in Sabah, one adult was observed to remain in the nest whilst its mate flew to and fro from the waters edge, collecting material from the debris along the edge of the river over which the nest was constructed (A. Owen *in litt.*). In some instances, the entrance hole may be partially hidden by a sort of hood built into the nest above the hole (R. Timmins *in litt.*).

Eggs Clutch-size is reported to be 2-3 (Büttikofer 1900, Baker 1926, Delacour and Jabouille 1931, Medway and Wells 1976), but in south-east Thailand and Tenasserim clutch-size sometimes reaches four (Hume 1880, Riley

1938). Eggs are laid at the rate of one per day (Hume 1880). Twenty-four eggs had average dimensions 26.8 x 18.8mm with a range of 25.0-29.3mm x 18.2-20.7mm (Baker 1926). In shape they vary from broad short ovals to long blunt ones. The texture is not fine, and the surface is quite glossless, whilst they are very variable in colour. Baker (1926, 1934) describes three types: the first, and most frequent, has the ground colour pale dull pink, covered all over with dull pale reddish-brown freckles that may form ill-defined blotches in some cases; the second form is white or almost white, with the marks claret or purplish-red; the third type also has the ground colour white, but the markings are less numerous and almost purple-black. In appearance, the eggs of Black-and-Red Broadbill are similar to those of Dusky Broadbill, but are much smaller (Baker 1934).

The male probably shares the task of incubation, since one was shot that was disturbed from a nest (Baker 1934).

DESCRIPTION *C. m. macrorhynchos*
Adult Forehead, crown, mantle and back black with distinctive glossy dark greenish sheen, this colour extending across the upper breast to form a conspicuous band of variable width. Feathers of lores, and below eye, black, lacking green sheen, the black extending around bill and base of bill to chin and upper throat. Rump and uppertail-coverts bright maroon. Upperwing-coverts black with glossy blue-green sheen to feather edges; scapulars largely pure white on outer web, considerably elongated and pointed, these forming a long white line from near the bend of the closed wing to half-way along the folded tips to secondaries. Flight feathers blackish above, greyish-brown below, with concealed white bar formed by white basal spot on inner web of all but outermost two primaries, and broad white basal bar along edge of outer web of the secondaries; this white patch being shortest on the outer primaries, narrowest and longest (reaching three-quarters towards tip) on the innermost secondaries. Tertials black, slightly glossy. Narrow bright orange-yellow line at bend of wing, usually visible in field, formed by orange-yellow outer webs to feathers along leading edge of closed wing. Some individuals have a small, indistinct reddish-maroon patch at the tip of a single lesser or median covert feather. Underwing-coverts blackish to blackish-brown, with some white in feathers of inner part; axillaries white with yellow-orange tinge. Underparts neatly marked: chin and upper throat, and broad breast-band black (the latter glossy); rest of underparts deep maroon. Feathers of tarsus black. Tail feathers blackish, strongly graduated, with or without white patch in terminal part of inner web of outermost (shortest) 1-2 pairs of feathers. The white in the tail may be tinged with buff. Adults of all subspecies very occasionally have extensive patches of orange in the plumage. Although such patches are most consistently on the flanks and belly, some specimens have orange patches on the rump and ear-coverts. Iris iridescent emerald-green; legs and feet bright blue, sometimes tinged violet; upper mandible bright turquoise-blue, lower mandible yellow-orange with blue cutting edge and tip, and bordered by a narrow line of greenish where the two colours meet. Bill tip hooked, but not strongly; rictal bristles well developed, usually with two or more of 20-25mm extending from base of bill at lores, and shorter bristles on chin.
Immatures Older immatures are only distinguished from adults by browner upperwing-coverts, tertials and flight feathers, and by the presence of small white or yellowish-white spots in the tips of the median coverts, these forming two diffuse lines of spots in the closed wing. Slightly older, subadult birds, can only be identified by the presence of the remnants of these white spots. Younger immatures have variable amounts of brown on the throat, chin and ear-coverts, blackish upperparts without extensive greenish gloss (this gloss first appears on the mantle and breast-band) and more conspicuous yellowish spots in the wing-coverts.
Juvenile Upperparts brown with scattered blacker feathers on mantle and back, and reddish-maroon patches on rump and uppertail-coverts. Breast-band well developed at very young age, being black and acquiring general greenish gloss before black of upperparts. Rest of underparts brown, with red feathers gradually appearing, firstly on lower throat and centre of lower breast to belly and undertail-coverts. Juveniles have the bright yellow-orange leading edge to the wing of adults but the rest of the wing is brown, with white patches formed by white in outer webs of scapulars, but these are not elongated or pointed as in the adult. Iris bronze, upper mandible blackish with dirty blue base, lower mandible dirty blue. The inside of the mouth of fledglings is orange.

MEASUREMENTS Wing 100-108; tail 83-94; bill length 30.5-33, bill width 18.5-22; tarsus 23-25.5; weight 50-76.5.

PSARISOMUS

A monotypic genus, with sexes similar, distributed from the Himalayas to Indochina and the Sunda region. A slimmer, longer-tailed species than other broadbills, with green plumage and beautifully marked black and yellow head. It is an inhabitant of tropical and subtropical forests, occurring in lowlands in some parts of its range, but more often frequenting submontane and montane regions. Bill hooked, like other broadbills, but lacks rictal bristles. The nest is an elongated, rather elegant structure, usually suspended from the tip of a rattan tendril or branch. Nests are invariably attended by helpers as well as parents.

40 LONG-TAILED BROADBILL
Psarisomus dalhousiae   Plate 20

Eurylaimus dalhousiae Jameson, Edinburgh New Phil. J., 18, 1835, p.589: northern India, probably near Simla.

FIELD IDENTIFICATION Length 23-26 (nominate subspecies). Long-tailed Broadbill is widely distributed from the eastern Himalayas through South-East Asia to Sumatra and Borneo. Sexes are very similar: where differences are apparent they are not consistent in all individuals. Plumage green with striking black cap, bright yellow throat, face and collar, and long, graduated blue tail (black below). Top of crown blue. In flight, the wings appear blue, and the white patch on the underside of the blue flight feathers is prominent. Immatures have green crowns but pale yellow on the lores and rear of ear-coverts, and greenish-yellow, rather than bright yellow, chin and throat. Some birds from northern Burma and the Himalayas may have bluish rather than green underparts. The wing-beat is rapid, and the wings whir in flight.

Similar species Long-tailed Broadbill is a very distinctive species, and unlikely to be confused with any other broadbills within its range. Sympatric barbets in the genus *Megalaima* and Fire-tufted Barbet *Psilopogon pyrolophus* are superficially similar, with green bodies and colourful heads, but barbets are stocky, short-tailed birds with large heads and massive bills, and can therefore be distinguished on morphology alone.

VOICE Long-tailed Broadbills are most vocal in flight or when about to move. At such times they may give a series of loud sharp whistles, *tseeay* or *pseew*, repeated 5-8 times, with little change in pitch, but falling in tone. The call is reminiscent of calls given by Dusky Broadbill, but sounds less frantic, and is downwards inflected, not upwards, as in Dusky. Whilst feeding, this species is relatively silent, although individuals occasionally utter a single sharp *tseeay* or *seweet* (Ali and Ripley 1948, 1983), and occasionally a short, rasping *psweep*.

DISTRIBUTION The Long-tailed Broadbill has the largest range of any Asian broadbill, being distributed from the foothills of the eastern Himalayas east through Nepal, Sikkim and Bhutan into north-east India and Burma, and discontinuously to Indochina, southern China, South-East Asia and the Sunda Region, where present in Peninsular Malaysia, Sumatra and north Borneo.

In the east of its range, it occurs in the Himalayan foothills and adjacent terai plains from Garhwal, Uttar Pradesh (c.30°25'N, 78°10'E), and Kumaon through Nepal (where the majority of the records come from the centre and east of the country: Inskipp and Inskipp 1991); Darjeeling; extreme north of West Bengal; Sikkim (Stevens 1915); Bhutan (in the centre and east of the country and on the southern border, as well as in the Royal Manas National Park: Ludlow and Kinnear 1937, Inskipp and Inskipp 1993, Ali *et al.* in press); to the north-east Indian states of Arunachal Pradesh, Assam, Meghalaya, Nagaland, Mizoram and Manipur; the Chittagong Hill Tracts in Bangladesh; and Burma in the hills of the Chindwin Valley, Arakan Hills, Pegu Hills, Karen Hills, and the hills and mountains of the Shan States and Tenasserim (Hopwood 1919, Inglis *et al.* 1920, Baker 1926, Ali and Ripley 1948, 1983, Abdulali 1976, Ripley 1982, Smythies 1986, Harvey 1990).

From the Shan States its range extends southwards and eastwards through northern Thailand and Laos to Tonkin and North Annam in northern Vietnam and to southern China in western, southern and south-east Yunnan Province, south-west Guizhou Province (Kweichow) and in south-west Guangxi Zhuang Autonomous Region (Kwangsi) (Delacour 1930, Bangs and van Tyne 1931, Delacour and Jabouille 1931, Delacour and Greenway 1940, Cheng 1987).

Another population, apparently isolated, occurs in south-east Thailand, Cambodia and South Annam (Robinson and Kloss 1919). In this part of Thailand it is known from Khao Yai to the Dangrek Range bordering Cambodia, from relict plains woodlands in Nong Khai Province (P. Round *in litt.*), and in the mountains of the south-east from Khao Chamao and Khao Soi Dao through the mountains to the Cambodian border (Round 1988, Lekagul and Round 1991), and presumably through the Cardamom Mountains to the Elephant Mountains, although it has only been recorded near Kampot in Cambodia (Thomas 1964).

In Peninsular Malaysia it is found in the Larut Hills, the Main Range from Mt Korbu to Ulu Langat, and on Mt Tahan, in the east (Medway and Wells 1976). In Sumatra, the species is known from the Gayo Highlands, Aceh Province, to c.5°S in the Barisan Range, Lampung Province (van Marle and Voous 1988, Holmes in prep.). In Borneo, it apparently has a very restricted range, being known only from the north-west, where distributed from Mt Kinabalu, Sabah, through the Crocker Range and on Mt Trus Mardi, south to Mt Dulit and Mt Mulu (Banks 1935, 1937, Sheldon in prep.). There are no records from Kalimantan, Indonesia (Andrew 1992).

GEOGRAPHICAL VARIATION Five subspecies have been recognised, but examination of large series of specimens shows that variation within subspecies is considerable, suggesting that some may not be valid. The geographic range of the species in Indochina is poorly known, and whilst populations in south-east Thailand and Cambodia appear to be isolated from others, the range limits of nominate

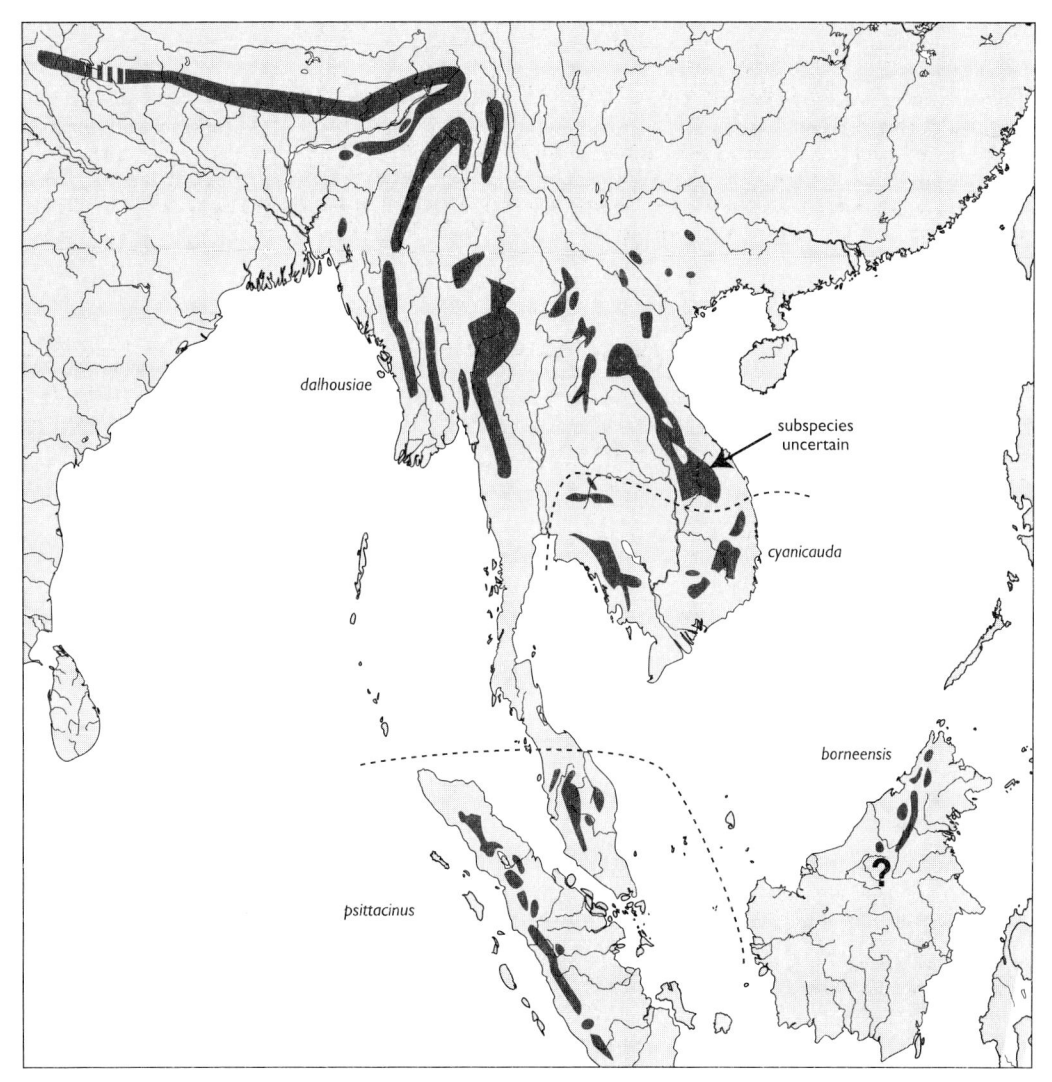

Distribution based on approximate extent of forest remaining within the altitudinal range of the species. Former range more widespread.

dalhousiae and *cyanicauda* in the Annamitic Chain of Vietnam and Laos is not clear. No consistent differences were found when skins of *cyanicauda* and *davinus* were examined, and these taxa are therefore lumped below.

P. d. dalhousiae (Himalayas to NE India, SE Bangladesh, Burma south to Tenasserim, N Thailand and Laos, to N Vietnam – Tonkin, North Annam – and S China. Birds in Central Annam may be this subspecies or *cyanicauda*) Broad bright yellow collar almost meets at rear of neck, where there is usually a blue patch, which sometimes extends along fringe of the black feathers of nape. Birds from east of range (northern Indochina) often have whitish or pale yellow feathers at bottom edge of collar, especially nearer nape, but in general there is little white in the collar of this subspecies. See Description. Wing of male 99-105; tail 116-133.

P. d. psittacinus (Peninsular Malaysia and Sumatra) Shorter-winged but longer-tailed than the nominate subspecies, but populations on the mainland and in Sumatra differ, with Sumatran birds being the most distinctive (Peninsular Malaysia: wing of male 101-105, of female 101-104; tail of male 125.5-144, of female 119-143. Sumatra: wing of male 94-97, of female 95-99.5; tail of male 130-153, of female 131-141). The collar is usually a mixture of yellow and white, the white usually predominating and often extending as a narrow border across the upper breast, bordering the yellow throat. Females usually differ from males in having a concealed or partially concealed yellow band across the nape.

P. d. borneensis (NW Borneo) Very similar to *psittacinus* but smaller (wing of male 93.5-98.5, tail of male 119-133, weight 52.8); tail slightly paler, more ultramarine-blue, than other subspecies.

P. d. cyanicauda (includes *davinus* Deignan 1947) (SE

Thailand, Cambodia and Vietnam in South Annam) Tail darker blue than that of nominate and *borneensis*, similar in shade to *psittacinus*, but shorter. Males and females differ as in *psittacinus*. Wing of male 97-101, female 93-99; tail 116-124.

HABITAT In the Indian subcontinent, Long-tailed Broadbills occur in tropical and subtropical forests in the foothills of the Himalayas and adjacent terai and plains, up to an altitude of c.2,000m (Ali and Ripley 1983), although apparently most numerous between c.600-1,200m (Baker 1895, 1934). Baker (1934) also reported that, in the Indian subcontinent, they show an apparent preference for ravines running through high forest with little undergrowth and with a stream in the centre, or ephemeral pools, over which they can build their nest.

In Nepal, Long-tailed Broadbills frequent broadleaved forests in the foothills (Inskipp and Inskipp 1991), whilst in Bhutan they have been apparently found in the undergrowth of dense tropical forest from c.245-1,525m (Ludlow and Kinnear 1937, Inskipp and Inskipp 1993, Ali *et al.* in press). In the east of their Himalayan range, in the foothills of Arunachal Pradesh, Singh (1994) found Long-tailed Broadbills at various sites at 200-1,800m in primary and secondary tropical semi-evergreen and evergreen forests and in subtropical broadleaved forests. In the Mishmi Hills of eastern Arunachal Pradesh, Ali and Ripley (1948) found them in bamboo clumps and forest at 200-700m. Stevens (1925) also noted that groups of Long-tailed Broadbills are often encountered in areas dominated by bamboo in Sikkim, at altitudes up to 1,830m. Here, and in upper Assam, this species usually keeps to dense forest during the breeding season, but gregarious parties often venture into more open habitats during cold weather in the non-breeding season: in Sikkim, for example, Stevens (1915, 1925) found groups in open scrub interspersed with trees in the vicinity of habitations. Oates (1883) noted that in Burma they sometimes visited gardens in cold weather.

In Burma, Long-tailed Broadbills are widely distributed in forest at low to moderate elevation. Thus, Hume (1875a) found them throughout the forests of the Pegu Hills, at a variety of altitudes, whilst Mayr (1938) encountered them below 200m further north in the region of the Chindwin River. However, as elsewhere, they also occur in submontane forests, and Heinrich (Stresemann and Heinrich 1939) collected them at 1,300m on Mt Victoria in the Chin Hills. In the Karen Hills of Burma, Smith *et al.* (1943) observed them in evergreen forest and high regrowth, and discovered a nest where pine forest merged into evergreen forest, suggesting that this species must sometimes venture into areas of pine or mixed forest. In northern Thailand, Long-tailed Broadbills also usually occur in evergreen forest, but are occasionally encountered in mixed deciduous forests, as well as in bamboo, from 650-2,000m (Deignan 1945, Round 1988). Deignan (1946) reported that the species had been collected as high as 2,230m on Doi Pha Hom Pok, although this record may be erroneous. In Tenasserim, which represents the southern part of the range of nominate *dalhousiae*, Hume and Davison (1878) reported that it occurs from the low hills to 1,830m, whilst Hopwood (1919) found the species breeding there at altitudes of 610-915m.

Further east, in southern China, Long-tailed Broadbills occur in subtropical mixed forest with bamboo in Yunnan (Cheng 1987). Populations in Indochina occur through a broad altitudinal range. In North Annam, Long-tailed Broadbill has been recorded as low as 50m (Robson *et al.* 1989), whilst in Tonkin it occurs at 150-1,525m (Delacour 1930, Bangs and van Tyne 1931). In Laos, it has been recorded at numerous localities throughout the hills and mountains (Oustalet 1903, Delacour and Jabouille 1931, Delacour and Greenway 1940, R. Timmins *in litt.*). In extreme southern Laos, R. Thewlis (*in litt.*) found birds in the canopy of primary forest and in tall bamboo in lowland forest, whilst Bangs and van Tyne (1931) found them in Laos at 580-1,340m and R. Timmins (*in litt.*) observed them at 1,200m at Nam Theun (Nakai Plateau, central Laos).

In Central Annam, birds have been observed as low as 120m (pers. obs.), whilst specimens of *cyanicauda* have been collected at 915-1,370m in South Annam (Robinson and Kloss 1919, Chasen and Kloss 1932). Wildash (1968) reported that, in southern Vietnam, the species occurs in evergreen forests at low altitudes, although this needs verification. In southern Cambodia, Delacour (1929) recorded *cyanicauda* in areas dominated by pine at 730m near Kampot, but there are no records from elsewhere in the country (Thomas 1964).

In the Sunda Region, where this species apparently prefers higher altitudes, small groups frequent middle and upper levels of montane rainforests. In Peninsular Malaysia *psittacinus* has been found at 800-1,550m in the Larut Hills and the Main Range, and on Mt Tahan, in the east, where possibly occurring as low as 150m (Medway and Wells 1976). M. Chong (*in litt.*) has found it at 205m at Belum, in the north of the peninsula (upper Perak), in March. In Sumatra, the species is known virtually throughout the mountains at 700m to at least 2,440m (Meyer de Schauensee and Ripley 1940, van Marle and Voous 1988). In Borneo, where Long-tailed Broadbills appear to have a very restricted range, *borneensis* has been recorded only in Sabah and northern Sarawak, at altitudes of 900-1,650m (Banks 1935, 1937, Sheldon in prep.).

STATUS Long-tailed Broadbills are relatively common in much of their range, although in the north-western part they appear to be rarer. Populations of this species are possibly declining in Nepal: in the central duns, for example, it was reported to be common in dense forests in 1947, but there has been only one record (1970) from this area since (Inskipp and Inskipp 1991).

In the eastern Himalayan region, Hume (1888) described the status as very common throughout the entire high forest regions of Assam, Cachar and north Sylhet hills and plains of north-eastern India. Oates (1883) considered this to be a very common species in Burma, although, in contrast, Hume (1875) found it uncommon in the Pegu Hills, whilst in the Karen Hills it was described as fairly common (Smith *et al.* 1943). The status of Long-tailed Broadbill in northern parts of its range in the Indian subcontinent and Burma may have changed significantly, although there are no recently published comments to this effect. There are no recent records from Bangladesh, although the species may survive in the Chittagong Hill Tracts (Harvey 1990), while in parts of northern Thailand it is now considered to be uncommon, although still common in some areas where there is good forest (Round 1984, P. Round *in litt.*). Although probably not threatened by the bird trade, young birds are still supplied to the domestic cagebird markets in Thailand (P. Round *in litt.*).

Further east, Delacour *et al.* (1928) and Delacour

(1929) found Long-tailed Broadbill to be common in parts of Tonkin, North Annam and south-west Cambodia, whilst Delacour and Greenway (1940) reported that it was very numerous in the vicinity of the Mekong in central-western Laos. It was apparently already rare in South Annam (Beaulieu 1939) at that time, and in view of major changes in habitat that have occurred in northern Vietnam, it is likely to now be rare there, and populations must be rather isolated. In China, the species is reported to be very rare (Cheng 1987).

Long-tailed Broadbills are fairly common in suitable habitat in Peninsular Malaysia (pers. obs.), but generally scarce in Sabah and Sarawak. Banks (1937) described the status in Sarawak as rather local and uncommon, although on Mt Kinabalu Whitehead found it fairly common at 925-1,220m (Sharpe 1889). Sheldon (in prep.) classifies it as a scarce species in Sabah.

MOVEMENTS In the Indian subcontinent, Long-tailed Broadbill may make local altitudinal movements during cold weather (Ripley 1982). Inglis et al. (1920) noted that it moved from the hills of northern West Bengal into the plains during the winter months, whilst Stevens (1925) described its occurrence in the Rungbong Valley of Sikkim as sporadic.

FOOD This is an insectivorous species. In the Indian subcontinent Long-tailed Broadbills have been recorded eating small beetles, cicadas, hemipteran bugs, cockroaches, green grasshoppers, crickets, spiders, butterflies, large black and large brown ants, grubs and insect larvae (Baker 1895, Ali and Ripley 1948, 1983). The stomach of a female collected off a nest in the Indian subcontinent was full of beetles and other insects (Hume 1890). Specimens from northern Sumatra contained a cicada, an acridid grasshopper, a chrysomelid beetle and a spider (Chasen and Hoogerwerf 1941). Specimens from Sarawak (SarM) have contained large grasshoppers and other orthopterans, cicadas and beetles. Robinson (1927) reported that Long-tailed Broadbills sometimes eat berries and other fruit, although there is no evidence to support this.

HABITS Long-tailed Broadbills are very sociable except when breeding, travelling around in small, noisy parties which can usually be located by their regular shrill calls and chattering notes. Flocks in the Indian subcontinent are said typically to comprise 15 to 30 birds (Ali and Ripley 1983), but may contain as many as 40 birds. Large flocks are occasionally recorded in other parts of its range, but flocks usually number less than 15. During the breeding season, pairs separate from these flocks and small parties and are much more secretive. Ali and Ripley (1948) noted that the flocking behaviour was rather similar to that of minivets *Pericrocotus* spp. Groups or pairs of Long-tailed Broadbills also join mixed species flocks, associating with such birds as drongos *Dicrurus* spp., babblers Timaliidae, flycatchers Muscicapidae, bulbuls Pycnonotidae, trogons *Harpactes* spp. and nuthatches Sittidae (pers. obs.). McClure (1974a) noted, however, that such associations are often only temporary, with the broadbills breaking away after a short time.

Long-tailed Broadbills often sit motionless on exposed branches of the lower canopy, occasionally jerking the tail upwards in the manner of a shrike *Lanius* sp. Insects are gleaned from foliage or from the air during fluttering, often clumsy, sallies from perch to perch. Cicadas are frequent prey items of Long-tailed Broadbills in parts of their range, in particular in the Indian subcontinent, and the 'hysterical' noises made by cicadas after capture often help in locating this species (pers. obs.). Long-tailed Broadbills also hunt for prey along branches and trunks and, occasionally, climb trailing creepers or pendant branches (Robinson 1927, Wells 1985, Smythies 1986). Baker (1895) found prey items in the stomachs of Long-tailed Broadbills that suggested that they must sometimes feed by clinging to trunks and branches.

BREEDING
Seasonality From the Indian subcontinent to northern Thailand (subspecies *dalhousiae*), breeding is seasonal, beginning in March, April, May or June. In the Indian subcontinent, the young usually fledge by June or July (Hume 1890). Although this is a relatively common and widespread species, remarkably few breeding records are documented.

In Nepal, a pair was found nest building and three old nests were found at the end of April (Inskipp and Inskipp 1991). Breeding has been recorded in Sikkim at 760m in April (Stevens 1925), and a probable nest was found on 18 May (Ali et al. in press). Hopwood (1912, 1919) found nests between 7 and 9 May in Arakan, Burma, and in April and May in Tenasserim, whilst Smith et al. (1943) recorded nest building in the Karen Hills on 15 April. Darling found a nest in Tenasserim on 3 April (Hume 1880). Baker (1934) gave May and June as the principal breeding months, but he found fresh eggs as early as 3 April and as late as 24 August, although he noted that the latter probably related to a second brood. Normally, Long-tailed Broadbill is not double-brooded, but if the first clutch of eggs or brood is lost, replacement nests are generally built and a second clutch may be laid (Baker 1934).

In Thailand, nest building has been observed at Khao Yai on 3 February and at Kaeng Krachen on 30 April-1 May; occupied nests have been found at Khao Yai on 22 February, 11 March and in April, and on Doi Suthep on 10 April; recently fledged young were observed in Khao Yai in two seperate parties on 10 July (P. Round *in litt.*). At an active nest was found c.30m above the forest floor in a heavily forested ravine in Khao Yai National Park in April (McClure 1974a). The female was incubating whilst the male remained nearby. Immatures have been collected in northern Thailand on 2 August and 2 September (Riley 1938). The only records of breeding from Vietnam are of a bird observed constructing a nest high in an open tree at Cuc Phuong in east Tonkin on 6 July, and of a family party, including juveniles, in South Annam (subspecies *cyanicauda*) on 5 June (C. Robson *in litt.*). Elsewhere in Indochina, the only breeding records are from Laos, where R. Timmins (*in litt.*) found a nest under construction on 22 March and observed immatures in June and July.

Wells (1986) documents finding active nests of *psittacinus* at Maxwell's Hill in Peninsular Malaysia on 18 April, and on Fraser's Hill in April and May. Nest construction at the latter site has been observed as early as late February (S. Duffield *in litt.*). In Sumatra, nest building has been observed in February and eggs collected twice in April (van Marle and Voous 1988), whilst Meyer de Schauensee and Ripley (1939) collected a male with enlarged testes on 1 April, and Chasen and Hoogerwerf (1941) collected a juvenile during the same month. Immatures have been collected in Sumatra between 30 May and 22 August (RMNH). These records indicate that *psittacinus* breeds during the driest part of the year.

In Sabah, Phillipps (1970) observed five Long-tailed Broadbills (*borneensis*) attending a nest in the lower branches of a forest tree, and hanging over a steep slope, on 25 April. An adult carrying nesting material was observed on 3 May, whilst a female with a brood patch has been collected in mid-March on Mt Kinabalu (Sharpe 1889, Sheldon in prep.). A juvenile was collected in Sarawak on 13 August (SarM).

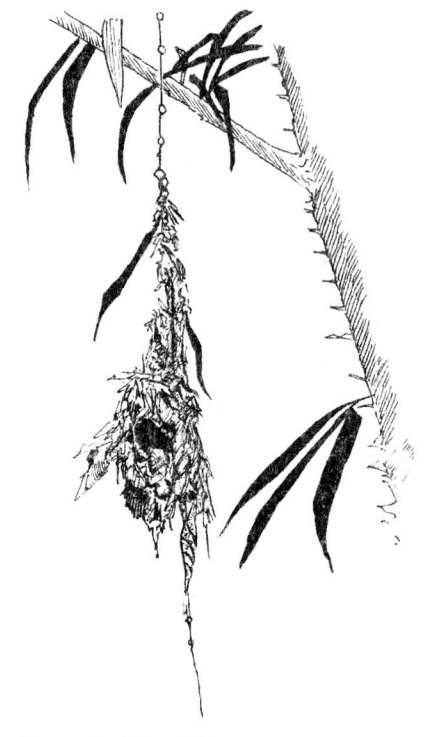

Nest of Long-tailed Broadbill

Nests Both adults participate in nest building, incubation and feeding the young. Although both adults build the nest, construction may take as long as three weeks (Baker 1934).

The nest is usually suspended from the tip of a branch or creeper up to 10m above the ground, so that it hangs well away from vegetation. Nests are frequently suspended above streams, roads, paths or open spaces. The nest is attached by a narrow extension of the nest ball to the tip of an isolated branch of a tree or tree-fern, or the tip of a rattan stem, presumably providing maximum protection against nest predation by arboreal mammals and snakes. In India, nests have even been found attached to telegraph wires strung across hilltops in forest (Peile 1914). The same nest site may be used every season, so that several old nests may be found hanging nearby, often suspended from the same branch. Bingham found six nests in a single, spine-covered tree in Burma (Hume 1880). Nests are reported to be very robust, withstanding very strong winds without serious damage (Peile 1914).

The nest is a conspicuous bulky ball of closely woven plant material. It is pear-shaped, but variable from short and stumpy to long and thin (Baker 1934), with a long pendant tail. The entrance, which occasionally has a small porch over the entrance (Hume 1880, Peile 1914), is a hole in the side, usually situated in the lower half about a third of the way from the base. The overall length of the structure may reach a metre, or occasionally more (one was c.1.4m in length: Baker 1934). The dimensions of the bulky ball that contains the nest chamber is much smaller, although apparently extraordinarily variable in size, and may be anything from 23.0 x 12.7cm to 35.6 x 20.3cm in the Indian subcontinent (Baker 1934). The chamber itself is of internal diameter 10-12.5cm, and the entrance hole c.4-5cm in diameter.

In the Indian subcontinent, nests have been made of fine roots, moss and the stems and tendrils of creepers and other stringy plant material, intermingled with some leaves and with a grass interior and lining of bamboo spathes (Hume 1880). They are sometimes adorned with pieces of green moss and the egg cases of spiders (Baker 1934). In Malaysia, a nest was made with black palm fibre, fern fronds and stems, bryophytes (mosses, liverworts), woody stems, dead leaves and twigs, and the nest chamber lined with dead bamboo leaves, grass blades or roots overlaid with green leaves (Medway and Wells 1976). Baker (1934) found eggs that were almost hatching lying on perfectly fresh leaves, suggesting that the green leaves must have been renewed frequently during incubation.

Eggs The nominate subspecies usually lays 5-6 eggs (Hume 1880, Ali and Ripley 1983), though occasionally only four eggs, or as many as eight, have been found in nests in the Indian subcontinent (Baker 1934). In Peninsular Malaysia clutch-size is generally three. The eggs are rather long ovals, slightly compressed at the smaller end but never really pointed. Two hundred eggs from the Indian subcontinent had an average size of 27.4 x 19.4mm, with a range of 25.0-29.6 x 17.0-20.5mm (Baker 1934), whilst three from Peninsular Malaysia measured 27.5-28.5 x 18.5-19.0mm (Medway and Wells 1976).

Egg colour is very variable, ranging from pure white to deep pink, marked with blotches of shades of pink, red and reddish-brown, and smaller lilac-grey markings (BMNH). Markings are usually most dense at the broader end. Hume (1880) notes that eggs with markings are almost glossless, whilst those that are pure white are fairly glossy. The period of incubation is unrecorded, although Baker (1934) stated that it was more than 14 days.

DESCRIPTION *P. d. dalhousiae*

Adult Head striking: narrow yellow band across forehead at base of bill, broadening across lores and joining yellow below eye and of throat, sometimes tinged with green on forehead; black crown broadening behind eye to cover ear-coverts and extending to nape, marked with a large bright pale blue central crown-patch extending from above or just in front of eye to rear of crown, but not onto nape, and a prominent bright yellow patch above the ear-coverts. Throat and chin bright pure yellow, extending as a broad collar that almost meets at rear of neck, where there is usually another blue patch, this blue sometimes extending along fringe of the black feathers of nape. Birds from east of range (northern Indochina) often have whitish or pale yellow feathers at bottom edge of collar, especially nearer nape. Mantle, back, rump and uppertail-coverts dark bright green, occasionally with a few indistinct blue fringes, in small patches, to feathers of back. Tertials concolorous with back, dark bright green. Inner web of primaries black, with concealed white band about one-third of distance from base, across all primaries; outer web of basal two-thirds of primaries bright shining azure-blue, forming a prominent wing-panel, this colour changing to dark blue and then dark green along outer web of distal third to one-quarter of the feathers, usually at the point where feathers are emarginated. All primaries except outer are strongly emarginated. Secondaries have black inner web, dark green outer web (as mantle), this green extending to tip on innermost secondaries; tertials dark green. Greater coverts black with turquoise-blue leading edge, broadest on outer feathers. Other wing-coverts dark green. Underwing-coverts green, mottled by darker feather-bases. Underparts green, much paler than upperparts, often with bluish tinge, especially on belly, but occasionally extending to entire underparts. Some specimens from Nepal, Bhutan and northern Burma have the entire underparts turquoise, except for green undertail-coverts. Many individuals have a very narrow, indistinct, band of blue-fringed feathers across upper breast, bordering yellow of throat. Tail bright blue above, black below, strongly graduated, with the outer feathers being only about a third of the length of the central ones, and with narrow blackish edge to innermost three pairs, and, frequently, green edges near the base. Bill tip hooked; lacks rictal bristles. Iris colour reported to be very variable (e.g. Robinson and Kloss 1924), being iridescent green, or iridescent pink mixed with green or greyish-brown or greenish-grey; eye-ring fine, greenish-yellow. Bill pale green, or greenish-yellow, tipped bluish, and with blue base to upper mandible and, sometimes, pinkish-orange basal two-thirds of underside of lower mandible. Occasionally, some individuals may have yellow-greenish tip to mandible. Legs pale green, or greenish-yellow.

Juvenile Young birds (Sumatra) have the crown green, or green and black, and the throat greenish-yellow. Blackish or dark grey upper mandible, yellowish-green cutting edge and light greyish-green lower mandible; feet dirty olive-green.

MEASUREMENTS Wing of male 103-116, of female 96-105; tail of male 116-132, of female 116-127; bill length 30-31.5, depth 19-21.5; tarsus 28-30; weight 64-67 (*borneensis* 52.8).

SERILOPHUS

Serilophus is a monotypic Asian genus. It is a small, unobtrusive, yet gregarious species with one of the largest distributions of any broadbill species; within this range eight subspecies are recognised. However, the limits of these subspecies' ranges are poorly known and merit further study. Within its range, this species is found in a wide variety of forested habitats, in both lowland and montane forests. Populations in the north of its range may make altitudinal movements in response to harsh weather in winter. Although this species is insectivorous and feeds on relatively large prey items, its bill tip is hardly hooked, in contrast to most other insectivorous broadbills; a further dissimilarity is that *Serilophus* hangs its nest by weaving the materials across a relatively broad section of a branch, rattan or other structure. Observations suggest that helpers may be involved in nest building and feeding young. The six innermost primaries and seven adjacent secondaries are unusual in that they have distinctly notched tips.

41 SILVER-BREASTED BROADBILL
Serilophus lunatus　　　　　　Plate 23

Eurylaimus lunatus Gould, 1833 (1834), Proc. Zool. Soc. London, pt 1, p. 133: near Rangoon; type locality restricted to the hills of the Pegu district, Pegu Division, Burma, by Deignan (1948).

Alternative names: Gould's Broadbill, Collared Broadbill, Hodgson's Broadbill

TAXONOMY Early authors (e.g. Baker 1934, Garthwaite and Ticehurst 1937, Deignan 1948) treated *rubropygius* of the Indian subcontinent as a separate species from *lunatus*. However, more recent authorities include *rubropygius* as a subspecies of *lunatus* (e.g. Peters 1951). The limits of the ranges of *rubropygius*, *lunatus* and *elisabethae* in Burma are unclear, and further field study in central Burma, in particular in the northern Shan States and the upper Irrawaddy River valley and in the north-east, might show that there is indeed an area of sympatry between *rubropygius* and one of the other subspecies, and therefore a basis for recognising two species. *S. l. rubropygius* has been collected in the upper Irrawaddy in the Myitkyina district, and Garthwaite collected it in two localities just to the south of there, close to but west of the Irrawaddy in the Shwebo Forest Division and on the Kaukkwe River, Bhamo Forest Division. *S. l. lunatus* has been collected close to the latter site, in Bhamo, and in view of the lack of intergradation, Ticehurst concluded that these taxa represented two species (Garthwaite and Ticehurst 1937).

FIELD IDENTIFICATION Length 16-17. This small, gregarious broadbill is widely distributed from the Himalayas to Indochina and the Sunda Region. The plumage is rather subdued, with grey underparts, becoming white on belly and undertail-coverts, and grey to grey-brown upperparts. The distinctive features are the broad black supercilium and striking blue, black and rufous markings on the wing, and a more or less rufous rump. The tail is black, with white-upped outer feathers. A white band across the base of the flight feathers is conspicuous in flight. Females of some subspecies differ from males in having a narrow, often broken, silver band across the upper breast. Immatures closely resemble adults from a young age, but may have very short wings and tails.

Similar species Silver-breasted Broadbill is not like any other broadbill and is unlikely to be confused with any bird species in its range.

VOICE Although Silver-breasted Broadbills are often silent, they may call frequently in the vicinity of their nest. The most commonly heard call is a melancholy *ki-uu*, or *pee-uu*, with a lower second syllable. Another call, perhaps given as an alarm, is a high-pitched, thin, grating, staccato trill of 5-7 notes that sounds like an insect, given in flight (when described as *kitikitikit*) or when perched. In southern Laos, Silver-breasted Broadbills frequently utter a distinctive contact call described as *pri-iip* (R. Thewlis *in litt.*). Ali and Ripley (1948) report that birds caught in a net made a loud mouse-like squeaking when disturbed.

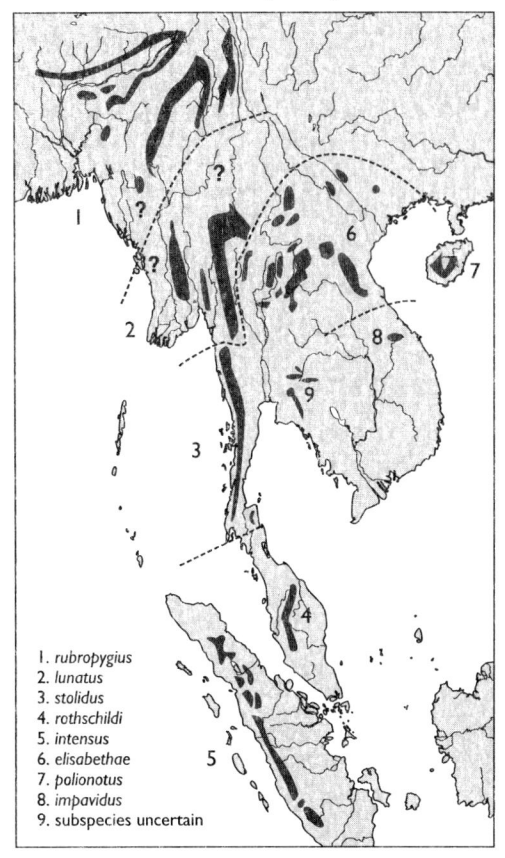

1. rubropygius
2. lunatus
3. stolidus
4. rothschildi
5. intensus
6. elisabethae
7. polionotus
8. impavidus
9. subspecies uncertain

Distribution based on approximate extent of forest remaining within the altitudinal range of the species. Former range more widespread.

DISTRIBUTION The Silver-breasted Broadbill has one of the largest distributions of any broadbill species, from the eastern Himalayas, Burma and south China to South-East Asia and the Sunda Region, where it occurs in Sumatra. An isolated population also occurs on Hainan.

Populations in the Himalayas occur from Nepal east through Darjeeling, northern West Bengal and southern and central Bhutan to the north-east Indian states of Arunachal Pradesh, Nagaland, Mizoram, Manipur, Assam, Meghalaya and Tripura (Ripley 1982); and south to north-east and south-east (Chittagong Hill Tracts) Bangladesh (Harvey 1990). In Burma, Silver-breasted Broadbill is widely distributed (Garthwaite and Ticehurst 1937, Smythies 1986). In China, it is known from southern Yunnan Province (south-west Mengding, southern Xishuangbanna and south-eastern Hekou) and the southern part of Guangxi Zhuang Autonomous Region (Cheng 1987).

In Indochina, the species occurs in Vietnam, Laos and Cambodia. Although an isolated population was thought to be confined to the Bolovens Plateau of southern Laos, recent records suggest that the species is widely distributed in central and southern Laos (J.W. Duckworth *in litt.*). Another population that might be isolated occurs in the mountain ranges of south-east Thailand and presumably throughout the adjacent mountains of southern Cambodia. In Cambodia, the only record of this species comes from the region of Kampot, at the eastern extremity of the Cardamom Mountains (Engelbach 1938).

In Peninsular Thailand, there is a population that ranges from the extreme south through the Larut Hills and main range of Peninsular Malaysia in northern Perak south to Jelebu, Negri Sembilan (Robinson 1927, Medway and Wells 1976, Lekagul and Round 1991). In Sumatra, it occurs throughout the mainland ranges, where known from Gunung Leuser National Park, Aceh Province, to Mt Dempu, South Sumatra Province (van Marle and Voous 1988).

GEOGRAPHICAL VARIATION Deignan (1948) recognised eleven subspecies of Silver-breasted Broadbill, excluding *rubropygius*, which he believed was a separate species (see Taxonomy). However, more recent authorities include *rubropygius* as a subspecies of *lunatus* and recognise only eight or nine subspecies (e.g. Peters 1951). Differences between these subspecies are generally minor.

Meyer de Schauensee (1984) stated that *lunatus* occurs in south-west Yunnan Province, China, but Cheng (1987) recognises *elisabethae* as being the subspecies in this area. In Tenasserim, the range of *lunatus* appears to be almost parapatric with that of *stolidus*, though the exact boundary between these two subspecies has not been determined: Riley (1938) reported that the southern limit of *lunatus* was Thasan, Chumphon, in Peninsular Thailand. Silver-breasted Broadbills in the mountains of southeast Thailand and adjacent Cambodia are perhaps isolated, and the subspecies to which they belong is not clear, although this population is included here in *elisabethae*, following Engelbach (1938). The subspecies *moderatus* (from Mt Leuser) was not recognised by van Marle and Voous (1988), and is included in *intensus*.

S. l. lunatus (Burma in the S Chin Hills, Pegu Hills, Karenni and S Shan States, and adjacent NW Thailand, south to N Tenasserim, probably at c.15°N) Lores usually rusty, ear-coverts, crown and mantle ferruginous, mixed with some grey on mantle. Small area of forehead ashy in most birds. Uppertail-coverts and rump bright pale rufous. Lighter on rump and tertials than *stolidus*.

S. l. rubropygius (Nepal east through the Himalayas to NE India, south to the Chittagong Hill Tracts in Bangladesh; Arakan and the upper Chindwin Valley, Burma) Differs from all other subspecies in wing pattern, having narrower, darker blue secondary bar; broad dark blue tips to outer webs of flight feathers; and white wing-flash near tip, formed by white subterminal spots to edge of outer web of secondaries. Differs from *lunatus* in having a pale grey crown, nape and upper mantle, the rest of the upperparts being darker grey, tinged brownish. The silver breast-band of females is broken in the centre. Lores dark grey. Supercilium less strongly marked than in other subspecies, and underparts generally greyer.

S. l. elisabethae (includes *aphobus* and *astrestus* of Deignan 1948) (NE Burma and the Shan States, eastern Thailand, Tonkin east to the Red River, S Yunnan,

China [Baker 1921, Cheng 1987] and Indochina in N Annam [Vietnam], Laos [except southern] and S Cambodia) Very similar to nominate, differing in having blackish to blackish-rusty lores and in the chestnut-rufous of the rump and the rufous of the uppertail-coverts being deeper-coloured, less bright. Greyer and with deeper-coloured rump and inner flight feathers than *stolidus*.

S. l. impavidus (S Laos, where perhaps confined to the Bolovens Plateau) Doubtfully distinct from *elisabethae*. Deignan (1948) stated that *impavidus* differed from the latter in having an ashy hue to the scapulars and upper back. Birds in the lowlands of southern Laos (e.g. at Xe Piane) must either belong to this subspecies or to *elisabethae*.

S. l. rothschildi (Peninsular Malaysia and extreme south of Peninsular Thailand) Differs from other subspecies in having the lores pale ashy and sides of head and ear-coverts greyish, lacking any ferruginous or brownish tinge, more extensive bluish bend of wing and bluish upper edge of black supercilium in front of eye, and more extensive rufous on upperparts, extending from rump and uppertail-coverts to back or lower mantle. Females tend to have broader white breast-bands than nominate.

S. l. intensus (Sumatra) Very similar to *rothschildi*, from which it differs in having the lores blackish-ashy. There is some doubt as to the validity of this taxon.

S. l. polionotus (Mountains of Hainan) Lores black; sides of head, ear-coverts, crown and nape pale ashy, tinged with darker olive-brown; scapulars and upper back ashy-grey. Breast darker than in *lunatus*, contrasting with paler, ashy throat.

S. l. stolidus (S Tenasserim, Burma and Peninsular Thailand between 15°N and 8°30'N and in mountains of Nakhon Si Thammarat) Very similar to *lunatus*, differing in having more extensive and paler ashy on forehead, and darker, more chestnut lower back and rump.

HABITAT In the Indian subcontinent, *rubropygius* is found in semi-evergreen and evergreen forests from the plains to about 1,700m. In the Mishmi Hills and other parts of eastern Arunachal Pradesh this subspecies has been recorded at c.195-690m (Ali and Ripley 1948, Ghosh 1987), whilst in central Arunachal Pradesh, Betts (1956) found it at 460m in bamboo jungle. Singh (1994) found Silver-breasted Broadbills at 200-700m in primary tropical evergreen and semi-evergreen forests in south-west Arunachal Pradesh and in an area bordering the Mishmi and Patkai Hills, in the south-east of the state. Stevens (1925) reported that, in Sikkim, *rubropygius* was primarily a bird of the plains, although a nest was found at 915m on one occasion. Stanford (1938) reported that this subspecies was found to 1,220m in northern Burma, whilst Mayr (1938) encountered it below 200m in the region of the Chindwin River. Garthwaite and Ticehurst (1937) collected this species in evergreen forests in the foothills near Bhamo (upper Irrawaddy River). Baker (1926) reported that *rubropygius* was most common at c.305-915m, and that it prefers forest with areas of bamboo, though occurring in every kind of forest within its range.

Further south, Hume (1875a) found the subspecies *lunatus* throughout the Pegu Hills region but not on the plains. In the Karen Hills and Karenni, Smith *et al.* (1943) found it at 915-1,220m. Smith (1943) saw it in evergreen oak forest in Burma.

In Thailand, Silver-breasted Broadbills have been recorded in evergreen forests, mixed deciduous forests, including areas dominated by pines (pers. obs.), and in bamboo (Robinson and Kloss 1923). In northern Thailand the nominate subspecies occurs in hills and mountains, mostly between 790m and 1,800m, though it has been collected as high as 2,230m (Riley 1938, Deignan 1945, 1946, Round 1988), whilst *stolidus* has been recorded to at least 885m in Nakhon Si Thammarat (TIST). To the east, *elisabethae* generally occurs at lower altitudes: it has been collected in Tonkin at 365m and in Laos at 700m (Bangs and van Tyne 1931), and observed at below 100m in Cuc Phuong National Park, Tonkin (pers. obs.). Habitats in which this species occur in central Vietnam include primary and selectively logged evergreen forests in limestone hills (pers. obs.). In Cambodia, *elisabethae* has been collected at 800m in the eastern Cardamom Mountains (Engelbach 1938). In Peninsular Thailand, *stolidus* has been recorded as low as 140m (Jeyarajasingam 1983). Hume and Davison (1878) reported that Silver-breasted Broadbills are confined to the northern and central part of Tenasserim, where regularly encountered in agricultural land and even shady gardens, as well as forested habitats.

In southern parts of their range, such as in the Sunda region, Silver-breasted Broadbills are restricted to higher altitudes, frequenting the middle and lower levels of upper hill and montane forests, often in areas with stands of bamboo. Indeed, Robinson (1927) believed that *rothschildi* was only found in forest where there was a large amount of bamboo. In Sumatra, the species occurs at 800-2,000m (Chasen and Hoogerwerf 1941). Robinson (1927) noted that *rothschildi* generally occurred from 610-1,220m in Peninsular Malaysia, and nests have been found at c.600-1,000m (Wells 1984, 1986, Anon. 1988). However, in the northern part of its range in the Peninsula it has been found at c.150m, whilst a small group was observed at 250m at Ulu Gombak in Selangor (pers. obs.) and it has also been found at this altitude at Belum (M. Chong *in litt.*). Wells (1986) suggested that, in Peninsular Malaysia, this species inhabits a narrow altitudinal zone at the upper limit of dipterocarp forest on hill-slopes; although it is usually found in this zone, it also clearly wanders to much lower altitudes.

STATUS In Nepal, Silver-breasted Broadbill is apparently now extinct, but was collected by Hodgson in the first half of the nineteenth century (Inskipp and Inskipp 1991): its loss there can be attributed to loss of suitable habitat (C. Inskipp *in litt.*). In north-east India, Hume (1888) described it as not rare in north-east Cachar, and the Khasi and Garo hills. In Burma, Hume (1875a) found it to be very common in the Pegu Hills and Hopwood (1912) described it as common in Arakan. Although still fairly common in Thailand, it is uncommon in parts of the north (Round 1984, P. Round *in litt.*). It was reported to have been common in parts of Tonkin and North Annam by Delacour (1929, 1929a), whilst Delacour and Jabouille (1927) reported that this species was locally numerous in Central Annam, and Delacour and Greenway (1940) described it as very common throughout northern Laos. On Hainan, although well distributed, *polionotus* is reported to be very rare (Cheng 1987). In the Sunda Region, this is a locally common to uncommon resident.

In Thailand, young birds of this species are still being supplied for the illegal, burgeoning domestic cagebird market (P. Round *in litt.*).

MOVEMENTS In the Indian subcontinent, *rubropygius* is apparently an altitudinal migrant that moves to lower altitudes during the cold winter months. Cripps noted it as a visitor to the Dibrugarh district of north-east Assam during cold weather (Hume 1888), whilst Inglis *et al.* (1920) stated that Silver-breasted Broadbills moved further into the plains of northern West Bengal during the winter. Stevens (1915) also noted that it was possibly a local migrant in north-east Assam.

FOOD Grasshoppers, mantises and other insects, insect larvae, grubs and land snails have been recorded in the diet of *rubropygius* in the Indian subcontinent (Ali and Ripley 1983). The stomach of one individual collected in north-east India was crammed with the shells of tiny land snails (Hume 1888). Hume (1875a) reported that specimens from the Pegu Hills of Burma had fed principally on grasshoppers. Hume and Davison (1878) found small insects in the stomachs of specimens of *lunatus* that they examined in Tenasserim, and reported that one individual had swallowed, head first, a green lizard more than 10cm long. R. Thewlis (*in litt.*) observed one catch a small beetle in flight in southern Laos.

HABITS Silver-breasted Broadbills are often quiet and unobtrusive and are sometimes very tame and rather sluggish; in the northern part of their range this may be a consequence of the rather crepuscular activity peak that has been reported (Baker 1926). Near the nest, however, they may call frequently. They occur in pairs or small, loosely associated groups, feeding in the lower branches of canopy trees and bushes. In northern Thailand, groups of up to 20 can sometimes be found, but further south, in Peninsular Malaysia, groups are rarely this large. Delacour and Jabouille (1927) noted that, during February and March in Indochina, small groups of Silver-breasted Broadbills joined understorey bird flocks comprised of species such as Asian Paradise Flycatcher *Terpsiphone paradisi*, Black-naped Monarch *Hypothymis azurea* and fulvettas *Alcippe* spp.. In the Indian subcontinent, and occasionally in Peninsular Malaysia, these broadbills are also found in mixed species flocks (Ali *et al.* in press, pers. obs.).

Occasionally this broadbill may take insects by flycatching from a perch, but more usually it gleans prey off the branches and foliage whilst sallying from one perch to another. Such perches may only be a few metres apart. Baker (1926) reported that *rubropygius* extract larvae and grubs from the bark of trees and that they will also capture insects on the wing, even though the flight is rather heavy and awkward. They perch very erect, often for long periods, with the tail often twitched. In Vietnam, a flying bird was observed either to drink or to take an insect from a puddle on a road (pers. obs.).

Both sexes (and possibly helpers) participate in nest building, incubating eggs and feeding young (Oates 1879, Baker 1934, Ali and Ripley 1983). In the Indian subcontinent, nest building activities are carried out only during the first three and last three hours of daylight, with frequent breaks for feeding (Baker 1934).

BREEDING
Seasonality Baker (1934) reported that, in the Indian subcontinent, *rubropygius* breeds mostly between May and June, whilst *lunatus* breeds from March to July in Burma, although records from various parts of Burma (see below) perhaps indicate a longer breeding season in the south (Tenasserim) than elsewhere. The breeding season in Peninsular Malaysia, where many nests have been found, and probably in Indochina, is apparently similar. Hence breeding in the colder, northern part of its range occurs in spring, whilst in warmer parts of its range, to the south, breeding coincides with the south-west monsoon: in the northern parts of its range this brings rain, but in the south this timing corresponds to the drier season. Breeding records are as follows:

Hopwood (1912, 1919) found nests in the Arakan Hills, Burma, in early May, and on 12 March found seven nests in Tenasserim on a single day, all suspended from bushes in or near a stream. Stanford and Ticehurst (1930) found nests in Burma in the Arakan Hills in April and under construction near Prome on 14 April. Smith (1943) found eggs of Silver-breasted Broadbills in Burma between 25 March and 7 June. Dissection of birds from the Pegu Hills of Burma suggests that Silver-breasted Broadbills lay eggs there at the end of April or in early May (Hume 1875a). Oates (1879) found one nest with four eggs near Pegu in dense evergreen jungle on 12 May. Many nests have been found in Tenasserim from March to May, and eggs have been found there as late as 11 July (Baker 1926). In northern Thailand, Deignan (1945) collected immatures of *lunatus* in post-juvenile moult in mid-August and early September, and adults in post-nuptial moult in mid-August and late December. At Khao Yai, Silver-breasted Broadbills have been seen carrying nest material or building nests on 6 March, 24 April and 13 May, and a nest with two eggs was found on 6 March (P. Round *in litt.*). A female *stolidus* with brood patch was collected in Nakhon Si Thammarat, Peninsular Thailand, on 10 May (TIST).

Further south, occupied nests of *rothschildi* have been found in Peninsular Malaysia between 11 April (with nestlings) and 19 June (incubating) (Jeyarajasingam 1983, Wells 1984, 1986, 1990b, Anon. 1988). In Sumatra, a nest of *intensus* containing two eggs was discovered on 22 March, and a male with enlarged gonads collected on 30 March (Robinson and Kloss 1918, Meyer de Schauensee and Ripley 1940).

In Indochina, a nest of *elisabethae* was found in March (Delacour and Jabouille 1931) but contained no eggs. All other breeding records from Indochina are from mid-June, during which five nests, including one under construction, have been found in North Annam. One of these nests contained a single chick (C. Robson *in litt.*).

Nests The majority of nests found have been suspended from the outer branches of smaller trees, or from the tips of palm fronds or bamboo, but also occasionally from bushes or tree-ferns (Baker 1934). They invariably hang over open spaces, in particular roads, paths or small streams, usually 3-5m above the ground. Occasionally, nests have been discovered as low as 1m above water and 1.5m above the ground, and as high as 7m. Active nests have been observed within 70m of each other.

The nest is suspended from a suitable branch by a mass of fine plant fibres that are interwoven across the branch over a width similar to that of the nest or slightly shorter (pers. obs.).

It is an untidy pendant ball with a circular entrance hole, usually somewhat sheltered by a protruding hood. The entrance is usually on one side, exactly in the centre. A long beard of loose material dangles below the nest. Nevertheless, the nest of Silver-breasted Broadbill is notably neater than that of most other Asian broadbills, with the exception of *Calyptomena*. Baker (1934), who saw many nests of *rubropygius*, estimated that the adults took 5-10 days to build them, although many of the 'decorations'

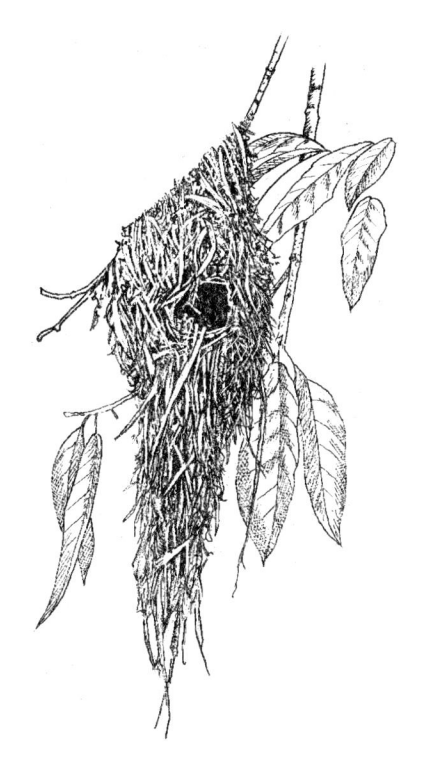

Nest of Silver-breasted Broadbill

on the tail, such as cocoons and moss, may be added after the eggs are laid.

Nest size of *lunatus* is reported to vary from around 25.5cm from top to bottom and 10cm in diameter to 38 x 20.5cm, excluding the tail, which may add considerably to the total length: a nest from Tenasserim had a total length of 60cm (Oates 1879). Nests of *rubropygius* tend to be larger, some being as large as 76cm from top to bottom and 30.5cm in diameter, although the majority are about the same size as the largest nests of *lunatus*.

A nest of *rothschildi* collected in Peninsular Malaysia was a solid, elongated structure, c.27.5cm long and with a more ragged tail c.14cm long (UM). The entrance hole, also elongated, about 4cm from top to bottom, and 2.7cm wide, was situated 16cm from the top, and protected above by a hood, made of frayed palm leaves, which extended about 2cm outwards from the nest. This hood presumably provides some protection from the sun and rain to the occupants of the nest cavity.

In Burma, nests of *lunatus* are made from a variety of materials including coarse grass and the outer bark of elephant-grass and weeds. These are bound together by fine black hair-like roots and lined with various types of broad grass blades or bamboo. Numerous dead leaves and thin twigs are incorporated into the interior of some nests (Hume 1880). Seven nests found by Hopwood (1919) in Tenasserim were unusual in apparently being made of moss. A nest of *lunatus* described by Oates (1879) from Tenasserim was firmly lined with broad leaves of elephant or thatch grass and a few green leaves spread over the nest cup, whilst Baker (1926) describes the usual nest of *lunatus* as being lined with bamboo leaves and grass with a layer of green leaves over them. Smith (1943) found green leaves lining some, but not all, nests that he investigated in Burma. Nest materials used by *rubropygius* from the Indian subcontinent differ little from that described above (Ali and Ripley 1983): they are bound together with fine black hair-like roots or fungal fibres, and lined with bamboo and grass leaves. Oates (1879) noted that the nest is very similar to, but larger than, the nest of Olive-backed Sunbird *Nectarinia jugularis*.

In Peninsular Malaysia, the principal material used by *rothschildi* is often green moss rather than grasses, although this presumably depends on the availability of materials, since some nests are made of straw-like fibres or grasses. One nest found at Genting by J. Howes (*in litt.*) was apparently entirely constructed from Tiger Grass *Thysanolaena maxima* (W. Geisen *in litt.* to J. Howes). Examination of photographs shows that the nest may be hung from near the end of the twig by a matrix of fine roots or other finely woven fibres. A characteristic of the nest of Silver-breasted Broadbill is the adornments attached to the untidy beard of loose material that usually dangles below: these may include spiders' egg-cocoons, caterpillar excreta, and pieces of moss or lichen, and, in one instance, bits of wool (Hume 1880, Baker 1934).

The outside of a nest from Peninsular Malaysia primarily consisted of fine black roots (some of which may have been fungal hyphae) and moss, with a few frayed pieces of palm. The cavity was lined almost entirely with pieces of palm leaf, although there were a few entire, but small, green leaves, presumably of forest trees (UM). M. Chong (*in litt.*) found green leaves with a silvery tinge of c.2.5cm length in a nest in Malaysia.

Eggs According to Baker (1926, 1934) *rubropygius* may lay from 4-7 eggs, although 4-5 is more normal. *Lunatus* lays 4-5 eggs (Hopwood 1919, Ali and Ripley 1983). In contrast, clutch-size of *rothschildi* in Peninsular Malaysia is usually three or less (UM).

The eggs are short, blunt glossy ovals, rarely rather lengthened but never pointed. Those of *lunatus* are white, faintly tinged cream, with a sparse speckling of tiny reddish-purple spots, most dense at the broader end. Eggs of *rubropygius* are very similar but they are more variable, some being pure spotless white, others being warm pink spotted sparsely with claret. Typically, those of *rubropygius* are pinker and have larger spots (Baker 1926). Eighty eggs of *rubropygius* on average measure 17.3 x 23.6mm with a range of sizes from 22.3-25.0 x 16.2-18.1mm, whilst 38 eggs of *lunatus* averaged 23.9 x 17.4mm with a range of 22.25-26.7 x 16.3-18.0mm (Baker 1926). Two eggs of *intensus* from Sumatra were slightly larger, being 17.2-18.0mm wide and 26.5-26.7mm long (Robinson and Kloss 1918).

Both adults are reported to take turns at incubating the eggs (Smythies 1986), although Jeyarajasingam (1983) only observed one adult attending a nest, believed to contain eggs, during two days of observation. At this nest, an adult took a solitary green leaf into the nest each time it entered and then stayed sitting there for periods of up to 75 minutes. In the Indian subcontinent it has also been noted that the incoming bird often brings a green leaf or grass-blade (Ali and Ripley 1983).

Both sexes also feed the young. Baker (1934) noted that, in the Indian subcontinent, although double-brooding is not usual, a second nest is immediately made if the first clutch is robbed. Second nests are usually built in very close proximity to the original.

In Sumatra, a nest of Silver-breasted Broadbill containing three eggs was discovered on Mount Kerinci at 900m: one of the eggs, however, apparently belonged to an uni-

219

dentified cuckoo. The cuckoo's egg measured 14.7 x 18.8mm and was completely white. A cuckoo egg was also found in the nest of a Silver-breasted Broadbill in Tenasserim (Robinson and Kloss 1918).

DESCRIPTION *S. l. lunatus*

Adult male Lores rusty, or occasionally blackish-rusty; sides of the head and ear-coverts ferruginous, forehead pale ashy, this colour changing gradually to the light ferruginous of the crown and nape; a bold black supercilium, starting above the lores, extends to the nape. Orbital skin yellow, forming bold eye-ring. Mantle ashy-brown; lower back rufescent changing to pale chestnut-rufous on the rump and uppertail-coverts. The tail is graduated, black, with the innermost three pairs broadly tipped white, though this is hidden from above. Upperwing-coverts are blackish-brown. Feathers at bend of wing with pale blue tips. Outermost primaries blackish, becoming greyish or blue-grey towards tip: tips of longest two primaries (3 and 4) narrowing sharply from broad inner web to form a distinct, white, sharp triangular point. Primary 4 pointed, with blue-grey inner web; primaries 5-10 distinctly notched at tip, with outer web pointed and inner web rounded, the inner webs being brown with ferruginous tip, this being broadest on outermost and narrowest on innermost feathers. Primaries 4-10 black on outer web with narrow blue tip, and narrow white margin to notch. Inner web of all primaries with white spot near base, forming narrow white bar on underwing extending across secondaries. Outer web of base of all but outermost primary pale blue, this extending along edge of feathers towards tip to form prominent blue wing-panel that is mimicked by pattern of blue in outer web of outermost seven secondaries, so that closed wing is largely blue and black. Outermost seven secondaries similar in shape to innermost primaries, being distinctly notched at tip. Outer webs of these feathers black, with very narrow blue and white tips (the white nearest the notch); inner webs basally white, rest grey-brown, with very narrow blue-grey margin at tip to outermost 3-4 secondaries, and ferruginous tip or spot near notch at tip of secondaries 5-6. Secondaries 7-9 with grey base to inner web, but ferruginous inner web to distal half; outer web of these feathers almost entirely rufous-buff. Tertials black with rufous-buff tips. The throat and breast are silky ashy-grey, fading to white on the belly and undertail-coverts; thighs black. Iris colour variable, from iridescent emerald-green to sapphire-blue; orbital skin bright pale yellow to greenish-yellow; bill colour variable, usually with extreme base orange-yellow, and narrowly edged with dark green which grades into sky-blue of rest of the bill, but sometimes greenish-blue with greyish tinge, basally orange-yellow; feet yellowish-green or olive, soles yellow. Bill tip hardly hooked, rictal bristles present but mostly rather short.

Adult female Differs from male in having a gorget of silvery-white across the upper breast and sides of the neck behind the ear-coverts, formed by white tips to feathers in narrow band across breast. The extent and thickness of this band varies between individuals.

Immature Closely resembles adult from a very early age, so that even when only half-grown it has the appearance of a miniature adult, but with short, rounded wings and short tail. However, the head is generally darker than that of adult, in particular the forehead may be buffy, rather than grey, whilst the black coronal bands are proportionally wider and may extend across the nape. The upperparts may be more rufous than in adult, the underparts are marginally darker, most notably on the breast, whilst colours in the wing may be very slightly duller. *S. l. intensus*: iris pinkish-grey, bill dark sooty, tinged green and basally orange-yellow.

MEASUREMENTS Wing 82-91; tail 67-80; bill length 22-24, width 14.5-15.5; tarsus 19-20.5; weight (*intensus* 25.2-34, *rubropygius* 33-35).

EURYLAIMUS

Four species of brightly plumaged South-East Asian broadbills, including two endemic to the Philippines. The Philippine species, however, are characterised by their possession of a periocular wattle, and may be better placed in a separate genus. *Eurylaimus* broadbills are usually encountered in pairs or family groups, but the density of individuals in a forest may be high: the characteristic and regularly uttered, explosive insect-like trills of *E. ochromalus* and *E. javanicus* are invariably answered by a chorus of others. These are colourful, rather squat, foliage-gleaning insectivores. They have large, brightly coloured and strongly hook-tipped bills with which they prey on large arthropods and occasionally lizards. The two Philippine species, which are sometimes considered to be conspecific, are probably seriously threatened by habitat loss.

42 BANDED BROADBILL
Eurylaimus javanicus   Plate 21

Eurylaimus javanicus Horsfield, 1821, Trans. Linn. Soc. London, 13, pt. 1, p.170: Java.

Alternative name: Purple-headed Broadbill

FIELD IDENTIFICATION Length 21.5-23. Banded Broadbill is a relatively large, distinctive species, found from Indochina to the Sunda Region. The combination of purple-red head and underparts, and yellow streaking and spotting on the dark upperparts, serves to identify adult Banded Broadbill from all other species in its range. Males differ from females in having a narrow black band across the upper breast, although this is indistinct or lacking in males on Borneo and Java. Juveniles have pale brown heads and upperparts marked with yellow, most notably on their wings, where there are rows of large yellow spots. The throat is pale yellowish-white, the breast to belly buffy and, in recently fledged birds, there is a sooty band across the upper breast. Immatures have crowns strongly washed with shades of yellowy-green, and purple markings on the sides of head and neck and belly.

Similar species Banded Broadbill can easily be distinguished from its smaller, sympatric congener, Black-and-Yellow Broadbill, by the coloration of the head (purple in Banded, black in Black-and-Yellow), and from the lack of a white neck collar. Black-and-Red Broadbill is more similar but, instead of yellow on the upperparts, has a prominent white wing-flash. It also has a bicoloured bill, and is a deeper, richer maroon colour below.

VOICE The most frequently given call, often as a duet with one bird starting shortly after the other, is a brief, sharp, whistle, *wheeoo*, followed immediately by a loud, rising, rather cicada-like, frantic trill, 5-6s long. The trill increases in tempo and rises in pitch until the last two or three notes, which tail away quietly and at a noticeably lower pitch. The sequence may be frequently repeated during bouts of calling. Very often, birds making this long call make other vocalisations to each other in-between, most frequently a brief nasal disyllabic *whee-u* or *whee-oo*, with emphasis on the second part of the first note. Other calls are a falling-tone *kyeeow*, which could be confused with a call of Blue-winged Pitta or Blue Pitta, a rolling *koewrr* and a yelping *keek-eek-eek*. Some of these calls are also similar to those of the Black-and-Yellow Broadbill.

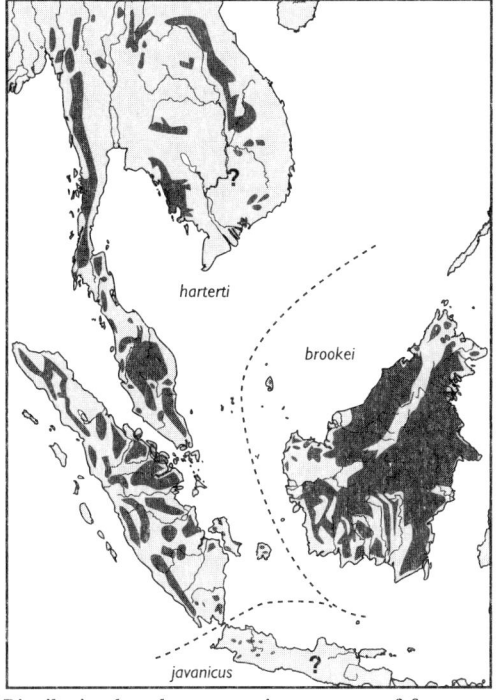

Distribution based on approximate extent of forest remaining within the altitudinal range of the species. Former range more widespread.

DISTRIBUTION The Banded Broadbill is distributed from Indochina and south-eastern Burma to the Sunda Region. Where suitable habitat exists, it occurs throughout much of Thailand (except central), and is widespread in Peninsular Malaysia and on the islands of Java, Borneo and Sumatra and some of their satellite islands. Relatively small forested offshore islands are also often colonised by this species, such the Tiga and Sebatik Islands of Sabah. A population also occurs on the Natuna Islands. Although probably formerly widespread in parts of Java, its distribution there is now very fragmented.

Delacour and Jabouille (1931) could document few localities where they found Banded Broadbill in Indochina, these being at several sites in Cochinchina and southern Annam and near Vientiane, Laos. More recently, R. Timmins and J.W. Duckworth (*in litt.*) found this species at many sites in central and southern Laos. In Burma, it is known from Karenni, the Karen Hills and Tenasserim.

GEOGRAPHICAL VARIATION Five subspecies were listed by Mayr (1979), but only three are recognised here. Kloss (1931) stated that *billitonis* (of Bangka and Belitung [Billiton] Islands, to the east of south Sumatra) was an intermediate form between *harterti* and *brookei*, differing in the coloration of the crown, throat, foreneck and breast, and in the thickness of the black breast-band. However, examination of a series of skins in BMNH of *harterti* shows that there is a great deal of variability in these features, with some birds being identical to specimens from Bangka and Belitung, as noted by Chasen (1937). Hence, following Mees (1986), *billitonis* is treated here as a synonym of *harterti*. Similarly, no clear differences could be found in specimens from the South-East Asian mainland and from Sumatra. Chasen (1935) separated these populations into *harterti* (Sumatra and the Riau Archipelago) and *pallidus*, describing the latter as being paler on the head and underparts. In the following account, *pallidus* is included in *harterti*.

E. j. javanicus (Endemic to Java) Small. Undertail-coverts pure yellow. Male lacks black breast-band. Head and throat of female glossed with grey.

E. j. harterti (Sumatra and the Riau Archipelago, Burma, Thailand, Laos, Cambodia?, Vietnam and Peninsular Malaysia including Penang and Singapore) Differs from nominate subspecies in large size (wing 101-116, weight 74-84) and in having maroon, not brown, upper mantle. Underparts darker (dark pink to pinky-maroon) than nominate birds. Undertail-coverts pinky-maroon. Males have distinct black breast-bands. Iris ultramarine-blue to emerald-green.

E. j. brookei (Borneo and the North Natuna Islands) Males separable from similar-sized *harterti* by pinker, obsolescent black pectoral band, darker more blackened forehead and often pinker, less leaden throat. Females differ only in having slightly darker forehead. Iris usually greyish, occasionally light blue. Weight 73-87, wing 104-112.

HABITAT In the northern part of its range, the Banded Broadbill is usually restricted to evergreen forest and mixed deciduous forest in the vicinity of streams and rivers, although Baker (1934) noted that it also frequents gardens, parks and the surroundings of villages. In Malaysia and Sumatra, this is a widely distributed species in forested habitats (Chasen and Hoogerwerf 1941, Medway and Wells 1976), and has occasionally been found in rubber plantations (pers. obs.). Habitats in which Banded Broadbills have been recorded in north Borneo include primary and logged lowland to lower montane forests, upland heath forest (*kerangas*), peatswamp forest, *Albizia* groves (plantation) and overgrown rubber estates (Mitra and Sheldon 1993). In Sumatra, this species occurs in freshwater swamp forest (Silvius and Verheugt 1986, Verheugt *et al.* 1993) as well as a variety of other forested habitats. In Java, Banded Broadbill is often found at forest edge, particularly on mountain slopes (B. van Balen verbally).

It is a fairly common resident throughout the lowlands and hills of Thailand, Peninsular Malaysia and Sumatra to c.1,050m and throughout Borneo, where it has been recorded from sea level up to 1,220m. In Java, the nominate subspecies occurs from sea level to 1,500m (Hoogerwerf 1948), although Kuroda (1933) gives the altitudinal range as 485-915m, suggesting that it is commonest at these altitudes. *Harterti* has been found at 915m in South-East Thailand (Riley 1938), and at 610m in Nakon Srithamarat, Peninsular Thailand (Robinson and Kloss 1924a). In Cochinchina (south Vietnam), Beaulieu (1932) found this subspecies below 170m, although Wildash (1968) reported, perhaps mistakenly, that it is a montane species in south Vietnam. In Laos, it is most common below 500m, but occurs to at least 1,100m on the Bolovens Plateau (R. Timmins and J.W. Duckworth *in litt*.). On Banguran, in the Natuna Islands, Oberholser (1932) reported Banded Broadbill at 305m.

STATUS In northern Thailand, Banded Broadbill was described as a bird of extreme rarity by Deignan (1945), whilst on Java the nominate subspecies is now confined to scattered patches of forest, although still widespread and not uncommon where suitable habitat occurs in West Java (B. van Balen *in litt*.). However, it may have always been rare in central and eastern Java since Bezemer (1929) wrote that Horsfield collected it only once in East Java and did not encounter it there again, whilst he knew it himself from only a single locality in central Java, despite making numerous field trips there.

It is a relatively common species in suitable habitat elsewhere in the Sunda Region. Data from a study carried out near Kuching, Sarawak, suggests that the population density of this species in primary rainforest was around 10 per km² (Fogden 1976). In other parts of Borneo, such as in the lowlands of East Kalimantan (Pearson 1975) and parts of West Kalimantan (B. van Balen *in litt*.), it also appears to be common, but in Brunei it is apparently rather rare (Mann 1987). Delacour and Jabouille (1931) reported that it was uncommon in the lowlands of southern Indochina.

MOVEMENTS None recorded.

FOOD Observations in Sarawak (Fogden 1970) suggest that Banded Broadbills feed largely on orthopteran prey (grasshoppers, crickets, katydids). The size of prey taken was generally found to be larger than that eaten by its smaller sympatric congener, Black-and-Yellow Broadbill. Fogden estimated that the mean size of orthopterans eaten was about 55mm. However, Sharpe (1877a) found only beetles in a specimen that he collected, and Coleoptera indeed appear to be important in the species's diet. Specimen labels from birds collected in Sabah record various beetles including members of the Curculionidae (weevils, bark beetles), Tenebrionidae (ground beetles) and Carabidae (predatory ground beetles) in stomach contents, as well as orthopterans, hemipteran bugs and a snail. Smythies (1981) also lists spiders in the diet of birds on Borneo. Banded Broadbills have also been observed flycatching in a forest clearing in Borneo.

In Java, the diet is reported largely to comprise insects, although the fruits of *Ficus* (figs) have also been recorded. Small fruits have been found in the stomach of a bird collected in Sumatra (Chasen and Hoogerwerf 1941). Insects eaten in Java are reported to include grasshoppers, beetles, homopteran and heteropteran bugs and katydids (Tettigoniidae: *Scambophyllum* spp.) (Sody 1989). Banded Broadbills in Burma have been reported to eat beetles, caterpillars and insect larvae, and one bird shot by Davison contained a 10cm-long lizard (Baker 1926). An adult was observed feeding a newly fledged young with a large green katydid at Semengo by Fogden (1965).

HABITS Banded Broadbill, like its congeners, is most frequently encountered in the mid-level of forest, in pairs or small groups, where it is active all day. The calls of this species are given frequently, often during the heat of the day when few other species are calling. The cicada-like trill may be taken up in quick succession by a number of birds, as they answer each other in the forest.

Although not agile, Banded Broadbills hunt by gleaning invertebrates whilst in flight. Typically they may remain motionless, apart from the head, which is rotated as the bird searches for food, gleaning their prey during irregular fluttering flight, and then alighting on a new perch. Unlike the majority of foliage-gleaning insectivores, Banded Broadbill, and the closely related Black-and-Yellow Broadbill, frequently look upwards whilst searching for invertebrates from their perch, and pluck prey from the underside of leaves (Fogden 1972). This species has also been observed catching flying insects.

BREEDING

Seasonality Evidence suggests that mainland populations of Banded Broadbill breed mainly during the dry period of the year that follows the heavy rains associated with the north-east monsoon. Nests have been found in Peninsular Malaysia on 4 February and 3 March, whilst immatures have been recorded between 8 April and 2 September (Medway and Wells 1976, UM). In addition, Baker (1934) was sent several nests from Malacca and Perak and was able to give laying dates as being from 23 March to 7 May. Eggs have been taken in Peninsular Thailand on 19 July (Medway and Wells 1976) and an immature was collected in the south-east on 29 December (Riley 1938). Davison collected a nest containing two fresh eggs in Tenasserim on 21 March (Hume 1880). In Laos, however, an immature was observed on 11 June (R. Timmins *in litt*.), suggesting that breeding had taken place at the beginning of the wet season, rather than during the dry season.

The main breeding season of Banded Broadbill on the Greater Sunda Islands may coincide with that for mainland populations, but evidence suggests that the breeding season is longer on both Sumatra and Java. In Sabah, an adult was observed stripping leaves and plant fibres, presumably for nesting material, from low vegetation in primary forest on 16 March (A. Owen *in litt*.) and birds with enlarged sexual organs have been collected between 16 March and 25 July (Sheldon in prep.). A newly fledged young was observed at Semengo in Sarawak on 25 September (Fogden 1965) and an immature was trapped in Sabah on 10 August. In Sumatra, an immature was collected in Lampung on 23 November, and eggs have been collected on the nearby island of Belitung on 6 April, whilst an immature was collected in Lampung on 23 November (Kloss 1931a, van Marle and Voous 1988).

In Java (*juvanicus*), nests have been found in April, June and December (Hellebrekers and Hoogerwerf 1967), one of which was suspended about 2.5m above the ground (J. Houwing *in litt*.), and two recently fledged juveniles were collected on 1 April. Immatures of similar age were collected in Java on 9 November, 24 December, 1 March and 26 June, whilst similar-aged immatures have been collected

in Sumatra on 4 March, 4 July and 28 September (RMNH).
Nests In Peninsular Malaysia, only two nests have been well documented (UM). Both were at the building stage, one in logged forest at c.250m altitude on Mount Benom, and the other at the edge of a clearing in primary forest. The former was suspended from a twig in a dead tree, and hung very close to the trunk. The nest in primary forest, which was described as a ragged structure similar to that of a Black-and-Red Broadbill, was hanging between the main trunk of a tall *Koompassia excelsa* tree and a small lateral branch, some 21m above the ground. Two birds were observed to be building the nest. Nests are not always suspended from trees, however. A nest in Tenasserim was suspended from the tip of a very tall bamboo and overhanging a stream (Hume 1880).

The nest is a ragged, globular structure with a lateral entrance. Baker (1926, 1934) described it, based on several collected in Malaysia last century, as being shaped like a huge pear with an ample porch over the entrance, measuring anything from 75-90cm long, and being built from a miscellaneous mass of materials (almost anything that can be picked up in a forest being made use of). Nests are very strong and compact, and primarily made from small twigs, roots, leaves, grass, moss (both dead and fresh) and other materials, interwoven together with bamboo leaves and leaf skeletons, and with a lining of leaves in the nest cavity. A nest from Java was made of grass, roots and twigs, and lined with leaves (J. Houwing *in litt.*), whilst one from Tenasserim was described as being lined with bamboo leaves (Hume 1880). The entrance is generally about two-thirds of the distance from the bottom, and Baker (1934) noted that the roughly made porch, of twigs, leaves and grass, is often huge and may completely conceal the entrance.

Scraps of lichen and green moss, caterpillar excreta and cocoons are invariably fastened to the outside with cobwebs, tendrils and plant stems; these also hang down to form the distinctive long tail typical of broadbill nests and presumably provide camouflage (Baker 1934).

Clutch-size is usually 2-3 throughout the species's range (Baker 1926, 1934, Hellebrekers and Hoogerwerf 1967), although it is possible that it is sometimes larger. The eggs are a narrow oval with almost equal ends and a smooth, faintly glossy surface. Eggs from West Java are dirty white or greyish-cream, occasionally pinkish-clay, and fairly densely covered with irregular, small and rather pale markings of which the predominant are dull rusty-brown and the others lavender-grey. The markings are concentrated at the blunt end, where lavender-grey spots predominate (Hellebrekers and Hoogerwerf 1967). Eggs from elsewhere are reported to be white to creamy-white, spotted and speckled with deep purple, dark reddish-brown, or, occasionally, with pale reddish-lavender: although very distinctive, these eggs may resemble those of some drongos Dicruridae (Baker 1934). Ten eggs measured by Baker (1926) varied in size from 26.1-28.0 x 17.1-20.0mm, with an average size of 27.0 x 18.8mm. Seven eggs from West Java measured 27.6-31.5 x 20.1-22.2mm (Hellebrekers and Hoogerwerf 1967).

Observations in Sarawak showed that whilst about 70-80% of food was provided to young birds by their parents 13 weeks after fledging, this had dropped to around 20-30% by the time they were 20 weeks old (Fogden 1972).

DESCRIPTION *E. j. javanicus*
Adult male Forecrown to nape dark purple, becoming almost black at base of bill. Lores and narrow rim of feathers around front half of eye black; chin, throat and ear-coverts pale purple, concolorous with underparts. Mantle dark brown, tinged maroon, merging into blacker back. Back streaked with light yellow, most prominently in centre, the streaks formed by yellow inner webs of feathers; concealed basal half of central back feathers white; rump a mixture of light yellow and black, the yellow primarily in centre, where streaks join to form a single long broad streak; uppertail-coverts black, broadly fringed with yellow at tip. Tail feathers black, all but central pair with white (or buffy-white) spot near tip of inner web, this spot becoming larger towards innermost feathers. Spots hidden from above, conspicuous from below. Entire underparts from chin to breast and belly purple-pink. Undertail-coverts yellow. Primaries dark brown, with very narrow yellow leading edge to base of outer primaries, this yellow extending along leading edge and bend of wing. Primary coverts dark brown, rest of wing-coverts blackish. Primaries 4-7 with small yellowish spot in outer web at base, usually visible as a small spot in wing at rest. Secondaries dark brown with bright yellow fringe to outer web, extending from a quarter from tip to half-way down feather, narrowest on outer secondaries, broadest on innermost feathers, and forming a distinct yellow panel in flight feathers. Inner tertials blackish; outer tertials black and yellow, the yellow either confined to distal inner half of inner or outer web, or, in some feathers, the entire distal third to half may be yellow. Underside of flight feathers greyish, underwing-coverts pale yellow. Iris sapphire-blue; bill turquoise-blue with green tip and green or black edges to both mandibles; feet lilac-grey, pink or violet-tinged. Bill strongly hooked; rictal bristles well developed.

Adult female Differs from male in coloration of head and underparts. Head has distinctive greyish sheen to rear of crown, edged by slightly contrasting purple-pink collar across top of mantle. Forecrown distinctly darker than rear of crown. Chin dark purple, throat purple-pink, washed with greyish, the intensity slightly paler than on crown, this grey wash extending onto upper breast; central breast greyish, lower breast and belly purple-pink (without grey wash), becoming mixed with buffy-yellow on belly and lower flanks. Undertail-coverts yellow.

Juvenile (male) Forehead pale brown with bases of feathers yellow, most prominent on feathers of sides of forehead; rest of crown pale brown, some feathers having yellowish inner webs at base. Feathers of narrow supercilium brown with broad yellow centres and tips forming a distinct yellow supercilium that extends from the sides of forehead to sides of nape, where broadening and extending laterally to form a narrow, broken yellow collar. Lores and feathers around eye dark brown. Ear-coverts paler brown with long, narrow, yellow steaks. Mantle pale brown with droplet-like yellow spots formed by yellow patches in most feather-centres. Feathers of central back and rump mostly yellow, those of sides of back mostly brown. Tail blackish with white spot in inner web near tip of all but innermost pair. Feathers of wing, including retrices, dark brown, marked with three lines of yellow spots formed by yellow droplet-like spots in the centre and tip of feather tracts in the lesser and median coverts and by broad yellow fringes to tips of greater coverts. Tertials and secondaries dark brown with narrow yellow patch in outer web of innermost part of feathers forming, with the yellow tips of greater coverts, a distinct yellow wing-patch. Primaries dark brown, with feathers 2-6 strongly emarginated. Fringe to inner web of innermost part of all remiges white; underwing-coverts white. All remiges appear grey from below. Chin yellow; throat yellow, slightly streaky in appearance due to brown basal parts of feathers, and demarcated from similarly marked breast by narrow solid yellowy-white band across upper breast. Belly and flanks yellow to yellowy-white with pinky tinge to centre of belly. Undertail-coverts yellow. Bill orangy-horn; dried legs and feet yellow. Iris greyish-blue.

Immature As juveniles get older, the crown becomes strongly washed with shades of yellowy-green, whilst purple-pink patches appear on the sides of head and neck, and belly. Iris bluish.

MEASUREMENTS Wing 99-101; tail 52.5; bill length 33, width 22-24; tarsus 23-24; weight (*brookei*) 73-87, (*harterti*) 74-84.

43 BLACK-AND-YELLOW BROADBILL
Eurylaimus ochromalus   Plate 22

Eurylaimus ochromalus Raffles, 1822, Trans. Linn. Soc. London, vol. 13, pt. 2, p.297: Singapore Island.

FIELD IDENTIFICATION Length 13.5-15. A small, distinctive Sundaic broadbill, with black head and upperparts separated by a white collar, and conspicuous bold yellow markings on the back and wings; the breast is vinaceous-pink, fading into pale yellow on the belly and undertail-coverts. The bill is bright turquoise to cobalt-blue with green basal part of upper mandible and black cutting edge. The male has a complete black band across the breast, whilst the female has a broken band. Juveniles lack this breast-band and have greyish-white underparts and a pale yellow supercilium.

Similar species Within its range, Black-and-Yellow Broadbill is unlikely to be confused with any other species. However, the call is very similar to that of Banded Broadbill and confusion is therefore possible if the calling bird is not seen. For differences in call, see under Voice.

VOICE Black-and-Yellow Broadbill calls regularly, at all times of the day, and is more frequently heard than seen. The most commonly heard call is a cicada-like trill. This begins slowly, with a few sharp, down-slurred notes, then gradually gains speed until ending as a loud quivering trill. The call is confusingly similar to that of Banded Broadbill, but can be fairly easily distinguished. The rising, bubbling trill of Black-and-Yellow Broadbill lacks the introductory *wheeoo* of Banded Broadbill, accelerates more gradually, and does not tail away at the end, but stops rather abruptly. The trill is also slightly longer than that of Banded Broadbill, lasting some 8-12s. Often more than one bird of a pair will call at once, but alternately, with the end of one bird's call overlapping with the beginning of the others.

The throaty *keowrr* and squeaky *kyeeow* given by Black-and-Yellow Broadbill are also similar to calls of Banded Broadbill. In Brunei, a Black-and-Yellow Broadbill in the company of a pair that were making the bubbling trills was heard to utter a plaintive, rather shrill *peep* (D. Holmes *in litt.*). At the nest, males have been heard to make a call described as *kor kor kor*, but adults have also been observed making the more usual cicada-like call when perched within 50cm of their nest.

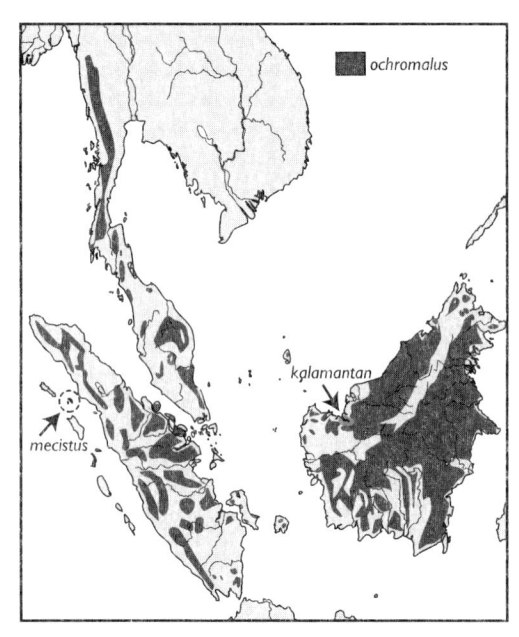

Distribution based on approximate extent of forest remaining within the altitudinal range of the species. Former range more widespread.

DISTRIBUTION The Black-and-Yellow Broadbill is distributed virtually throughout the Sunda Region, although not present on Java. It has been recorded from Peninsular Thailand and Tenasserim, Burma, south to Johore, Peninsular Malaysia, and on the islands of Borneo and Sumatra. Northernmost populations are reported to occur at c.15°N at Ye in Tenasserim (Robinson 1927).

Populations also occur on a number of offshore islands between Peninsular Malaysia and Sumatra, including Penang, the Riau Archipelago, Lingga Archipelago, Belitung and Bangka (van Marle and Voous 1988). To the west of Sumatra, the species is only known from Pini and Tana Masa in the Batu Islands (Meyer de Schauensee 1940). Oberholser (1932) reported it from Bunguran Island in the Natuna Islands, half-way between Borneo and Peninsular Malaysia. Gibson-Hill (1949) reported that it was formerly present in Singapore, although noting that it was unlikely that the species still occurred there. Its extinction in Singapore was confirmed by Lim (1992).

GEOGRAPHICAL VARIATION Three subspecies were recognised by Peters (1951), but examination of series of specimens (BMNH) reveals their great variability. Peters's three subspecies, two of which have very restricted ranges, are all very similar, and it might be more appropriate to treat Black-and-Yellow Broadbill as monotypic. However, since it has not been possible to examine sufficient material from the Banyak Islands or the Saribas District of Sarawak, all three are included in this account. Nevertheless, it seems rather unlikely that a distinct subspecies should occur in part of Sarawak, and this form may warrant inclusion with the nominate. Van Marle and Voous (1988) recognised *mecistus*, but provided no discussion on taxonomy.

E. o. ochromalus (Peninsular Thailand, Tenasserim, Peninsular Malaysia, Sumatra, Borneo, the Riau Archipelago, Lingga Archipelago, Belitung, Bangka, Batu

Islands and North Natuna Islands) See Description. Apparently slightly smaller than the other subspecies.

E. o. mecistus (Tuangku, in the Banyak Islands, to the west of north Sumatra, at 2°10'N, 97°17'E) Apparently larger than the nominate subspecies (Riley 1938, Ripley 1944). Wing 87.5-88.

E. o. kalamantan (Saribas District, Sarawak) Supposedly consistently longer-winged (Robinson and Kloss 1919a) and with more yellow on the mantle and back than nominate *ochromalus*, and with a broader white bar across the nape. Wing of male 81-89, of female 76-78 (Chasen and Kloss 1930, Kloss 1930).

HABITAT In Thailand Black-and-Yellow Broadbill is a species of evergreen forests, forest edge, logged forests and secondary growth. In Peninsular Malaysia, Sumatra and Borneo, it occurs in lowland to lower montane rainforest, freshwater and coastal peatswamp forest and upland heath forest (kerangas). Wells (1976) found it in lowland primary peatswamp forest in western Sabah, whilst in Peninsular Malaysia it has been observed in selectively logged peatswamp forest (Prentice 1988, Kang and Lee 1993). In Sarawak, it has been found in mixed dipterocarp forest approaching heath forest in character and also in tidal swamp forest (Duckworth and Kelsh 1988). B. van Balen (*in litt.*) found it to be a common species throughout the freshwater swamp forests of Danau Sentarum (West Kalimantan). Black-and-Yellow Broadbills are rather adaptable, and are frequently found in logged forests, secondary habitats where trees persist and, occasionally, overgrown plantations including cocoa, rubber and *Albizia* (Thompson 1966, Holmes 1969, Mitra and Sheldon 1993).

The species has a relatively wide altitudinal distribution. In Borneo, it generally occurs from sea level to 1,000m, and occasionally higher: in Sarawak, Wells *et al.* (1979) found it up to 980m on Mt Mulu, and specimens have been collected at c.1,130m in the Kelabit Uplands (Smythies 1957) and 1,220m on Mt Dulit (UMZC). In Sabah, it has been recorded from sea level to 1,200m on Mt Lumaka (Sheldon in prep.). It occurs throughout Sumatra and its eastern satellite islands to 900m (van Marle and Voous 1988); and to 700m in Peninsular Thailand and Peninsular Malaysia. Robinson (1915) found it at 365-460m in Bandon Province, Thailand. Hume and Davison (1878) reported that, in southern Tenasserim, this species is absent from high hills.

STATUS Robinson (1915) stated that it was not very common in Bandon Province, Thailand, but in most parts of its range it seems to be a relatively common bird. Pfeffer (1960) described it as locally abundant in Borneo, whilst Pearson (1975) found it common in the forested lowlands of East Kalimantan; B. van Balen found it common there upto c.1,000m, and common in West Kalimantan, and Gore (1968) reported that it was common in the lowlands of Sabah. In Brunei, it is also very common and widely distributed (Holmes 1969). In Peninsular Malaysia this is still a relatively common species in forested parts of the lowlands and hills (pers. obs.). The status in Sumatra is undocumented, but in view of some large areas of forest that remain, it must still be common where appropriate habitat exists.

There may have been no observations of the subspecies *mecistus* from the Banyak Islands since at least 1970, but the islands were still heavily forested in 1989 (Holmes 1994) and it seems likely that Black-and-Yellow Broadbills must still occur there.

Data from a study carried out in an isolated patch of primary rainforest near Kuching, Sarawak, suggest that the population density of this species there was around 10 per km^2 (Fogden 1976). Evidence suggests that this species survives well in logged forests in Sabah (Lambert 1990, 1992).

MOVEMENTS None recorded.

FOOD In Sarawak, the main food of this species is reported to be orthopterans (grasshoppers, crickets and katydids), mostly of about 35mm length (Fogden 1970), although other invertebrates that have been recorded in the diet of Black-and-Yellow Broadbill in Borneo include caterpillars, mantises, beetles including Scarabaeidae (chafers, dung beetles) and Cerambycidae (long-horned beetles), unspecified hymenopterans (wasps, bees, ants) and molluscs (SarM, Smythies 1980, Sheldon in prep.). In Peninsular Malaysia, birds have also been observed catching flying termites from perches high in the upper canopy: sallies into the swarm of emerging termites were made, with the termites snatched in mid-air (pers. obs.). Birds shot in *Albizia* plantations in Sabah frequently contained lepidopteran larvae. Pfeffer (1959) found flies in the stomach of this species in Borneo.

One bird collected in Sarawak had the remains of orange-red berries, with large dark pips, in its stomach (SarM), but there is little evidence that this species regularly takes fruit. Thus while Robinson (1927) reported that Black-and-Yellow Broadbills feed largely on fruit, especially figs, Lambert (1989a) did not record them eating figs on any occasion during a three-year study of fig-eating and seed dispersal by rainforest birds in Malaysia.

HABITS Small groups frequent the middle and upper canopy and emergent trees of forests, and occasionally taller trees within plantations. Although sometimes noisy, members of such groups, which may be rather dispersed, generally sit quietly scrutinising their surrounding for invertebrate prey for considerable periods of time. When something is spotted, the bird generally poises before launching into the air, catching the prey in mid-flight, usually from the surface of leaves, and then alighting at a new hunting station. Hunting stations are usually perches in relatively open parts of the lower to upper canopy (pers. obs.). Flocks of up to 10-15 birds have been reported (Thompson 1966). Individuals and groups sometimes also join mixed bird flocks, although these associations may be rather short-lived.

During calling close to its nest, one adult was observed to hold its wings slightly, where flexed and vibrating, whilst the tail was slightly cocked and fanned (F. Sheldon *in litt.*). Thompson (1966) noted that the head was usually bowed slightly during calling.

Prentice (1988) observed mating in this species in logged peatswamp forests of Peninsular Malaysia in the early afternoon, after frequently hearing birds calling in the morning. A calling male was observed to alight on a small horizontal branch near the trunk, just above another branch occupied by a silent female. The male slowly stretched its wings out horizontally, almost to full stretch, where it held them whilst wagging its tail conspicuously from side to side. Simultaneously, the female half-spread its wings and shivered them. After a short time, the male flew down and copulated with the female. Both birds then flew c.3m to another perch where copulation resumed, after which the male sat c.30cm from the female slowly raising and lowering his wings, before both birds flew off once more.

Nuptial behaviour of Black-and-Yellow Broadbill

BREEDING Mating has been observed on 7 February (Prentice 1988) in Peninsular Malaysia, but the eight nesting records documented for Peninsular Thailand and Peninsular Malaysia all come from between 10 May and 14 July (UM, Anon. 1988), whilst two juveniles with adults have been observed in Peninsular Malaysia on 21 August at Gombak and 19 October at Sungai Buloh (M. Chong *in litt.*). A juvenile was collected in Thailand at the end of May (Meyer de Schauensee 1934).

In Sumatra, mating has been witnessed in mid-March (Chasen and Hoogerwerf 1941), nest building has been observed in May, June and July (Nash and Nash 1985, Holmes in prep.), eggs collected in May (van Marle and Voous 1988) and immatures collected in June. On Pulau Tuangku, off the Sumatran coast, a partly finished nest of *mecistus* was obtained on 30 January, hung from the tip of a branch of a lime tree in a clearing, some 9m off the ground (Richmond 1903).

In Sabah and Sarawak, nest construction has been observed in mid-March whilst occupied nests have been found in April, May, June and July, and recently fledged young observed between early May and mid-June (Fogden 1965, Smythies 1981, Sheldon in prep.). In Sarawak, an immature male was collected on 9 April. Birds with enlarged gonads have been collected there between mid-March and 19 August.

All these records suggest that, in general, breeding coincides with the onset of the dry season, but that breeding may start earlier in the Greater Sunda Islands than in Peninsular Malaysia and Thailand.

Nests of Black-and-Yellow Broadbill are usually suspended from a branch or other suitable vegetation. Those found in Malaysia and Peninsular Thailand have been suspended 5-18m above the ground, in open situations. One of these was in a rubber tree, another two in remnant forest trees in a rubber plantation. Nests are not always suspended in conspicuous places: one in Sabah was being built on some branch tips at the lower edge of the canopy in a dense tree and was only discovered when an adult was observed carrying about six thin twigs, c.15-20cm in length, into the tree (F. Sheldon *in litt.*). Occasionally, nests may be suspended above water (M. Chong *in litt.*).

Nests found in Malaysia have been large, untidy, pear-shaped structures composed of looped grasses or twigs and adorned with lichens. One from Tenasserim was made of moss, fungal mycelia and skeleton leaves, with a lining of rough grass roots, bamboo leaves and leaf stalks. This nest was supported in a loop of a cane and measured about 17.8cm long, 12.7cm broad one way and 10cm the other, with walls some 3.8cm thick and an entrance hole of 5 x 6.4cm in dimensions. There was no porch, but a sort of step projected below the entrance hole (Baker 1934).

Of four nests discovered in Sabah, two have been placed close to bee hives, and, according to tribal people in Sarawak, this is also very typical of the species there. One nest, discovered at Gombak, Peninsular Malaysia, on 14 May 1972, contained the egg of an unidentified cuckoo, in addition to three broadbill eggs (UM). The eggs were described as being mushroom-pink, flecked all over with brown and purple-brown, concentrated in a darker ring around the broader end, and 23.7 x 17.4mm in dimensions. This species has also been observed feeding the young of Indian Cuckoo *Cuculus micropterus* (Wells 1984), and, on 16 September 1984 at Gombak, an unidentified recently fledged cuckoo was being fed by two adult Black-and-Yellow Broadbills (pers. obs.).

The clutch-size in Malaysia is reported to be three, although a nest found in Tenasserim contained only two eggs (Baker 1934). Baker (1926) gave the measurements of these as 23.5 x 17.0mm and 23.5 x 16.5mm, noting that they were exactly like those of Banded Broadbill in appearance.

DESCRIPTION *E. o. ochromalus*
Adult male Head, including throat, black, separated by more or less distinct narrow white collar across nape from the black of mantle, this white collar broadening sharply and extending across sides of neck and upper breast, where often tinged with pink or pinky-yellow tones. Mantle and back black, variably patterned with bold yellow markings. Feathers of upper mantle have broad white bases that become yellow in exposed part of many feathers and extend towards feather tip, although in some feathers the terminal third to two-thirds is black; in other feathers the yellow may extend along entire inner or outer web so that the mantle appears to be boldly streaked with yellow. The black of the mantle extends across sides of lower neck and across upper breast to form a distinct band separating the white (or pinkish) lower throat and upper breast from the pink lower breast. Lower back, rump and uppertail-coverts similarly patterned to back, with streaks formed by yellow in outer margins of feathers near tip. Lower breast pink, sometimes with buffy tinge, becoming darker towards belly and flanks, where mixed with yellow feathers of inner flanks (especially on lower part) and yellow undertail-coverts. Scapulars black, boldly marked with yellow as on back; some scapular feathers may be entirely yellow. Lesser and median-coverts blackish. Tertials and secondaries blackish, marked with yellow band along outer edge of outer web, this extending from close to feather tip to about half-way down feather on innermost feathers, but progressively longer towards outer wing. These yellow marks form conspicuous yellow wing-panel.

Primaries 4-6 have white spot in outer web near base of feather, the position of these spots coinciding with the base of yellow margin in outer secondaries. Bend of wing mixed yellow and white, particularly on underwing. Inner web of all primaries greyish-white at base, extending progressively further towards tip from outer to inner wing so that underwing appears greyish-white. Underwing-coverts mostly yellow, with a few blackish feathers at base of outer primaries. Tail blackish, feathers graduated so that inner feathers are longest. Innermost pair of tail feathers marked with yellowish-white spot on inner web close to tip, this spot white on other tail feathers and becoming progressively larger so that it extends across both webs of outermost tail feather, where yellowish on outer web and white on inner. These spots are prominent from below, but mostly hidden from above. Rictal bristles poorly developed, most being short, with a few reaching 10mm in length. Eye-ring black; iris sulphur-yellow; bill bright turquoise to cobalt-blue, the distal two-thirds of the cutting edge black, the distal half of the upper mandible bright blue-green to green; extreme gape sometimes fleshy-yellow; feet colour variable, from bluish-pink to pinkish-horn or brownish-grey.

Adult female Differs from adult male in having incomplete black band across breast.

Juvenile Upperparts dark brown to black, with similar yellow markings to an adult but with the white collar tinged yellow and with distinctive yellow patches on the sides of the forehead that extend across the top of the lores, and usually a few indistinct yellow feathers in a line running towards the nape starting behind and above the eye. Mantle rather grey at first, later becoming darker and developing more yellow. Less extensive yellow in upperwing-coverts when compared to adult, this being confined to the tips of median wing-coverts, but also present is the yellow leading edge to secondaries that is found in adult birds. Chin black; throat yellowish-white; flanks grey with pale centre of breast and belly, later becoming yellow on the flanks to undertail-coverts prior to turning pink. An immature male from Sarawak had iris pale grey, beak pale ultramarine-blue, feet pale flesh-blue.

MEASUREMENTS Wing 74.5-85; tail 43-57; bill length 26-30, width at base 17-19; tarsus 19-23; weight 31.0-39.0.

44 MINDANAO WATTLED BROADBILL
Eurylaimus steerii Plate 22

Eurylaemus [sic] *Steerii* Sharpe, 1876, Nature 14, p.297: Basilan.

Alternative name: Steere's Broadbill.

TAXONOMY The allopatric taxa of *Eurylaimus* broadbills endemic to the Philippines were originally described as two species, *Sarcophanops samarensis* and *Eurylaemus steerii*. Although early texts on Philippine birds (e.g. McGregor 1909-10, Hachisuka 1934-35) also recognised two species, these taxa were subsequently treated as one species by Delacour and Mayr (1945, 1946) and Peters (1951) and more recently by Dickinson *et al.* (1991). No discussion of taxonomic status was provided by these authors, Delacour and Mayr (1945) simply stating that 'although differing clearly in size and color, the two Philippine forms *steerii* and *samarensis* are in our opinion subspecies of one species...'. These Philippine taxa do indeed differ considerably, both in plumage and in biometrics. Differences in plumage are described in the appropriate sections of the species accounts. Furthermore, *steerii* is subdivided into two subspecies, *E. s. steerii* and *E. s. mayri*.

McGregor (1909-10) provided biometric data from an unpublished manuscript by F.S. Bourns and D.C. Worcester, based on ten male and ten female specimens of *steerii* from Basilan and five male *samarensis* from Samar, which suggested a considerable difference in size between the two forms: average wing lengths of 84mm (male) and 85mm (female) for *steerii* and 78mm for male *samarensis*, and average total length being 2cm longer in *steerii*. Average tail length for the males was 60mm for *steerii* and 62mm for *samarensis*, suggesting that *samarensis* is proportionally longer-tailed. Salomonsen (1953) gave average wing measurements of 87.5 (male, n=16, range 85-91) and 87.2 (female, n=14, range 84-92) for *steerii*, and 82.4 (male, n=6, range 80-86) and 80.8 (female, n=5, range 79-82) for *mayri*, and average tail lengths of 60.1 (male *steerii*), 53.9 (female *steerii*), 54.8 (male *mayri*) and 52.8 (female *mayri*). Rand and Rabor (1960) gave wing lengths of six male and nine female *samarensis* from Samar as 78-85.5mm (average 80.2) and 76.5-82mm (average 79), respectively, and of a female from Bohol as 78mm. Rabor (1938) also provided wing length measurements for three specimens of *samarensis* from Leyte: 77mm, 78mm (two males) and 77mm (female).

Fully grown adult specimens examined during the course of researching this book were limited to 18 (10 female, eight male) of *steerii* from Zamboanga (Mindanao) and Basilan; 27 (14 female, 13 male) *mayri* from Mindanao and Dinagat; and 15 *samarensis* (nine male, six female) from Samar and Leyte. Wing lengths of these specimens were as follows: *steerii* male 84-90mm, female 84.5-90mm (average all specimens 86.4); *mayri* male 79.5-86mm, female 79-85mm (average all specimens 81.9); *samarensis* male 75.5-81mm, female 76.5-80mm (average all specimens 78.0). These measurements support the assertion of early authors that *samarensis* is a smaller taxon than *steerii*, but indicate some overlap in wing length between *samarensis* and *mayri*.

Eight adult (four male, four female) *steerii* with full-grown tails had tail length of 55-62mm, average 58mm; eight specimens of *samarensis* (four adult male, two subadult male, two female) had tail lengths of 46-50mm, average 48.25mm. These measurements do not indicate that *samarensis* has a proportionally longer tail, as suggested by measurements given by McGregor (1909-10), and the possibility of an error in McGregor's measurements cannot be excluded.

In view of these, and of the very different appearance of *samarensis* from the other two taxa in the Philippines, *steerii* and *samarensis* are treated in the following account as separate species, thus reinstating the views of early authors. These two species are kept in the genus *Eurylaimus* in the following accounts, but they are quite distinct from other members of the genus, as noted by Olson (1971), and they may be better placed in the genus *Sarcophanops*, first used when *samarensis* was described.

One final point perhaps worth noting is that, as a rule, plumage differences between recognised subspecies of broadbills in Asia are usually rather small. If the Philippine taxa were kept as one species, the differences in plumage between *steerii*/*mayri* and *samarensis* would be the most extreme in the whole Eurylaimidae.

FIELD IDENTIFICATION Length 16.5-17.5. Endemic to the Philippines, where confined to Mindanao and its satellite islands. The male has a black face and throat, maroon-purple forehead and crown, and white nuchal collar. The eyes are surrounded by a blue fleshy wattle. The mantle is dark grey, the lower back and tail are rufous with a purple wash to the rump and uppertail-coverts. The breast to belly are lilac with the centre of the belly yellowish-white. The wings are black with a prominent narrow white-and-yellow bar across the secondaries and tertials. Females differ from males in having pure white underparts. Immatures differ from adults in having white throat (this gradually becoming black), olive-green crown and wash to upperparts, and ill-defined wing-bar with buffish-pink rather than white.

Similar species The only broadbill in its range, this species is unlikely to be confused with any other. It is slightly larger than the closely related Visayan Wattled Broadbill, which differs in having a mottled grey nuchal collar, purple mantle and white-and-lilac wing-bar. The crown of Visayan Wattled Broadbill tends to be slightly paler than that of Mindanao Wattled Broadbill.

VOICE The voice of Mindanao Wattled Broadbill is poorly-known. McGregor (1909-10) stated that the call is a plaintive whistle.

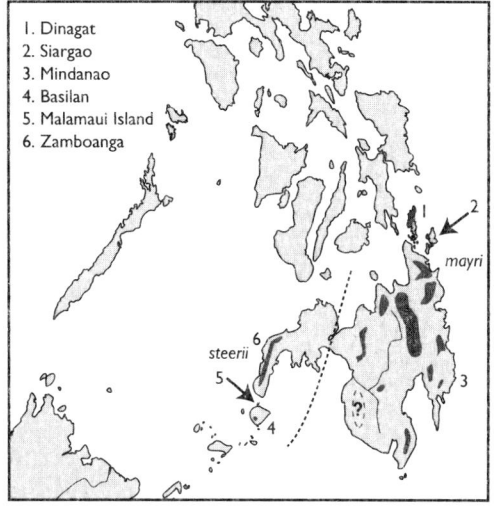

Distribution based on approximate extent of forest remaining within the altitudinal range of the species. Former range more widespread.

DISTRIBUTION Mindanao Wattled Broadbill is endemic to the Philippines, where known only from Mindanao, Basilan, Melamaui, Dinagat and Siargao. Records suggest that it is widespread in Mindanao, since it has been collected or observed in the following provinces: Agusan del Norte, South Cotabato (east of Sarangani Bay), North Cotobato, Surigao del Sur, Davao Oriental, Misamis Oriental and Lanao del Norte.

GEOGRAPHICAL VARIATION Two subspecies are recognised, differing primarily in size.

E. s. steerii (Zamboanga Peninsula of SW Mindanao, and on the adjacent islands of Basilan and Malamaui) Forehead and crown bright maroon-purple; white nuchal collar.

E. s. mayri (Dinagat, Siargao and Mindanao, except Zamboanga) Slightly smaller than nominate subspecies (wing 79-86, bill tip to gape 25.2-27.9, bill width 17.7-18.3). Underparts of male usually paler than in nominate.

HABITAT The species frequents the understorey and middle layers of rainforests. Recently, Mindanao Wattled Broadbills (*mayri*) have been observed in lowland secondary forest close to more pristine riverine forest on Mindanao (Evans *et al.* 1993, G. Dutson *in litt.*). On Dinagat and Siargao, duPont and Rabor (1973a) found this species in dense patches and areas of remnant original dipterocarp forests and in mixed dipterocarp and secondary forests on hillsides in the interior of these islands. Occasionally, pairs were also encountered in scrub forests on both islands. McGregor (1909-10) reported that Mindanao Wattled Broadbill has also been found in mangrove.

Dickinson *et al.* (1991) reported that wattled broadbills occurred up to 1,000m, but specimens have been collected from sea level to 1,220m on Mindanao (FMNH, BMNH). On Mt Apo, Mindanao, this species has recently been observed at c.800m (J. Scharringa *in litt.* to BirdLife), but elsewhere there are no recent records away from the lowlands.

STATUS DuPont and Rabor (1973a) reported that Mindanao Wattled Broadbill was a fairly common species in suitable habitat on Dinagat and Siargao. There do not appear to be any records from Maguindanao or Sultan Kudarat Provinces, or from South Cotabato Province to the west of Sarangani Bay, suggesting perhaps that Mindanao Wattled Broadbill is scarce in the western part of southern Mindanao.

Dickinson *et al.* (1991) considered that Philippine broadbills were uncommon, and most active field workers who have recently visited the country believe that broadbills must now be threatened by habitat loss. Collar *et al.* (1994) treated Wattled Broadbill – following the taxonomy of Sibley and Monroe (1990), and hence lumping *steerii* and *samarensis* – as a Vulnerable species in their treatise on threatened birds of the world. In view of the revised taxonomy presented here, however, it would be pragmatic to reevaluate the threat status of broadbills in the Philippines.

MOVEMENTS None recorded.

FOOD Little information on the diet has been published, although the species has been observed gleaning insects off vegetation and catching them in flight (Delacour and Mayr 1946, duPont and Rabor 1973a). The allopatric Visayan Wattled Broadbill is also insectivorous (see p.230).

HABITS The Mindanao Wattled Broadbill is unobtrusive but tame. Whilst often silent and easily overlooked, in flight its wings often make a distinctive whirr, drawing attention to its presence. This species has also been reported to make a distinctive whirring bill-snapping noise when perched, which is audible for some distance (McGregor 1909-10, Wells 1985).

It is usually observed feeding at low levels of the forest, either singly, in pairs or in small flocks of up to six, and sometimes joins mixed bird parties (Hachisuka 1934-35, duPont and Rabor 1973a). DuPont and Rabor (1973a) reported that the birds would usually perch motionless on the lower branches of trees, shrubs or even bushes in the darker parts of the forest. From time to time, they would make short sallies to glean insects off nearby vegetation, before returning to the same perch or one close by. Delacour and Mayr (1946) also reported that this species catches insects on the wing. Insects are reportedly rapped on a branch several times before being swallowed (McGregor 1909-10).

In mixed bird flocks, Wattled Broadbills typically move perch every few minutes, occasionally giving quiet calls (G. Dutson *in litt.*). A specimen collected in dipterocarp forest in Davao was observed following a flycatcher and chasing insects by jumping from branch to branch (FMNH).

BREEDING Few records of breeding have been documented, and it is not possible to identify any seasonality that exists. Most passerines, however, breed from April to June in the Philippines (Dickinson *et al.* 1991), and the little evidence available suggests that the breeding season of Mindanao Wattled Broadbill also spans this period.

On Basilan, an egg was found in the oviduct of a female collected in May (Dickinson *et al.* 1991), whilst another female with an egg in the oviduct was collected on Mt Hilong-hilong, Agusan Province, Mindanao, on 16 April. Two immature males were collected on 25 August 1891 on Basilan (USNM), although it is not clear whether these were siblings. On Mindanao, well-grown immatures have been collected on Mt Hilong-hilong on 15 and 16 April, at Ayala on 3 December, and in Davao on 2 January (USNM).

DESCRIPTION *E. s. steerii*
Adult male Centre of forehead, crown and upper nape dark maroon-purple, sometimes mixed with a few pale grey marks; sides of forehead at base of bill white, extending along the edge of lores to form distinctive patches; lores, ear-coverts, chin and throat black. Eye surrounded by distinctive light blue wattle, which is narrowest in front (c.2-4mm on dried specimens) and broadest behind and below eye (c.4-6mm). Distinct white collar (c.5mm wide) extends across top of mantle, usually narrowly bordered above by blackish formed by bases of feathers of lower nape. Collar extends to sides of neck, bordering black ear-coverts, but does not extend onto breast. Mantle and back dark grey, extending across side of upper breast as a narrow band; lower back, rump and uppertail-coverts purple-red; tail bright rufous. Breast to belly and flanks pink, becoming white on lower belly and undertail-coverts. Occasionally, pinky-buff feathers are found on the breast and flanks (in *mayri*: not clear if this feature is common to both subspecies). Feathers of tarsus grey. Primaries and secondaries blackish, with extremely narrow outer web to outer primaries. Basal two-thirds of outer web of outer primary white, the white extending onto bend of wing. Inner web of all primaries greyish-white at base, extending progressively further towards tip from outer to inner wing. Primary coverts blackish; inner and median coverts black; tertials black with broad white band across centre of both webs (the extreme outer margin of white patch on innermost tertial may be yellow), forming a bold patch in wing, this extending as band across secondaries as a bold yellow patch, formed by yellow band across outer margin of outer web of secondaries, slightly nearer base than tip. Flight feathers grey from below with white base; underwing-coverts mixed white and grey. Iris golden yellow, light green or blue, depending on the way the light strikes (Hachisuka 1934-35). Bill, legs and feet light blue.

Cutting edge of mandible and extreme tip dries yellow. Bill strongly hooked; rictal bristles well developed, both from lores and chin.
Adult female Similar to male, but underparts white, not pink.
Immature (*steerii*: described from McGregor 1909-10, Hachisuka 1934-35) Male: differs from adult male in having white throat (this gradually becoming black), white breast and belly, with some feathers usually tipped with pale lilac; crown olive-green, brightest on forehead; back and wing-coverts washed with olive-green, brightest on lower back; wing-bar ill-defined and buffish-pink rather than white as in the adult; slightly duller maroon on rump than adult. Bill as in adult, except centre of upper mandible, which is black. Wattle yellow. Immature females differ from immature male in having no lilac on underparts. Subadult females have black patches on throat and distinctive olive sheen to lower back, and usually. A yellow-olive wash to upper rump.

MEASUREMENTS (*steerii*) Wing 84-90; tail 55-62; bill tip to gape 27-30, width 18.5-19; tarsus 20-23.5; weight (*mayri*) 33.7-44.4.

45 VISAYAN WATTLED BROADBILL
Eurylaimus samarensis   Plate 22

Sarcophanops samarensis Steere, 1890, List of Birds and Mammals, Steere Expedition, p.23: Catbalogan, Samar.

Alternative name: Samar Broadbill.

FIELD IDENTIFICATION Length 14.5-15.0. Endemic to the Philippines where found only on Leyte, Samar and Bohol. The male has a black throat, lores and ear-coverts, purple crown and mottled grey nuchal collar. The eyes are surrounded by a pale, sky-blue, fleshy wattle. The mantle is purple, the lower back, rump and tail are rufous with a purple tinge. The underparts are lilac with the centre of the belly yellowish-white. The wings are black with a prominent white-and-lilac bar across the tertials and secondaries. Females differ from males in having pure white underparts. Juveniles have dark brown heads washed with grey on crown; a whitish collar extending across sides of neck; upperparts mixed olive and grey-brown with rufescent tinge in patches; breast and flanks greyish-brown; and belly and undertail-coverts white.
Similar species In its Philippine range Visayan Wattled Broadbill should not be mistaken for any other species. It is slightly smaller than the closely related Mindanao Wattled Broadbill, which differs in having a white nuchal collar, dark grey mantle and white-and-yellow wing-bar.

VOICE Observations suggest that this species rarely calls, but the voice has been described as typical of the genus, an insect like *tik, tik, t-rrrrrrrrr*, usually given twice in quick succession with a two to three second gap (Hornskov 1995). It should be noted, however, that the call of Blue Fantail *Rhipidura superciliaris*, which is sympatric, is very reminiscent of that of *Eurylaimus* broadbills, and could easily be mistaken for a broadbill (pers. obs.). Delacour and Mayr (1946) stated that the call is a plaintive whistle, and calls on Bohol thought to be the contact call of Visayan Wattled Broadbill have been described as short, rather quiet, mid-pitched whistles (D. Allen *in litt.*).

1. Samar
2. Leyte
3. Bohol

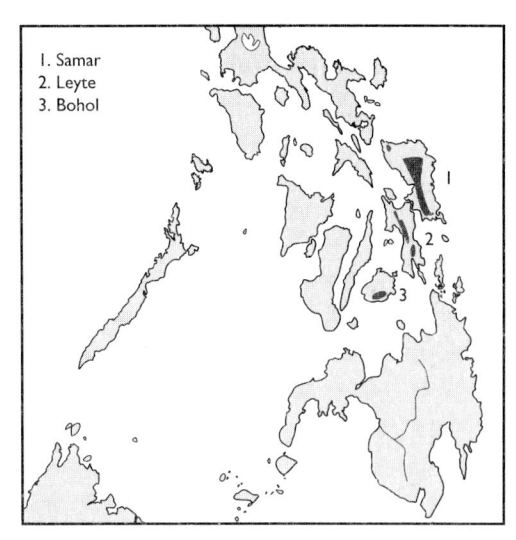

Distribution based on approximate extent of forest remaining within the altitudinal range of the species. Former range more widespread.

DISTRIBUTION Known only from Leyte, Samar and Bohol in the Visayan islands of the Philippines. Visayan Wattled Broadbill has been collected at Caliwag, Patok and Buri on Leyte (Parkes 1973); at Cantaub, Sierra Bullones, and near Pamilacan on Bohol; and at Matuguinao, San Isidro and on Mt. Capoto-an on Samar (Rand and Rabor 1960).

GEOGRAPHICAL VARIATION None recorded.

HABITAT On Bohol, Samar and Leyte, this species has only been found in primary forest (Rabor 1938, Rand and Rabor 1960). Some of these areas, if not all, were characterised by limestone outcrops. On Samar, Rand and Rabor (1960) found Visayan Wattled Broadbills at 100-600m altitude, but on Bohol they found them at 700-750m. In Rajah Sikatuna National Park, Bohol, Visayan Wattled Broadbill occurs at altitudes of c.300-400m (pers. obs.). Rabor (1938), who observed five together in deep forest on Leyte, noted that this species was rare in the highlands of the island.

STATUS Bourns and Worcester reported than Visayan Broadbills were abundant on Samar, but irregularly distributed (McGregor 1909-10), although Rand and Rabor (1960) reported that it was rare on this island. In parts of its range, such as on Bohol, it is reported to be locally common (Hornskov 1995), but is unobtrusive and certainly under-recorded (Brooks et al.1995). Nevertheless, this species is probably threatened throughout its range through continuing deforestation, and Wattled Broadbill – following the taxonomy of Sibley and Monroe (1990), and hence lumping *steerii* and *samarensis* – is listed as Vulnerable in the World List of Threatened Birds (Collar et al. 1994). In view of the revised taxonomy presented here, in which *samarensis* and *steerii/ mayri* are recognised as distinct species, the threat status of broadbills in the Philippines should be reevaluated.

MOVEMENTS None recorded.

FOOD A specimen collected on Bohol contained 'small insects', whilst one collected on Samar had a green caterpillar in its bill (Rand and Rabor 1960).

HABITS Visayan Wattled Broadbill is shy and easy to overlook. It is apparently an unusually silent species that perches on horizontal boughs, branches and vines in relatively dense vegetation (pers. obs.). In contrast to other species in the genus, it apparently prefers the lower levels of the forest, where it frequents the understorey or lower canopy of smaller forest trees. However, Hornskov (1995) notes that, on Bohol, it usually calls from a well-concealed canopy perch. It sometimes joins mixed bird flocks, and on Bohol, has been observed with flocks that contain Blue Fantail, Black-crowned Babbler *Zosterornis (Stachyris) nigrocapitata*, Philippine Leaf Warbler *Phylloscopus olivaceus*, Rufous-tailed Jungle Flycatcher *Rhinomyias ruficauda*, Yellow-bellied Whistler *Pachycephala philippinensis* and Spangled Drongo *Dicrurus hottentottus* (pers. obs.). Usually, Visayan Wattled Broadbill is encountered in pairs or small groups: up to five have been observed together (Rabor 1938).

BREEDING Evidence points strongly towards seasonal breeding between February-March and May-June. A female with enlarged gonads was collected on Samar in April, whilst four of seven males collected were moulting their primaries in April and May (Rand and Rabor 1960). A female collected on Samar on 28 March was breeding (FMNH). A female observed on Bohol on 26 March 1990 was carrying a leaf (A. Greensmith *in litt.*) which may have been nest material. A recently fledged juvenile was collected on Mt Capoto-an, Samar, on 12 May, and immature males have been collected at Catbalogan and Paranas (=Wright), Samar, on 24 April, 12 July and 8 August (FMNH, USNM). On Bohol, an adult and juvenile were seen together on two occasions between 23 and 27 July 1994 (Brooks et al. 1995).

The nest and eggs of this species remain undescribed.

DESCRIPTION

Adult male Feathers of crown dark purple, sometimes tinged with reddish tones and with grey bases that may show through, giving a mottled appearance. Feathers of forehead similar, but stiffer and with whiter bases. Band of feathers across nape and to behind ear-coverts grey, with narrow whitish fringes, especially behind the ear-coverts and sides of nape, forming narrow-barred appearance (the white fringes are sometimes completely absent from centre of nape); feather tips at rear edge of this band are purple, forming a diffuse indistinct purple band across the nape and sides of neck at rear of the grey (barred white) band. Lores, ear-coverts, throat and chin black. Mantle and back purplish-red, but the colour not uniform, giving a somewhat mottled appearance; rump, uppertail-coverts and tail brighter and more reddish than mantle, with less purplish tone. Black throat clearly demarcated from deep pink breast and flanks. Centre of belly pink, merging into white of undertail-coverts and lower belly. Feathers of tarsus blackish mixed with dull rufous at feather tips, and occasionally forming bars. Tertials pale grey basally, greyish-white in central portion and with broad black terminal bar; scapulars and upperwing-coverts black; secondaries black with conspicuous broad band across base, this being formed by white bar with narrow pink edge in outer web of feathers, this bar extending across whole inner web of innermost two secondaries and lacking pink in these feathers. Primaries blackish above, greyish below; outer primaries with very narrow white leading edge that extends onto white feathers at bend of wing.

Inner primaries with greyish-white to outer edge (basal half) of inner web. Alula black; underwing-coverts greyish. Wattle pale sky-blue. Eye light emerald-green, with bronze often present in ring around iris, but sometimes appearing yellow to orange-yellow or even blue, depending on light (these colour changes may occur when the bird is exposed to bright light: P. Morris verbally); legs and feet pale blue to dull grey; bill light bluish, deeper in colour at base and more silvery towards tip and along cutting edge.

Adult female Differs from male in having pure white instead of pink on underparts.

Juvenile male Forehead to rear of crown, lores, ear-coverts and sides of upper throat dark brown with greyish wash to crown formed by greyish basal two-thirds of feathers. Rest of throat white, joined to broad whitish collar extending across sides of neck, this collar gradually fading and not meeting on the back of neck. Small grey wattle. Mantle and back mixed olive and grey-brown with rufescent tinge in patches; uppertail-coverts and rump dull pinky-maroon; tail dull maroon. Breast and flanks greyish-brown; belly and undertail-coverts white. Upperwing-coverts blackish, with narrow olive fringes to median coverts. Primaries brownish above, greyish-brown below, with greyish-white fringe to basal half to two-thirds of inner web, most conspicuous from below. Secondaries and tertials brownish with bold pinky band across basal two-thirds of tertials, this band extending as a patch across secondaries, which have a pinky-maroon patch in the outer web of their basal two-thirds. Dried bill with yellowish gape and underside of lower mandible. Rictal bristles on chin and lores well developed, the latter reaching 14mm in length; bill tip strongly hooked.

MEASUREMENTS Wing 75.5-81; tail 46-50; bill tip to gape 24.4-27.6, width 16-18; tarsus 19.0-20.5; weight 33.5-41.5.

CORYDON

A monotypic genus; sexes similar. Subspecies are not strongly differentiated. *Corydon sumatranus* is a chunky, sociable, drab species of the forest canopy in South-East Asia. It is permanently gregarious and strictly insectivorous, and most notable for its extraordinary, disproportionately large, prominently hooked pink bill – reported to be the widest of any of the passeriformes. Nests, which are conspicuous bulky pendant structures, often attached to the tip of a rattan spike or the end of a branch, are usually attended by more than a pair, suggesting cooperative breeding.

46 DUSKY BROADBILL
Corydon sumatranus Plate 20

Coracius Sumatranus Raffles, 1822, Trans. Linn. Soc. London,13, pt. 2., p.303: interior of Sumatra.

FIELD IDENTIFICATION Length 24-27.5. Dusky Broadbill is an inhabitant of tall forests from Indochina and Tenasserim to Malaysia, Sumatra and Borneo. It is a blackish, thickset, gregarious species, with an obvious massive pinkish bill and orbital skin. A small white patch in the primaries is another good feature. Most subspecies have a fairly obvious off-white or buffy-brown throat-patch, although this is lacking in juveniles. An orange or flame-coloured streak on the mantle is virtually never visible in the field.

Similar species Unlikely to be confused with any other species in its range.

VOICE Groups of Dusky Broadbills can often be located, high in the canopy, by voice. The most commonly heard call is a series of around six to eight screaming notes, rising up the scale *hi-ky-ui, ky-ui, ky-ui...* (also described as *pee-u* and the last notes often more like *ki-ip*) of rising inflection and accented on the *ky* note, is perhaps the most distinctive. Other frequently heard calls are a shrill thin *psseeoo* of falling pitch, and a piercing, high-pitched *tsiu*. A carrying croaking note or rattling laugh is also sometimes given whilst at rest. In flight this species sometimes gives a clear whistle, *pee-u* (Deignan 1945). In Laos, birds have been heard to utter a repeated quavering *ch whit* in flight and an accelerating sequence of 4-7 *chwoo* or *phwoo* notes and occasionally a *pepepepepe...* call near a nest (R. Timmins *in litt.*).

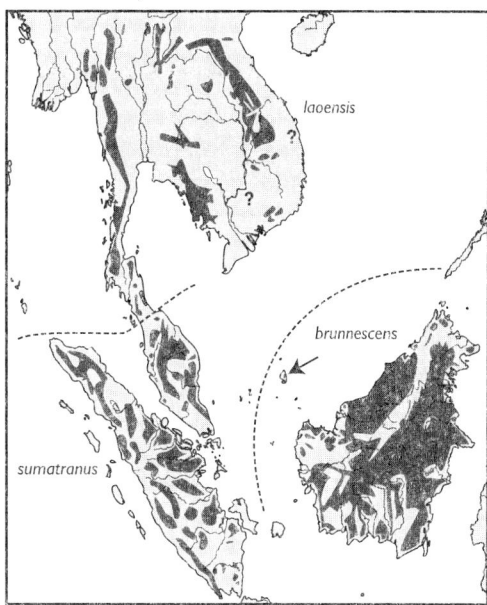

Distribution based on approximate extent of forest remaining within the altitudinal range of the species. Former range more widespread.

DISTRIBUTION Dusky Broadbills occur from Indochina and Tenasserim (Burma) to the Sunda Region. In

231

Indochina, it is found as far north as Khebon, 80km north of Phuqui in North Annam (Delacour 1929). However, the exact distribution of Dusky Broadbill in Indochina is poorly known. For example, its occurrence in Central Annam has not been documented, although it seems likely that it must occur there since it occurs in both North and South Annam (Delacour and Jabouille 1931), whilst in Cambodia the only records are from forest near Angkor (Oustalet 1903, Engelbach 1948) and from the base of hills at Kep, within 200m of the sea (Thomas 1964) in the extreme south-east (Elephant Mountains). In Laos, Dusky Broadbill has been found from near Vientiane and Saravane (Delacour and Jabouille 1931) to the extreme south at Phou Xang He and Xe Piane (R. Timmins and J.W. Duckworth *in litt.*). To the west, the range of Dusky Broadbill extends to the Karen Hills of Burma and from northern Thailand south through the peninsula and Tenasserim (Burma) to the Sunda Region.

In the Sunda Region, Dusky Broadbill is widespread in suitable habitats, occurring from the north of Peninsular Malaysia south to the state of Johore, Penang Island, and the islands of Sumatra and Borneo. A population also occurs on the North Natuna Islands between Borneo and Peninsular Malaysia. Gibson-Hill (1949) reported an old record for Singapore, although noting that it was unlikely that the species still occurred there. Its extinction in Singapore was confirmed by Lim (1992).

GEOGRAPHICAL VARIATION Up to eight subspecies have been described, but most are rather similar. Examination of series of skins indicates that the features used by various taxonomists to differentiate between subspecies (primarily the colour of the throat, the amount of olive in the plumage and the coloration of the concealed spot on the back) are so variable that clear differences are rarely apparent. In particular, the four subspecies described by Deignan (1947) are not recognised here. As a consequence, only three subspecies are described in the following account.

C. s. sumatranus (includes *pallescens* of Deignan 1947) (Sumatra and Peninsular Malaysia, from Pattani, Peninsular Thailand, south to the state of Johore, Malaysia, and on Penang Island) Concealed pale orange and white dorsal spot.

C. s. laoensis (includes *khmerensis*, *ardescens* and *morator* of Deignan 1947) (Burma in the Karen Hills, and northern Thailand south to Chanthaburi Province, Trang at about 12°40'N, and Tenasserim, Burma; Laos, Cambodia and Vietnam) Distinguished from the nominate subspecies by generally darker coloration and whiter throat and more scarlet-coloured concealed dorsal spot. Wing 132-143. Facial skin and bill perhaps with more purple tones than *sumatranus*: facial skin fleshy-purple; bill purple, tipped light blue-grey.

C. s. brunnescens (includes *orientalis* of Mayr 1938a) (Endemic to Borneo and the North Natuna Islands) Darker than *sumatranus*, with sooty-black upperparts distinctly tinged with olive-green, particularly on the rump and uppertail-coverts. Throat usually a much deeper rufous-brown than other subspecies. Concealed dorsal spot scarlet. Wing 130-137.

HABITAT Groups of Dusky Broadbill frequent the upper canopy of primary and logged deciduous and evergreen forests and rainforests, mostly in the lowlands and hills but locally in mossy forests at higher elevations. In Borneo and Vietnam, its habitats include forests on limestone, and in Sumatra it occurs in freshwater peatswamp forests (Delacour 1929, Silvius and Verheugt 1986, Verheugt *et al.* 1993). In Peninsular Malaysia, it occasionally visits gallery forest and small patches of tall trees at the edge of rubber plantations (pers. obs.), but a comprehensive study of birds using *Albizia* groves in Sabah showed that it does not usually use these man-made forests (Mitra and Sheldon 1993). Baker (1934) also stated that, unlike some of the other mainland species of broadbill, Dusky Broadbill was never found in cultivation or in the vicinity of villages.

In Indochina, Dusky Broadbills have been found in tall lowland forest in limestone-dominated hills in North Annam (Delacour 1929). Wildash (1968) states that, in southern Vietnam, it occurs to 1,830m, and Delacour and Jabouille (1931) concur that it can occur up to 2,000m in the mountains of Indochina. However, such high altitudes are probably unusual, and it is more likely to be found at lower altitudes. Beaulieu (1932) found Dusky Broadbill only below 170m in Cochinchina whilst Robinson and Kloss (1919) collected it at 200m in South Annam.

In northern Thailand and in the adjacent Karen Hills of Burma, Dusky Broadbills frequent evergreen forests to at least 1,000m (Deignan 1936), whilst in south-east Thailand it has been claimed that they occur to at least 1,070m (Meyer de Schauensee 1946), although this needs verification. Further south, in Peninsular Thailand and in Peninsular Malaysia, this species favours dipterocarp forests of the lowlands and hills, in both primary and logged forests. Although occasionally found in submontane forests in this part of its range, it is generally rarer at higher altitudes, although Baker (1934) stated that nests from Perak, Peninsular Malaysia, had been found between 610m and 1220m, and that this species perhaps occurred as high as 1,525m. Hume and Davison (1878) reported that in Tenasserim it is absent from high hills, although Hopwood collected a nest at 1,220m on Mt Nwalabo.

In the Greater Sunda Islands, it is an uncommon resident. In Borneo, the subspecies *brunnescens* occurs from the lowlands to at least 1,220m (Mt Dulit, Sarawak) in both primary and selectively logged forests (pers. obs.), whilst in Sumatra *sumatranus* has been recorded to c.1,000m (van Marle and Voous 1988). Although one specimen from Mount Kinabalu, Sabah, was said to have been collected at 1,830m (Gore 1968, Jenkins and deSilva 1978), the occurrence of the species there at this altitude needs confirmation. In the Natuna Islands it occurs to at least 300m (Oberholser 1932).

STATUS The Dusky Broadbill has been described as uncommon in northern Thailand (Meyer de Schauensee 1946) and in the Greater Sundas, but in Peninsular Thailand and Peninsular Malaysia it is a locally common species in suitable habitat. Deforestation in Vietnam and parts of Thailand must have restricted the range of this forest bird and isolated populations. However, there is much forest in its Malaysian and Indonesian range, and it must still have large populations there.

Deignan (1945) described its status as rare in the Doi Suthep-Pui region of northern Thailand (at 750-1,000m), and Round (1984) failed to find it there more recently. In north Borneo it has long been considered a rather scarce species (Sharpe 1889, Gore 1968) but in some areas, such as the Kutai National Park, East Kalimantan, it is reported to be common (Pearson 1975). Delacour and Jabouille (1931) found it to be fairly rare and local in

Indochina. Wildash (1968) noted that it was sporadic and localised in southern Vietnam, and the only record from Vietnam during a recent series of forest bird surveys was from Nam Bai Cat Tien National Park in southern Annam (Robson *et al.* 1989, 1991, 1993, Lambert *et al.* 1994, J. Eames verbally).

Although based on few data, studies on the effects of logging on bird populations in Sabah, Malaysia, suggest that populations of Dusky Broadbill declined following logging, although a nest was found in forest that had been recently logged (Lambert 1990, 1992).

FOOD The diet of Dusky Broadbill is reported to include large items such as lizards, large brown hemipteran bugs and grasshoppers up to 8-10cm in length (R. Thewlis *in litt.*), as well as a variety of invertebrates, including a variety of grasshoppers and crickets, scarabid beetles, ants and 'soft insects'. Large flying insects are occasionally caught by making upward leaps from their perch.

HABITS Dusky Broadbills are invariable found in small groups of up to ten birds, but as many as 20 individuals have been observed together in northern Thailand. Groups feed in the upper parts of the forest, usually in the canopy or emergent trees at 15m-30m. They are somewhat crepuscular, usually being most active in the early morning and evening. Groups of Dusky Broadbill are frequently very vocal, but the birds spend a considerable proportion of their time sitting quietly whilst scanning for prey. Prey is often gleaned from nearby branches of leaves after a short flight. Since sympatric species of broadbill generally feed at lower heights in the forest, there is little overlap in foraging station.

Even whilst breeding, Dusky Broadbill is a very social species. All members of a group may engage in nest building, as noted by Meyer de Schauensee (in Deignan 1945), who collected an incomplete nest in northern Thailand in mid-March that was being built by about ten birds. In Sabah, at least three adults were observed building a nest in logged forest (pers. obs.), whilst in Laos R. Timmins (*in litt.*) saw at least three adults attending a nest. Other observers have also noted that more than two individuals may help build the nest (see Breeding). Whether helpers are also involved in feeding young is unclear, but fledged young have been observed being fed by two adults (M. Chong *in litt.*), suggesting that at least both parents must take part in this activity.

BREEDING There are relatively few recent breeding records for Dusky Broadbill, although Baker (1926, 1934) documented a number of nests and had access to at least 20 eggs. However, recent nests have been well-documented, and there is therefore a surprisingly large amount of information on the breeding of a species that must otherwise be considered poorly known.

Seasonality Evidence suggests that the nesting season in the northern part of its range coincides with the latter part of the dry season, so that immatures are being fed by adults during the rains of the south-west monsoon. Further south, in Peninsular Malaysia, the nesting season apparently starts earlier, during the wettest part of the year, and the known nesting season is much longer, perhaps spanning 10 months. Breeding records are documented below.

In the northern part of its range, Deignan (1945) collected a male with enlarged gonads in late March in northern Thailand, and eight adults in postnuptial moult between late May and early December. Two adults carrying nesting material (beaks full of moss) were observed in Khao Yai National Park, Thailand, on 3 April (B. Gee *in litt.*). Meyer de Schauensee (1946) collected immatures between 9 August and 20 September. Hume and Davison (1878) obtained a fledgling that had just left the nest on 1 May in Tenasserim, whilst Hopwood (1919) found a nest there on 12 March. Two immatures were collected in Peninsular Thailand on August 6 (Riley 1938).

In Cambodia, Engelbach (1948) observed a group of Dusky Broadbills with young at Angkor on 17 September, and regarded this as an indication that this species bred there in August. In Laos, a nest under construction by six adults was found on 29 March in the Phou Xang He National Biodiversity Conservation Area and two other nests were observed in the Xe Piane NBCA on 9 May and 14 May (R. Timmins *in litt.*).

In Peninsular Malaysia and Tenasserim, nests containing eggs have been found from mid-December to late June (Baker 1934, BMNH, M. Chong *in litt.*). However, the breeding season must be longer than this suggests, since very young birds were collected near Klang, Peninsular Malaysia, in early September (Robinson 1927), and other recently fledged birds have been observed on 12 September and in December (M. Chong *in litt.*). However, there is perhaps a peak in nesting activity, since five nests discovered in Peninsular Malaysia in recent years were all found between 4 March and late May (M. Chong *in litt.*, S. Duffield *in litt.*, pers. obs.).

There are very few breeding records from the Greater Sunda Islands, but these suggest that the timing of breeding in Borneo and Sumatra is similar to that than on the mainland, although nesting in Borneo has been recorded as early as mid-November (Holmes and Burton 1987). In Sabah, nests under construction have been discovered on 12 April and 20 May, and in Kalimantan on 12 and 29 November. In Sarawak, a juvenile was collected on 16 January (SarM), and in Sabah one was observed on 6 May. Nests have been found in Sumatra on 24 February and 28 June (van Heyst 1919, Nash and Nash 1985) and a recently fledged juvenile was collected on 21 November in Lampung (MZB). Immatures have been collected in Sumatra in May and on 6 July (RMNH).

Nests Dusky Broadbills typically suspend their nests from the ends or tips of strong plant parts so that they hang in rather open places, making nest predation difficult. Nests have all been found 4-13m above the ground, usually over open spaces. Three nests found by R. Timmins (*in litt.*) in Laos were suspended over dry river beds: one from a tree, one from a climber and one from bamboo. Three nests received by Baker (1934) were hanging from pendant branches of trees, two over a stream and the other over a pool of water. In Peninsular Malaysia, three nests have been found suspended from the tendrils of rattan palms (one identified as a species of spiny *Calamus*), one so that it hung about 4.5m above a river and another over a more open area of forest (Anon. 1988, M. Chong *in litt.*, M. Strange *in litt.*, pers. obs.). A fourth nest was suspended at mid-canopy height over a clearing where a mature forest tree had fallen (S. Duffield *in litt.*). One nest in Sabah was suspended 15m above the floor of primary forest from a long drooping branch, whilst another was suspended from a hanging rattan tendril about 4m above a bare logging road in recently logged forest (pers. obs.). A nest from Kalimantan was suspended from a rattan about 8m above the ground in peatswamp forest (Nash and Nash 1985).

Nest of Dusky Broadbill

The nest is a large, bulky, domed, pear-shaped pendant structure about 30cm in diameter, and often two or more metres long. Hopwood (1919) describes a nest in Tenasserim, suspended some 12m above the ground from a bamboo cane tip, as being built of coarse dry moss with pieces of sticks, leaves and cocoons added. Baker (1926, 1934) noted that grass, twigs, leaves, moss, roots, plant stems and other vegetable matter are used to make the nest, whilst green leaves are used as a lining and cobwebs are used to attach all sorts of material, including cocoons, caterpillar excreta, spiders' egg-bags, leaves, etc., to the outside and bottom, so that it hangs down to form a long tail that may reach 70cm in length. A nest photographed in Malaysia (M. Strange) was adorned with the runners of an epiphytic plant (possibly a fig) so that lines of small, living green leaves formed a partial outer layer, providing excellent camouflage. The entrance is about one-third down the body of the nest and is overhung with and vitually concealed by a large porch constructed of the same materials as the body of the nest (Baker 1934, R. Timmins *in litt.*).

A nest collected on 16 December in Peninsular Malaysia (BMNH) was suspended from a rattan, and was a compact, elongated, pear-shaped structure, 60cm from top to bottom and of maximum width c.29cm. The entrance hole was 5cm wide and 6cm from top to bottom, and situated 16cm from the bottom of the nest. Although the nest was presumably complete, since it contained three eggs, it did not have the long tail that is reportedly usually present. The main component of the structure was small roots and twigs (1-2mm diameter) with regular nodules, these almost certainly derived by stripping them from epiphytic plants growing on tree trunks. Intermixed with these finer materials were larger twigs, up to 10mm wide and 52cm long. A few small leaves in the structure were attached to the small root-like twigs and were perhaps those of a climbing *Ficus*. Also present were pieces of moss. The nest cavity was made almost entirely of long, interwoven strips of palm leaf and lined with entire, fleshy dicotyledonous leaves of up to 8cm long and 3.5cm wide. The external structure of roots and twigs did not extend to cover the bottom of the nest chamber, so that the palm strips were exposed at the bottom.

Nest materials used in Laos also included strands of green, succulent epiphytic climbers, as well as pieces of vine, bamboo leaves, and roots (R. Timmins *in litt.*). Occasionally, strips of palm leaf may be used as a nest lining (Baker 1934), whilst in Peninsular Malaysia, M. Chong (*in litt.*) observed three adults bringing dried bamboo leaves to a nest; these were used in the nest cavity, presumably as a lining to the cup. During half an hour of observation, eight visits of less than one minute were made to the nest.

Eggs In Peninsular Malaysia, clutch-size is reported to be three (Medway and Wells 1976), but a nest found in Tenasserim contained four newly hatched young (Hopwood 1919), and Baker (1926) gives clutch-size as 2-4. The ground colour varies from pale reddish to pale dull cream. The markings consist of numerous small blotches and freckles of reddish-brown that may almost obscure the ground colour. Some eggs have fewer markings, and two collected by Hopwood have rather larger blotches and show underlying marks of pale lavender. Twenty eggs obtained by Baker (1926) measured on average 29.4 x 22.2mm, with minimum and maximum measurements of 27.2-34.9 x 20.0-24.0mm.

DESCRIPTION *C. s. sumatranus*
Sexes similar.
Adult Head blackish-brown, occasionally with slight olive tinge to crown and nape, and dominated by massive, strongly hooked bill that is almost as broad as long, and wider than the skull. Bare skin around eye and across lores pinkish, most prominent below eye. Body blackish-brown with strong olive tinge to feathers of rump and uppertail-coverts. Wings brown with white bar across base of primaries, this formed by narrow spot on outer web of outer primary that is adjacent to a broader bar on outer web of adjacent primaries and is longest on innermost primary. Mantle has concealed patch of pale orange feathers mixed with white and, occasionally, some brighter orange and scarlet feathers. Feathers around edge of lower mandible usually dark brown, this occasionally extending onto chin; rest of throat and feathers of upper breast whitish with variable extent of dull orange-brown edges (and occasionally whole feathers) giving pale white or orangy-brown patch that contrasts strongly with rest of underparts, which are brown. Flanks often tinged with olive. Central two pairs of tail feathers all brown, rest brown with broad white subterminal bar visible from below at rest. Tail graduated, with innermost feathers longest and outermost shortest, so that the white bars on undertail appear as a line of bars. There is considerable variation in the amount of white in the tail, possibly related to age. Iris dark brownish-crimson. Bill colour very variable, upper mandible usually yellowish-horn stained with pinkish tones, cutting edge greyish, lower mandible pinkish with yellowish base

(Robinson and Kloss 1924, Smythies 1986). Rictal bristles well developed but inconspicuous due to size of bill.
Juvenile Much browner than adult. Lacks the orange patch on mantle and has brown throat with a few pale patches. Size of white spots in wings much reduced, and white in tail limited to small spots in the outer webs of outer tail feathers. Iris pale grey to brownish-grey, bill pale pinkish-horn with vinaceous stains, gape orange-yellow to reddish. Subadults have upper mandible dark sepia, tip yellowish-horn (Robinson and Kloss 1924).

MEASUREMENTS Wing 132-142; tail 90-104; bill length 36-40, width at base 29-33; tarsus 25-28.5; weight c.140.

PSEUDOCALYPTOMENA

The single species of this genus was originally thought to be a flycatcher. It is very poorly known due to the relative inaccessibility of its very restricted range. It is an inhabitant of central African montane forests. *Pseudocalyptomena* is more closely related to Madagascan asities than to African broadbills or *Calyptomena* broadbills, despite similarities to the latter in shape, coloration and diet. It is highly frugivorous, but also eats insects.

47 AFRICAN GREEN BROADBILL
Pseudocalyptomena graueri   Plate 18

Pseudocalyptomena graueri Rothschild, 1909, Ibis, p.690, pl. 10: 50 miles west of Russisi, north of Lake Tanganyika, 2,000m.

Alternative name: Grauer's Broadbill

FIELD IDENTIFICATION Length 13.6-15.6. African Green Broadbill has a restricted montane range in eastern Zaïre and western Uganda. It is a relatively small, short-tailed broadbill that is often silent and easily overlooked. Adults are identified by the combination of green upperparts and belly, pale blue throat and undertail-coverts. A narrow black eye-stripe extends a short distance behind the eye, whilst the moustachial area is narrowly streaked with black. The crown is buffish, and also finely streaked with black. Immatures are duller, with green undertail-coverts. The flight is direct and rather slow, often with some gliding.
Similar species African Green Broadbill is not likely to be confused with other species, although its coloration resembles that of some bee-eaters. However, African Green Broadbill is more stocky and upright compared to any species of bee-eater, and has a short tail.

VOICE The call of this species is described as a feeble *tsi-tsi* or *cree-cree* repeated 3-8 times at a rapid rate, of about four notes a second. In Uganda, African Green Broadbills were heard to utter a one-syllabled high-pitched *prrrp* at intervals of about 30s (Friedmann 1970). The *tsi-tsi* call is reminiscent of the voice of Oriole Finch *Linurgus olivaceus* (Keith *et al.* 1992). When breeding, a very high-pitched, prolonged bell-like ringing noise has been heard. Although this noise may have been a vocalisation, the possibility that it was mechanically produced cannot be ruled out since it started when the bird took flight, and stopped as it landed (E. Smith verbally).

DISTRIBUTION The African Green Broadbill is known only from a small region in extreme eastern Zaïre and in adjacent west Uganda. These localities are in the Itombwe Mountains between 3°41'S, 28°31'E and 3°04'S, 28°48'E (Ibachilo, Karungu, Luvumba, Miki and Muusi: Prigogine 1971) and Mount Kahuzi to the west of Lake Kivu (Nyavaronga, 2°00'S, 28°49'E) in Zaïre, and from the Bwindi and Ruhizha areas of the Impenetrable Forest, Uganda (Kigezi Province).

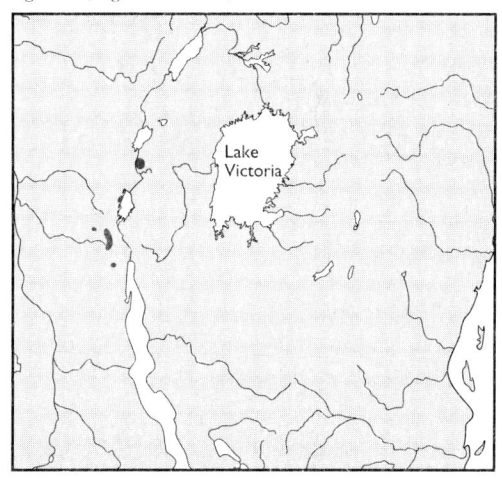

Distribution based on known sites.

GEOGRAPHICAL VARIATION No subspecies are described: the suggestion that the population in the Impenetrable Forest, Uganda, may have narrower bills (Friedmann and Williams 1968) was subsequently disproved by the collection of further specimens (Friedmann 1970).

HABITAT African Green Broadbills inhabit primary montane forests, forest edge, bamboo-dominated forest and, occasionally, isolated trees in clearings and agricultural areas, at altitudes of 1,760-2,480m in Zaïre (Rockefeller and Murphy 1933, Prigogine 1971) and 2,060-2,285m in the Bwindi-Impenetrable National Park in Uganda (Friedmann and Williams 1968, T. Butynski *in litt.*).

In Zaïre, Rockefeller and Murphy (1933) noted that it occurred mainly below the bamboo zone, and appeared to prefer forest edge in the vicinity of cultivation. African Green Broadbills in Uganda appear to be associated with stands of the dominant tree species, *Chrysophyllum gorungosanum*, on steep slopes, and they have not been observed away from good stands of this tree or in valley bottoms (T. Butynski *in litt.*). Prigogine (1971) noted that they tended to occur in areas with dense foliage. Within forest, the species has been observed at all levels: whilst reports of some observers suggest that it prefers the upper branches of fairly tall trees at 7.5-23m above the ground in Zaïre (Rockefeller and Murphy 1933, Chapin 1953), field notes accompanying specimens from Uganda indicate that it occupies the upper portions of the undergrowth, about 2.5m from the ground (Friedmann 1970). Hence there may be some difference in the ecology of these two isolated populations.

STATUS The species has been reported as quite common but localised in the Itombwe Mountains of Zaïre, but rare in the Inpenetrable forest in Uganda. Experienced observers usually only encounter it about once in two days within its range in Uganda (T. Butynski *in litt.*). Based on a careful analysis of distribution and potential threats, Collar and Stuart (1985) classified it as Rare in their documentation of the threatened birds of Africa, and it is considered as Vulnerable by Collar *et al.* (1994). Forest clearance around villages, commercial logging and mining activities were all identified as potential threats to the species, although the population in the mountains west of Lake Kivu is probably safeguarded in the Kahuzi-Biega National Park, and montane habitat in Itombwe is at present largely intact (Wilson and Catsis 1990).

MOVEMENTS None recorded.

FOOD The African Green Broadbill is apparently omnivorous. Whilst early observers reported a flycatcher-like habit (Chapin 1953), the stomachs of five specimens collected in Uganda revealed no indication of any aerial feeding habits (Friedmann 1970). The stomach of one contained a single small beetle, some small seeds and a small flower bud; the second contained only flower buds; the third the remains of small fleshy fruits; the fourth only small white flowers and the fifth a small snail, a beetle, several insect larvae and a white seed. Three stomachs examined by Prigogine (1971) contained orange and green fruit and insects.

The presence of seeds in stomachs suggest that, like broadbills in the Asian genus *Calyptomena*, this species is highly frugivorous and acts as a seed disperser. Rockefeller and Murphy (1933) first observed the species in a large tree with abundant juicy berries.

HABITS Comparatively little is known of this species, which was originally thought to be a flycatcher. It has been encountered singly or in small flocks of up to ten individuals. Evidence suggests that African Green Broadbills prefer lower levels of the forest, often occurring in the upper part of the understorey at less than 3m above the ground, but occasionally foraging in the canopy at elevations of up to 25m. The species is noted to be fairly tame, as it sometimes feeds close to human habitation (Prigogine 1971).

Rockefeller and Murphy (1933) saw this species behaving like a flycatcher: apparently catching insects on the wing and then returning to the same perch between flights, and subsequently flying between perches some 30-90m apart. The flight was noted to be on a straight, level course, with vigorous rapid, but silent, wing-beats. In Uganda, however, the flight is reported to be slow and of a gliding nature. Even on short flights, the birds rarely flapped their wings after starting and seemed to glide by preference (Friedmann 1970). The original observations of birds flycatching from a perch high in the forest perhaps, therefore, relate to some sort of display, similar to that of species of *Smithornis* broadbill, rather than to feeding behaviour.

Sometimes the species joins mixed-species flocks (Prigogine 1971): at Ruhizha, an African Green Broadbill was observed to join a mixed species flock that included White-headed Wood-hoopoes *Phoeniculus bollei*, Black-backed Puffbacks *Dryoscopus cubla* and Black-faced Apalis *Apalis personata*.

Like other broadbills, the African Green Broadbill often perches without moving for prolonged periods, and gleans insects from leaves during upward sallies. However, unlike other broadbills, this species has also been observed climbing along the underside of boughs of trees, and along vertical boughs in the manner of a woodpecker (Prigogine 1971, Chapin 1978), perhaps whilst searching for invertebrates. Birds observed in the forest understorey and in *Neoboutonia* trees in Uganda were reported to feed in the manner of a waxbill *Estrilda* sp., and in flight were reminiscent of a crombec *Sylvietta* sp. because of the short tail and small size (Friedmann 1970).

Although no display by this species has ever been documented, Lowe (1931) notes that the pectoralis major muscle of African Green Broadbill is remarkable for its size and thickness. These muscles are responsible for the wing's downstroke, and as noted by Webster (1991) these are the muscles that the Green Broadbill uses to remain in the spinning ('beak pirouette') display that has been observed in captivity. Given the development of these muscles in African Green Broadbill it may well perform a similar display.

BREEDING Chapin (1953) reported that the breeding season must be long, since four adults with enlarged gonads were collected in late July in Zaïre at the same time as an immature. Rockefeller and Murphy (1933) report that dissected specimens collected between 26 July and mid-August appeared to be approaching breeding condition. In Uganda, a nest was discovered in April in rather open scrubby vegetation with many 20m-tall trees (Keith *et al.* 1992). The nest, suspended from the outermost branches of a tree in mid-canopy, c.11m above the ground, was directly overhanging a stream. It was a spherical structure estimated to be c.20-25cm in diameter, with a c.5cm-wide side entrance and an outside layer of green lichen ('spanish moss'). A bird, presumed to be the female, was observed sitting in the nest whilst another adult, presumed to be the male, sat quietly nearby (E. Smith verbally). When this latter bird flew off, it produced a prolonged, very high-pitched metallic noise like a ringing bell.

DESCRIPTION
Sexes alike.
Adult female Forehead to top of crown buffy, marked by broad dark brown streaks. Rear of crown mixed brown and yellow-green, sometimes with some bluish tinge. Narrow black eye-stripe extending from lores, where broadest, through eye to rear of ear-coverts. Moustachial area and ear-coverts blackish-brown, with pale blue feathers overlaid in the ear-coverts and dominating behind eye. Upperparts bright green (paler than Asian *Calyptomena*), often with a few scattered pale bluish marks in some feathers. Chin and throat white, merging into pale blue lower throat and breast. Belly and flanks pale green to lime-green, slightly paler than upperparts, and clearly demarcated from blue breast. Undertail-coverts variable, being

pale blue, or green with pale blue tips. Innermost pair of tail feathers green above, with distinct blue tinge to basal two-thirds. Rest of tail feathers with brown on inner webs, this increasing in extent to cover whole web of outermost three pairs and extending across tip of outermost feather, so that green is confined to the basal two-thirds of the outer web. From below the tail appears brown. Upperwing-coverts and tertials pale green, concolorous with upperparts, with hidden brown bases and inner webs. Flight feathers dark brown with green leading edges, narrowest on primaries (absent on outermost) and broadening towards innerwing to cover outer half of outer web. Primaries 2-5 strongly emarginated, with the narrow tips lacking green edge. Underwing-coverts mixed white and pale green. Tail short and rounded, the outermost feathers some 6mm shorter than the central pair. Eye blackish-brown, bill black, legs and feet light greyish-green. Rictal bristles very short.

Immature Slightly duller plumage than adult, with undertail-coverts entirely green. Iris dark grey with dark brown undertone; legs dark bluish-green. Lower mandible dark yellowish shaded with brown; skin at gape yellow.

MEASUREMENTS Wing 74-76.5; tail 35.5-39.5; bill 11.8-15; gape width 8.5-13.6; weight 29.0-32.5.

PHILEPITTA

An endemic Madagascan genus, until recently included, together with sunbird asities, in an endemic family. *Philepitta* contains two sexually dichromatic species, one an inhabitant of Madagascar's dry western forests, the other an eastern, wet forest species. Plumage varies seasonally. Males have elaborate, brilliantly coloured caruncles (wattles) during the breeding season. *P. castanea* is highly frugivorous. *P. schlegeli* also feeds on fruit, but is perhaps more specialised towards a nectarivorous diet. Both species have narrow bills in comparison to typical broadbills. *P. castanea* is a polygamous species in which males lek and perform elaborate displays.

48 VELVET ASITY
Philepitta castanea   Plate 24

Turdus castanea P.L.S. Müller, 1776, Natursyst., suppl., p.143: Madagascar.

FIELD IDENTIFICATION Length 14-16.5. Endemic to the wet forests of eastern and northern Madagascar. A short-tailed, plump species with a medium length, slightly curved bill. In breeding plumage, males are entirely velvety-black with a prominent, brilliant deep lime-green caruncle over and in front of the eye that protrudes above the forehead, and a concealed golden-yellow 'shoulder' patch. In non-breeding plumage, the body becomes scaled with yellow, and the forehead to nape is spotted yellow, leaving only the sides of head and throat unmarked black. The caruncle is absent or vestigial, forming a narrow strip over the eye. The female is generally olive-green with prominent yellow-green eye-ring and moustachial streak and less distinct yellow supercilium in front of the eye. The upperparts are unmarked, the underparts are scaled with yellow and green, although this is indistinct or absent on central belly and undertail-coverts. Immatures and subadults resemble adult females, but subadult males may have a bare whitish patch of skin above the eye. Wings whirr in flight at all ages. Floppy flight through the forest understorey is distinctive.

Similar species The female is similar to female Schlegel's Asity, which is sympatric in the Sambirano region at the north-western end of Madagascar's central mountain range. Female Velvet Asity, however, differs in having a slightly curved, longer bill; a yellow patch below the back of the eye; more or less spotted or streaked rather than mottled underparts, these markings extending to the sides of belly and, often, the sides of undertail-coverts (in contrast, Schlegel's Asity has virtually unmarked yellow undertail-coverts and lower belly); and dull olive-green edges to the flight feathers: in female Schlegel's Asity these are bright yellowish-green.

VOICE Territorial males give a weak, very quiet but high-pitched, rather metallic and squeaky advertisement vocalisation. The call is an amorphous series of squeaky *weee-cheew* or *weee-dooo* notes, with a clear emphasis on the first syllable, and somewhat resembling a squeaking door. When males interact with each other, these vocalisations may be extended into a long continuous call that lasts several seconds, *weetweetweet...weet*. At other times, males may utter very quiet metallic single *tsip* notes, a brief *tsip-tsip-ou*, or a long thin *seeeee* call on a single frequency (Prum and Razafindratsita in prep., pers. obs.).

DISTRIBUTION Velvet Asity is endemic to Madagascar where it is distributed almost the entire length of the island in the east, from Taolanaro (Fort Dauphin, 25°01'S) in the south to Sambava (14°16'S) in the north, and in the north-west of the island in the Sambirano and Tsaratanana Massif at the northern end of the central mountain range. It is not found in the extreme north of the island, however, where the humid forest that it inhabits is replaced by dry forests, nor in rainforests of Montagne d'Ambre in the north. Areas inhabited by Velvet Asity have year-round rain.

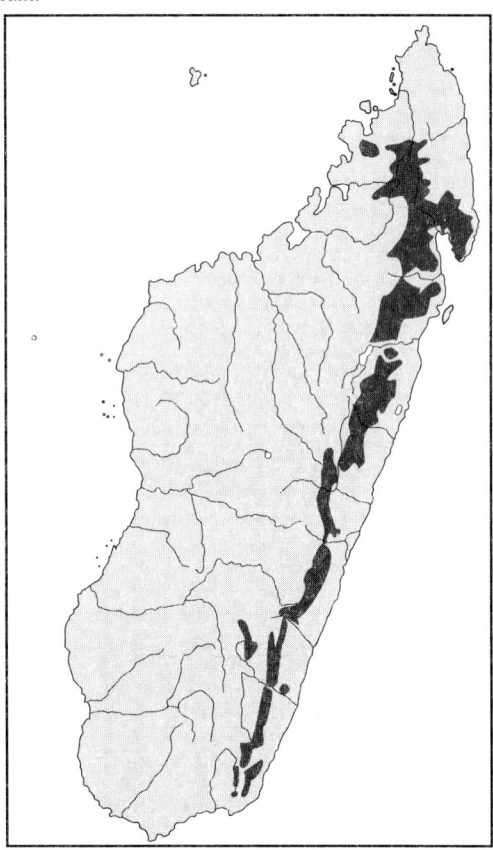

Distribution based on approximate extent of forest remaining within the altitudinal range of the species. Former range more widespread.

GEOGRAPHICAL VARIATION None reported.

HABITAT The Velvet Asity inhabits primary rainforest as well as bushes and secondary growth in degraded and logged forests and secondary woodland. At Ranomafana, observations suggest that it is most common in secondary growth (R. Prum *in litt.*). However, F. Hawkins (*in litt.*) reports that Velvet Asity only occurs in secondary habitats where these are immediately adjacent to primary forest. In general, it is reported to be most common on forested mountain slopes, and although it occurs to sea-level on such slopes it is a rare species in flat lowland forest (Rand 1936). In montane regions, it has been recorded in mossy forest to 1,800m (Rand 1936). Although most frequently encountered in the understorey and lower levels of the forest, it regularly ventures into the mid-levels and occasionally the canopy, particularly to visit nectar sources (R. Prum *in litt.*). Birds have been seen foraging at c.30m above the ground in primary forest on the Masoala Peninsula (pers. obs.).

STATUS This is probably the most frequently encountered species of asity in Madagascar. Delacour (1932b) described its status as very common to abundant, having collected 153 specimens between 1929 and 1931. It is still a relatively common species throughout most of its range, although, as with other forest birds in eastern Madagascar, forest loss is a threat in the long term (see under Common Sunbird Asity).

MOVEMENTS None recorded.

FOOD The Velvet Asity is highly frugivorous, with fruit forming the bulk of its diet, but there is some evidence that it also feeds on nectar. In the Andringitra Strict Nature Reserve, Goodman and Putnam (in press) identified a number of plants with fruits eaten by Velvet Asities, as well as observing them eat small orange-coloured buds of *Macaranga oblongifolia* (Euphorbiaceae). The fruits that Goodman and Putnam identified were those of two species of melastome, *Gravesia caliantha* and *Medinilla ericarum* (orange-red fruit); those of *Pittosporum* sp. (Pittosporaceae); and those of two species of Rubiaceae, *Pauridiantha lyallii* (cranberry-coloured) and *Chassalia* sp. (bright red). At nearby Ranomafana National Park, Velvet Asity has been observed eating red fruits from a small tree in the family Piperaceae, and fruits of a number of species in the Rubiaceae. These fruits were up to c.8mm in diameter, and were swallowed whole (pers. obs.). Fruits are eaten from a perched position rather than being snatched in flight in the manner of many frugivores.

Small seeds, up to 3 x 2mm in size, have been found in the stomachs of specimens, indicating that the birds probably act as dispersal agents for the plants on which they feed. Stomachs of eight specimens examined by Rand (1936) and another eight examined by Benson *et al.* (1976) contained only fruits and seeds. A female Velvet Asity that had been feeding on small, unidentified pink fruits of a vine in the mid-canopy of forest at Ranomafana was subsequently observed regurgitating seeds (pers. obs.).

Evidence from nectarivory comes from the structure of the tongue, and the observation of this species feeding frequently at the flowers of a mistletoe *Bakerella* sp. in the mid to upper canopy (Prum and Razafindratsita in prep., R. Prum *in litt.*) and from a root parasite, *Ditepalanthus* sp. (Balanophoraceae) (Safford and Duckworth 1990). F. Hawkins (*in litt.*) also records observing a male feeding at a small cluster of flowers over at least 48hrs. R. Prum (*in litt.*) reports that the tongue of pickled specimens of this species is bifid for the distal 3-4mm and that each side is divided into many fine brushy tips: this clearly suggests some degree of specialisation for nectarivory.

Insects are occasionally also eaten, and Thompson and Evans (1991) observed Velvet Asity sally-gleaning and flycatching for insects. As in the case of most other frugivores, insects are likely to be fed to nestlings.

HABITS The Velvet Asity is sometimes solitary, but often observed in pairs or small groups in the understorey and lower levels of the forest. Although it is rather unobtrusive, often sitting motionless in dense vegetation for prolonged periods between feeding bouts, it is also a rather tame species. When feeding, Velvet Asities move from perch to perch almost by hopping, and may assume a crouched posture with the head jerked and bobbed backwards and forwards whilst apparently searching for fruit (pers. obs.). Rand (1936) and Petter (1969) noted that it occasionally joins mixed bird flocks. On the Masoala Peninsula, a female Velvet Asity was observed accompanying a group of Common Newtonias *Newtonia brunneicauda* and White-headed Vangas *Leptopterus viridis* (pers. obs.). O. Langrand (*in litt.*) has also observed this species with mixed bird parties of insectivores, but it was feeding on fruits.

During the breeding season, Velvet Asities may interact relatively frequently. F. Hawkins (*in litt.*) observed at least four male Velvet Asities together in Zahamena on 20 September. Two males were fully black (breeding plumage), one had some yellow fringing on the back, and the other was in non-breeding plumage. One of the black males was observed to chase the other black male and, between chasing, returned to the vicinity of a female-plumaged bird. Later, the same black male chased off the other males and was observed to perform the erect wing-flap display (described below) to the female. In October, two males in breeding plumage were observed interacting in the understorey of forest at Ranomafana. Brief chases were interrupted by very frequent, rapid wing-flicking whilst perched and the birds uttered very quiet, metallic *tsip* notes (pers. obs.).

At Ranomafana National Park, detailed observations of more than ten male Velvet Asities during two breeding seasons indicated that males occupy very small territories from at least mid-November until late January (Prum and Razafindratsita in prep.). The territories were either solitary, being isolated from others by more than 100m, or clumped into groups of 2-4 adjacent males. Two males were observed to occupy the exact same territories in subsequent breeding seasons. Male territories, of no more than 20-30m in diameter, were too small to provide sufficient food resources to sustain the male or potential mates and offspring, and, furthermore, none of five territories examined in detail contained any nest trees (Velvet Asity nests exclusively in *Tambourissa* trees: see under Breeding).

On territory, males perch on small branches c.2-4m off the ground where they call and display. At the beginning of a delayed breeding season in November 1995, three males were in attendance in more than 80% of five-minute observation periods between 0600hrs and 0900hrs, but were rarely present later in the day. By January, midway through the breeding season, three males were observed in attendance in more than 80% of five-minute observation periods from 0600hrs to 1600hrs.

Within these territories, Prum and Razafindratsita (in prep.) were able to witness a number of male display behaviours. Four males were seen to perform three highly stereotyped displays. These were an erect wing-flap display, a gape display and a hanging gape display.

During the erect wing-flap display, the male assumes

an erect posture with the neck and body elongated, and proceeds to lean forward over the branch. The brilliant green caruncles are raised, exposing the light blue stripe that is usually hidden above the eye, to form two straight erect planes on the side of the head like the front of a three-cornered hat. Maintaining this rigid posture, the male quickly jumps up and down on his long tarsi. Two males that were observed performing this display differed slightly in detail at this point: one pumped up and down once and then, during the second vertical pump, rapidly opened both wings simultaneously at the body sides, flashing the bright yellow spots at the bend of the wing in the process. When the second male performed the pumping action, this was accompanied by a series of 2-5 asynchronous, single wing-flaps, in an apparently random order, so that either left or right wing was flapped first, followed by the other side or repeated. Each of these two males always performed the display in the same fashion.

During the gape display, males perch with the head pulled in and the mouth open wide and held up at a slight angle, prominently exposing the yellow gape. When perched in this posture, an extended and energetic series of calls are given. On one occasion, the gape display was followed by a more elaborate hanging gape display, in which the male suddenly falls forward and hangs from the perch. In this hanging position, with closed wings, the head is held horizontal and the gape wide as the bird continued to call. After half to several seconds, the bird flies off from the perch. The suddenness of the hanging gape display makes it look like the bird is about to fly off but trips over its perch.

The erect wing-flap display was observed on more than 30 occasions by Prum and Razafindratsita (in prep.), and on most of these it was apparently made in response to a second, unobserved bird. On several occasions, other species of bird, entering the male's territory, elicited this display. These included Blue Coua *Coua caerulea* and Madagascar Pygmy Kingfisher *Ispidina madagascariensis*. On at least five occasions, however, the erect wing-flap display was performed repeatedly during a visit to a male's territory of a presumed female. These female-plumaged birds did not call, posture or act aggressively towards resident males, as did other apparently immature male birds in female-like plumage. Both the gape and hanging gape displays were performed during competitive interactions between males at the boundaries of their territories, or during visits by non-resident males. Males typically increased calling in response to vocalisations of a neighbouring or intruding male, and then began to perform the gape and hanging gape displays as vocal excitement increased. The males frequently approached one another at territorial boundaries, perched less than 5m apart, vocalised energetically, chased one another and performed these displays. Visiting immature males in female-like plumage also frequently perform these displays, behaving identically to mature, black-plumaged males.

Territorial males observed by Prum and Razafindratsita (in prep.) frequently preened their plumage with their bills and feet. This preening was also often extensive: one male preened on a single perch for c.25 minutes. On one occasion, a male was seen to seize a 1cm-long ant and rub it on the body plumage for c.10s.

BREEDING Observations of displaying male Velvet Asities, documented above, and those on breeding documented below, suggest that this species is polygynous (Prum and Razafindratsita in prep.: see Introduction). Breeding is highly seasonal although season apparently depends on latitude.

Seasonality In the northern part of its range, breeding starts earlier than in the south. Males in breeding plumage can be found from 7 July to mid-September in northern Madagascar and to at least 20 November in east-central Madagascar (Benson *et al.* 1976, Langrand 1990). Rand (1936), working at Andapa (14°39'S, 48°17'E), found a nest with eggs on 29 August, and noted that all adults collected between 15-30 August were in breeding condition. Between 1-7 September most of the birds were in breeding condition, including one female that was ready to lay eggs. In contrast, specimens collected in May-July were not breeding. In the central part of its range, O. Langrand (*in litt.*) found an old nest at Périnet (18°56'S, 48°25'E) on 15 August and a male was reported to be collected from a nest near Tamatave (18°10'S, 49°23'E) on 23 November (UMZC). F. Hawkins observed a lone female feeding a newly fledged juvenile at Zahamena (17°44'S, 48°42'E) on 27 September. In the southern part of the range of Velvet Asity, all evidence clearly points to a much later breeding season. At Ranomafana (21°16'S, 47°28'E), nest construction usually starts in November or early December, and fledging occurs in January to February (R. Prum and T. Razafindratsita *in litt.*).

Occasionally, birds may breed earlier or later than these times: a female was observed collecting yellowy bamboo leaves at Ranomafana on 20 October (pers. obs.), and O. Langrand (*in litt.*) found a nest there on 16 January. All adult males observed at Ranomafana in November 1995, during unusually dry weather, still had 20-50% of the yellow edges to their body plumage, yet in the five previous Novembers, during normal weather conditions, S. Zack had never observed an adult male with yellow feather edges (R. Prum *in litt.*). Hence, the timing of breeding varies slightly according to prevailing weather conditions. To complicate the picture further, it should be noted that some birds may breed in non-breeding plumage, since there is ample evidence of delayed plumage maturation in this species (Rand 1936, R. Prum *in litt.*; see under Description).

Nests Evidence collected at Ranomafana by Prum and Razafindratsita (in prep.) indicates that Velvet Asities nest exclusively in a single species of tree, *Tambourissa obovata*. The nest is a pear-shaped structure that is somewhat compressed laterally (Rand 1936) and hung from a low branch, c.2-5m off the ground, and usually in a shady part of the forest. J. Colebrook-Robjent found several nests at Marojejy along fast-flowing streams, and another on a ridge far from water (Benson *et al.* 1976).

Rand (1936) provided a relatively detailed description of a nest found on 29 August. Outside, it was constructed of moss and long, threadlike palm fibres. Inside, there was a complete and thick lining of broad dead leaves, c.1cm thick except on the bottom where this layer was c.3cm thick. There was no other lining. Rand (1936) reported that this nest was very well camouflaged, hanging c.2m above a rocky streambed from a sapling. It measured 28cm from top to bottom, 22.5cm from back to front and was 15cm wide. The entrance was placed about 5cm from the top and was protected by a projection of the roof that extended 7.5cm beyond the 6cm-wide nest entrance. The nest cavity was 16cm high and 9.5cm wide. One nest found in dense forest was made almost entirely of moss, and was suspended from a 2m-tall sapling so that it hung above the rocks in a dry river bed (Benson et al. 1976). A nest observed under construction by Prum and Razafindratsita (in prep.) was lined with dry bamboo leaves.

As in the case of Common Sunbird Asity, it appears that females alone build the nest. Furthermore, and adding intrigue, Prum and Razafindratsita (in prep.) observed more than one individual in female plumage (both assumed to be female) assisting in the construction of two nests, strongly suggesting that the breeding system employed by this fascinating, polygynous species might also involve cooperative breeding. One nest was built entirely in the absence of any male, but the other was briefly visited by a single male that appeared to be loosely associated with the pair of females at the nest. It is therefore possible that multiple flexible breeding strategies are being pursued (R. Prum *in litt.*, Prum and Razafindratsita in prep.). A single female-plumaged bird was observed to feed the recently fledged young during the Ranomafana study. Contrary to observations by Prum and Razafindratsita (in prep.), Langrand (1990) reported that both parents participate in feeding the young, although noting that only the female incubates eggs (O. Langrand *in litt.*).

Nest construction is conducted in a manner that is almost unique among birds, although shared with Common Sunbird Asity (Prum and Razafindratsita in prep.): initially, a complete, hollow sphere is woven, without any entrance. When it is complete, the female(s) then poke a hole into the side and construct the doorway. The final stage in nest construction is the addition of the nest lining which, in the case of the nests observed by Prum and Razafindratsita (in prep.), consisted of bamboo leaves. Many other eurylaimids do not construct nests in this fashion but there are no observations available for Schlegel's Asity or Yellow-bellied Sunbird Asity, or the apparently related African Green Broadbill.

Eggs Clutch-size is reported to be three (Rand 1936, Langrand 1990), although an egg in BMNH (collected before 1870) was reported to have come from a clutch of nine. The eggs are pure white, elongated, rather pointed ovals with smooth, slightly glossy shells. Three eggs collected by Rand (1936) measured 28.3 x 18.4mm, 28.7 x 18.6mm, and 28.5 x 19mm. Two eggs in BMNH measure c.25.5 x 19mm and c.28.5 x 19mm.

DESCRIPTION
A short-tailed, plump, sexually dimorphic species with a medium-length, slightly curved bill. Although male plumage has been neatly categorised into 'breeding' and 'non-breeding' plumage, males exhibit delayed plumage maturation and may perhaps breed in plumages that are not the 'breeding plumage' described below. Rand (1936), for example, caught a subadult male in immature plumage on 2 September that had considerably enlarged testes and was apparently breeding, and Goodman (1995) reports that numerous individuals collected during 1990-95 in 'female' plumage proved to be males, with large testes and convoluted epididymis. Indeed birds were identical in plumage to adult female but had well-developed wattles and testes of equivalent size to males in breeding plumage. R. Prum and T. Razafindratsita have observed such birds singing and displaying (R. Prum *in litt.*). Nothing is known about the amount of time that males spend in female-like subadult plumage before they acquire adult male plumage and caruncles. Subadult males develop adult male plumage and wattles simultaneously (in contrast to Schlegel's Asity). At Ranomafana, adult males moult into fresh black plumage with yellow edges ('non-breeding') in April and May and these edges become abraded by the breeding season, usually during late October and November (R. Prum *in litt.*).

Adult male non-breeding Plumage black with bold yellow markings formed by narrow yellow tips to feather edges, except on throat and face, which is unmarked. Crown appears streaked or scaled; rest of upperparts and breast marked with distinctive narrow bands of yellow. Belly to undertail-coverts mixed yellow and black. Tertials, primaries and secondaries brownish-black. The caruncle is absent or vestigial, forming a narrow strip over the eye (usually during the months of April to August, but occasionally as late as October: Benson et al. 1976, Prum et al. 1994). Eyes dark brown; bill black with pale yellow gape; legs and feet olive.

Adult male breeding Breeding plumage is obtained when the yellow feather fringes wear off through abrasion (Benson et al. 1976, Benson 1985). Body entirely velvety-black with a concealed golden-yellow 'shoulder' patch formed by yellow marginal underwing alular coverts. A few very fine yellow tips to feathers of the forehead and above the caruncle may also be present. Tertials, primaries and secondaries brownish-black. Head distinctively marked by the presence of a prominent bright lime or apple-green, fleshy supraorbital wattle (caruncle) that extends along the sides of the head from above the ear-coverts to the bill, and protrudes upwards to form a laterally compressed wattle at the sides of the forehead. The lower part of the caruncle, extending as a narrow stripe from the bill to just behind the eyes, is bright blue, although the blue stripe is usually concealed except when displaying. The caruncle is covered in small erect papillae and can be extended by internal muscles so that they stand erect (Prum et al. 1994, R. Prum *in litt.*). Eyes dark brown; bill black with extreme tip and cutting edge of distal half yellowy-green; legs and feet olive.

Males are often individually identifiable by the shape of the caruncle or by the last remnants of the yellow feather edgings at the end of feather wear.

Adult female Forehead, crown, mantle, back, rump and uppertail-coverts dark olive-green. Head marked by a few whitish-green or yellow-olive feathers above the ear-coverts; ear-coverts dark olive-green, darker than mantle but concolorous with the broad moustachial streak, which widens towards end. Area around eye, extending backwards as a broad patch below ear-coverts, yellow-green. Tail brown with olive tinge and yellow-olive edges, especially near base. Chin and throat pale yellow-green to yellowy-white, indistinctly mottled darker (caused by darker, olive-yellow bands at feather tips) and becoming more spotted and often mixed with whitish-yellow on breast. Belly to undertail-coverts largely yellow; sides of breast olive-green (as upperparts), marked with whitish-yellow to yellow-olive spots that coalesce to form streaks along the lower, dark olive, flanks; lower flanks and sides of belly whitish-yellow with dark olive streaks. Lesser coverts brownish with broad dark olive-green tips; tertials brown with yellow-green edge to outer web and tips of outer web forming distinctive wing-panel that crosses the secondaries; secondaries brown with the same yellow-green edge to outer web; primaries similar but yellow-green edges narrower. Bare skin around eyes dull olive to olive-grey; eyes dark brown, bill black with base of lower mandible yellow or pinky, gape dull yellow or yellowy-pink; legs and feet olive.

Immature Similar to female, but subadult males have a bare whitish patch of skin above the eye that is the rudimentary caruncle. Subadult males gradually acquire black feathers with yellow tips as in non-breeding adult male; those first to appear are on belly, rump, mantle sides and greater and lesser coverts.

MEASUREMENTS Wing male 77-88, female 80-91; tail male 46.5-52, female 45-48.5; bill (from skull) male 19.1-23.2, female 19.6-23.7; tarsus 22-27.5; weight 31.5-38. Wing, tarsus and bill measurements from Goodman 1995.

49   SCHLEGEL'S ASITY
Philepitta schlegeli   Plate 24

Philepitta schlegeli Schlegel 1867, Proc. Zool. Soc. London, p.422: Madagascar.

FIELD IDENTIFICATION Length 12.5-14. Schlegel's Asity inhabits the rainforests of Sambirano and the seasonally dry forests of the western half of Madagascar. In breeding plumage, males are easily identified by the black head with conspicuous large blue and pale green wattle around the eyes, the bright yellow underparts and bright olive-green upperparts with yellow and black markings on the mantle and back. In non-breeding plumage, the crown and sides of the head become green, flecked with yellow and white, whilst the wattle is reduced in size or disappears completely. The upperparts are olive-green with some black spotting on the mantle and back, and usually some yellow at the top of the mantle. The throat is largely scaled white and yellow, whilst the rest of the underparts become yellowy-green with diffuse yellow or white mottling, although the lower belly and undertail-coverts remain largely yellow. As birds change back into breeding plumage, black feathers appear on the head. Subadult males may have almost fully developed wattles before acquiring solid black heads, and may breed in such plumage. Females resemble males in the non-breeding plumage, but have a bold yellow eye-ring and olive-green upperparts, usually with narrow, short yellowish streaks on the crown to nape. The belly and undertail-coverts are more or less unmarked yellow. Immature females and some subadult males are indistinguishable from adult females.

Similar species Males in breeding plumage are unmistakable. Non-breeding males and females are only likely to be confused with Velvet Asity: both species occur in the forests of north-west Madagascar (Sambirano). However, Schlegel's Asity is slightly smaller, lacks a distinct yellow patch below the eye, has clear yellow belly and undertail-coverts and bright yellowish edges to the flight feathers, and the bill is more slender, shorter and straight.

VOICE The call has been described as an insignificant down-scale whinny, easily overlooked (A. Greensmith *in litt.*). The song of the male is a quiet jingle of seven notes that lasts some 2-3 seconds, with the first five notes rising in pitch and the last two descending. The song may be delivered from a relatively hidden perch low in a bush, or an exposed perch high in the canopy (pers. obs.).

DISTRIBUTION Schlegel's Asity is widely distributed in western Madagascar, although the distribution may be somewhat scattered. It is known from the following localities (Dee 1986, Langrand 1990, *in litt.*, Hawkins 1994): Anaborano, Sambirano (including the Manongarivo Special Reserve), Maromandia, Bora, Ankarafantsika, Ampijoroa, Namoroka, Tsiandro, Tsimafana (19°42'S, 44°33'E), 5km north of Beroboka (c.19°50'S, 44°34'E) and Ankazoabo (the most southerly part of its known range, at 22°18'S, 44°30'E). There is a specimen recorded as originating in north-east Madagascar (Dee 1986), but the presence of the species there requires confirmation.

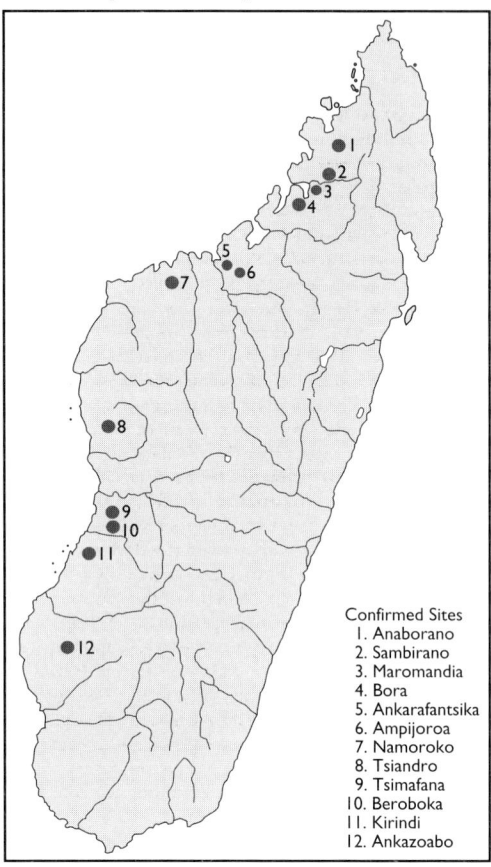

Confirmed Sites
1. Anaborano
2. Sambirano
3. Maromandia
4. Bora
5. Ankarafantsika
6. Ampijoroa
7. Namoroko
8. Tsiandro
9. Tsimafana
10. Beroboka
11. Kirindi
12. Ankazoabo

Map shows location of known sites. Probably widespread in western Madagascar in areas where suitable habitat occurs.

GEOGRAPHICAL VARIATION None recorded.

HABITAT Schlegel's Asity usually inhabits the bushy lower stratum of dry deciduous forests of western Madagascar and the lower strata of rainforests in the Sambirano area (north-west Madagascar). It has been found from sea level to at least 800m. In the western dry forests its distribution appears to be linked to that of calcareous soils, although not exclusively so (Delacour 1932b).

Hawkins (1994) notes that outside of the Sambirano rainforest, Schlegel's Asity occurs only in gallery forests in river valleys, where it has a rather patchy distribution, and in humid forest fringes in limestone areas. In November, at Ampijoroa, this species was observed visiting flowers in the canopy of rather open, dry forest dominated by leafless deciduous trees at that time, indicating that it will leave preferred dense leafy forest to feed in drier habitats nearby (pers. obs.). O. Langrand (*in litt.*) observed Schlegel's Asity in a coffee tree at the limit of rainforest in the Manongarivo Special Reserve, at 500m.

STATUS Schlegel's Asity is reported to be common in the rainforests of Sambirano and Bora in the north-west of Madagascar, but less common in the western savanna areas of the country (Rand 1936, Milon *et al.* 1973, F. Hawkins *in litt.*). In the west, it is reported to be commoner in woodlands on calcareous soils, such as those at Anaborano, Namoroka and Tsiandro (Delacour 1932b). It seems to be rather rare in the dry forests at Ampijoroa (pers. obs.).

MOVEMENTS None recorded.

FOOD Eight Schlegel's Asities collected by Rand (1936) had fruit in their stomachs, indicating that this is a highly frugivorous species. Like other asities, however, Schlegel's also appears to be highly nectarivorous (F. Hawkins verbally, pers. obs.). F. Hawkins (*in litt.*) has observed it feeding on the nectar of a brilliant red-flowering climbing shrub *Combretum* sp. (Combretaceae). At Ampijoroa, a male was observed feeding on the white flowers in the canopy of a flowering *Albizia* tree (Leguminosae) in November 1993 (probably either *A. lebbeck* or *A. fastigiata*; F. Hawkins *in litt.*, D. Reid *in litt.*). This bird was seen hanging from inflorescences and probing the flowers, and although it is possible that it was searching for small insects, the behaviour was very reminiscent of that of Asian leafbirds *Chloropsis* spp. when they feed on nectar. One of the eight specimens examined by Rand (1936) also contained a large spider, and another, small insects. Although fruit are included in the diet, the species utilised are not documented, with the exception of the dark red-orange fruits of an understorey shrub, *Cabucala* sp. (Apocynaceae) (F. Hawkins *in litt.*). Clearly, the diet of this species is very varied, and this may be an adaptation to its rather seasonal habitat, where reliance on one type of food source, such as fruit, may not be possible year round. Examination of specimens preserved in spirit at BMNH indicates that the tongue of this species is specialised, although both tongues appear damaged. The tongue is attenuated near the tip, with the distal portion very narrow, perhaps even tubular, whilst the tip has a small number of short, fine, brush-like protrusions (c.1-2mm long).

HABITS Schlegel's Asity, in contrast to Velvet Asity, is primarily a bird of the canopy, where it visits flowering and fruiting trees. Birds tend to be solitary, although they have occasionally been observed in pairs. P. Morris (verbally), encountered five males calling simultaneously at Ampijoroa on 18 November from separate perches along 150m of trail. As pointed out by Prum and Razafindratsita (in prep.), this pattern is suggestive of dispersed leks. Of four of these individuals that were observed, only two were in full adult male breeding plumage: the other two were subadults in female-like plumage but had fully developed wattles. All were probably singing, since they were located initially on call. At one time, two males (one subadult) were observed in the same tree.

Males may call from exposed perches high in the canopy, but also call from hidden perches at various levels in the forest. Sometimes, they may call from close to the ground. Unlike Velvet Asity, Schlegel's Asities tend to fly rather long distances between feeding or calling bouts, and in savanna usually fly above the level of the tallest trees (F. Hawkins *in litt.*, pers. obs.).

Hawkins (1994) observed a male Schlegel's Asity display to a female near a nest at Namoroka in mid-October. The male perched next to the female with its wings drooping and simultaneously fluffed up its breast and lifted its tail so that it almost touched its rump. During this display, the male made a series of quiet sibilant squeaks. Later, in the same area, an adult male was observed to attack a younger or partially moulted male which had a mixture of adult male and female plumage and a half-sized wattle. The latter individual had perched within 30cm of the adult pair, and after a few minutes the adult male attacked. The two birds then fought in the air and, briefly, on the ground, before the young male flew off.

An adult male feeding at a relatively large flowering *Albizia* tree at Ampijoroa was observed to visit the tree repeatedly, at intervals of c.15-30 minutes, during mid-morning. Between feeding bouts, this male periodically called from within the flowering tree as well as from various hidden perches in the vicinity, and also flew off to call from sites c.100m away (pers. obs.).

Schlegel and Pollen (1868; in Dee 1986) noted that Schlegel's Asity was found in mixed bird parties, and O. Langrand (*in litt.*) has observed two pairs of Schlegel's Asities in a mixed bird party in mixed deciduous forest at Bora on 9 June, and two males in a mixed bird party in rainforest at Manongarivo on 18 June. Such birds may associate with various vangas Vangidae, Crested Drongo *Dicrurus forficatus*, Common Newtonia *Newtonia brunneicauda*, Long-billed Greenbul *Phyllastrephus madagascariensis*, Ashy Cuckoo-shrike *Coracina cinerea*, Common Jery *Neomixis tenella*, Madagascar Paradise Flycatcher *Terpsiphone mutata* and other typical flocking species.

BREEDING Very little is known about the breeding biology of Schlegel's Asity. Males in breeding plumage have been collected or observed between late October and mid-July. Until 1993, when two nests were found, nothing had been documented about the nesting of this species. Both nests found in 1993 were under construction: one was discovered in Namoroka Strict Nature Reserve on 15 October (Hawkins 1994), the other at Ampijoroa on 5 December (S. Harrap *in litt.*).

The nest at Namoroka was in a small, 10ha, forest patch isolated by savanna and about 1km from a more continuous forest tract. The nest was half-built, suspended 3m above the ground from a 6cm-thick horizontal branch which was the lowest in a 4.5m-tall tree. It hung over an area with only limited herbaceous growth beneath it, forming a clear area for at least 3m in all directions except towards the tree, being positioned c.1.5m from the trunk (F. Hawkins *in litt.*). Both adults were building the nest, using strips of leaf, up to 10cm long, to construct the frame and to hold the frame to the branch. Pieces of moss and bark, collected from shrubs nearby and held together with spiders' webs, formed the outer covering. The frame was an oval, some 20cm tall, and with an entrance hole evidently being left near the top. Although not observed, the completed nest would probably have been pear-shaped (Hawkins, 1994).

At Ampijoroa, both adults were also building the nest, which was hung 3m off the ground. The nest was close to the tip of a branch of a broadleaved tree near the forest edge, in a shallow gully. It was a rather scruffy bag, measuring some 22-24cm long and 12-14cm wide, and attached by weaving the material over the branch along a length of

Schlegel's Asity at nest

c.12-14cm. In contrast to the Namoroka nest, the exterior was made of long, dried, grass-like fibres, with a few pale brown, dried broad leaves in the vicinity of the nest entrance. A trailing mass of material protruded from below one side of the nest bottom. A few narrow twigs or petioles were sticking out of the dangling mass (S. Harrap *in litt.*). Clutch-size is unknown and the eggs are undescribed.

There is no evidence that this species is polygynous, unlike its rainforest-inhabiting congener, Velvet Asity for which there is some evidence of polygyny. Nevertheless, as documented under Habits, some observations suggest the existence of dispersed leks, and further study of the breeding system may show that this species exhibits considerable plasticity in its breeding ecology.

DESCRIPTION
Although male plumage has been categorised into 'breeding' and 'non-breeding' plumage, males apparently exhibit delayed plumage maturation and may perhaps breed in plumages that are not the 'breeding plumage' described below (see below under Subadult male).

Adult male breeding Forehead, crown to nape and ear-coverts black, the sides of the crown and lores adorned with a bold, colourful, fleshy wattle (caruncle). Eye-ring emerald-green, encircled above, in front and partly below (front half), by the bicoloured, multi-lobed wattle. Wattle colour above eye bright dark blue, this colour also present in front of the eye and extending down across lores to the base of the bill, where bordered behind by a shining emerald-green crescent that points back towards the ear-coverts; the upper part of the wattle protrudes as a single, shiny broad emerald-green lobe towards the forehead, whilst a similar but bifurcated shiny emerald-green lobe extends towards the nape. The caruncle surface is apparently covered with broad, round, bubble-like tubercles (Prum *et al.* 1994). Upperparts olive-green with variable amount of yellow at top of mantle, adjoining the black of nape and often forming a patch. The mantle, and occasionally back, are marked with irregular black spots, although sometimes these marks are missing. Upperwing-coverts olive-green with brighter fringes; flight feathers dark brown, broadly fringed with yellow-green, although this does not extend to the emarginated tips of outer five primaries. Tail olive-green. The underparts are golden-yellow. Short, wide, black bill, with distinct orange-red gape and interior of mouth; eyes brown; legs and feet slaty.

Adult male non-breeding The plumage of non-breeding males is variable, depending on whether or not they are moulting into breeding plumage. The non-breeding plumage is reminiscent of that of adult female, differing primarily in having denser, broader and more prominent whitish streaks on the top of the mantle (mostly at the sides). The crown may be similar to that of the female for a short time, but there is usually a variable amount of black on the crown and ear-coverts. Black spots, and some yellow at the top of the mantle, may also be present.

Subadult male There appear to be two distinct subadult male plumages or age-classes. One is female-like, and the other is female-like with wattles: one female-like specimen in BMNH (collection date unrecorded) is a male with fully formed wattles (larger than that of most birds that are apparently in 'breeding plumage'). The existence of this specimen and observations of singing males with female plumage and wattles (see Habits) suggests the possibility that males may exhibit delayed plumage maturation, like Velvet Asity, and perhaps sometimes breed in 'non-breeding' plumage. In addition to differences mentioned above, the plumage of this specimen also differs from that of an adult female in having a more or less white throat and a larger number of scattered white feathers on the breast and belly. Full adult male plumage is perhaps acquired only after more than one moult cycle, the number of years for transition to full adult plumage being unknown.

Adult female Eye large, surrounded by prominent broad fleshy yellow ring. Forehead to crown bright olive-green, with variable amount of creamy and yellowy-green mottles on forehead and sparsely scattered narrow creamy streaks on crown to nape, these being formed by cream centres of feathers. A line of cream or yellowish-white in feather tract behind eye forms a more or less distinct line from top of eye at rear to back of ear-coverts. Lores blackish; ear-coverts very variable in colour, being a mixture of olive-green, yellowy-green and cream. Upperwing-coverts brownish on inner webs and mostly olive-green on outer webs, usually being brightest along the feather edge; feathers at bend of wing mostly bright yellow, forming narrow yellow streak at bend of closed wing. Flight feathers dark brown above, grey-brown below, with prominent narrow bright yellowy-green leading edges, broadest on secondaries; these edges reach tip of all but outermost five primaries, which are strongly emarginated. Innermost edges of all flight feathers narrowly edged with greyish-white. Underwing-coverts mostly silky white. Tail feathers basally dark brown, but with broad olive-green edges to outer webs, extending across both webs of innermost pairs. Chin and upper throat pale yellow; lower throat to upper breast mottled, with feathers white to yellowish-white narrowly edged with olive-green; this pattern continues on lower breast and flanks but white tones are generally replaced by yellowish-white and become increasingly yellow, so that

upper belly is mottled yellow and olive-green. Lower belly and undertail-coverts unmarked bright yellow. Eye brown; bill black; interior of mouth orange-yellow; legs slate-grey.
Immature Full-grown immature males are indistinguishable from adult females until they start to acquire elements of breeding plumage (BMNH).

MEASUREMENTS Wing 70-82; tail 38-46; tarsus 18-21.8; bill from skull 15.2-17.3 (measurements, except tail, from S. Goodman *in litt.*). Weight unrecorded, but probably smaller than Velvet Asity (i.e. less than 38).

NEODREPANIS

This Madagascan genus comprises two tiny species that have a radically different appearance to all their broadbill relatives. They resemble sunbirds in size, diet, the shape of the long, decurved bill, and in having a long tubular tongue. One species is an inhabitant of moist montane forests, the other is more widely distributed at lower altitudes in the eastern rainforests. Like *Philepitta*, breeding males have brightly coloured plumage and wattles around the eyes. Recent observations suggest that these species are polygamous, with lekking males which perform elaborate ritual displays.

50 COMMON SUNBIRD ASITY
Neodrepanis coruscans Plate 24

Neodrepanis coruscans Sharpe, 1875, Proc. Zool. Soc. London, p.76: Madagascar.

Alternative names: False Sunbird, Wattled Asity, Wattled Sunbird, Wattled Sunbird Asity, Sunbird Asity.

FIELD IDENTIFICATION Length 9.5-10.5. Common Sunbird Asity is endemic to the forests of eastern Madagascar. It is a tiny species with a distinctive, long, strongly downcurved bill, reminiscent of a sunbird. The tail is very short, hardly extending past the wing tips. Males have separate breeding and non-breeding plumages. In breeding plumage, the head is bright dark metallic blue and adorned with a conspicuous broad rectangular pale azure-blue wattle that extends over and behind the eyes. The upperparts, including the wings, are metallic dark blue except for the greater coverts and secondaries, which are yellowish, and the primaries, which are greyish-brown with yellow outer margins. The underparts from the throat to undertail-coverts are yellow to yellow-olive, being brightest on the flanks and undertail-coverts, more olive on the breast and marked with dark greyish to blackish mottles on the breast. Females and immatures are olive-green above, paler, more mottled olive-grey below with bright yellow flanks and undertail-coverts, and have greenish heads. Non-breeding males are like females but retain metallic blue tails and feathers on the rump, uppertail-coverts and inner wing-coverts. Subadult males resemble females.

Similar species Common Sunbird Asity can be confused with Yellow-bellied Sunbird Asity, which is sympatric at some sites at higher altitudes, or with the non-breeding males and females of the two true sunbirds of Madagascar: Souimanga Sunbird *Nectarinia souimanga* and Long-billed Green Sunbird *N. notatus*.

Breeding male Common Sunbird Asity is distinguished from similar-plumaged Yellow-bellied Sunbird Asity by larger size, more sharply decurved bill, possession of olive-yellow fringes to the flight feathers, olive-yellow breast marked with dark grey dots, and in the shape and colour of the orbital wattle, which extends backwards as an elongated oval behind and above the eye in Common Sunbird Asity and is a paler, more uniform blue colour. For other differences with respect to the wattle, see under Yellow-bellied Sunbird Asity. Males in non-breeding plumage and females differ from Yellow-bellied Sunbird Asity in having yellow largely restricted to the flanks and undertail-coverts. Voice and the noise made by the wings in flight are also good identification features.

Both true sunbirds are larger than Common Sunbird Asity and have a more elongated shape, longer tails and less curved bills. In contrast to Common Sunbird Asity, the sunbirds do not make a distinctive buzzing whir with their wings during short bursts of flying, whilst Common Sunbird Asity does not hover in the manner typical of the Souimanga Sunbird, which does so frequently. Common Sunbird Asities also flick their wings frequently when perched, unlike sunbirds. Male sunbirds of both species are quite different to Common Sunbird Asity, and unlikely to be confused. Non-breeding male and female sunbirds are very similar to female Common Sunbird Asity. Apart from structural differences mentioned above, female Souimanga Sunbird differs in having much darker green upperparts, has more yellow on average below, and lacks the bright olive-green edges to the flight feathers (which form a distinct panel in the wing of female Common Sunbird Asity). Female Long-billed Green Sunbird is less likely to be confused, since it is much larger, with a long, robust bill is darker above without a bright yellow-green wing-panel and the breast is heavily mottled with dark brown.

VOICE The call of Common Sunbird Asity is a thin, high-pitched vigorous *seeseeeseeeseeeeseee* delivered very rapidly and sometimes trailing off into more separated series of notes. It is loud and may be audible from around 20m. Single, slightly drawn-out *szeee* or *szee-u* notes may also be given from a perch. When flying short distances, as for example between flowers, the wings of Common Sunbird Asity make a very distinctive mechanical buzz, audible at at least 10m. These noises apparently differ between the sexes (R. Prum *in litt.*).

DISTRIBUTION Endemic to Madagascar, where restricted to the east and north-west. In the east it is present from Andohahela in the south (24°40'S) to the Masoala Peninsula and Marojejy Strict Nature Reserve (c.14°18'S) in the north, and extends westwards to Tsaratanana on the edge of the Sambirano Floristic Zone. It has not been found in the extreme north in the isolated mountains of

Montagne d'Ambre National Park (Langrand 1990), even though suitable habitat apparently occurs there.

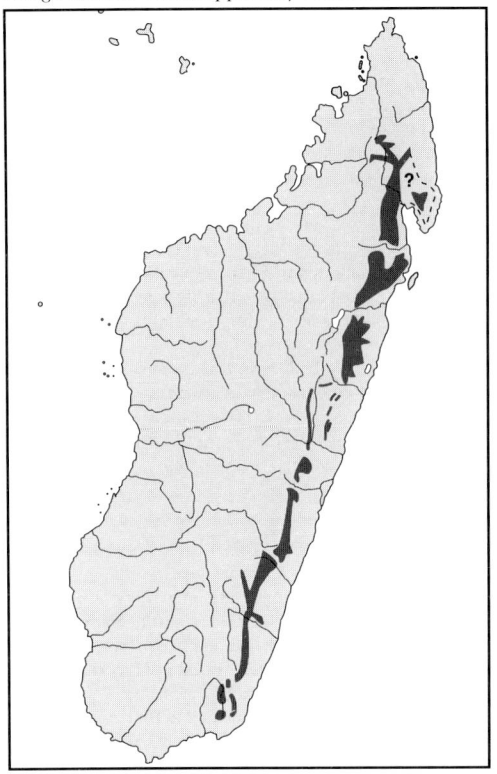

Distribution based on approximate extent of forest remaining within the altitudinal range of the species. Former range more widespread.

GEOGRAPHICAL VARIATION None reported.

HABITAT The Common Sunbird Asity is generally distributed throughout the eastern rainforests of Madagascar, and in the north-west at Tsaratanana. These forests are characterised by having no dry season, although precipitation is greatly reduced between mid-September and the end of November. It occurs at all levels in the forest, and from sea level to at least 1,200m. A specimen collected by Rand (1936) at 1,800m was most likely prepared at camp at that altitude but collected at a lower elevation (F. Hawkins *in litt.*). In areas where both sunbird asities occur, they apparently replace each other altitudinally, as is the case in the Andringitra Strict Nature Reserve, where Goodman and Putnam (in press) did not find Common Sunbird Asity above 900m, but found Yellow-bellied Sunbird Asity at elevations above 1,100m. In some parts of its range Common Sunbird Asity does not appear to occur, or is very rare, at low altitudes. For example, in the Masoala Peninsula there have been few observations below 500m (Hawkins *et al.* in prep.).

It inhabits various types of forest within the rainforest belt in eastern Madagascar, including mossy forest on Tsaratanana (Milon 1951), and also frequents adjacent dense secondary forests and brush (Rand 1936). Benson (1976) collected specimens in tangled undergrowth and saplings in degraded forest. Although the species has been recorded in plantations, this has only been observed where there is adjacent forest and all evidence suggests that the Common Sunbird Asity requires relatively intact forest (Hawkins *et al.* in prep.).

STATUS This species is fairly common throughout most of its range and in some areas, such as at Fanovana and Maroantsetra (Sihanaka Forest), it has been described as abundant (Dee 1986). In general, it is commoner at higher altitudes within its range (Rand 1936, Dee 1986). As with all forest-dependent species on Madagascar, habitat loss through clearance and fire is of serious concern, and populations of forest species are becoming more and more isolated. Although it is present in various protected areas, the long-term survival of this and other specialised forest bird species outside of such areas is questionable. Green and Sussman (1990) calculated that, at present rates of moist forest loss on Madagascar, only the most inaccessible areas, such as forest on very steep slopes, would remain by about 2015.

MOVEMENTS None recorded, but it seems possible that this species may make altitudinal movements.

FOOD The diet of Common Sunbird Asity is poorly known. However, the structure of the bill, and tubular tongue, clearly support observations suggesting that it is a nectarivorous species (Salomonsen 1965). Like other nectarivores, it is also reported to eat insects and spiders (Rand 1936, Benson *et al.* 1976). According to Salomonsen (1965), Common Sunbird Asities are attracted to flowers, such as *Impatiens humblotiana* (Balsaminaceae), to feed on nectar. The long red corolla of this species is curved and almost in the same size and form as the bill of Common Sunbird Asity. Surprisingly, Langrand (1990), asserted that Common Sunbird Asity visits this species and *Hedychium coronarium* (Zingiberaceae) to feed on small insects that are attracted to their flowers: it seems more likely that visits to such flowers are actually related to feeding on nectar or pollen.

At Ranomafana National Park, Common Sunbird Asity has been observed feeding at a variety of flowers in the understorey and in the mid to lower canopy of rainforest. In October (pers. obs.), these included: the red flowers of an epiphytic mistletoe *Bakerella* (Loranthaceae), the small yellow-flowered inflorescences of a species of *Psychotria* (Rubiaceae), about 1.5m off the ground; the large white flowers on terminal inflorescences of a small tree which was probably *Mussaenda* (Rubiaceae) (B. Lewis *in litt.*); the medium-sized pinky-white cauloflorous flowers of *Rhedia* sp. (Clusiaceae); the small white flowers in terminal inflorescences of *Gaertnera* trees (Rubiaceae), and the flowers of *Gravesia* (Melastomataceae). At Périnet, in November (pers. obs.), Common Sunbird Asities regularly fed at the small red-based pink flowers on the branches of an epiphytic melastome (probably a species of *Medinilla*: B. Lewis *in litt.*), at 1,150m altitude, and at lower altitudes, at the long head of white flowers of *Dracaena*, in the forest understorey. O. Langrand (*in litt.*) has observed pairs of Common Sunbird Asities feeding at the flowers of *Grevillea alba*, whilst Goodman and Putnam (in press) observed them visiting the dull reddish-pink flowers of *Bakerella clavata* and the white flowers of *Liparis* sp. (Orchidaceae). Safford and Duckworth (1990) describe Common Sunbird Asities feeding at the terrestrial root-parasite flowers of *Ditepalanthus* sp. (Balanophoraceae) and the flowers of a ginger *Aframomum* sp. (Zingiberaceae). Rand (1936) collected 16 Common Sunbird Asities that were feeding on flowers with corollas that were too long to attract the shorter-billed Souimanga Sunbird, although he noted that Long-billed Green Sunbird occasionally fed at them.

A pair observed at Périnet in 1973, originally identified as Yellow-bellied Sunbird Asities but now reidentified as Common Sunbird Asities (R. Safford verbally, Hawkins et al. in prep.), were feeding flying termites to young in a nest sited in the mid-level of the forest patch (A. D. Forbes-Watson in Dee 1986). Safford and Duckworth observed this species probing into arboreal moss, presumably to feed on invertebrates.

HABITS The Common Sunbird Asity is a rather restless species that usually stops only briefly whilst moving through the understorey and lower levels of the forest. Occasionally, individuals will perch, calling, for more prolonged periods in the canopy, though even during calling bouts there is a tendency to move perch frequently (pers. obs.). Rand (1936) noted that they may also sit quietly in dark parts of the forest. They are usually found alone or in pairs in dense vegetation within forest. Whilst feeding, Common Sunbird Asities regularly flick their wings, and generally make regular, short flights between flowers during which the loud insect-like buzz of the wings is readily heard. Common Sunbird Asities sometimes associate with nectarivorous species such as the Souimanga Sunbird and Madagascar White-eye *Zosterops maderaspatana* which are attracted to the same types of flowers, and occasionally several Common Sunbird Asities will independently visit the same flowering tree: Rand (1936) collected 16 specimens from a single tree over a collecting period of around five weeks, and noted that other birds visiting were not collected. At other times, they may search for small insects on the bark of twigs and branches (Rand 1936).

BREEDING The breeding season coincides with the dry season in Madagascar, probably starting in August or September, although fledglings may not emerge until the onset of the rainy season. Rand (1936) collected a male with enlarged testes on 23 August, whilst Salomonsen (1965) noted that specimens of males in breeding plumage collected in September had swollen testes. Goodman and Putnam (in press) collected three males in the Andringitra Strict Nature Reserve with well-developed wattles in late September; two in breeding plumage, the third moulting into breeding plumage. Langrand (1990) reported that nesting has been recorded between September and mid-January. R. Prum and T. Razafindratsita (*in litt.*) observed the construction of two nests in early November at Ranomafana, during a late dry season, and fledging during the early rains in December. P. Morris (*in litt.*) also found this species breeding at Vohiparara, Ranomafana in November-December.

The nest is reported to be made of dead leaves and covered with moss (Langrand 1990), although an incomplete nest observed at Ranomafana in mid-October (pers. obs.) was made almost entirely of green, drying moss with a few small dried bamboo leaves, a few other leaves, a few twigs of up to 3mm thick and 5cm long, and some small black rootlets. The nest is a rather ragged-shaped oblong structure some 8-10cm long and up to 35cm long, becoming narrower and scrappier nearer the end. It is usually hung from the end of a branch in the lower storey or understorey, with the moss being threaded over the top of a narrow branch. The incomplete nest observed at Ranomafana was about 4.5m off the ground, suspended from near the tip of a hanging branch over a path. Another nest found by O. Langrand (*in litt.*) at Ranomafana contained two pale green eggs.

Initial studies of the species suggest that only the female constructs the nest (R. Prum *in litt.*) and that only the female incubates the eggs (O. Langrand *in litt.*). A nest found by P. Morris (verbally), thought to contain young, was visited frequently by a female: a breeding-plumaged male also visited the vicinity of the nest, but never entered during the brief period of observation. The female was seen to enter the nest and stay there for c.7 minutes, was away from the nest for c.7 minutes, returned and spent four minutes in the nest followed by four minutes away, returned again briefly, flew off and then almost immediately came back again. It is of interest to note that the observer had the impression that the last visit was possibly made by a different female.

Observations by T. Razafindratsita and R. Prum have shown that nest construction is conducted in an almost unique manner, as in both *Philepitta* asities. The female first constructs a complete, hollow orb, then pokes a hole in it to create a doorway, and finally builds the overhanging porch (R. Prum *in litt.*).

DESCRIPTION
Bill long and downcurved, being broad at the base but extremely fine and pointed at the tip.
Adult male breeding Forehead and lores black, tinged with iridescent blue; crown to nape, mantle, back, rump, uppertail-coverts and scapulars bright, iridescent, metallic dark blue, in some lights appearing blue-green. The head is adorned by a large, smooth, rectangular wattle (caruncle) that is distinctly rounded at the rear. The wattle is very narrow around the lower half of the eye, with a short rounded lobe extending forward from the upper half of the wattle at the front, towards the upper lores and forehead, although the lores are not obscured by the wattle. The eye itself is bordered by a narrow turquoise-green eye-ring. This wattle reaches some 6.5mm in width and 11mm in length. Tail feathers black, with broad metallic dark blue edges and tips to innermost pairs and progressively less blue on outer feather, the blue being restricted to the outer edge of the outer pairs and usually absent completely from the outermost pair. Tertials black with bright iridescent dark blue terminal bands that extend along the outer edge; lesser coverts iridescent metallic dark blue; median coverts brown with broad yellow-olive tips, greater coverts brown with broad yellow-olive tips and leading edge. Flight feathers brown with narrow yellow edge to all but outermost primary, narrowest in outer primaries and broadest on inner secondaries. Primary coverts dark brown with narrow iridescent dark blue outer edges. Chin, throat and breast bright yellow with olive tinge and greyish-brown speckling formed by exposed bases to feathers. Undertail-coverts and flanks brighter, purer yellow. Underwing-coverts yellow, with white bases to feathers of axillaries. Outer primary very narrow and pointed. The bill is black with an ultramarine-blue base to the upper part of the upper mandible that extends to the nostril and is bordered in front and below by a light, almost lime, green. The entire basal quarter of the lower mandible is bright pale green. Iris dark brown; legs olive-grey. Males remain in non-breeding plumage until at least the end of April, with some birds acquiring breeding plumage as early as May but others not until August. Moult is discussed in some detail by Salomonsen (1965) and

Benson *et al.* (1976): males apparently moult twice a year, and have an eclipse (non-breeding) plumage that lasts for 4-5 months, from January-March to May-August.

Adult male non-breeding Males in non-breeding plumage are variable, depending on state of moult. In eclipse plumage, males are dull, resembling females, but they moult again within 4-5 months to acquire their nuptial plumage. All non-breeding males retain metallic blue tails and uppertail-coverts, but lack the caruncle of the breeding plumage and blue on upperparts, although some metallic blue-green feathers may be present in the lesser coverts. Development of metallic blue in the upperparts when moulting out of eclipse occurs in patches, usually starting with the lower back and scapulars and then progressing to the tertials and mantle. At this stage the wattle remains undeveloped.

Adult female Forehead to nape olive-green, with mottling formed by brown centres to all feathers. Wattle absent at all ages, but there is a narrow band of yellow feathers around the lower margin of the eye. Ear-coverts olive-green, usually mottled with grey or whitish. Rest of upperparts uniform olive-green, glossed with a bright green iridescence that is only visible in bright sunlight. Tail feathers brown with narrow yellowy-green edges to all but outermost pair. Tertials brown with narrow olive-yellow outer edge and tip. Lesser and median coverts brown with broad yellow-green terminal bands; greater coverts brown with olive-green leading edge; primary coverts dark brown. Flight feathers similar to, but slightly paler brown than, adult male, but basal edge to inner web grey. Chin, throat and breast mottled, being mixed olive-green and grey, occasionally with some yellow. Belly variable, being olive-green to yellow-olive; lower flanks and undertail-coverts bright, unmarked, yellow. Bill black without colourful base.

Immature Sexes similar and virtually indistinguishable from adult female except that they have a slightly swollen pale base to the bill. Immature males are usually indistinguishable from adult females, but may have one or two metallic blue feathers or patches in feathers on the uppertail-coverts or rump, and may have more yellowy feathers on the centre of belly and lower breast.

MEASUREMENTS Wing 47-51; tail 22-28; tarsus 13-15; bill long and strongly decurved, measured in straight line from base to tip, male 24-26 (n=7), female 26-27 (n=3); weight 6.2-6.6.

51   YELLOW-BELLIED SUNBIRD ASITY
Neodrepanis hypoxantha   Plate 24

Neodrepanis hypoxantha Salomonsen, 1933, Bull. Brit. Orn. Club, 53, p.182: east of Tananarive (now Antananarivo), eastern Madagascar.

Alternative names: Small-billed False Sunbird, Small-billed Asity.

FIELD IDENTIFICATION Length 9-10. Yellow-bellied Sunbird Asity is endemic to eastern Madagascar, where confined to montane forest. It is one of the two smallest members of the Eurylaimidae, with sexually dimorphic plumage that also varies with season. The bill is long and curved, but the tail is very short, barely extending past the tail tip. The underparts are entirely bright yellow in all adult plumages. Breeding males have metallic dark blue head, upperparts and tail, and brilliant golden-yellow underparts. The head is adorned by a large, rather rectangular, green-and-blue wattle over and behind the eyes. Fringing on the primaries, secondaries and greater and median upperwing-coverts is iridescent blue and concolorous with rest of the upperparts. Non-breeding males have an undeveloped wattle that may or may not be visible, are slightly duller yellow below than breeding males, and have no metallic blue on the head and mantle. Some males may have fully developed wattles but female-type plumage. Such birds usually have indistinct olive streaking on the sides of the yellow breast and flanks. Females have a greenish crown and greenish upperparts and bright yellow cheeks and underparts, although at some times of the year (around September) females may have a dull olive wash over the yellow, particularly on the upper breast and throat. Subadults and immatures resemble adult females. The flight is swift and direct.

Head of male Yellow-bellied Sunbird Asity.

Similar species This species is very similar to the Common Sunbird Asity from which it was not distinguished until 1933 (Salomonsen 1933). Both species of sunbird asity have been observed together at Vohiparara, near Ranomafana, and they may also be partially sympatric in Sihanaka Forest. Elsewhere, however, they are usually apparently separated altitudinally (Goodman and Putnam in press), although at some sites bith species may occur. Yellow-bellied Sunbird Asity is also very similar to the non-breeding males and females of Souimanga Sunbird *Nectarinia souimanga*.

Yellow-bellied Sunbird Asity can be distinguished from Common Sunbird Asity in all plumages by the shorter, finer and less sharply decurved bill (although, if the tongue is protruding, bill length and shape may appear to be similar), and, in most cases, by unmarked yellow underparts. Males are further distinguished from similar-plumaged Common Sunbird Asity by markedly smaller size, lack of yellow edges to the secondaries, the absence of dark grey dots on the breast or mottled appearance to the underparts, and, in breeding plumage, by the shape of the wattle. The fully formed wattle is squarer than that of male Common Sunbird Asity, and notched at the rear (that of Common is rounded), is relatively broad below the eye (very narrow in Common) and, when fully developed, extends across the lores to the base of the bill. The wattle is bicoloured, being green around the eye whilst the remainder is blue (a darker blue than that in the wattle of Common Sunbird Asity). The lower border or the wattle of Yellow-bellied is marked with a line of small iridescent paler blue spots that are absent from the wattle of Common. Males with developed wattles and female-type plumage usually have some light olive streaking on the flanks and sides of breast, but this is neater and less distinctive than that shown by male Common Sunbird Asity. In non-breeding plumage, male Yellow-bellied Sunbird Asities may also differ from Common in having yellow in the feathers of the uppertail-coverts and rump that may form a narrow band, although not all birds show this character. In the hand, male Yellow-bellied Sunbird Asity can also be distinguished from its congener by the more attenuated outer primary: the emargination of the outer primary of Yellow-bellied extends c.7mm from the feather tip whilst that of Common is 3-5mm long (see Salomonsen 1933). Voice is also a very good character for distinguishing the species (P. Morris *in litt.*).

Female Yellow-bellied Sunbird Asity is best distinguished from female Common by bill shape, clean yellow underparts and the lack of a bright yellow panel in the flight feathers (these are a duller olive-green in Yellow-bellied). Both true sunbirds are notably larger than Yellow-bellied Sunbird Asity and have a more elongated shape and longer tails. However, neither have the unmarked clear yellow underparts of female Yellow-bellied Sunbird Asity, and this is the best feature for separating these species. Female Souimanga Sunbird, which is the most similar sunbird in both size and structure, also differs in having darker upperparts.

VOICE Hildebrandt (1881) noted that the voice was a barely audible soft whistle. R. Safford (*in litt.*) confirms that the call is a very quiet, insignificant *si*. This call, given singly or repeated at intervals, is often barely audible, and could be dismissed as a frog or insect (Hawkins *et al.* in prep.). During short flying bouts, the wings of males in breeding plumage make a distinctive, clear mechanical whirring, almost trilling noise, apparently reminiscent of the drumming of Common Snipe *Gallinago gallinago*, and perhaps derived from the attenuated outer primary, which is confined to males in breeding plumage. Yellow-bellied Sunbird Asities in other plumages are reported to be much quieter (Hawkins *et al.* in prep.). This noise is notably louder than the mechanical buzzing noise made by flying Common Sunbird Asity. In flight, it is also reported to utter a shrill *tseee* (Langrand 1990).

DISTRIBUTION Endemic to the highlands of Madagascar, where patchily distributed in the east of the island in the zone that does not experience distinct dry seasons. It was thought to be extremely localised, occurring in forest relicts near the top of the eastern escarpment of the Central Domain (see Langrand 1990) and outlying massifs rising out of the eastern lowlands (Hawkins *et al.* in prep.). However, recent field research has suggested that Yellow-bellied Sunbird Asity is more widespread than formerly realised. Whilst the exact locality of some early collections that include this species are unknown, specimens are known to have been collected at the following locations: east of Antananarivo, east Imerina (near Antananarivo), Sihanaka Forest (including Fito), and Andrangoloaka (Andrangolesaka) (Collar and Stuart 1985). Stresemann (1937) and Greenway (1967) reported that Andrangoloaka was on the eastern slope of the plateau east of Antsirabe, some 150km south of Antananarivo, but Collar and Stuart (1985) provide evidence that this site was much closer to Antananarivo, on a forested escarpment east of Mantasoa Reservoir. Neither site, however, is still forested; nor is east Imerina.

Confirmed sightings have more recently been made at Marojejy (Safford and Duckworth 1990), at the summit of Maharira and at Vohiparara in Ranomafana National Park; in Andringitra Strict Nature Reserve (Goodman 1994, R. Safford *in litt.*, P. Morris *in litt.*); and in Anjanaharibe-sud Special Reserve (14°42S, 49°32'E) (F. Hawkins *in litt.*). These sightings suggest that this species is found along the whole length of the island in the east, although there are no records to the south of Andringitra (22°20'S), despite the presence of high forested mountains as far south as Andohahela (24°40'S).

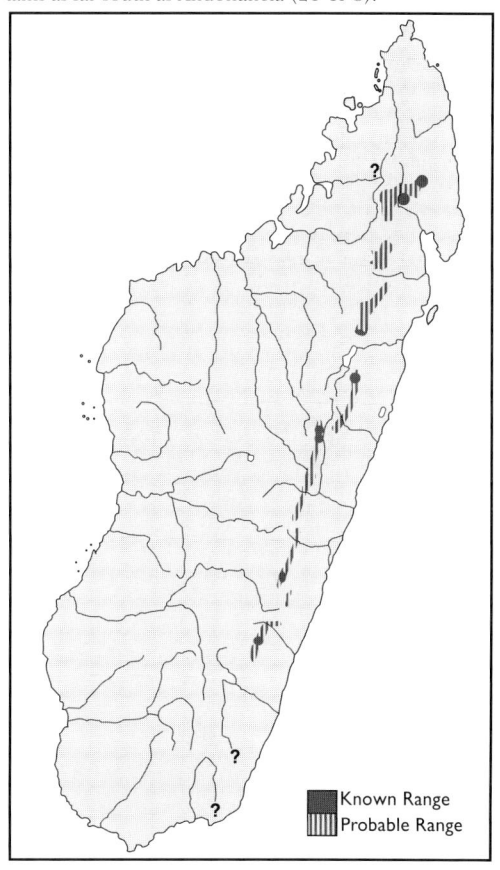

Map shows known sites and probable distribution based on extent of forest in appropriate altitudinal range.

GEOGRAPHICAL VARIATION None described.

HABITAT This is a very poorly known species that appears to be endemic to the humid montane forests of eastern Madagascar.

In the Marojejy Strict Nature Reserve in the northern part of the eastern rainforests, this species was only observed in a narrow altitudinal zone, between 1,500m and 1,700m, in sclerophyllous montane forest: it was the only species observed exclusively in this habitat (Evans *et al.* 1992). In Ranomafana National Park, it has been confirmed to occur around the highest peak, at 1,375m. Here, where R. Safford (*in litt.*) found the species in June 1993, all sightings were within 2m of the summit ridge. The habitat at this elevation is described as low, mossy forest (height 5-15m) with much bamboo (Hawkins *et al.* in prep.). In December 1994, several Yellow-bellied Sunbird Asities, including an adult male in full breeding plumage, were observed at c.1,100m at Vohiparara, Ranomafana (P. Morris *in litt.*). Reports of this species from 1,050-1,100m at

Maromiza, on the edge of the Analamazoatra Reserve near Périnet, require confirmation. The pair of sunbird asities photographed here, and reported to be Yellow-bellied Sunbird Asities in Collar and Stuart (1985), were evidently Common Sunbird Asities (Hawkins *et al.* in prep.). In Andringitra Strict Nature Reserve, a pair was observed in mid-January at 1,950m in 12-15m-tall forest along a stream (Hawkins *et al.* in prep.), and Goodman and Putnam (in press) observed birds daily at 1,100-1,765m. Specimens from Andrangoloaka are also believed to have been collected at high altitudes, at 1,000-1,400m, but since the exact location of this site is uncertain, this cannot be verified.

Available evidence suggests that the two species of sunbird asity are usually altitudinally separated, although there is a narrow contact zone where both may occur, and indeed even breed, e.g. at Vohiparara in Ranomafana, and (apparently) at Fito, between Maroantsetra and Fanovana (Benson 1974). In November 1995 at Vohiparara P. Morris (*in litt.*) observed adult Yellow-bellied Sunbird Asity feeding a newly fledged juvenile within 150m of a nesting pair of Common Sunbird Asities. At Zahamena Strict Nature Reserve, F. Hawkins (*in litt.*) found them to be altitudinally separated, with Common Sunbird Asity at 500-1,200m and Yellow-bellied at 1,200-1,500m. Here there was an abrupt change in species composition coincidental with the vegetation change from mid-altitude forest to mossy forest on ridges. However, at Anjanaharibe-sud, Yellow-bellied Sunbird Asity was found in both mossy forests and in the upper margins of mid-altitude forest at 1,400m, in an area where Common Sunbird Asity did not occur (F. Hawkins *in litt.*).

Nevertheless, the preferred habitat of Yellow-bellied Sunbird Asity would appear to be sclerophyllous montane forest. This vegetation type, typical of the upper elevation of the central domain of the eastern Malagasy region, lies between vegetation at higher altitude comprised of low scrub and the taller trees of moist montane forest below (White 1983). Sclerophyllous montane forest is characterised by a low canopy, of 3-10m height, the absence of a shrub layer, a dense mossy layer on the forest floor and, usually, hanging lichen and moss and an abundance of bamboo (Hawkins *et al.* in prep.).

STATUS The Yellow-bellied Sunbird Asity is believed to be one of the numerically rarest and most threatened of all Madagascar's bird species. Although it presently occurs in at least four protected areas (Hawkins *et al.* in prep.), an analysis of distribution and threats by BirdLife International led to its listing as Endangered (Collar *et al.* 1994). Hildebrandt (1881), describing birds now believed to have been this species (Collar and Stuart 1985), noted that it was uncommon around Andrangoloaka, whilst the Mission Franco-Anglo-Americaine failed to find it, reinforcing the view that it has always been rare (Collar and Stuart 1985).

Nevertheless, the status of Yellow-bellied Sunbird Asity may vary with locality. At Marojejy, for example, Safford and Duckworth (1990) reported that it was very sparse, but in the Andringitra Strict Nature Reserve it was observed daily by Goodman and Putnam (in press), and the mean number of detections at census points at 1,100-1,765m did not differ significantly from the mean number of detections of Common Sunbird Asity at points between 720 and 960m. Hence, Yellow-bellied Sunbird Asity is apparently as numerous at Andringitra within its restricted altitudinal range as is Common Sunbird Asity at lower elevations. Further north, Hawkins *et al.* (in prep.) reported that Yellow-bellied Sunbird Asity seemed to be one of the commonest species in the stunted forests at the top of Maharira, in Ranomafana National Park, during surveys carried out in June 1993. It should be noted, however, that this type of forest usually holds few birds of any species (Evans *et al.* 1992).

Many of the forests of the high plateau of Madagascar within the known range of Yellow-bellied Sunbird Asity have been seriously affected by deforestation, as have the majority of forests in the adjacent eastern escarpment. Hence populations of Yellow-bellied Sunbird Asity are probably now more fragmented than in the past, and there is less hope that recolonisation could occur if isolated populations become extinct through stochastic events. Although the forests at the altitudes at which Yellow-bellied Sunbird Asity occur are reported to be partly buffered from the devastating effects of slash-and-burn agriculture that has been responsible for so much forest loss in Madagascar, these forests are nevertheless susceptible to serious fires and to cattle-grazing (Sayer *et al.* 1992, Hawkins *et al.* in prep.).

MOVEMENTS Recent observations suggest that both species of sunbird asity occur at sites such as Vohiparara, but not all year round. The possibility therefore exists that Yellow-bellied Sunbird Asities could undertake altitudinal movements, moving to lower altitudes in times of food scarcity at higher elevations, or perhaps to take advantage of food abundance (perhaps related to flowering seasons) at lower altitudes, although some evidence would suggest that it might even breed at these lower altitudes (see above). Such movements would not be unexpected for a species that relies on an ephemeral food resource such as nectar.

FOOD The Yellow-bellied Sunbird Asity, like Common Sunbird Asity, appears to be a specialised nectarivorous species, feeding on nectar and presumably pollen (Salomonsen 1965, Hawkins *et al.* in prep., P. Morris *in litt.*). Hildebrandt (1881) reported that it fed at flowering bushes in forest clearings. More recently, observers have seen this species feeding on the nectar of mistletoes (Loranthaceae). These included an orange-flowered mistletoe growing on a treetrunk, the vermilion flowers of *Bakerella clavata* and the dull reddish-pink flowers of *B. tandrokensis* (Goodman and Putnam in press, Hawkins *et al.* in prep.). Birds have also been observed visiting smaller white flowers of an unidentified species (Hawkins *et al.* in prep.). Like sunbirds, Yellow-bellied Sunbird Asities also feed opportunistically on invertebrate prey: Hawkins *et al.* (in prep.) report observations of them eating tiny invertebrates, in two cases small spiders, whilst Goodman and Putnam (in press) observed a male glean a small green caterpillar that was passed backwards and forwards through its mandibles before swallowing. F. Hawkins (*in litt.*) observed this species catching tiny flies (see below).

HABITS This species is very poorly known, since it was not until the 1990s that its existence at several sites was confirmed. It is evidently very furtive, rarely remaining still for long and therefore often difficult to observe. However, it is also inquisitive, and may often perch within one metre of an observer while calling vigorously, and even displaying, during the breeding season (F. Hawkins *in litt.*). Most observations are of birds observed singly, although individuals may congregate in areas where particular flowering plants occur and hence may form loose associations. P. Morris (verbally) observed up to three together at a mistletoe plant at Vohiparara on 4 December: at least five different individuals appeared to be in this area, including two birds that were in a female-like plumage but had wattles and some blue flecks on the mantle.

On one occasion, a Yellow-bellied Sunbird Asity was seen to rattle dead bamboo leaves, perhaps in an attempt to dislodge invertebrate prey (Hawkins *et al.* in prep.). On another occasion, F. Hawkins (*in litt.*) observed birds flycatching during the late afternoon from the tops of canopy shrubs, c.3-5m off the ground: they were flying 2-3m upwards, apparently to catch tiny flies.

Males perform a display that is reminiscent of the hanging gape display of Velvet Asity. During this display, the male leans forward with its body held below the perch and the tail held above, whilst simultaneously fluffing out the throat feathers and holding the bill and head horizontally: in this posture, the male flicks his wings and vocalises energetically (Prum and Razafindratsita in prep.). Hawkins (in prep.) has observed a similar display, apparently given in response to the presence of a human observer, with the male sunbird asity perched only 1-2m away. During these 'throat displays', birds observed by Hawkins leant forward and fluffed out the yellow throat feathers. In this position, not only is the bright face-wattle displayed prominently, but the angle of posture also means that the iridescent blue back is clearly visible. On more than one occasion, birds displaying in this way turned a complete somersault around a branch, the bird dropping forward and down, and swinging rapidly upright while gripping the branch throughout the movement. Hawkins (in prep.) also noted that female-plumaged birds may also display to human intruders in this way, although more rarely. P. Morris (*in litt.*) has also observed a male with fully developed wattles, but otherwise in non-breeding plumage, perform a somersaulting display with quivering, open wings, in response to playback of its call.

Hawkins (in prep.) also observed a number of interspecific displays. Males in breeding plumage were seen to perform displays to female-plumaged birds. On one occasion, a male perched far forward on the branch, performing the throat display described above, whilst a female-plumaged bird, perched in a similar way c.0.5m away, flapped its wings vigorously whilst making an intense quiet twittering noise, before flying off. The male remained, in a normal sitting posture, and after a few seconds fluffed up its body plumage and drooped its wings. On 23 November, a similar display was observed, except that the female-plumaged bird was clearly an immature or non-breeding male, since it had a poorly developed wattle and a few blue feathers on the mantle. As previously, only the female-plumaged individual performed the wing-flapping. The following day, the same female-plumaged male displayed in the presence of at least two males in full breeding plumage. Shortly afterwards, all three birds were involved in a chase. During the following few days, until 2 December, similar displays were again observed. Hawkins (in prep.) suggests that the throat display may be aggressive, since it was made both to human observers and to immature or non-breeding males, whilst the wing-flapping display may be submissive.

On 17 September, a completely different type of display was witnessed by Hawkins (in prep.) at Zahamena Strict Nature Reserve. A male, perched within 10cm of a female, pulled his head and neck back whilst depressing the bill, so that it touched the breast ('hunched display'). Simultaneously, the crown feathers were erected at the rear and the tail was held downwards. During the display, the male called vigorously and continuously and the female touched the male's back feathers with the tip of her bill whilst slightly spreading and fluttering her wings. On 21 September, four birds – two breeding-plumage males and two in female plumage that were probably females – were observed performing a display. One male performed the hunched display whilst the adjacent female plumaged bird wing-flapped. Immediately following this, Hawkins (in prep.) saw the two female-plumaged birds fighting vigorously in mid-air, falling several metres and separating just above the ground. The two female-plumaged birds were observed to chase the males from perch to perch, and whilst not fighting or performing other displays, all four birds would wing-flap very rapidly and frequently call.

BREEDING A male specimen in full breeding plumage, with bare skin on the sides of the head encircling the eye, was probably collected in October or November 1895, whilst another specimen in near-breeding plumage was collected in November 1880 (Salomonsen 1965, Benson 1974, 1976). Other male specimens which were apparently coming into breeding plumage have been collected in July and August (Benson 1974). One or two breeding-plumage males and two moulting males were observed at Marojejy during 16-22 October (Hawkins *et al.* in prep.), whilst a male observed at Vohiparara on 4 December was in full breeding plumage, with large wattles (P. Davidson and P. Morris *in litt.*). During late September, F. Hawkins (*in litt.*, in prep.) observed males in full breeding plumage actively displaying in Zahamena Strict Nature Reserve. Two males collected at Andringitra Strict Nature Reserve in late October were also apparently in full breeding condition, with well developed wattles, whilst a female collected at this time had a slightly enlarged oviduct (Goodman and Putnam in press). In contrast, of six birds observed at Maharira on 3 June, none were males in breeding plumage. F. Hawkins (*in litt.*) observed newly fledged juveniles at 1,600-2,000m at Anjanaharibe-sud on 14 November, 27 November and 2 December. A bird believed to be a juvenile male was collected in June 1880 (Benson 1976) and a recently fledged juvenile was observed at Vohiparara by P. Morris (*in litt.*) in November 1995. These observations suggest that breeding occurs in late September, October and November, during the main peak of breeding for birds in Madagascar (Langrand 1990), and therefore spanning the driest months of the year. A male observed in Andringitra Strict Nature Reserve on 14 January was reported to have blue upperparts (O. Langrand *in litt.*), although it is not clear whether this bird was in full breeding plumage, since nothing is recorded about the wattle.

Only one nest of the species has ever been found, and the eggs are unknown. The nest was discovered on 29 October in Andringitra Strict Nature Reserve, at 1,600m (Goodman and Putnam in press). The female was observed to make frequent visits to the nest, lining the dome's interior with dry leaves of a climbing bamboo. Although the male perched nearby much of the time, and displayed when the female was present, he was not observed to enter the structure or assist in nest construction. When collected, on 30 October, the nest was empty and lacked any sort of roof over the entrance, but was probably incomplete.

Bill and tip of first primary of Common Sunbird Asity, above, and Yellow-bellied Sunbird Asity, below.

DESCRIPTION
Sexually dimorphic and plumage varies with season. Tail very short, barely extending past tail tip. Bill long and strongly downcurved.

Adult male breeding Head metallic dark blue, adorned by a large, smooth, rather rectangular wattle (caruncle) that is notched at the rear. The inner part of the wattle, surrounding a narrow pale lime-green eye-ring, is a darker brilliant lime-green, this colour narrowest above and below the eyes and extending backwards being the eye as a broad band that narrows to a blunt point at the rear of the wattle. The rest of the wattle, encircling the lime-green, is a bright cobalt-blue, this colour extending around the rear of the wattle as lobes that do not quite meet at the rear, so that the wattle is notched, and extending forward to cover the lores (from a side view, but in fact slightly raised above the feathers of the lores, so the lores may be visible from the front). In front of the eye, the wattle almost meets the blue and lime-green of the base of the bill. The surface of the wattle is not smooth, and the lower border, from the lores to the rear, is distinctly marked by a line of small swollen areas that catch the light and appear as a line of 8-9 bright, pale azure-blue dots. Mantle, back, rump, uppertail-coverts and scapulars bright, iridescent, metallic dark blue, in some lights appearing blue-green. Tail feathers black, with narrow metallic dark blue edges and tips to innermost pairs and progressively less blue on outer feather, the blue being restricted to the outer edge of the outer pairs and usually absent completely from the outermost 1-2 pairs. Tertials black with bright iridescent dark metallic blue terminal bands that extend along the outer edge, concolorous with mantle; lesser and median coverts iridescent metallic dark blue with hidden blackish bases; greater coverts blackish with broad metallic dark blue leading edge. Flight feathers dark brown with very narrow metallic blue leading edge to basal half to two-thirds of secondaries. Underparts, from chin to undertail-coverts, brilliant uniform golden-yellow with dark bases hidden by overlapping feathers, hence usually appearing unmarked but may occasionally show a slight mottling on the breast, or fine olive tips to the throat feathers (such features may be the result of feather abrasion as plumage changes to non-breeding). Underwing-coverts yellow, with white bases to feathers of axillaries. Tenth (outer) primary strongly emarginated, the tip being very narrow, and measuring from the notch 7mm Iris dark brown; bill black with slightly swollen cobalt-blue base to upper mandible that extends to above nostril, bordered in front and below nostril by narrow swollen area of lime-green that extends to the gape and is adjacent to a similar-coloured swollen area on the lower mandible that extends across about one third of the bill; legs brown (Hildebrandt 1881, Salomonsen 1933). As with other asities, males with fully developed wattles but female-type plumage have also been recorded (P. Morris *in litt.*). The underparts of such birds are a relatively bright yellow, but have light olive streaking on the sides of the breast and flanks.

Adult male non-breeding Males in non-breeding plumage are variable, depending on state of moult, but all lack the head wattle. All birds probably retain metallic blue edges to tail and secondaries and have blue feathers in uppertail-coverts and upperwing-coverts but lack blue on head and mantle. Some individuals have a narrow yellow band across the rump.

Adult female Forehead to nape olive-green, with mottling formed by brown centres to all feathers. Lacks wattle, but there is a narrow band of yellow feathers around the lower margin of the eye. Ear-coverts olive-green. Rest of upperparts uniform olive-green, glossed with a bright green iridescence that is only visible in bright sunlight. Tail feathers brown with very narrow olive-green edges to basal half of all but outermost pair. Tertials brown with narrow olive-yellow outer edge and tip. Lesser and median coverts brown with broad olive-green terminal bands; greater coverts brown with olive-green leading edge. Flight feathers brown with narrow olive-green leading edge to secondaries; basal edge to inner web of all flight feathers greyish-white. Underparts golden-yellow, occasionally with indistinct narrow dark olive tips to feathers of chin. Bill blackish-brown without colourful base.

Immature Sexes similar and virtually indistinguishable from adult female except that they have a slightly swollen pale base to the bill. Immature males may have one or two metallic blue feathers or patches in feathers on the uppertail-coverts or rump, and may have more yellowy feathers on the centre of belly and lower breast. Some individual males occasionally have some very fine yellowish edgings to secondaries. Underparts duller yellow below; less metallic sheen to upperparts and head. Immature males moulting into breeding plumage have mottled iridescent dark blue on the upperparts and head. Longmore (1985) described the plumage of an immature male moulting into adult plumage, although the description could equally relate to an adult male moulting into breeding plumage.

Juvenile Dull olive above with dull clear yellow underparts, more olive on flanks. The bill is weaker, shorter and straighter than that of adults, and has conspicuous pale horn basal colour to both mandibles (Benson 1976, F. Hawkins *in litt.*). A recently fledged bird observed by P. Morris (*in litt.*) resembled a female, but had a short bill with the basal half pinky-orange.

MEASUREMENTS (n=2). Wing 49.5,50; tail 25.5,26.5; tarsus 14,14.5. Bill long and curved, with exposed culmen, measured in straight line 18,22 mm (compared to 24-27 mm for Common Sunbird Asity). Weight 6.4,8.1.

BIBLIOGRAPHY

Abdulali, H. 1976. A catalogue of the birds in the collection of the Bombay Natural History Society - 18 (Eurylaimidae, Pittidae, Alaudidae). *J Bombay Nat. Hist. Soc.* 72: 299-327.

Aksornkoae, S. 1993. *Ecology and Management of Mangroves.* IUCN, Bangkok.

Alexander, S. 1994. Pittas in the Chizarira National Park. *Honeyguide* 40: 27-28.

Ali, S. 1962. *The Birds of Sikkim.* Oxford University Press, Madras.

— & Ripley, S.D. 1948. The birds of the Mishmi Hills. *J. Bombay Nat. Hist. Soc.* 48: 1-37.

— & — 1983. *Handbook of the Birds of India and Pakistan. Volume 4: Frogmouths to Pittas* (Second edition). Oxford University Press, Delhi, London, New York.

Ali, S., Biswas, B. & Ripley, S.D. in press. The birds of Bhutan. *Rec. Zool. Surv. India.* Occ. Paper 136.

Allen, F.G.H. 1955. The nesting of the Bluewinged Pitta (*Pitta megarhyncha* Schleg.). *Malay Nat. J.* 10: 67-68.

Allen, G.M. 1930. *The Birds of Liberia.* Harvard University Press, Harvard.

Allport, G., Ausden, M., Hayman, P.V., Robertson, P. & Wood, P. 1989. The conservation of the birds of Gola Forest, Sierra Leone. International Council for Bird Preservation, Cambridge. (Study Report 38).

Amadon, D. 1951. Le pseudo-soumanga de Madagascar. *L'Ois. Rev. Franc. Ornithol.* 21: 59-63.

— 1979. Philepittidae, in Traylor, M.A. (ed.) *Check-list of Birds of the World.* Vol. 8: 330-331. Museum of Comparative Zoology, Cambridge, Massachusetts.

Andrew, P. 1992. *The Birds of Indonesia: a Checklist* (Peters' sequence). *Kukila* Checklist No. 1. Indonesian Ornithological Society, Jakarta.

Anon. 1976. *Complete Book of Australian Birds.* Reader's Digest, New South Wales.

— 1988. Recent Reports. *Enggang* 1. (Newsletter of the MNS Selangor Bird Group, Kuala Lumpur).

— 1988a. Recent reports. *Bangkok Bird Club Bull.* 5(6).

— 1988b. Recent reports. *Bangkok Bird Club Bull.* 5(8).

— 1994. Recent Reports: Cameroon. *Bull. African Bird Club* 1(2): 105.

Auber, L. 1964. The possible evolutionary significance of differences in feather structure between closely related Pittidae (Passeres: Mesomyodes). *J. Linn. Soc. Lond. (Zool.)* 45: 245-250.

Audley-Charles, M.G., Hurley, A.M. & Smith, A.G. 1981. The vertebrate faunas. Pp. 9-23 in Whitmore, T.C. (ed.) 1981. *Wallace's Line and Plate Tectonics.* Clarendon Press, Oxford.

Austin, O.L. 1948. The birds of Korea. *Bull. Mus. Comp. Zool.* 101 (1): 1-301.

— & Kuroda, N. 1953. The birds of Japan, their status and distribution. *Bull. Mus. Comp. Zool.* 109(4): 279-637.

Baker, E.C.S. 1895. The birds of North Cachar, Part V. *J. Bombay Nat. Hist. Soc.* 10: 161-168.

— 1913. Zoological results of the Abor Expedition, 1911-1912. Birds. *Rec. Indian Mus.* 9: 251-254.

— 1919. Notes on a collection of bird-skins formed by Mr E.G. Herbert C.M.Z.S., M.B.O.U. *J. Nat. Hist Soc. Siam* 3: 409-443.

— 1921. New birds from SE Yunnan, SW China. *Bull. Brit. Orn. Club* 42: 13-18.

— 1926. *The Fauna of British India, including Ceylon and Burma. Birds Vol. 3.* Taylor and Francis, London.

— 1934. *The Nidification of Birds of the Indian Empire. Vol. 3: Ploceidae-Asionidae.* Taylor and Francis, London.

Baker, N.E. & Baker, E.M. 1993. Four Afrotropical migrants on the East African coast: evidence for a common origin. *Scopus* 15: 122-124.

Ballman, P. 1969. Die Vögel aus der altburdigalen Spaltenfüllung von Wintershof (West) bei Eichstätt in Bayern. *Zitteliana* 1: 5-60.

Bangs, O. & van Tyne, J. 1931. Birds of the Kelley-Roosevelts expedition to French Indo-China. *Field Mus. Nat. Hist., Zoology* 18(3): 33-119.

Banks, E. 1935. A collection of montane mammals and birds from Mulu in Sarawak. *Sarawak Mus. J.* 4: 327-341.

— 1937. Birds from the highlands of Sarawak. *Sarawak Mus. J.* 4: 497-518.

— 1949. Noteworthy birds from Mt. Dulit. *Sarawak Mus. J.* 5: 147-148.

— 1952. Mammals and birds from the Maga Mountains in Borneo. *Bull. Raffles Mus.* 21: 160-163.

Bannerman, D.A. 1936. *The Birds of Tropical West Africa,* Vol. 4. Crown Agents, London.

Barnard, H.G. 1911. Field notes from Cape York. *Emu* 11: 25.

Basilio, A. 1963. *Aves de la Isla de Fernando Poo.* Coculsa, Madrid.

Batchelor, D.M. 1959. North Borneo bird notes. *Sarawak Mus. J.* 9: 263-266.

Bates, G.L. 1905. Field notes on the birds of Efulen in the West-African colony of Kareun. *Ibis* 8(5): 89-98.

— & Ogilvie-Grant, W.R. 1911. Further notes on the birds of southern Cameroon. *Ibis* 9(5): 479-545.

Beakbane, A.J. & Boswell, E.M. 1984. Nocturnal Afrotropical migrants at Mufindi, southern Tanzania. *Scopus* 8: 124-127.

de Beaufort, L. F. & de Bussy, L.P. 1919. Vogels can de oostkust van Sumatra. *Bijdr. Dierk* 21: 229-276.

Beaulieu, D.A. 1932. Les oiseaux de la région de Honquan (Province de Thudaumot, Cochinchine). *L'Ois. Rev. Franc. Ornithol.* 2: 133-154.

— 1939. Les oiseaux de la région de Pleiku. *L'Ois. Rev. Franc. Ornithol.* 9: 163-182.

— 1944. *Les Oiseaux du Tranninh.* Université Indochinoise, Hanoi.

Beehler, B. 1981. Ecological structuring of forest bird communities in New Guinea. *Monographie Biologicae* 42: 837-861.

Bell, H.L. 1968. The Noisy Pitta in New Guinea. *Emu* 68: 92-94.

— 1982a. A bird community of lowland rainforest in New Guinea. 1. Composition and density of the avifauna. *Emu* 82: 24-41.

— 1982b. A bird community of lowland rainforest in New Guinea. 2. Seasonality. *Emu* 82: 65-74.

— 1982c. A bird community of lowland rainforest in New Guinea. 4. Birds of secondary vegetation. *Emu* 82: 217-224.

Benson, C.W. 1970. A Blue-winged Pitta on Christmas Island, eastern Indian Ocean. *Bull. Brit. Orn. Club* 90: 24-25.

— 1971. The Cambridge collection from the Malagasy Region (part II). *Bull. Brit. Orn. Club* 91: 1-7.

— 1974. Another specimen of *Neodrepanis hypoxantha*. *Bull. Brit. Orn. Club* 94: 141-143.

— 1976. Specimens of *Neodrepanis hypoxantha* in Dresden. *Bull. Brit. Orn. Club* 96: 144.

— 1981. Les oiseaux: des espèces uniques au monde. Pp. 63-74 *in* Oberlé, P. (ed.) *Madagascar, Un Sanctuaire de la Nature.* Lechevalier, Paris.

— 1985. Asity. P.25 *in* Campbell, B. & Lack, E. (eds). *A Dictionary of Birds.* T & A.D. Poyser, Calton, England.

— & Benson F.M. 1977. *The Birds of Malawi.* Montfort Press, Limbe, Malawi.

—, Colebrook-Robjent, J.F.R. & Williams, A. 1976. Contribution à l'ornithologie de Madagascar. *L'Ois. Rev. Franc. Ornithol.* 46: 103-134, 209-242, 367-386.

— & Irwin, M.P.S. 1964. The migrations of the pitta of eastern Africa (*Pitta angolensis longipennis* Reichenow). *N. Rhodesia J.* 5(5): 465-475.

Beruldsen, G. 1980. *A Field Guide to Nests and Eggs of Australian Birds.* Rigby, Adelaide.

Beruldsen, G.R. & Uhlenhut, K. 1995. Cape York in the wet. *Aust. Bird Watcher* 16(1): 3-10.

Berry, P.S.M. & Ansell, P.D.H. 1978. African Pitta (*Pitta angolensis*) displaying in the upper Luangwa Valley. *Bull. Zamb. Orn. Soc.* 10(1): 29.

Betham, R.M. 1922. Breeding of the Indian Pitta (*Pitta brachyura*) and the Streaked Wren Warbler (*Prinia lepida*). *J. Bombay Nat. Hist. Soc.* 28: 1135.

Betts, F.N. 1956. Notes on birds of the Subsansiri area, Assam. *J. Bombay Nat. Hist. Soc.* 53: 396-414.

Bezemer, K.W.L. 1929. Enkele waarnemingen van *Eurylaimus javanicus* (Horsf.) in Midden-Java. *Ardea* 18: 2-8.

Bingham, C.T. 1903. A contribution to our knowledge of the birds occurring in the southern Shan States, Upper Burma. *Ibis* (8)3: 584-606.

Bishop, K.D. 1992. New and interesting records of birds in Wallacea. *Kukila* 6: 8-34.

— & Coates, B.J. in prep. *A Field Guide to the Birds of Wallacea.* Dove Publications, Alderley, Queensland.

Blaber, S.J.M. & Milton D.A. 1994. The distribution of nests of the Black-and-Red Broadbill *Cymbirhynchus macrorhynchos* along a river in Sarawak. *Forktail* 10: 182-184.

Blakers, M., Davies, S.J.J.F & Reilly, P.N. 1984. *The Atlas of Australian Birds.* Melbourne University Press, Melbourne.

Bô Khoa Hoc, Công Nghê & Môi Truòng 1992. [*Red Data Book of Vietnam.* Vol 1. Animals] Science and Technics Publ., Hanoi. [In Vietnamese].

Bolster, R.C. 1921. Breeding of the Indian Pitta. *J. Bombay Nat. Hist. Soc.* 28: 284.

Bouet, G. 1931. Contribution à la répartition des oiseaux en Afrique Occidentale (Fin). *L'Ois. Rev. Franc. Ornithol.* 1: 487-502.

Bowden, C.G.R. in press. The birds of Mount Kupe, southwest Cameroon. *Bird Conserv. Internatn.*

Bradley, J. 1994. African Pitta *Pitta angolensis* at Jadina, Kenya coast. *Scopus* 18: 56-57.

Brazil, M.A. 1991. *The Birds of Japan.* Christopher Helm, London.

Britton, P.L. (ed.) 1980. *Birds of East Africa: their Habitat, Status and Distribution.* East African Natural History Society, Nairobi.

— & Rathbun, G.B. 1978. Two migratory thrushes and the African Pitta at the Kenya coast. *Scopus* 2: 11-17.

Brooks, T., Dutson, G., King, B. & Magsalay, P.M. 1995. An annotated check-list of the forest birds of Rajah Sikatuna National Park, Bohol, Philippines. *Forktail* 11: 121-134.

Brosset, A. & Erard, C. 1986. *Les Oiseaux des Régions Forestières du Nord-est du Gabon,* Vol 1. Société Nationale de Protection de la Nature, Paris.

Brown, G. 1931. Notes on the hill-migrating birds of Ceylon. *Bull. Brit. Orn. Club* 51: 20-24.

Brüggemann, F. 1878. Weitere Mitteilungen über die Ornithologie von Central-Borneo. *Abhandl. Naturwis. Ver. Bremen* 5: 525-537.

Buckingham, D.L., Dutson, G.C.L. & Newman, J. in prep. Birds of Manus, Kolombangara and Makira (San Cristobal) with notes on mammals and records from other Solomon Islands.

Bucknill, J.A.S. & Chasen, F.N. 1927. *The Birds of Singapore Island.* Government Printing Office, Singapore.

Butchart, S.H.M., Brooks, T.M., Davies, C.W.N., Dharmaputra, G., Dutson, G.C.L., Lowen, J.C. & Sahu, A. 1993. Preliminary report of the Cambridge Flores/Sumbawa Conservation Project 1993. Unpublished.

Büttikofer, J. 1887. Contribution to the ornithology of Sumatra. On a collection of birds, made by Dr. C. Klaesi, in the highlands of Padang (W. Sumatra) during the winter 1884-1885. *Notes Leiden Mus.* 9: 1-96.

— 1900. Zoological results of the Dutch scientific expedition to Central Borneo. The Birds. *Notes Leiden Museum* 21: 145-276.

Caldwell, H.R. & Caldwell, J.C. 1931. *South China Birds.* Hester May Vanderburgh, Shanghai.

Carey, G.J. (ed.) 1994. *The Hong Kong Bird Report 1993.* Hong Kong Bird Watching Society, Hong Kong.

Chalmers, M.L. 1986. *Annotated Checklist of the Birds of Hong Kong.* Hong Kong Bird Watching Society, Hong Kong.

Chapin, J.P. 1939. The birds of the Belgian Congo I. *Bull. Amer. Mus. Nat. Hist.* 65: 1-756.

— 1953. The birds of the Belgian Congo III. *Bull. Amer. Mus. Nat. Hist.* 75A: 1-821.

Chasen, F.N. 1931. Birds from Bintang Island in the Rhio Archipelago. *J. Raffles Mus.* 5: 114-120.

— 1935. Four new races of Malaysian birds. *Bull. Raffles Mus.* 10: 43-44.

— 1935a. A handlist of Malaysian birds. *Bull. Raffles Mus.* 11: 1-389.

— 1937. The birds of Billiton Island. *Treubia* 16: 205-238.

— & Hoogerwerf, A. 1941. The birds of the Netherlands Indian Mt. Leuser Expedition 1937 to north Sumatra. *Treubia* 18 (Supplement): 1-125.

— & Kloss, C.B. 1930. On a collection of birds from the lowlands and islands of North Borneo. *Bull. Raffles Mus.* 4: 1-112.

— & — 1932. On birds from Doi Suthep, 5,600 feet, north Siam. *J. Siam Soc.* 8(4): 231-248.

Cheng Tso-hsin 1987. *A Synopsis of the Avifauna of China.* Science Press, Beijing.

Choudhury, A. 1986. Bird killing at Jatinga. *Guwahati Sentinel* 7 Sept. (Assam, India).

— 1990. *Checklist of the Birds of Assam.* Sofia Publishers, Guwahati.

Clancey, P.A. 1963. Miscellaneous taxonomic notes on African birds 20, 1. The south African races of the Broadbill *Smithornis capensis* (Smith). *Durban Mus. Novit.* 6(19): 231-241.

— 1970. On *Smithornis capensis suahelicus* Grote. 1926. *Bull. Brit. Orn. Club* 90: 164-166.

— 1971. A handlist of the birds of southern Moçambique. *Mems. Inst. Invest. Cient. Moçamb.* 10, Série A. 1969-1970.
— 1980. *S.A.O.S. Checklist of Southern African Birds.* South African Ornithological Society, Pretoria.
— 1985. *The Rare Birds of Southern Africa.* Winchester Press, Johannesburg.
Coates, B.J. 1990. *The Birds of Papua New Guinea.* Vol. 2. Dove Publications, Alderley, Queensland.
Coffin, M.F. & Rabinowitz, P.D. 1987. Reconstruction of Madagascar and Africa: evidence from the Davie fracture zone and western Somali Basin. *J. Geophys. Res.* 92(B9): 9385-9406.
Collar, N.J. & Andrew, P. 1988. *Birds to Watch: the ICBP World Checklist of Threatened Birds.* International Council for Bird Preservation, Cambridge. (Technical Publication 8).
—, Gonzaga, L.P., Krabbe, N., Madroño Nieto, A., Naranjo, L.G., Parker, T.A. & Wege, D.C. 1992. *Threatened Birds of the Americas.* International Council for Bird Preservation, Cambridge.
—, Crosby, M.J. & Stattersfield, A.J. 1994. *Birds to Watch 2: the World Checklist of Threatened Birds.* BirdLife International, Cambridge.
— & Stuart, S.N. 1985. *Threatened Birds of Africa and Related Islands.* International Council for Bird Preservation, Cambridge.
— & Stuart, S.N. 1988. *Key Forests for Threatened Birds in Africa.* International Council for Bird Preservation, Cambridge. (ICBP Monograph 3).
—, Round, P.D. & Wells, D.R. 1986. The past and future of Gurney's Pitta *Pitta gurneyi. Forktail* 1: 29-51.
Collins, N.M., Sayer, J.A. & Whitmore, T.C. (eds). 1991. *The Conservation Atlas of Tropical Forests: Asia and the Pacific.* Macmillan, London.
Colston, P.R. & Curry-Lindahl, K. 1986. *The Birds of Mount Nimba, Liberia.* British Museum (Natural History), London.
Coomans de Ruiter, L. 1938. Oölogische en biologische aanteekeningen van eenige pitta's of prachtlijsters in der westerafdeeling van Borneo. *Limosa* 11: 35-46.
Counsell, D. 1986. The Royal Air Force Ornithological Society's expedition, May-June 1984. *Brunei Mus. J.* 6(2): 164-204.
Cracraft, J. 1983. Species concepts and speciation analysis. *Current Ornithol.* 1: 159-187.
— 1987. Species concepts and the ontology of evolution. *Biol. Philos.* 2: 63-80.
Cranbrook, Earl of 1981. The vertebrate faunas. Pp. 57-69 *in* Whitmore, T.C. (ed.) 1981. *Wallace's Line and Plate Tectonics.* Clarendon Press, Oxford.
Crawford, D.N. 1972. Birds of the Darwin area, with some records from other parts of the Northern Territory. *Emu* 72: 131-148.
Dahl, F. 1899. Das Leben der Vögel auf den Bismarckinseln. *Mitt. Zool. Mus. Berlin* 1:107-222.
Daniels, R.J.R. 1986. The Indian Pitta. *Newsletter for Birdwatchers* 26(5&6): 15-16.
Danielsen, F. & Heegaard, M. 1995. The birds of Bukit Tigapuluh, southern Riau, Sumatra. *Kukila* 7: 99-120.
—, Balete, D.S., Christensen, T.D., Heegaard, M., Jakobsen, O.F., Jensen, A., Lund, T. & Poulsen, M.K. 1994. *Conservation of Biological Diversity in the Sierra Madre Mountains of Isabela and southern Cagayan Province, the Philippines.* Department of Environment and Natural Resources (Philippines), BirdLife International, Zoological Museum of Copenhagen University and Danish Ornithological Society, Copenhagen.
Davidson, N.C. 1978. Additions to local avifaunas: Bia National Park, Ghana. *Bull. Nigerian Orn. Soc.* 14(46): 88.
Davidson, P. & Stones, T. 1993. Birding in the Sula Islands. *Oriental Bird Club Bull.* 18: 59-63.
—, Lucking, R.S., Stones, A.J., Bean, N.J., Raharjaningtrah, W. and Banjaransari, H. 1994. Report on an ornithological survey of Taliabu, Indonesia, with notes on the Babirusa pig. BirdLife International - Indonesia Programme, Bogor.
—, Stones, T. & Lucking, R. 1995. The conservation status of key bird species on Taliabu and the Sula Islands, idonesia. *Bird Conserv. Internatn.* 5: 1-20.
Davison, G.W.H. 1980. A survey of terrestrial birds in the Gunung Mulu National Park, Sarawak. *Sarawak Mus. J.* 27: 281-289.
Dean, W.R.J., Macdonald, I.A.W. & Vernon, C.J. 1974. Possible breeding record of *Cercococcyx montanus. Ostrich* 45: 188.
Decoux, J.P. & Fotso, R.C. 1988. Composition et organisation spatiale d'une communauté d'oiseaux dans la région de Yaondé. Conséquences biogéographiques de la dégradation forestière et de l'aridité croissante. *Alauda* 56: 126-152.
Dee, T.J. 1986. *The Endemic Birds of Madagascar.* International Council for Bird Preservation, Cambridge, England.
Deignan, H.G. 1936. A revised hand-list of the birds of the Chiengmai region. *J. Siam Soc.* 10(2): 71-129.
— 1945. The Birds of Northern Thailand. *U.S. Natn. Mus. Bull.* 186.
— 1946. Notes on birds of northern Siam. *Not. Nat.* 173.
— 1946a. A new pitta from the Malay Peninsula. *Proc. Biol. Soc. Washington* 59: 55.
— 1947. New races of Asiatic broadbills (Eurylaimidae). *Proc. Biol. Soc. Washington* 60: 119-122.
— 1948. The races of the Silver-breasted Broadbill, *Serilophus lunatus* (Gould). *J. Washington Acad. Sci.* 38: 108-111.
Delacour, J. 1928. A collection of living birds from central Annam. *Avic. Mag.* 4: 212-216.
— 1929. On the birds collected during the fourth expedition to French Indochina. Part II. *Ibis* (12)5: 193-220.
— 1929a. Three new subspecies from southern Indo-China. *Bull. Brit. Orn. Club* 49: 49-50.
— 1930. On the birds collected during the fifth expedition to French Indo-China. *Ibis* (12)6: 564-599.
— 1932a. On the birds collected in Madagascar by the Franco-Anglo-American Expedition, 1929-1931. *Ibis* 13(2): 284-304.
— 1932b. Les oiseaux de la Mission Franco-Anglo-Americaine à Madagascar. *L'Ois. Rev. Franc. Ornithol.* 2: 1-96.
— 1934. Breeding of the Hooded Pitta (*Pitta cucullata*). *Avic. Mag.* (4)12: 222-226.
— 1947. *Birds of Malaysia.* MacMillan, New York.
— 1970. The contribution of Gilbert Tirant to Indochinese ornithology. *Nat. Hist. Bull. Siam Soc.* 23(3): 325-329.
— & Greenway, J. 1940. Liste des oiseaux recueillis dans la Province du Haut-Mekong et le Royaume de Luang-Prabang. *L'Ois. Rev. Franc. Ornithol.* 11 (Suppl.): 1-21.
— & —1940a. Notes critiques sur certains oiseaux

Indochinois. *L'Ois. Rev. Franc. Ornithol.* 10: 60-77.
— & Jabouille, P. 1925. On the birds of Quangtri, Central Annam; with notes on others from other parts of French Indo-China. *Ibis* (12)1: 209-260.
— & — 1927. Recherches ornithologiques dans les provinces du Tranninh (Laos) de Thua-Thien et de Kontoum (Annam) et quelques autres régions de L'Indonchine Française. *Archives d'Histoire Naturelle.* Paris: Société Nationale d'Acclimation de France.
— & — 1929. Les brèves de l'Indochine Française (Pittidae). *L'Ois. Rev. Franc. Ornithol.* 10: 113-122.
— & — 1931. *Les Oiseaux de l'Indochine Française.* Exposition Coloniale Internationale, Paris.
—, — & Lowe, W.P. 1928. On the birds collected during the third expedition to French Indochina. Part II. *Ibis* 12(4): 285-317.
— & Mayr, E. 1945. Notes on the taxonomy of the birds of the Philippines. *Zoologica* 30: 105-117.
— & — 1946. *Birds of the Philippines.* MacMillan, New York.
Demey, R. & Fishpool, L.D.C. 1991. Additions and annotations to the avifauna of Côte d'Ivoire. *Malimbus* 12: 61-86.
— & — 1994. The birds of Yapo forest, Ivory Coast. *Malimbus* 16: 100-121.
Dharmakumarsinhji, R.S. 1955. *Birds of Saurashtra, India.* Times of India Press, Bombay.
Diamond, J.M. 1972. *Avifauna of the Eastern Highlands of New Guinea.* Nuttall Ornithological Club, Cambridge, Massachusetts.
—, Bishop, K.D. & Balen, S. van. 1987. Bird survival in an isolated Javan woodland: island or mirror. *Cons. Biol.* 1: 132-142.
Dickinson, E.C. & Chaiyaphun, S. 1968. On a small collection of birds from in or near Nakhorn Ratchasima province, Eastern Thailand. *Nat. Hist. Bull. Siam. Soc.* 22(3&4): 309-315.
—, Kennedy, R.S. & Parkes, K.C. 1991. *The Birds of the Philippines.* BOU Check-list 12. British Ornithologists' Union, Tring.
Donald, C.H. 1918. The occurrence of the Indian Pitta (*Pitta brachyura*) in the Kangra District, Punjab. *J. Bombay Nat. Hist. Soc.* 25: 497-499.
Dowsett, R.J. & Dowsett-Lemaire, F. 1989. Avifaune de Congo: additions et corrections. *Tauraco Res. Rep.* 2: 17-19.
—- & — (eds.) 1993. A contribution to the distribution and taxonomy of Afrotropical and Malagasy birds. *Tauraco Res. Rep.* 5. Tauraco Press, Liège, Belgium.
— & Forbes-Watson, A.D. 1993. *Checklist of Birds of the Afrotropical and Malagasy Regions. Volume 1: Species Limits and Distribution.* Tauraco Press, Liège.
Dowsett-Lemaire, F. & Dowsett, R.J. 1991. The avifauna of the Kouilou Basin, Congo. *Tauraco Res. Rep.* 4: 189-239.
Draffen, R.D.W., Garnett, S.T. & Malone, G.J. 1983. Birds of the Torres Strait: an annotated and biogeographical analysis. *Emu* 83: 207-234.
Driscoll, P.V. & Kikkawa, J. 1989. Bird species diversity of lowland tropical rainforests of New Guinea and northern Australia. Pp. 123-152 *in* Harmelin-Vivian, M.L. and Bourlière, F. (eds.) *Vertebrates in Complex Tropical Systems.* Springer-Verlag, New York.
Duckworth, J.W. & Kelsh, R. 1988. A bird inventory of Similajan National Park. International Council for Bird Preservation, Cambridge. (Study Report 31).
—, Timmins, R.J. & Cozza, K. 1993. A wildlife and habitat survey of Phou Xang He proposed protected area. IUCN, Vientiane.
Dunning, J.B. 1992. *CRC Handbook of Avian Body Masses.* CRC Press, Boca Raton, Florida.
duPont, J.E. & Rabor, D.S. 1973. South Sulu Archipelago birds: an expedition report. *Nemouria* 9.
— & — 1973a. Birds of Dinagat and Siargao, Philippines. *Nemouria* 10.
Dutson, G. 1995. The birds of Salayar and the Flores Sea Islands. *Kukila* 7: 129-141.
Dutson, G.C.L. & Newman, J.L. 1991. Observations on the Superb Pitta *Pitta superba* and other Manus endemics. *Bird Conserv. Internatn.* 1: 215-222.
Eames, J.C. & Ericson, P.G.P. in prep. The Björkegren expeditions to French Indochina: a collection of birds from Vietnam and Cambodia. (*Forktail*).
Edden, R & Boles, W. 1986. *Birds of the Australian Rainforests.* Reed, New South Wales, Australia.
Elgood, J.H., Heigham, J.B., Moore, A.M., Nason, A.M., Sharland, R.E. & Skinner, N.J. 1994. *The Birds of Nigeria: an Annotated Check-list.* BOU Check-list 4 (2nd ed.). British Ornithologists' Union, Tring.
Elliot, D.G. 1870. Remarks on some lately-described *Pittae*, with a synopsis of the family as now known. *Ibis* (2)6: 408-421.
— 1893. *Monograph of the Pittidae.* Second Edition. Bernard Quaritch, London.
Engelbach, P. 1932. Les oiseaux du Laos Méridional. *L'Ois. Rev. Franc. Ornithol.* 2: 439-498.
— 1936. Notes de la région de Kampot (Cambodge). *L'Ois. Rev. Franc. Ornithol.* 6: 347-348.
— 1938. Notes sur quelques oiseaux du Cambodge. *L'Ois. Rev. Franc. Ornithol.* 8: 384-394.
— 1948. Liste complémentaire aux oiseaux du Cambodge. *L'Ois. Rev. Franc. Ornithol.* 18: 5-26.
— 1953. Les oiseaux d'Angkor et leur identification sur le terrain. *La Terre et La Vie* 100: 148-166.
Evans, G. H. 1904. The nidification of the Little Blue Winged Pitta (*Pitta cyanoptera*) in upper Burma. *J. Bombay Nat. Hist. Soc.* 16: 171-172.
Evans, M.I., Duckworth, J.W., Hawkins, A.F.A., Safford, R.J., Sheldon, B.C. & Wilkinson, R.J. 1992. Key bird species of Marojejy Strict Nature Reserve, Madagascar. *Bird Conserv. Internatn.* 2: 201-220.
Evans, T.D., Dutson, G.C.L. & Brooks, T.M. (eds) 1993. Cambridge Philippines Rainforest Project 1991. BirdLife International, Cambridge. (Study Report 54).
Everett, C. 1974. The breeding of the Bengal Pitta. *Avic. Mag.* 80: 33-35.
Favaloro, N.J. 1931. Notes on a trip to the Macpherson Range, south-eastern Queensland. *Emu* 31: 48-59.
Finsch, O. 1900. Systematische Ueberstict der Vögel der Südwest-inselns. *Notes Leiden Mus.* 22: 225-309.
Fogden, M.P.L. 1965. Borneo bird notes, 1963-65. *Sarawak Mus. J.* 12: 395-413.
— 1970. Some aspects of the ecology of bird populations in Sarawak. D.Phil. thesis, Oxford University.
— 1972. The seasonality and population dynamics of equatorial forest birds in Sarawak. *Ibis* 114: 307-342.
— 1976. A census of a bird community in tropical rainforest in Sarawak. *Sarawak Mus. J.* 24: 251-267.
Forbes, W.A. 1880a. Contributions to the anatomy of passerine birds. Part II. On the syrinx and and other points in the anatomy of the Eurylaemidae. *Proc. Zool. Soc. Lond.* 1880: 380-386

— 1880b. Contributions to the anatomy of passerine birds. Part IV. On some points in the structure of *Philepitta* and its position amongst the Passeres. *Proc. Zool. Soc. Lond.* 1880: 387-391.

Fraser, L. 1842. On two new species of birds from western Africa. *Proc. Zool. Soc. London* Part I: 189-190.

— 1843. On eight new species of birds from western Africa. *Proc. Zool. Soc. London* Part II: 16-17.

Friedmann, H. 1970. The status and habits of Grauer's Broadbill in Uganda (Aves: Eurylaimidae). *Los Angeles County Mus. Contr. Sci.* 176.

— & Williams, J.G. 1968. Notable records of rare or little-known birds from western Uganda. *Rev. Zool. Bot. Afr.* 57: 11-36.

Frith, H.J. & Calaby, J.H. 1974. Fauna survey of the Port Essington District, Cobourg Peninsula, Northern Territory of Australia. *CSIRO Wild. Res. Tech. Paper* 28.

Fuggles-Couchman, N.R. 1984. The distribution of, and other notes on, some birds of Tanzania. *Scopus* 8: 1-17.

Garrod, A.H. 1876. On some anatomical characters which bear upon the major divisions of the passerine birds. Part 1. *Proc. Zool. Soc. London* 1876: 506-519.

— 1877. Notes on the anatomy of passerine birds. Part 2. *Proc. Zool. Soc. London* 1877: 447-453.

— 1878. Notes on the anatomy of passerine birds. Part 4. *Proc. Zool. Soc. London* 1878: 143.

Garthwaite, P.F. & Ticehurst, C.B. 1937. Notes on some birds recorded from Burma. *J. Bombay Nat. Hist. Soc.* 39: 552-560.

Gartshore, M.E., Taylor, P.D. & Francis, I.S. 1995. Forest Birds in Côte d'Ivoire. BirdLife International, Cambridge (Study Report 58).

Gatter, W. & Gardner, R. 1993. The biology of the Gola Malimbe. *Bird Conserv. Internatn.* 3: 101.

Germain, M., Dragesco, J., Roux, F. & Garcin, H. 1973. Contribution à l'ornithologie du sud Cameroun II. Passeriformes. *L'Ois. Rev. Franc. Ornithol.* 43: 212-259.

Ghosh A.K. 1987. *Qualitative analysis of faunal resources of proposed Namdapha Biosphere Reserve, Arunachal Pradesh.* Zoological Survey of India, Calcutta.

Gibson-Hill, C.A. 1949. An annotated checklist of the birds of Malaya. *Bull. Raffles Mus.* 20: 1-299.

— 1950. A checklist of the birds of Singapore Island. *Bull. Raffles Mus.* 21: 132-183.

— 1950a. A collection of birds eggs from north Borneo [V. W. Ryves coll.]. *Bull Raffles Mus.* 21: 106-115.

— 1952. The apparent breeding seasons of land birds in North Borneo and Malaya. *Bull. Raffles Mus.* 24: 270-293.

Gilliard, E.T. 1950. Notes on a collection of birds from Bataan, Luzon, Philippine Islands. *Bull. Amer. Mus. Nat. Hist.* 94: 461-504.

— & LeCroy, M. 1966. Birds of the middle Sepik Region, New Guinea. Results of the American Museum of Natural History Expedition to New Guinea 1953-54. *Bull. Amer. Mus. Nat. Hist.* 132: 247-275.

Ginn, P.J. McIlleron, W.G. & Milstein, P.le S. (eds.) 1989. *The Complete Book of Southern African Birds.* Struik Winchester, Cape Town.

Glydenstolpe, N. 1916. Zoological results of the Swedish Zoological Expedition to Siam 1911-1912 and 1914-1915: Birds. *Kungl. Svenska Vet.-Akad. Handl.* 56(2): 1-160.

Gonzales, P.C. 1983. The birds of Catanduanes. *Zool. Papers* No. 2, National Museum, Manila, Philippines.

Goodman, S. 1994. Recent expedition to the Andringitra Reserve. *Working Group on Birds in the Madagascar Region Newsletter* 3(2): 10.

— 1995. Morphoilogical variation in *Philepitta* spp.(Aves: Eurylaimidae). *Working Group on Birds in the Madagascar Region Newsletter* 5(2): 1-3..

Goodman, S.M. & Gonzales, P.C. 1990. The birds of Mt. Isarog National Park, southern Luzon, with particular reference to altitudinal distribution. *Fieidiana, Zool.* 60: 1-39.

— & Putnam, M.S. in press. The birds of the eastern slopes of the Réserve Naturelle Intégrale d'Andringitra. *Fieldiana Zool.*

Gore, M.E.J. 1968. A check-list of the birds of Sabah, Borneo. *Ibis* 110: 165-196.

— & Won, Pyong-Oh 1971. *The Birds of Korea.* Royal Asiatic Society, Seoul.

Gray, G.R. 1860. List of birds collected by Mr. Wallace at the Molucca Islands with descriptions of new species, etc. *Proc. Zool. Soc. London* 1860: 341-366.

Green, G.M. & Sussman, R.W. 1990. Deforestation history of the eastern rainforests of Madagascar from satellite images. *Science* 248: 212-215.

Greenway, J.C. 1967. *Extinct and Vanishing Birds of the World.* Second Revised Ed. Dover Publications, New York.

Gretton, A. 1988. Gurney's Pitta and the lowland forests of southern Thailand. International Council for Bird Preservation, Cambridge. Unpublished report.

—, Kohler, M., Lansdown, R.V., Pankhurst, T.J., Parr, J. & Robson, C. 1993. The status of Gurney's Pitta *Pitta gurneyi* 1987-1989. *Bird Conserv. Internatn.* 3: 351-367.

Grimes, L.G. 1987. *The Birds of Ghana: an Annotated Checklist.* BOU Check-list 9. British Ornithologists' Union, Tring.

Guillemard, F.H.H. 1885. Report on the collection of birds obtained during the voyage of the Yacht Marchesa. Part V. The Molucca Islands. *Proc. Zool. Soc. London* 1885: 561-576.

Hachisuka, M. 1934-1935. *The Birds of the Philippine Islands.* Vol. 2. Witherby, London.

Hachisuka, M. & Udagawa, T. 1950. Contributions to the ornithology of Formosa. Part 2. *Quarterly J. Taiwan Mus.* 4: 1-180.

Haddon, D. 1981. *Birds of the North Solomons.* Wau Ecology Institute, Papua New Guinea. (Handbook No. 8).

Hall, B.P. 1952. Notes on races of *Pitta soror* Wardlaw-Ramsay, in southern Indochina. *Bull. Brit. Orn. Club* 72: 102-104.

— (ed.) 1974. *Birds of the Harold Hall Australian Expedition 1962-1970: a Report on the Collection made for the British Museum (Natural History).* British Museum (Natural History), London. (Publ. No. 745).

— & Moreau, R.E. 1970. *An Atlas of Speciation in African Passerine Birds.* British Museum (Natural History), London.

Harrison, T. 1964. Food capacity of a Green-breasted Pitta. *Sarawak Mus. J.* 11: 611-615.

Hartert, E. 1896. An account of the collections of birds made by Mr William Doherty in the eastern archipelago. *Novit. Zool.* 3: 537-590.

— 1897a. Mr William Doherty's bird collections from Celebes. *Novit. Zool.* 4: 153-166.

— 1897b. On the birds collected by Mr Everett in south Flores. Part I. *Novit. Zool.* 4: 513-528.

— 1898a. List of a collection of birds made in the Sula Islands by William Doherty. *Novit. Zool.* 5: 125-136.

— 1898b. On the birds of Lomblen, Pantar and Alor. *Novit. Zool.* 5: 455-465.
— 1901. On the birds of the Key and South-east Islands and of Ceram-laut. *Novit. Zool.* 8: 1-5, 93-101.
— 1903. The birds of Batjan. *Novit. Zool.* 10: 43-64.
— 1909. *Pitta schneideri*, sp. n. *Bull. Brit. Orn. Club* 25: 9-10.
— 1924. Notes on some birds from Buru. *Novit. Zool.* 31: 104-111.
Harvey, W.O. 1938. The East African Pitta (*Pitta angolensis longipennis* Reichenow). *Ibis* 14: 335-337.
Harvey, W.G. 1990. *Birds in Bangladesh.* University Press, Dhaka.
Hawkins, F. 1994. The nest of Schlegel's Asity *Philepitta schlegeli. Bull. African Bird Club* 1(2): 77-78.
— in prep. The display of the Yellow-bellied Sunbird Asity *Neodrepanis hypoxantha.*
Hawkins, A.F.A., Safford, R.J., Duckworth, J.W., Evans, M.I. and Langrand, O. in prep. Field characters and conservation status of the Yellow-bellied Sunbird Asity *Neodrepanis hypoxantha.*
Hazevoet, C.J. 1995. *The Birds of the Cape Verde Islands: an Annotated Check-list.* BOU Check-list 13. British Ornithologists' Union, Tring.
Heath, P. 1992. A report on the birds and mammals seen on a trip to the Greater Sundas between 7th June and 2nd August 1991. Unpublished report.
Heinrich, G. 1956. Biologische Aufzeichnungen über Vögel von Halmerhera und Batjan. *J. Orn.* 99: 31-40.
Hellebrekers, W. Ph.J. & Hoogerwerf, A. 1967. A further contribution to our oological knowledge of the island of Java (Indonesia). *Zool. Verhandelingen, Leiden* 88: 1-164.
Herbert, E.G. 1924. Nests and eggs of birds in central Siam (part 3). *J. Nat. Hist. Soc. Siam* 6: 293-311.
Henry, G. M. 1971. *A Guide to the Birds of Ceylon.* Oxford University Press.
Hildebrandt, J.M. 1881. Skizze zu einem Bilde centralmadagassischen Naturlebens im Frühling. *Zeit. Gesell. Erdk. Berlin* 16: 194-203.
Hindwood, K.A. & McGill, A.R. 1958. *The Birds of Sydney.* Royal Zoological Society of New South Wales, Sydney.
Holmes, D.A. 1969. Bird notes from Brunei: December 1967–September 1968. *Sarawak Mus. J.* 17: 399-402.
— 1973. Bird notes from southernmost Thailand, 1972. *Nat. Hist. Bull. Siam Soc.* 25: 39-66.
— 1994. A review of the land birds of the west Sumatran islands. *Kukila* 7: 28-46.
— in prep. Sumatra bird report. *Kukila.*
— & Burton, K. 1987. Recent notes on the avifauna of Kalimantan. *Kukila* 3: 2-32.
Holyoak, D. 1970. Observations on the behaviour of a pair of Green Broadbills. *Avic. Mag.* 76: 16-18.
Hoogerwerf, A. 1948. Contribution to the knowledge of the distribution of birds on the island of Java. *Treubia* 19: 83-137.
— 1949. *Een bijdrage tot de oölogie van het Eiland Java.* Van de Kon. Plantentuin van Indonesia, Buitenzorg, Java.
Hopwood, C. 1912. A list of birds from Arakan. *J. Bombay Nat. Hist. Soc.* 21: 1196 1221.
— 1919. Notes on some nests recently found in south Tenasserim. *J. Bombay Nat. Hist. Soc.* 26: 853-859.
Hornskov, J. 1987. More birds from Berbak Game Reserve, Sumatra. *Kukila* 3: 58-59.
— 1995. Recent observations of birds in the Philippine Archipelago. *Forktail* 11: 1-10.

Hose, C. 1898. On the avifauna of Mount Dulit and the Baram District in the Territory of Sarawak. *Ibis* (6)5: 381-424.
Howe, R.W. 1986. Bird distributions in forest islands in north-eastern New South Wales. Pp. 119-129 in Ford, H. A. and Paton D. C. (eds). *The Dynamic Partnership: Birds and Plants in Southern Australia.* The Flora and Fauna of South Australia Handbooks Committee, South Australia.
Hume, A.O. 1875a. A first list of the birds of upper Pegu. *Stray Feathers* 3: 1-194.
— 1875b. Novelties. *Stray Feathers* 3: 262-303.
— 1888. The birds of Manipur, Assam, Sylhet and Cachar. *Stray Feathers* 11: 1-353.
— 1890. *The Nests and Eggs of Indian Birds.* Vol.II, R.H. Porter, London. (2nd ed., edited by E.W.Oates).
— & Davison, W. 1878. A revised list of the birds of Tenasserim. *Stray Feathers* 6: 1-524.
Hurrell, P. 1989. Schneider's Pitta rediscovered in Sumatra. *Kukila* 4: 53-55.
Husain, K.Z. & Sarkar, S.U. 1973. Notes on a collection of birds from Pabna (Bangladesh). *J. Asiatic Soc. Bangladesh* 18(1): 83-90.
ICBP 1992. *Putting Biodiversity on the Map: Priority Areas for Global Conservation.* International Council for Bird Preservation, Cambridge.
Ingalhalikar, S. 1977. Indian Pitta in captivity. *Newsletter for Birdwatchers* 17(7): 8-9.
Inglis, C.M., Travers, W.L., O'Donel, H.V. & Shebbeare, E.O. 1920. A tentative list of the vertebrates of the Jalpaiguri District, Bengal. Part II. *J. Bombay Nat. Hist. Soc.* 26: 988-999.
Inskipp, C. & Inskipp, T. 1991. *A Guide to the Birds of Nepal.* 2nd Ed. Croom Helm, London and Sydney.
— & — 1993. Birds recorded during a visit to Bhutan in spring 1993. *Forktail* 9: 121-142.
Irwin, M.P. S. 1981. *The Birds of Zimbabwe.* Quest Publishing, Salisbury.
IUCN 1992. Analysis of proposals to amend the CITES Appendices. IUCN, Cambridge, UK.
— 1994. *Guidelines for Protected Area Management Categories.* CNPPA with the assistance of WCMC. IUCN, Gland, Switzerland, and Cambridge, UK.
Jackson, F.J. (1938). *The Birds of Kenya Colony and the Uganda Protectorate.* Vol. II. Gurney and Jackson, London.
Jakobsen, O.F. & Andersen, C.Y. 1995. New distributional records and natural history notes on the Whiskered Pitta *Pitta kochi* of the Philippines. *Forktail* 11: 111-120.
Jensen, A., Poulsen, M.K., Accos, J., Jakobsen, O.F. & Andersen, C.Y. in prep. *Conservation of Biological Diversity of Mount Pulog National Park, The Philippines.* Department of Environment and Natural Resources (Philippines)–BirdLife International, Manila and Cambridge.
Jepson, P. 1993. Recent ornithological observations from Buru. *Kukila* 6: 85-109.
Jeyarajasingam, A. 1983. Observations of the nest of the Silver-breasted Broadbill (Aves, Eurylaimidae). *Malayan Naturalist* 37: 14-16.
Johnson, J.M. 1972. Indian Pitta *Pitta brachyura* behaviour of choosing to fly through holes in the bushes. *Indian Forester* 98: 450.
Jones, A.E. 1943. On the occurrence of the Green-breasted Pitta (*Pitta cucullata* Hartl.) at Simla. *J. Bombay Nat. Hist. Soc.* 43: 658.
Jones, M., Juhaeni, D., Banjaransari, H., Banham, W., Lace,

L., Linsley, M. & Marsden, S. 1994. The status, ecology and conservation of the forest birds and butterflies of Sumba. Unpublished report. Department of Biological Sciences, Manchester Metropolitan University, Manchester, UK.

—, Linsley, M.D. & Marsden, S. J. 1995. Populations sizes, status and habitat associations of the restricted-range bird species of Sumba, Indonesia. *Bird Conserv. Internatn.* 5: 21-52.

Junge, G.C.A. 1958. *Pitta erythrogaster bernsteini* nov. subspec. *Ardea* 46: 88.

— & Kooiman, J.G. 1951. On a collection of birds from the Khwae Noi Valley, western Siam. *Zool. Verhandelingen* 15: 3-38.

Kang, N. & Lee, P.G. 1993. The avifauna of the North Selangor peat-swamp forest, West Malaysia. *Bird Conserv. Internatn.* 3: 169-179.

Keith, S., Urban, E.K. & Fry, C.H. (eds.) 1992. *The Birds of Africa*, Vol. IV. Academic Press, London.

King, B. 1978. A new race of *Pitta oatesi* from Peninsular Malaysia. *Bull. Brit .Orn. Club* 98(3): 109-113.

Kinnear, N.B. 1929. On the birds collected by Mr H. Stevens in Northern Tonkin in 1923-24. *Ibis* (12)5: 107-150.

Kloss, C.B. 1921. New and known oriental birds. *J. Fed. Malay States Mus.* 10: 207-213.

— 1930. An account of the Bornean birds in the Zoological Museum, Buitenzorg, with the description of a new race. *Treubia* 12: 395-424.

— 1931. Some birds of Billiton Island. *Treubia* 13: 293-298.

— 1931a. An account of the Sumatran birds in the zoological museum, Buitenzorg, with descriptions of nine new races. *Treubia* 13: 299-370.

Kobayashi, K. & Cho, H. 1977. *Avifauna of Taiwan.* Japan Bird Society, No. 16.

Koelz, W. 1939. New birds from Asia, chiefly from India. *Proc. Biol. Soc. Washington* 52: 61-82.

Ku Chang-tung. 1957. Blue-winged Pitta *Pitta brachyura nympha* Temminck et Schlegel, obtained at Beidai-he, in Hebei Province. *Scientia* 19: 594.

Kuroda, N. 1933. *Birds of the Island of Java.* Vol. 1 Passeres. N. Kuroda, Tokyo.

Lambert, F.R. 1987. Fig-eating and seed dispersal by birds in a Malaysian lowland rainforest. PhD thesis, University of Aberdeen.

— 1988. The status of White-winged Wood Duck *Cairina scutulata* in Sumatra. Asian Wetland Bureau, Bogor, Indonesia. (Report No. 4).

— 1989a. Fig-eating by birds in a Malaysian lowland rainforest. *J. Trop. Ecol.* 5: 401-412.

— 1989b. Daily ranging behaviour of three tropical forest frugivores. *Forktail* 4: 107-116.

— 1990. *Avifaunal changes following selective logging of a north Bornean rainforest.* Institute of Tropical Biology, University of Aberdeen, Scotland.

— 1992. The consequences of selective logging for Bornean lowland forest birds. *Phil. Trans. Royal Soc. Lond.* B. 335: 443-457.

— 1992a. The significance of sexual dimorphism in bird dispersed figs of gynodioecious *Ficus*. *Biotropica* 24: 214-216.

— 1993. Some key sites and significant records of birds in the Philippines and Sabah. *Bird Conserv. Internatn.* 3: 281-297.

— 1994. Notes on the avifauna of Bacan, Kasiruta and Obi, North Moluccas. *Kukila* 7: 1-9.

—, Eames, J.C. & Nguyen Cu. 1994. *Surveys for Pheasants in the Annamese Lowlands of Vietnam.* IUCN, Cambridge.

—, Eames, J.C. & Nguyen Cu. 1995. The habitat, status, vocalizations and breeding biology of Blue-rumped Pitta *Pitta soror annamensis* in central Vietnam. *Forktail* 11: 151-155.

— & Yong, D. 1989. Some recent bird observations from Halmahera. *Kukila* 4: 30-33.

— & Howes, J.R. in prep. On the ranging behaviour of Black-headed Pitta and three species of terrestrial babblers in Sabahan rainforest.

Lan, Y. 1983. A taxonomic study on Chinese birds of the genus *Pitta* (Pittidae). *Zool. Res.* 4(3): 219-216.

Langrand, O. 1990. *Guide of the Birds of Madagascar.* Yale University Press, Newhaven and London.

Lansdown, R.V., Seale, W., & McLoughlin, J. 1991. Vocalisations and displays of the Hooded Pitta. *Nat. Hist. Bull. Siam Soc.* 39: 93-102.

Law, S.C. 1939. Haunt and habitat of *Pitta c. cucullata* Hartl. in West Bengal. *J. Bombay Nat. Hist. Soc.* 40: 759-762.

Lawson, W.J. 1961. Probable courtship behaviour of the Broadbill *Smithornis capensis*. *Ibis* 103a: 289-290.

Lee, G., Haeffner, S. & Zajicek, J. 1989. Breeding and handrearing of the Hooded Pitta *Pitta sordida* at the Denver Zoo (Colorado, USA). *Avic. Mag.* 95: 119-128.

Leighton, M. 1982. Fruit resources and patterns of feeding, spacing and grouping among sympatric Bornean hornbills (Bucerotidae). PhD thesis, University of California, Davis.

Lekagul, B. & Round, P.D. 1991. *A Guide to the Birds of Thailand.* Saha Karn Bhaet, Bangkok.

Lewis, A. & Pomeroy, D. 1989. *A Bird Atlas of Kenya.* Balkema, Rotterdam.

Lim, K.S 1992. *Vanishing Birds of Singapore.* The Nature Society, Singapore.

— 1994. Birds to Watch: Singapore. *Bull. Oriental Bird Club* 19: 52-54.

Longmore, N.W. 1985. A Sydney specimen of *Neodrepanis hypoxantha* (Philepittidae). *Bull. Brit. Orn. Club* 105: 85-86.

Louette, M. 1981. *The Birds of Cameroon: an Annotated Checklist.* Paleis der Académiën, Brussels.

Lowe, P.R. 1931. On the anatomy of *Pseudocalyptomena* and the occurrence of broadbills (Eurylaimidae) in Africa. *Proc. Zool. Soc. London* 1931: 445-461.

Ludlow, F. & Kinnear, N. 1937. The birds of Bhutan and adjacent territories of Sikkim and Tibet. *Ibis* (14)1: 467-504.

— & — 1944. The birds of south-eastern Tibet. *Ibis* 86: 348-389.

Mace, G. & Stuart, S. 1994. Draft IUCN Red List categories. *Species* 21-22: 13-14.

Madge, S.G. 1972. Some new distribution records for the Copperbelt and other notes. *Bull. Zambian Orn. Soc.* 4(1): 15-18.

Madoc, G.C. 1956. *An Introduction to Malayan Birds.* Malayan Nature Society, Kuala Lumpur.

MacGillivray, W. 1914. Notes on some north Queensland birds. *Emu* 13: 132-186.

MacKinnon, J. & Phillipps, K. 1993. *A Field Guide to the Birds of Borneo, Sumatra, Java and Bali.* Oxford University Press, Oxford.

Mackworth-Praed, C.W. & Grant, C.H.B. 1970. *Birds of West Central and Western Africa.* (Ser. 3, Vol. 1.). Longmans, London.

Mann, C.F. 1987. A checklist of the birds of Brunei Darussalam. *Brunei Mus. J.* 6(3): 170-212.
— 1988. Bird report for Brunei Darussalam, July 1986 to June 1988. *Brunei Mus. J.* 6(4): 88-111.
Manson, A.J. 1983. On the presence of the African Broadbill in the Vumba. *Honeyguide* 113: 33.
Mayr, E. 1935. Birds collected during the Whitney South Seas Expedition XXX: Descriptions of twenty-five new species and subspecies. *Amer. Mus. Novit.* 820.
— 1938. The birds of the Vernay-Hopwood Chindwin Expedition. *Ibis* (14)2: 277-320.
— 1938a. Notes on a collection of birds from south Borneo. *Bull. Raffles Mus.* 14: 5-46.
— 1944. The birds of Timor and Sumba. *Bull. Amer. Mus. Nat. Hist.* 83: 123-194.
— 1955. Notes on the birds of northern Melanesia, 3. Passeres. *Amer. Mus. Novit.* 1707.
— 1979. Family Pittidae, in Traylor, M.A. (ed.) *Check-list of Birds of the World*. Vol 8: 310-329. Museum of Comparative Zoology, Cambridge, Massachusetts.
— & Rand, A.L. 1937. Results of the Archbold Expeditions. No. 14. Birds of the 1933-1934 Papuan Expedition. *Bull. Amer. Mus. Nat. Hist.* 73: 1-248.
McClure, H.E. 1974. *Migration and Survival of the Birds of Asia*. US Army Component SEATO Medical Research Laboratory, Bangkok, Thailand.
— 1974a. Some bionomics of the birds of Khao Yai National Park, Thailand. *Nat. Hist. Bull. Siam Soc.* 24: 99-191.
— & Leelavit, P. 1972. Birds banded in Asia during the MAPS program, by locality, from 1963 through 1971. U.S. Army Research and Development Group, San Francisco.
McGill, A.R. 1960. *A Hand List of the Birds of New South Wales*. Fauna Protection Panel, Sydney.
McGregor, R.C. 1909-10. A Manual of Philippine Birds. *Bur. Sci. Manila.* No.2. (2 parts).
McKelvey, S.D. & Miller, B.W. 1979. Breeding of the Giant Pitta at the San Antonio Zoo. *Avic. Mag.* 85: 109-111.
McLachlan, G.R. & Liversidge, R. 1978. *Roberts Birds of South Africa*. John Voelker Bird Book Fund, Cape Town.
McLoughlin, J. 1988. Bird of the month - Giant Pitta. *Bangkok Bird Club Bull.* 5:(7).
Medway, Lord 1972. The Gunong Benom Expedition 1967: 6. The distribution and altitudinal zonation of birds and mammals on Gunong Benom. *Bull. Brit. Mus. (Nat. Hist.) Zool.* 23(5): 105-154.
— & Wells, D.R. 1976. *The Birds of the Malay Peninsula. Vol V: Conclusion, and Survey of Every Species*. Witherby, London.
Mees, G. F. 1971. Systematic and faunistic remarks on birds from Borneo and Java, with new records. *Zool. Mededelingen* 45(21): 225-244.
— 1977. Additional records of birds from Formosa (Taiwan). *Zool. Mededelingen* 51(15): 247-264.
— 1982. Birds from the lowlands of southern New Guinea (Merauke and Koembe). *Zool. Verhandelingen* 191: 1-188.
— 1986. A list of the birds recorded from Bangka Island, Indonesia. *Zool. Verhandelingen* 232: 1-176.
Meise, W. 1929. Die Vögel von Djampea und benachbarten Inseln nach einer Sammlung Baron Plessens. *J. Orn.* 77: 431-479.
— 1941. Ueber die Vogelwelt von Noesa Penida bei Bali nach einer Sammlung von Baron Viktor von Plessen. *J. Orn.* 4: 345-376.
Meyer, A.B. 1879. Field notes on the birds of Celebes. *Ibis* (4)3: 43-71, 125-147.
— & Wigglesworth, L.W. 1895. Bericht über die von den Herren P. und F. Sarasin in Nord-Celébes gesammelten Vögel. *Abh. ber. K. Zool. Mus. Dresden* (1894/95) 8: 1-20.
Meyer, O. 1906. Die Vögel der Insel Vuatom. *Natur und Offenbarung* 52.
Meyer de Schauensee, R. 1934. Zoological results of the third de Schauensee Siamese Expedition, part II: birds from Siam and the southern Shan States. *Proc. Acad. Nat. Sci. Philadelphia* 86: 165-280.
— 1940. The birds of the Batu Islands. *Proc. Acad. Nat. Sci. Philadelphia* 92: 23-42.
— 1946. On Siamese birds. *Proc. Acad. Nat. Sci. Philadelphia* 98: 1-82.
— 1984. *The Birds of China*. Oxford University Press, Oxford.
— & Ripley, S.D. 1940. Zoological results of the George Vanderbilt Sumatran Expedition, 1936-1939. Part 1 - Birds from Atjeh. *Proc. Acad. Nat. Sci. Philadelphia* 91: 311-368.
Milne-Edwards, A. & Grandidier, A. 1879. *Histoire Physique, Naturelle et Politique de Madagascar, 12. Histoire Naturelle des Oiseaux*. Tome I. Paris.
Milon, P. 1951. Etude d'une petite collection d'oiseaux du Tsaratanana. *Naturaliste Malgache* 3(2): 167-183.
Mitra, S. S. & Sheldon, F.H. 1993. Use of an exotic tree plantation by Bornean lowland forest birds. *Auk* 110: 529-540.
Mohan, D. & Chellam, R. 1990. New call record of Green-breasted Pitta *Pitta sordida* (P.S.L. Muller) in Dehra Dun, Uttar Pradesh. *J. Bombay Nat. Hist. Soc.* 87(3): 453-454.
Moreau, R.E. 1966. *The Bird Faunas of Africa and its Islands*. Academic Press, London.
Mori, T. 1939. 'Bird conservation in Korea' [in Japanese]. *Yacho* 6(1): 1-11.
Motley, J. & Dillwyn, L.L. 1855. Contributions to the natural history of Labuan and the adjacent coasts of Borneo. Part 1. ('Aves' Pp. 8-38, 53-62). Jan van Voorst, London.
Moulton, J.C. 1913. Notes on a collecting trip to Mt. Poi, Sarawak. *J. Straits Branch Royal. Asiatic. Soc.* 65: 1-12.
Müller, J.P. 1847. *Über die bisher unbekannten typischen Verschiedenheiten der Stimmorgane der Passerinen*. Akad. Wiss., Berlin.
Müller, S. & Schlegel, H. 1840. Overzigt der in den Indischen Archipel levende soorten van het geslacht Pitta. *Verhand. Natuurlijke Gesch. Nederl. overzeesche bezittingen. Zool. Pitta.*
Nash, A.D. & Nash, S. V. 1985. Breeding notes on some Padang-Sugihan birds. *Kukila* 2: 59-63.
— & —1987. An annotated checklist of the birds of Tanjung Puting National Park, Central Kalimantan. *Kukila* 3: 93-116.
Nash, S. (ed.) 1990. Project GREEN: Ghana Rainforest Expedition Eighty-Nine. Final Report.
Neelakantan, K.K. 1963. Regular slaughter of the Indian Pitta by crows. *Newsletter for Birdwatchers* 3(11): 1-2.
Nehrkorn, A. 1894. Zur Avifauna Batjan's. *J. Orn.* 42: 156-161.
Nguyen Cu 1990. Preliminary survey of some North Vietnam's forest and their avifauna. *Environment Newsletter* 2(3): 2-3.

Nicholson, F. 1883. On a second collection of birds made in the Island of Sumatra by Mr H. O. Forbes. *Ibis* (5)1: 235-257.

Norris, A.Y. 1964. Observations on some birds of the Tooloom Scrub, northern N.S.W. *Emu* 63: 404-412.

Noske, R. & Saleh, N. 1993. The status of lowland forest birds in West Timor. Unpublished report to Indonesian Institute of Science, Bogor.

Oates, E.W. 1879. Notes on the nidification of some Burmese birds. *Stray Feathers* 7: 164-168.

— 1882. A list of the birds of Pegu. *Stray Feathers* 10: 175-248.

— 1883. *A Handbook to the Birds of British Burmah including those found in the adjoining State of Karennee*. Vol. 1. Porter and Dulau, London.

Oberholser, H.C. 1932. The birds of the Natuna Islands. *Bull. U.S. Nat. Mus.* 159: 1-137.

Ogilvie-Grant, W.R. 1910. Additional notes on the birds of Hainan. *Proc. Zool. Soc. London* 1910: 572-579.

Olson, S. L. 1971. Taxonomic comments on the Eurylaimidae. *Ibis* 113: 507-516.

Oustalet, M.E. 1903. Oiseaux du Cambodge, du Laos, de L'Annam et du Tonkin. *Nouv. Arch. Mus. Hist. Nat. Paris* (4)5: 1-94.

Parker, V. 1992. *Swaziland Bird Checklist*. Conservation Trust of Swaziland, Swaziland.

Parkes, K.C. 1960. New subspecies of Philippine birds. *Proc. Biol. Soc. Washington* 73: 57-62.

— 1971. A new subspecies of pitta from the Philippines. *Bull. Brit. Orn. Club* 91: 98-99.

— 1973. Annotated list of the birds of Leyte Island, Philippines. *Nemouria* 11.

Parrish, R. 1983. Breeding the Noisy Pitta. *Int. Zoo News* 30(5): 17-19.

Paynter, R.A. 1970. Species with Malaysian affinities in the Sundarbans, East Pakistan. *Bull. Brit. Orn. Club* 90: 118.

Pearson, D.L. 1975. A preliminary survey of the birds of the Kutai Reserve, Kalimantan Timur, Indonesia. *Treubia* 28: 157-162.

Peile, H.D. 1914. Nest of the Long-tailed Broadbill on electric light wire. *J. Bombay Nat. Hist. Soc.* 23: 360-361.

Penry, H. 1994. *Bird Atlas of Botswana*. University of Natal Press, Pietermaritzburg.

Pérez del Val, J., Fa, J.E., Castroviejo, J. & Purroy, F.J. 1994. Species richness and endemism of birds in Bioko. *Biodiversity and Conservation* 3: 868-892.

Peters, J.L. 1951. Family Eurylaimidae, in *Check-list of Birds of the World*. Vol 7: 3-13. Museum of Comparative Zoology, Cambridge, Massachusetts.

Pfeffer, P. 1960. Etude d'une collection d'oiseaux de Borneo. *L'Ois. Rev. Franc. Ornithol.* 30: 191-218.

Poulsen, M.K. 1995. The threatened and near-threatened birds of Luzon, Philippines, and the role of the Sierra Madre Mountains in their conservation. *Bird Conserv. Internatn.* 5: 79-115.

Prentice, C. 1988. Observations of the nuptial behaviour of Black and Yellow Broadbill *Eurylaimus ochromalus*. *Enggang* (Newsletter of the MNS Selangor Bird Group, Kuala Lumpur) 1(4): 8.

Prescott, K.W. 1973. First report of *Pitta e. erythrogaster* from Leyte. *Bull. Brit. Orn. Club* 93: 32-33.

Prigogine, A. 1971. Les oiseaux de l'Itombwe et de son hinterland. *Mus. Roy. Afr. Cent. Ann.* Ser. 8 (185): 1-298.

Prum, R.O. 1993. Phylogeny, biogeography, and evolution of the broadbills (Eurylaimidae) and asities (Philepittidae) based on morphology. *Auk* 110: 304-324.

— 1994. A new hypothesis for the phylogenetic relationships of the asities: Madagascar loses a family of endemic birds, but not to extinction! *Working Group on Birds in the Madagascar Region Newsletter* 3(2) 5-8.

—, Morrison, R.L. & Eyck, G.R.T. 1994. Structural color production by constructive reflection from ordered collagen arrays in a bird (*Philepitta castanea*: Eurylaimidae). *J. Morphology* 222: 61-72.

— & Razafindratsita, T. in prep. Observations of the breeding behavior of the Velvet Asity (*Philepitta castanea*).

Pycroft, W.P. 1905. Contributions to the osteology of birds: Part 7: Eurylaimidae: with remarks on the systematic position of the group. *Proc. Zool. Soc. London* 1905: 30-56.

Rabor, D.S. 1938. Birds from Leyte. *Philippine J. Sci.* 66: 15-34.

Raikow, R.J. 1987. Hindlimb myology and evolution of the Old World suboscine passerine birds (Acanthisittidae, Pittidae, Philepittidae, Eurylaimidae). *Ornithol. Monogr.* 41. American Ornithologists' Union, Washington.

Rand, A.L. 1936. The distribution and habits of Madagascar birds. *Bull. Amer. Mus. Nat. Hist.* 72: 143-499.

— 1938. Results of the Archbold Expedition No. 20. On some Passerine New Guinea birds. *Amer. Mus. Novit.* 991.

— 1951. Birds from Liberia. *Fieldiana Zool.* 32(9): 560-653.

— & Gilliard, E.T. 1967. *Handbook of New Guinea Birds*. Weidenfeld and Nicholson, London.

— & Rabor, D.S. 1960. Birds of the Philippine Islands: Siquijor, Mount Malindang, Bohol, and Samar. *Fieldiana Zool.* 35(7): 225-441.

—, Friedmann, H. & Traylor, M.A. 1959. Birds from Gabon and Moyen Congo. *Fieldiana Zool.* 41(2): 221-410.

Rathbun, G.B. 1978. The African Pitta at Gede Ruins, Kenya. *Scopus* 2: 7-10.

Reichenow, A. 1903. *Die Vögel Afrikas*. Vol 2. Neumann, Neudamm.

Rensch, B. 1931. Die Vogelwelt von Lombok, Sumbawa und Flores. *Mitt. Zool. Mus. Berlin* 17: 451-637.

Richmond, C.W. 1903. Birds collected by Dr W.L. Abbott on the coast and islands of northwest Sumatra. *Proc. U.S. Nat. Mus.* 26: 485-524.

Riley, J.H. 1938. Birds from Siam and the Malay Peninsula in the United States National Museum collected by Drs. Hugh M. Smith and William L. Abbott. *U.S Nat. Mus. Bull.* 172.

Ripley, S.D. 1944. The bird fauna of the West Sumatran Islands. *Bull. Mus. Comp. Zool.* 44: 307-430.

— 1982. *A Synopsis of the Birds of India and Pakistan*. Bombay Natural History Society, Bombay.

— & Rabor, D.S. 1962. New birds from Palawan and Culion Islands, Philippines. *Postilla* 73.

Rippon, G. 1901. On the birds of the southern Shan States, Burma. *Ibis* (8)4: 525-561.

Roberts, T.J. 1992. *The Birds of Pakistan. Volume 2: Passeriformes*. Oxford University Press, Oxford.

Robinson, H.C. 1915. On a collection of birds from the Siamese province of Bandon, N.E. Malay Peninsula. *J. Fed. Malay St. Mus.* 5: 83-110.

— 1915a. Birds collected by Mr C. Boden Kloss on the

coast and islands of south-eastern Siam. *Ibis* (10)3: 718-761.
— 1927. *The Birds of the Malay Peninsula. I: the Commoner Birds*. Witherby, London.
— & Chasen F.N. 1939. *The Birds of the Malay Peninsula. IV: the Birds of the Low Country, Jungle and Scrub*. Witherby, London.
— & Kloss, C. B. 1918. On a collection of birds from the province of Phuket, peninsular Siam. *J. Nat. Hist. Soc. Siam* 3: 87-119.
— & — 1918a. Results of an expedition to Korinchi Peak, Sumatra (1914). II Birds. *J. Fed. Malay States Mus.* 8: 81-284.
— & — 1919. On birds from South Annam and Cochin China. *Ibis* (11)1: 392-625.
— & — 1919a. On new subspecies of Malay birds. *Bull. Brit. Orn. Club* 40: 11-28.
— & — 1921. The birds of south-west and peninsular Siam. *J. Nat. Hist. Soc. Siam* 5: 1-87.
— & — 1924. On a large collection of birds chiefly from West Sumatra made by Mr E. Jacobson. *J. Fed. Malay States Mus.* 9: 189-348.
— & — 1924a. The birds of south-west and peninsular Siam. *J. Nat. Hist. Soc. Siam* 5: 219-397.
— & — 1930. A second collection of birds from Pulo Condore. *J. Nat. Hist. Soc. Siam, Suppl.* 8: 79-86.
Robson, C.R. 1991. The avifauna of Nam Cat Tien National Park, Dong Nai. *Garrulax* 8: 4-9.
— 1993. From the field. *Oriental Bird Club Bull.* 18: 67-70.
— & Eames, J.C. 1992. *Forest Bird Surveys in Vietnam 1991*. International Council for Bird Preservation, Cambridge. (Study Report 51).
— , —, Wolstencroft, J.A., Nguyen Cu & Truong Van La 1989. Recent records of birds from Viet Nam. *Forktail* 5: 71-97.
— , —, Newman, M., Nguyen Cu & Truong Van La 1991. Forest bird surveys in Vietnam 1989/1990. Final report to ICBP. Unpublished report.
— , —, Nguyen Cu & Truong Van La 1993. Birds recorded during the third BirdLife/Forest Birds Working Group expedition in Viet Nam. *Forktail* 9: 89-119.
Rockefeller, J.S. & Murphy, C.B.G. 1933. The rediscovery of *Pseudocalyptomena*. *Auk* 50: 23-29.
Rodewald, P.G., Dejaifve, P.-A. & Green, A.A. 1994. The birds of Korup National Park, Cameroon. *Bird Conserv. Internatn.* 4: 1-68.
Rothschild, W. 1898. On three new species. *Bull. Brit. Orn. Club* 7: 33.
— 1901. On two new species from Isabel, Solomon Islands. *Bull. Brit. Orn. Club* 12: 22-23.
— 1904. On a new form of pitta from the Solomons. *Bull. Brit. Orn. Club* 15: 7-8.
— & Hartert, E. 1901. Notes on Papuan birds. *Novit. Zool.* 8: 55-88.
— & — 1903. Notes on Papuan birds. *Novit. Zool.* 14: 447-483.
— & — 1905. Further contributions to our knowledge of the ornis of the Solomon Islands. *Novit. Zool.* 12: 243-268.
— & — 1912. List of a collection of birds made by Mr. Albert Meek on the Kumasi River, north-eastern British New Guinea. *Novit. Zool.* 19: 187-206.
— & — 1914. The birds of the Admiralty Islands, north of German New Guinea. *Novit. Zool.* 21: 281-298.
— & — 1914a. On a collection of birds from Goodenough Island. *Novit. Zool.* 21: 1-9.
Round, P.D. 1983. Some recent bird records from northern Thailand. *Nat. Hist. Bull. Siam. Soc.* 31(2): 123-138.
— 1984. The status and conservation of the bird community in Doi Suthep-Pui National Park, north-west Thailand. *Nat. Hist. Bull. Siam. Soc.* 32(1): 21-46.
— 1988. *Resident Forest Birds in Thailand: their Status and Conservation*. International Council for Bird Preservation, Cambridge, U.K. (Monograph 2).
— 1992. Gurney's Pitta: the latest chapter. *Oriental Bird Club Bull.* 15: 29-32.
— 1995. On the seasonality and distribution of Gurney's Pitta *Pitta gurneyi*. *Forktail* 11: 155-158.
— & Treesucon, U. 1983. Observations on the breeding of the Blue Pitta (*Pitta cyanea*) in Thailand. *Nat. Hist. Bull. Siam Soc.* 31(1): 93-98.
— & — 1986. The rediscovery of Gurney's Pitta *Pitta gurneyi*. *Forktail* 2: 53-66.
— & — 1990-1992. The conservation and management of lowland rainforest at Khao Nor Chuchi, southern Thailand. Unpublished (BirdLife International Internal Annual Reports).
— , — & Eames, J.C. 1989. A breeding record of the Giant Pitta *Pitta caerulea* from Thailand. *Forktail* 5: 35-47.
Rozendaal, F.G. 1988. Pittas in Vietnam *Garrulax* 4: 2-5.
— 1989a. A field study of Indochinese pittas in Viet Nam, with special reference to Elliot's Pitta *Pitta elliotii* (Aves: Pittidae). Unpublished preliminary report.
— 1989b. Status of Elliot's Pitta in Viet Nam. *Dutch Birding* 11: 17-18.
— 1990. Report on surveys in Hoang Lien Son, Lai Chau and Nghe Tinh Provinces, Viet Nam. Unpublished report to WWF.
— 1991. *Pitta sordida* in northern Viet Nam. *Dutch Birding* 13: 16-17.
— 1993. New subspecies of Blue-rumped Pitta from southern Indochina. *Dutch Birding* 15: 17-22.
— 1994. Species limits within the Garnet Pitta-complex. *Dutch Birding* 16: 239-245.
— & Dekker, R.W.R.J. 1989. Annotated checklist of the birds of the Dumoga-Bone National Park, north Sulawesi. *Kukila* 4: 85-108.
Safford, R. & Duckworth, W. 1990 (eds). *A Wildlife Survey of the Marojejy Nature Reserve, Madagascar*. International Council for Bird Preservation, Cambridge. (Study Report 40).
Salomonsen, F. 1933. *Neodrepanis hypoxantha*, sp. nov. *Bull. Brit. Orn. Club* 53: 182-183.
— 1953. Miscellaneous notes on Philippine birds. *Vidensk. Medd. Dan. Naturhist. Foren.* 115: 205-281.
— 1965. Notes on the Sunbird-asitys (*Neodrepanis*). *L'Ois. Rev. Franc. Ornithol.* 35 (no. spéc.): 103-111.
Salvadori, T. 1887. Catalogo delle collezioni ornitologiche fatte presso siboga in Sumatra, e nell'isola Nias dal Signor Elio Modigliani. *Ann. Mus. Civ. Storia Nat. Genova* 14: 169-253.
— 1888. Viaggio de Leonardo Fea nella Burmania e nelle regioni vicini, III. Uccelli raccolti nel Tenasserim (1887). *Ann. Mus. Civ. Stor. Nat. Genova* (2a)5: 554-622.
Sayer, J.A., Harcourt, C.S. & Collins, N.M. (eds). 1992. *The Conservation Atlas of Tropical Forests: Africa*. Macmillan, London.
Schlegel, H. & Pollen, F.P.L. 1868. *Recherches sur la faune*

de Madagascar et de ses dependances, d'après les découvertes de François P. L. Pollen et D.C. van Dam. 2me partie. Mammifères et oiseaux. Leyde.

Schodde, R. & Mathews, S.J. 1977. Contribution to Papuasian Ornithology V. Survey of the birds of Taam Island, Kai Group. *CSIRO Aust. Div. Wildl. Res. Tech. Pap.* 33: 1-29.

Schouteden, H. 1966a. La faune ornithologique du Rwanda. *Doc. 10, Musée Royal de L'Afrique Centrale, Tervuren.*

— 1966b. La faune ornithologique du Burundi. *Doc. 11, Musée Royal de L'Afrique Centrale, Tervuren.*

— 1970. Quelques oiseaux du Libéria. *Rev. Zool. Bot. Afr.* 82(1-2): 187-192.

Sclater, P.L. 1863. Observations on the birds of south-eastern Borneo by the late James Mottley, Esq., of Banjarmassing. *Proc. Zool. Soc. London* 1863: 206-224.

— 1872. Observations on the systematic position of *Peltops, Eurylaimus* and *Todus. Ibis* (3)2: 177-180.

Sclater, W.L. & Moreau, R.E. 1933. Taxonomic and field notes on some birds of north-eastern Tanganyika Territory. *Ibis* 11(2): 399-439.

Serle, W. 1950. A contribution to the ornithology of the British Cameroons. *Ibis* 92: 343-376.

— 1954. A second contribution to the birds of the British Cameroons. *Ibis* 96: 47-80.

— 1959. Some breeding bird records at Ndian, British Southern Cameroons. *Nigerian Field* 24: 76-79.

Serventy D.L. 1968. Wanderings of the Blue-winged Pitta to Australia. *Bull. Brit. Orn. Club.* 88: 160-162.

Severinghaus, L.L., Liang, C.T., Severinghaus, S.R. & Lo, L.C. 1991. The distribution, status and breeding of Fairy Pitta (*Pitta nympha*) in Taiwan. *Bull. Inst. Zool., Academis Sinica* 30(1): 41-47.

Sharpe R.B. 1876. Prof. Steere's expedition to the Philippines. *Nature* 14: 297-298.

— 1877. Description of a new species of *Lobiophasis* and a new species of Pitta from the Lawas River, N.W. Borneo. *Proc. Zool. Soc. London* 1877: 93-94.

— 1877a. Contribution to the ornithology of Borneo, Part II. *Ibis* (4)1: 1-25.

— 1879. Contributions to the ornithology of Borneo, Part IV: on the birds of the Province of Lumbidan, north-western Borneo. *Ibis* (4)3: 233-272.

— 1888. Further notes on *Calyptomena whiteheadi. Ibis* 6(5): 231.

— 1889. On the ornithology of northern Borneo. *Ibis* 6(1): 409-443.

— 1892. On a collection of birds from Mount Dulit, in north-western Borneo. *Ibis* (6)4: 430-442.

— 1894. On a collection of birds from Mount Mulu in Sarawak. *Ibis* (6)6: 542-544.

— 1903. Remarks on *Pitta longipennis* Reichenowi and *Pitta reichenowi* Madarász. *Ibis* (8)3: 91-93.

— 1907. On further collections of birds from the Efulen district of Cameroon, West Africa (with notes by the collector, G.L. Bates). *Ibis* 9(1): 416-464.

Sheldon, F.H. in prep. *A Checklist of the Birds of Sabah, Malaysia.*

Shelford, R.W.C. 1900. The birds of Mt. Penrissen and neighbouring districts. *J. Straits Branch Royal Asiatic. Soc.* 33: 10-21.

Sibley, C.G. 1970. A comparative study of egg-white proteins of passerine birds. *Bull. Peabody Mus. Nat. Hist.* 32: 1-131.

— & Ahlquist, J.E. 1990. *Phylogeny and Classification of Birds.* Yale University Press, New Haven, Connecticut.

— & Monroe, B.L. 1990. *Distribution and Taxonomy of Birds of the World.* Yale University Press, New Haven, Connecticut.

— & Monroe, B.L. 1993. *A Supplement to Distribution and Taxonomy of Birds of the World.* Yale University Press, New Haven, Connecticut.

—, Williams, G.R. & Ahlquist, J.E. 1982. The relationships of the New Zealand Wrens (Acanthisittidae) as indicated by DNA-DNA hybridization. *Notornis* 29: 113-130.

—, Ahlquist, J.E. & Monroe, B.L. 1988. A classification of the living birds of the world based on DNA-DNA hybridization studies. *Auk* 105: 409-423.

Silvius, M.J. & Verheugt, W.J.M. 1986. The birds of Berbak Game Reserve, Jambi Province, Sumatra. *Kukila* 2: 76-84.

Singh, P. 1994. Recent bird records fom Arunachal Pradesh. *Forktail* 10: 65-104.

Skutch, A.F. 1987. *Helpers at Birds' Nests: a Worldwide Survey of Cooperative Breeding and Related Behavior.* University of Iowa Press, Iowa.

Smith, H.C. 1943. *Notes on Birds of Burma.* Liddells Printing Works, Simla.

—, Garthwaite, P.F. & Smythies, B.E. 1943. The birds of the Karen hills and Karenni found over 3000 ft. *J. Bombay Nat. Hist. Soc.* 43: 455-474.

Smythies, B.E. 1957. An annotated checklist of the birds of Borneo. *Sarawak Mus. J.* 7: 523-818.

— 1981. *The Birds of Borneo.* Third Ed. The Sabah Society with the Malayan Nature Society, Kuala Lumpur.

— 1986. *The Birds of Burma.* Third edition. Nimrod Press, Liss, U.K. with Silvio Mattacchione and Co., Pickering, Canada.

Snow, D.W. 1976. *The Web of Adaptation.* British Museum of Natural History, Tring.

Sody, H.J.V. 1989. Diets of Javanese birds. Pp. 164-221 in Becking, J.H. *Sody (1892-1959), His Life and Work, a Biographical and Bibliographical Study.* E. J. Brill, Leiden.

Stanford, J.K. & Ticehurst, C.B. 1930. The birds of the Prome District of lower Burma. *J. Bombay Nat. Hist. Soc.* 34: 901-915.

— & — 1938. On the birds of northern Burma - part IV. *Ibis* (14)2: 599-639.

Steere, J.B. 1890. *A List of the Birds and Mammals Collected by the Steere Expedition to the Philippines, with Localities, and with Brief Preliminary Description of Supposed New Species.* Courier Printers, Ann Arbor, Michigan.

Stevens, H. 1915. Notes on the birds of upper Assam. *J. Bombay Nat. Hist. Soc.* 23: 547-570.

— 1925. Notes on the birds of the Sikkim Himalayas. Part VI. *J. Bombay Nat. Hist. Soc.* 30: 664-685

Storr, G.M. 1973. *List of Queensland Birds.* Western Australian Museum, Perth. (Special Publ. 5.).

— 1984. Revised list of Queensland Birds. *Rec. W. Aust. Mus. Suppl.* 19.

Stresemann, E. 1914. Die Vögel von Seran (Ceram). *Novit. Zool.* 21: 25-153.

— 1923. Neue Formen aus Sud-China. *J. Orn.* 71: 362-365.

— 1937. Ein neuer Fund von *Neodrepanis hypoxantha, Salom. Orn. Monatsber.* 45: 135-136.

— 1938. Vögel von Fluss Kayan (Nordost Borneo) gesammelt von Baron Victor von Plessen. *Temminckia* 3: 109-136.

— 1940. Die Vögel von Celebes. Teil III. Systematik und

Biologie *J. Orn.* 88: 1-135.
— & Heinrich, G. 1939. Die Vögel des Mount Victoria. *Mitt. Zool. Mus. Berlin* 24: 151-264.
Stuart, S.N. (ed.) 1986. *Conservation of Cameroon Montane Forests.* International Council for Bird Preservation, Cambridge.
Takano, S. (ed.) 1981. *Birds of Japan in Photographs.* Tokai University Press, Tokyo.
Tarboton, W.R., Kemp, M.I. & Kemp, A.C. 1987. *Birds of the Transvaal.* Transvaal Museum, Pretoria.
Taylor, J.B. 1991. A status survey of Seram's Moluccan endemic avifauna. Unpublished report.
Taylor, P. & Taylor, P. 1995. Observations on nesting Noisy Pittas *Pitta versicolor. Aust. Birdwatcher* 16(1): 39-41.
Thewlis, R. 1993. Cambridge-Laos wildlife survey of Xe Pian Proposed Protected Area. Second interim report. Unpublished.
Thiollay, J.M. The birds of Ivory Coast. *Malimbus* 7: 1-59.
Thomas, J. 1991. Birds of the Korup National Park, Cameroon. *Malimbus* 13: 11-23.
Thomas, W.W. 1964. Preliminary list of the birds of Cambodia. Unpublished manuscript.
Thompson, M.C. 1966. Birds from North Borneo. *Univ. Kansas Publ., Mus. Nat. Hist.* 17(8): 377-433.
Thompson, P.M. & Evans, M.I. (eds.) 1991. *A survey of Ambatovaky Special Reserve.* Madagascar Environmental Research Group, London.
—, Harvey, W.G., Johnson, D.L., Millins, D.J., Rashid, M.A., Scott, D.A., Stanford, C. and Woolner, J.W. 1993. Recent notable bird records from Bangladesh. *Forktail* 9: 12-44.
Tirant, G. 1879. Les oiseaux de la Basse Cochin-Chine. *Bull. du Comité Agricole et Industriel de la Cochin-Chine* (3)1: 73-174.
Traylor, M.A. 1960. *Notes on the Birds of Angola, Non-passeres.* Museo do Dundo, Lisbon.
Urban, E.K. & Hakanson, T. 1971. An African Pitta, *Pitta angolensis longipennis,* from Ethiopia. *Bull. Brit. Orn. Club* 91: 9-10.
van Bemmel, A.C.V. 1939. Zwei neue Vögel von der Insel Morotai. *Treubia* 17: 125-126.
van Heyst, A.F.C.A. 1919. Aanteekeningen omtrent de avifauna van Deli (Sumatra's Oostkust). *Jaarber. Cl. Ned. Vogelk.* 9: 36-68.
van Marle, J.G. and Voous, K.H. 1988. *The Birds of Sumatra.* British Ornithologists' Union, Tring, England (BOU Check-list 10).
van Someren, V.G. 1919. On some new forms of bird from Africa. *Bull. Brit. Orn. Club* 40: 19-28.
van Someren, V.G. 1941. Two new races of *Smithornis capensis* (Eurylaimidae) from Kenya Colony. *Bull. Brit. Orn. Club* 62: 35-37.
van Tyne, J. 1933. Native bird traps of French Indo-china. *Sci. Monthly* 38: 562-565.
Verbelen, F. 1993. Observations of a major 'fall' of migrant Hooded Pittas (*Pitta sordida*) at Kaeng Krachan National Park, Phetchaburi Province. *Nat. Hist. Bull. Siam Soc.* 41: 121.
Verheugt, W.J.M., Skov, H. & Danielsen, F. 1993. Notes on the birds of the tidal lowlands and floodplains of South Sumatra Province, Indonesia. *Kukila* 6: 53-84.
Vernon, P. 1974. The breeding of van den Bosch's Pitta. *Avic. Mag.* 80(2): 42-45.
von Berlepsch, H. 1901. Systematisches Verzeichnis der von Herrn Professor Willy Kükenthal während seiner Reisen im Malayischen Archipel im Jahre 1894 auf den nörd-lichen Molukken-Inseln gesammelten Vogelbälge. *Abh. Senckenb. Naturf. Ges.* 25: 299-317.
Voous, K.H. 1961. Birds collected by Carl Lumholtz in eastern and central Borneo. *Nytt. Mag. Zool. Oslo* 10 (Suppl.): 127-180.
Vordeman, A.G. 1885. Bataviasche Vogels 1. *Natuuk. Tijdschr. Ned. -Indi.* 41: 182-211.
Wallace, A.R. 1860. The ornithology of Northern Celebes. *Ibis* 2: 140-147.
— 1861. Notes on the ornithology of Timor. *Ibis* 3: 347-351.
— 1869. *The Malay Archipelago.* Macmillan, London.
Wardlaw-Ramsay, R.G. 1877. Notes on some Burmese birds. *Ibis* (4)1: 450-473.
Watling, D. 1983. Ornithological notes from Sulawesi. *Emu* 83: 247-261.
WCMC 1995. *Checklist of Birds listed in the CITES Appendices.* JNCC Report 236 (Compiled by the World Conservation Monitoring Centre). Joint Nature Conservation Committee, Peterborough.
Webster, M.A. 1975. *An Annotated Checklist of the Birds of Hong Kong.* Hong Kong Bird Watching Society, Hong Kong.
Webster, R. 1991. The broadbills, an overview of the Eurylaimidae with an emphasis on the Lesser Green Broadbill (*Calyptomena viridis*) in the wild and in captivity. Unpublished report, San Antonio Zoo, Texas.
Wells, D.R. 1970. Bird Report: 1968. *Malay. Nat. J.* 23: 47-77.
— 1975. Bird Report: 1972 and 1973. *Malay. Nat. J.* 28: 186-213.
— 1976. Some bird communities in western Sabah, with distributional records, March 1975. *Sarawak Mus. J.* 24: 277-286.
— 1982. Bird Report: 1974 and 1975. *Malay. Nat. J.* 36: 61-85.
— 1983. Bird Report: 1976 and 1977. *Malay. Nat. J.* 36: 197-218.
— 1984. Bird Report: 1978 and 1979. *Malay. Nat. J.* 38: 113-150.
— 1985. Broadbill. Pp. 66-67 *in* Campbell, B. and Lack, E. (eds). *A Dictionary of Birds.* T & A.D. Poyser, Calton, England.
— 1986. Bird Report: 1980 and 1981. *Malay. Nat. J.* 39: 279-298.
— 1990a. Migratory birds and tropical forest in the Sunda region. Pp. 357-369 *in* Keast, A. (ed.) *Biogeography and Ecology of Forest Bird Communities.* SPB Publishing, The Hague.
— 1990b. Bird Report: 1982 and 1983; Bird Report 1984 and 1985; Bird Report 1986 and 1987. *Malay. Nat. J.* 43: 116-210.
— 1992. Night migration at Fraser's Hill, Peninsular Malaysia. *Bull. Oriental Bird Club.* 16: 21-25.
—, Hails, C.J. & Hails, A.J. 1979. A study of the birds of Gunung Mulu National Park, Sarawak, with special emphasis on those of lowland forest. Unpublished report to Royal Geographic Society, London.
Whistler, H. 1934. Occurrence of the Larger Blue-winged Pitta (*Pitta megaryncha* Schlegel) in Eastern Bengal. *J. Bombay Nat. Hist. Soc.* 37: 222.
White, C.M.N. 1961. *A Revised Check List of African Broadbills, Pittas, Larks, Swallows, Wagtails and Pipits.* Government Printers, Lusaka.
— & Bruce, M.D. 1986. *The Birds of Wallacea.* British Ornithologists' Union, London (Check-list 7).

White, F. 1983. *The Vegetation of Africa: a Descriptive Memoir to Accompany UNESCO/AETFAT/UNSO Vegetation Map of Africa.* UNESCO, Paris.

White, H.L. 1917. North Australian birds. *Emu* 16: 117-158, 205-231.

Whitehead, J. 1893. A review of the species of the family Pittidae. *Ibis* (6)5: 488-509.

— 1899. Field notes on birds collected in the Philippine Islands in 1893-6. *Ibis* (7)5: 81-111, 210-246, 381-399, 485-501.

Wildash, P. 1968. *Birds of South Vietnam.* Charles E. Tuttle Co., Rutland, Vermont and Tokyo, Japan.

Wiles, G.J. 1979. The birds of Salak Phra Wildlife Sanctuary, southwest Thailand. *Nat. Hist. Bull. Siam. Soc.* 28: 101-120.

Wilkinson, R., Dutson, G. & Sheldon, B. 1991a. *The Avifauna of Barito Ulu, Central Borneo.* International Council for Bird Preservation, Cambridge. (ICBP Study Report 48).

—, —, —, Darjono & Yus, R.N. 1991b. The avifauna of the Barito Ulu region, Central Kalimantan. *Kukila* 5: 99-116.

Williams, D.M. 1990. Blue-winged Pitta on Cheung Chau: a first record for Hong Kong. *Hong Kong Bird Report 1990*: 119-120.

Williamson, W.J.F. 1918. New or noteworthy bird-records from Siam. *J. Nat. Hist. Soc. Siam* 3(1): 15-42.

Wilson, J.R. and Catsis, M.C. 1990. A preliminary survey of the forests of the Itombwe mountains and the Kahuzi-Biega National Park extension, east Zaïre, July-September 1989. Unpublished report: World Wide Fund for Nature, Institut Zaïrois pour la Conservation de la Nature and the Fauna and Flora Preservation Society, London.

Woinarski, J.C.Z. 1993. A cut-and-paste community: birds of monsoon rainforests in Kakadu National Park, Northern Territory. *Emu* 93: 100-120.

—, Press, A.J. & Rusell-Smith, J. 1989. The bird community of a sandstone plateau monsoon forest at Kakadu National Park, Northern Territory. *Emu* 89: 223-231.

Wolters, H.E. 1982. *Die Vogelarten der Erde: eine Systematische Liste mit Verbreitungsangaben sowie Deutschen und Englischen Namen.* Paul Parey, Hamburg and Berlin.

Won, Pyong-Oh 1969. *An Annotated Checklist of the Birds of Korea.* [Korean with title page in English].

Wong, M. 1986. Trophic organization of understory birds in a Malaysian dipterocarp forest. *Auk* 103: 100-116.

Woodall, P.F. 1993. Measurements of the Noisy Pitta *Pitta versicolor* in Australia. *Corella* 17(4): 114-116.

— 1994. Breeding season and clutch size of the Noisy Pitta *Pitta versicolor* in tropical and subtropical Australia. *Emu* 94: 273-277.

Yang Lan. 1983. 'A taxonomical study on Chinese birds of the genus Pitta (Pittidae).' *Zool. Res.* 4: 219-226. [In Chinese].

Yen, K.Y. 1933. Les oiseaux du Kwangsi (Chine), (suite). *L'Ois. Rev. Franc. Ornithol.* 3: 755-788.

Zimmerman, D.A. 1972. The avifauna of the Kakamega forest, western Kenya, including a bird population study. *Bull. Amer. Mus. Nat. Hist.* 149: 257-339.

Zimmermann, U. in press. Displays and other behaviours of the Rainbow Pitta *Pitta iris.*

— in prep. The ecology of the monsoon-forest endemic Rainbow Pitta and a comparison with other Australian pittas. PhD thesis, Northern Territory University, Darwin, Australia.

INDEX OF SCIENTIFIC AND ENGLISH NAMES

Species are listed by their English vernacular name (e.g. Garnet Pitta), together with alternative names where relevant, and by their scientific names. Specific scientific names are followed by the generic name as used in the book (e.g. *granatina, Pitta*) and subspecific names are followed by both the specific and generic names (e.g. *coccinea, Pitta granatina*). In addition, genera are listed separately.

Numbers in italic type refer to the first page of the systematic entry and those in bold type refer to plate numbers.

abbotti, Pitta sordida 120
affinis, Cymbirhynchus macrorhynchos 72, *206*
affinis, Pitta guajana 105
African Green Broadbill **18**, 14, 17, 28, 66, 188, *235*
African Pitta **12**, 18, 54, *150*, 157
albigularis, Smithornis capensis 64, *187*
anerythra, Pitta 18, 58, *181*
anerythra, Pitta anerythra 182
Angolan Pitta, see African Pitta
angolensis, Pitta 18, 54, *150*
angolensis, Pitta angolensis 152
annamensis, Pitta soror 90
Anthrocincla, see Eared Pitta
aphobus, Serilophus lunatus 216
arcuata, Pitta, see *arquata, Pitta*
ardescens, Corydon sumatranus 232
arquata, Pitta 18, 52, *141*
aruensis, Pitta erythrogaster 136
astrestus, Serilophus lunatus 216
aurantiaca, Pitta cyanea 38, *102*
Azure-breasted Pitta **8**, 22, 28, 46, *129*

Banded Broadbill **21**, 21, 72, 205, *220*, 224, 226
Banded Pitta **5**, **16**, 15, 40, 62, *104*, 111, 114, 116
bangkana, Pitta sordida 44, *121*
Bar-bellied Pitta **6**, 8, 15, 42, 101, *108*
baudii, Pitta 42, *115*
Bengal Pitta, see Indian Pitta
bernsteini, Pitta erythrogaster 135
billitonis, Eurylaimus javanicus 221
Black-and-Crimson Pitta, see Black-headed Pitta
Black-and-Scarlet Pitta, see Graceful Pitta, Black-and-Crimson Pitta
Black-and-Yellow Broadbill **22**, 74, 207, 221, *224*
Black-backed Pitta, see Superb Pitta
Black-breasted Pitta, see Gurney's Pitta, Rainbow Pitta
Black-crowned Garnet Pitta, see Black-headed Pitta
Black-crowned Pitta, see Black-headed Pitta, Graceful Pitta
Black-faced Pitta **14**, 18, 28, 58, *181*
Black-headed Pitta **11**, 8, 18, 22, 52, 141, 143, 144, 145, 147
Black-throated Green Broadbill, see Whitehead's Broadbill
Blue Pitta **4**, 15, 27, 38, 83, 84, 92, *101*, 118, 221
Blue-banded Pitta **11**, 8, 15, 18, 52, *141*, 145, 147
Blue-breasted Pitta, see Red-bellied Pitta
Blue-headed Pitta **6**, 42, 104, *115*
Blue-naped Pitta **2**, **16**, 13, 15, 20, 27, 34, 62, 83, *85*, 89, 92, 101
Blue-rumped Pitta **2**, **16**, 15, 62, 85, 87, *88*, 92, 93

Blue-winged Pitta **13**, **16**, 18, 22, 27, 56, 62, 102, 110, 117, 150, 162, *166*, 170, 176, 177, 181, 221
bolovenensis, Pitta oatesi 32, *93*
borneensis, Psarisomus dalhousiae 211
brachyura, Pitta 18, 54, *159*, 163
brookei, Eurylaimus javanicus 221
brunnescens, Corydon sumatranus 70, *232*
budongoensis, Smithornis rufolateralis 66, *194*
Buff-breasted Pitta, see Noisy Pitta

caerulea, Pitta 20, 36, *97*
caerulea, Pitta caerulea 36, 62, *98*
caeruleitorques, Pitta eruthrogaster 135
Calyptomena 196
camerunensis, Smithornis capensis 187
capensis, Smithornis 64, *186*
capensis, Smithornis capensis 64, *187*
castanea, Philepitta 19, 78, *238*
castaneiceps, Pitta oatesi 32, *93*
celebensis, Pitta erythrogaster 50, *134*
Chinese Pitta, see Fairy Pitta
chyulu, Smithornis capensis 187
coccinea, Pitta granatina 52, *145*
coelestis, Pitta steerii 130
Collared Broadbill, see Silver-breasted Broadbill
Common Sunbird Asity **24**, 14, 78, 241, 243, *248*, 251, 252
concinna, Pitta elegans 58, *175*
conjunctus, Smithornis capensis 187
continentis, Calyptomena viridis 197
coruscans, Neodrepanis 14, 78, *245*
Corydon 231
cryptoleucus, Smithornis capensis 187
cucullata, Pitta sordida 18, 44, 62, *120*
cyanea, Pitta 38, *101*
cyanea, Pitta cyanea 38, *102*
cyanicauda, Psarisomus dalhousiae 211
cyanonota, Pitta erythrogaster 48, *135*
Cymbirhynchus 205

dalhousiae, Psarisomus 70, *210*
dalhousiae, Psarisomus dalhousiae 70, *211*
davinus, Psarisomus dalhousiae 211
deborah, Pitta oatesi 32, *93*
delacouri, Smithornis capensis 187
dohertyi, Pitta 50, *140*
Dusky Broadbill **20**, 21, 25, 26, 70, 209, 210, *231*

Eared Pitta **1**, 15, 22, 32, *83*, 101
elegans, Pitta 18, 58, *174*
elegans, Pitta elegans 58, *175*

269

Elegant Pitta **14**, 18, 22, 27, 58, 141, 162, *174*
elisabethae, Serilophus lunatus 216
elliotii, Pitta 42, *108*
Elliot's Pitta, see Bar-bellied Pitta
erythrogaster, Pitta 15, 18, 30, 48, 50, *133*, 140
erythrogaster, Pitta erythrogaster 48, *134*
Eucichla 15
eurylaemus, Smithornis sharpei 191
Eurylaimus 220
everetti, Pitta elegans 175
extima, Pitta erythrogaster 50, *136*

Fairy Pitta **13**, 11, 18, 27, 28, 29, 56, *162*, 166, 174
False Sunbird, see Common Sunbird Asity
finschii, Pitta erythrogaster 50, *136*
flynnstonei, Pitta soror 86, 90
forsteni, Pitta sordida 44, *119*
Fulvous Pitta, see Rusty-naped Pitta

Garnet Pitta **11**, 8, 18, 22, 52, 141, *143*, 147, 149
gazellae, Pitta erythrogaster 50, *136*
Giant Pitta **3**, **16**, 15, 20, 27, 36, 62, 94, 95, *97*
Gigantipitta 15
goodfellowi, Pitta sordida 119
Gould's Broadbill, see Silver-breasted Broadbill
Graceful Pitta **11**, 52, *143*, 144, *148*
granatina, Pitta 18, 52, *143*
granatina, Pitta granatina 52, *145*
graueri, Pseudocalyptomena 14, 66, *235*
Great Blue Pitta, see Giant Pitta
Great Pitta, see Ivory-breasted Pitta
Green Broadbill **19**, 21, 23, 24, 26, 68, *196*, 202, 207
Green-breasted Pitta **12**, 54, 153, 154, *160*. See also Hooded Pitta
Grey-headed Broadbill **18**, 66, 189, *193*, 197
guajana, Pitta 15, 40, 62, *104*
guajana, Pitta guajana 40, *105*
gurneyi, Pitta 42, 62, *110*
Gurney's Pitta **6**, **16**, 15, 20, 21, 27, 29, 42, 62, 101, 108, *110*, 117

habenichti, Pitta erythrogaster 136
harterti, Eurylaimus javanicus 72, *221*
hendeei, Pitta nipalensis 86
Hooded Pitta **7**, **16**, 15, 18, 20, 22, 25, 27, 29, 30, 44, 56, 62, 103, 107, 110, *117*, 125, 128, 129, 137, 168, 169, 178, 182
hosei, Pitta caerulea 36, *98*
Hose's Broadbill **19**, 10, 68, 196, *201*, 203
hosii, Calyptomena 201
hutzi, Pitta elegans 175
hypoxantha, Neodrepanis 78, *248*

impavidus, Serilophus lunatus 217
Indian Pitta **12**, 18, 22, 27, 54, 117, 124, *159*, 162, 170
inspeculata, Pitta erythrogaster 48, *135*
intensus, Serilophus lunatus 76, *217*
intermedia, Pitta soror 90
intermedia, Pitta versicolor 179

irena, Pitta guajana 40, *105*
Iridipitta 15
iris, Pitta 60, *183*
Ivory-breasted Pitta **8**, 18, 46, *126*

javanicus, Eurylaimus 72, *220*
javanicus, Eurylaimus javanicus 72, *221*

kalamantan, Eurylaimus ochromalus 225
kalaoenis, Pitta elegans 175
khmerensis, Corydon sumatranus 232
kochi, Pitta 18, 48, 62, *131*
Koch's Pitta, see Whiskered Pitta
kuehni, Pitta erythrogaster 135

laoensis, Corydon sumatranus 70, *232*
Larger Blue-winged Pitta, see Mangrove Pitta
lemniscatus, Cymbirhynchus macrorhynchos 206
Lesser Blue-winged Pitta, see Blue-winged Pitta, Fairy Pitta
Lesser Green Broadbill, see Green Broadbill
Long-tailed Broadbill **20**, 25, 26, 70, *210*
longipennis, Pitta angolensis 54, 150, *153*, 157
loriae, Pitta erythrogaster 136
lunatus, Serilophus 76, *215*
lunatus, Serilophus lunatus 76, *216*

macklotii, Pitta erythrogaster 50, *136*
macrorhynchos, Cymbirhynchus 72, *205*
macrorhynchos, Cymbirhynchus macrorhynchos 72, *206*
Magnificent Green Broadbill, see Hose's Broadbill
malaccensis, Cymbirhynchus macrorhynchos 206
Mangrove Pitta **13**, 22, 56, 117, 162, 166, 169, *170*
maria, Pitta elegans 58, *175*
Masked Pitta, see Black-faced Pitta
maxima, Pitta 18, 46, *126*
maxima, Pitta maxima 46, *127*
mayri, Eurylaimus steerii 228
mecistus, Eurylaimus ochromalus 225
medianus, Smithornis capensis 64, *187*
meeki, Pitta erythrogaster 50, *136*
mefoorana, Pitta sordida 119
megarhyncha, Pitta 56, 162, *170*
meinertzhageni, Smithornis capensis 187
Mindanao Wattled Broadbill **22**, 74, *227*, 229
Moluccan Pitta, see Blue-winged Pitta
moluccensis, Pitta 18, 56, 62, 162, *166*
morator, Corydon sumatranus 232
morotaiensis, Pitta maxima 46, *127*
muelleri, Pitta sordida 44, *119*

Neodrepanis 245
nigrifrons, Pitta anerythra 58, *182*
nipalensis, Pitta 34, 85
nipalensis, Pitta nipalensis 34, 62, *86*
Noisy Pitta **15**, 18, 20, 22, 27, 60, 138, 174, *177*, 183, 185
novaeguineae, Pitta sordida 44, *119*
novaehibernicae, Pitta erythrogaster 136
nympha, Pitta 18, 56, *162*

oatesi, Pitta 32, *92*

oatesi, Pitta oatesi 32, 62, *93*
obiensis, Pitta erythrogaster 135
oblita, Pitta erythrogaster 136
ochromalus, Eurylaimus 74, *224*
ochromalus, Eurylaimus ochromalus 74, *224*
orientalis, Corydon sumatranus 232

palawanensis, Pitta sordida 119
pallescens, Corydon sumatranus 232
palliceps, Pitta erythrogaster 135
pallida, Pitta anerythra 58, *182*
pallidus, Eurylaimus javanicus 221
petersi, Pitta soror 90
phayrei, Pitta 15, 32, *83*
Phayre's Pitta, see Eared Pitta
Philepitta 238
piroensis, Pitta erythrogaster 135
Pitta 83
plesseni, Pitta elegans 175
polionotus, Serilophus lunatus 76, *217*
propinqua, Pitta erythrogaster 134
Psarisomus 210
Pseudocalyptomena 235
psittacinus, Psarisomus dalhousiae 70, *211*
pulih, Pitta angolensis 54, *153*
Purple-headed Broadbill, see Banded Broadbill

Rainbow Pitta **15**, 20, 22, 60, 123, 146, 178, *183*
Red-bellied Pitta **6**, 15, 18, 22, 30, 48, 50, 117, 118, *133*, 140, 141, *182*
Red-breasted Pitta, see Red-bellied Pitta
Red-headed Scarlet Pitta, see Garnet Pitta
Red-sided Broadbill, see Rufous-sided Broadbill
reichenowi, Pitta 54, *156*
ripleyi, Pitta guajana 105
rosenbergii, Pitta sordida 30, 44, *119*
rothschildi, Serilophus lunatus 76, *217*
rubrinucha, Pitta erythrogaster 48, *135*
rubropygius, Serilophus lunatus 216
rufiventris, Pitta erythrogaster 135
rufolateralis, Smithornis 66, *193*
rufolateralis, Smithornis rufolateralis 66, *194*
Rufous-sided Broadbill **18**, 66, 186, 188, 189, 191, *193*
Rusty-naped Pitta **1**, **16**, 8, 15, 32, 62, 85, 86, 89, 92, 101

samarensis, Eurylaimus 74, *229*
sanghirana, Pitta sordida 30, *119*
schlegeli, Philepitta 78, *242*
Schlegel's Asity **24**, 13, 21, 26, 78, 238, *242*
schneideri, Pitta 36, 62, *95*
Schneider's Pitta **3**, **16**, 10, 13, 15, 28, 36, 62, *95*, 97, 149
schwaneri, Pitta guajana 40, *105*
Serilophus 215
sharpei, Smithornis 66, *190*
sharpei, Smithornis sharpei 66, *191*
shimba, Smithornis capensis 187
siamensis, Cymbirhynchus macrorhynchos 206
siberu, Calyptomena viridis 68, *197*
Silver-breasted Broadbill **23**, 76, *215*

simillima, Pitta versicolor 60, *179*
Small-billed Asity, see Yellow-bellied Sunbird Asity
Small-billed False Sunbird, see Yellow-bellied Sunbird Asity
Smithornis 186
sordida, Pitta 15, 18, 30, 44, *117*
sordida, Pitta sordida 44, *119*
soror, Pitta 34, *88*
soror, Pitta soror 34, 62, *90*
splendida, Pitta erythrogaster 136
Steere's Pitta, see Azure-breasted Pitta
steerii, Eurylaimus 74, *227*
steerii, Eurylaimus steerii 74, *228*
steerii, Pitta 46, *129*
steerii, Pitta steerii 130
stolidus, Serilophus lunatus 217
suahelicus, Smithornis capensis 187
Sula Pitta **10**, 50, 133, 134, *140*
sumatranus, Corydon 70, *231*
sumatranus, Corydon sumatranus 232
Sunbird Asity, see Common Sunbird Asity
Superb Pitta **8**, 28, 46, *128*, 183
superba, Pitta 46, *128*
Swinhoe's Pitta, see Fairy Pitta

tenebrosus, Cymbirhynchus macrorhynchos 206
thompsoni, Pitta erythrogaster 134
tonkinensis, Pitta soror 34, *90*

ussheri, Pitta (granatina) 18, 52, 143, *146*

Velvet Asity **24**, 19, 21, 24, 25, 26, 78, *238*, 242, 251
venusta, Pitta 52, 143, *148*
versicolor, Pitta 18, 20, 60, *177*, 183
versicolor, Pitta versicolor 60, *179*
vigorsii, Pitta elegans 58, *175*
virginalis, Pitta elegans 58, *175*
viridis, Calyptomena 68, *196*
viridis, Calyptomena viridis 68, *197*
Visayan Wattled Broadbill **22**, 74, *229*

Wattled Asity, see Common Sunbird Asity
Wattled Sunbird Asity, see Common Sunbird Asity
Whiskered Pitta **9**, **16**, 28, 29, 48, 62, *131*, 133
whiteheadi, Calyptomena 68, *203*
Whitehead's Broadbill **19**, 23, 24, 196, 201, 202, *203*
willoughbyi, Pitta cyanea 102

Yellow-bellied Sunbird Asity **24**, 28, 78, 245, 246, 248

zenkeri, Smithornis sharpei 66, *191*